Fluid Mixing and
Gas Dispersion in
Agitated Tanks

Other McGraw-Hill Chemical Engineering Books of Interest

AUSTIN ET AL. . *Shreve's Chemical Process Industries*

CHOPEY, HICKS . *Handbook of Chemical Engineering Calculations*

COOK, DUMONT . *Process Drying Practice*

DEAN . *Lange's Handbook of Chemistry*

DILLON . *Corrosion Control in the Chemical Process Industries*

FREEMAN . *Hazardous Waste Minimization*

FREEMAN . *Standard Handbook of Hazardous Waste Treatment and Disposal*

GRANT, GRANT . *Grant & Hackh's Chemical Dictionary*

KISTER . *Distillation Operation*

MILLER . *Flow Measurement Engineering Handbook*

PERRY, GREEN . *Perry's Chemical Engineers' Handbook*

REID ET AL. . *Properties of Gases and Liquids*

REIST . *Introduction to Aerosol Science*

RYANS, ROPER . *Process Vacuum System Design and Operation*

SANDLER, LUCKIEWICZ . *Practical Process Engineering*

SATTERFIELD . *Heterogeneous Catalysis in Industrial Practice*

SCHWEITZER . *Handbook of Separation Techniques for Chemical Engineers*

SHINSKEY . *Process Control Systems*

SHUGAR, DEAN . *The Chemist's Ready Reference Handbook*

SHUGAR, BALLINGER . *The Chemical Technician's Ready Reference Handbook*

SMITH, VAN LAAN . *Piping and Pipe Support Systems*

STOCK . *AI in Process Control*

YOKELL . *A Working Guide to Shell-and-Tube Heat Exchangers*

Fluid Mixing and Gas Dispersion in Agitated Tanks

Gary B. Tatterson, Ph.D.
Department of Chemical Engineering
North Carolina A&T State University
Greensboro, North Carolina

McGraw-Hill, Inc.
New York   St. Louis   San Francisco   Auckland   Bogotá
Caracas   Hamburg   Lisbon   London   Madrid
Mexico   Milan   Montreal   New Delhi   Paris
San Juan   São Paulo   Singapore
Sydney   Tokyo   Toronto

Library of Congress Cataloging-in-Publication Data

Tatterson, Gary B.
  Fluid mixing and gas dispersion in agitated tanks / Gary B.
Tatterson.
    p. cm.
  ISBN 0-07-062933-1
  1. Mixing. 2. Liquids. I. Title.
TP156.M5T38 1991
660'.2842—dc20   90-22595
                                           CIP

1 2 3 4 5 6 7 8 9 0 DOC/DOC 9 7 6 5 4 3 2 1

ISBN 0-07-062933-1

*The sponsoring editor for this book was Gail Nalven, the editing
supervisor was David E. Fogarty, and the production supervisor was
Pamela A. Pelton. This book was set in Century Schoolbook. It was
composed by McGraw-Hill's Professional Publishing composition unit.*

Printed and bound by R. R. Donnelley & Sons Company.

Contents

Preface

Mixing in agitated tanks is part of the infrastructure of the chemical, petrochemical, and biochemical industries. Practically every plant will contain some sort of tank mixing process. Unfortunately, such mixing is based upon fluid mechanics that are not taught in undergraduate and graduate fluid mechanics courses, and, as such, a fundamental understanding of tank mixing by plant personnel is seldom obtained. The general topic of mixing in agitated tanks should be properly developed in some manner since the understanding of mixing will directly aid commercialization of processes and retrofitting of existing processes. A specific example of an important and successful commercialization of a tank mixing process is penicillin production during World War II in which deep tank fermentation was utilized.

Mixing consumes a considerable portion of the processing time as material moves through a plant. The key issues in such processing are processing times and capacities. The convenient cycle times of 8 or 24 hours are no longer adequate. There is no justification to hold material which has been processed. Equally, many processes are undermixed and many are overmixed. Neither situation is particularly important in the sense of mixing. However, economically, these situations lead directly to unnecessary waste in processing capacity, time, and material. Processing of material has become too important in the present marketplace for mixing to be ignored, especially in processing high-technology materials and value-added materials.

Chemical reactions cannot take place between species unless they are mixed. If the reactions are fast, then the reaction rates, selectivity, and production rates are determined by the mixing rate. In heterogeneous systems, other considerations force the need for effective mixing. Mass transfer controlled processes are most often mixing controlled, "mass transfer controlled" being an inaccurate statement in many cases. Mixing lies at the heart of chemical reactor engineering since mixing times are necessary to be able to size chemical reactors in a rational manner. Safety considerations are also present in mixing. Accounts of industrial accidents where mixing was not maintained include "The Physicochemical Origins of the Sevesco Accident," by T. G. Theofanous (*Chem. Eng. Sci.*, **38**, 10, 1615–1625, 1983).

Mixing is no longer a unit operation. Mixing is the contacting mechanics of chemical processing. Unfortunately, mixing is usually taught as a unit operation and is, only occasionally included in undergraduate curriculum courses. Most commonly, a mixing experiment is included as part of a unit operations course. In such a situation, mixing cannot be properly understood by students. Mixing has basic principles and has contributions from many sectors. A major portion of fluid mechanics in chemical engineering is mixing, whether the flow is a circulation pattern on a tray of a distillation column or in a high-temperature twin screw extruder. However, mixing cannot be presented as simply as distillation because of its complexity. Much depends upon data correlations and evolving theory.

Whereas there are many specialized textbooks on such areas as distillation, extraction, and fluidization, books on mixing are sparse and are usually monographs written with rather limited reference listings. Most only explain the state of the art. No textbook is available from which a course on the fundamental principles of mixing can be taught. Further, only a few mixing courses are taught per year across the United States.

This book is designed to serve students, faculty, researchers, and practitioners. This book is meant to be an in-depth review of the literature, to present the fundamentals of mixing, to show what is known, and to provide the supporting documentation, references, and the different methods used to study mixing. The book establishes a course in mixing in which the fundamentals of laminar and turbulent fluid mechanics and transport are given at the intermediate undergraduate and graduate levels. Areas in need of research are noted or are made obvious by the discussion. The book provides the practitioners and professional engineers with the background necessary to be reasonably assured about design calculations, the limits of the design, and to assert the appropriateness of the methods used. Hopefully, they will appreciate the implications of mixing on design and plant operations.

In my opinion, design equations require some validation before their use, and the use of design equations without a fundamental understanding can lead to poor designs and mistakes. However, serious validation is difficult. One method of validation is in comparison of technical results in the open literature and with data. This book provides considerable technical references as would be needed. However, mixing is also a subject where individual contributions are important. As a result, one should recognize that results from author to author will vary due to geometric differences, different definitions of quantities, and different measurement techniques and fluid properties. The search for agreement and absolutes and the critical evaluation of the literature are difficult. Evaluation of the available studies from the cited literature can be accomplished in private and in a critical manner.

The acceptance of mixing as a subject worthy of study and research is also obvious. Mixing has progressed to the point where it can be taken as a serious subject warranting a place in curriculum of chemical and mechanical engineering as a second fluids course, much like a course on rheology or two-phase flow. The text can also be used to show the evolution of research in mixing. The different approaches, the different experimental techniques and results, and the basic principles which guide the research are included. This information would be useful in developing a sense for research which can be useful.

The individual chapters are structured such that the background and fundamentals are given first followed by specific literature. The book is divided into six chapters. Chapter 1 treats the general geometries involved in mixing. Differences in laminar mixing equipment and turbulent mixing geometries are discussed. Chapter 2 discusses the turbulent power numbers for various agitators from which mixing equipment can be sized. Chapter 3 provides the basic methods used to establish power numbers for laminar mixing. Chapter 4 discusses turbulence theories and turbulent mixing. Chapter 5 treats laminar mixing, and Chapter 6 treats gas dispersion in aerated agitated tanks.

The author wishes to thank Prof. Robert S. Brodkey, Department of Chemical Engineering, The Ohio State University, for his support and encouragement over the years. Drs. Dan Turner and Dara Childs, Department of Mechanical Engineering, Texas A&M University, Drs. Franklin King and Harold Martin, North Carolina A&T State University, and Drs. David Dickey and John Heibel are acknowledged for their help and support. The Linde Division of the Union Carbide Corporation is acknowledged for providing a grant which enabled the completion of the manuscript. Howard Harris and Dr. Jeffrey Kingsley are acknowledged. I wish to thank my students who have worked for and with me during my years at various universities. These individuals include: H. H. S. Yuan, A. M. Ali, Y. H. E. Sheu, P. C. H. Ouyang, Dr. T. P. K. Chang, A. McFarland, M. Tay, N. A. Wilcox, T. A. Sutter, R. Hsi, C. Ramsey, Terry Baggett, Scott Coe, William Usry, Roger King, Metin Gezer, Linda DeMore, Bill Pafford, Bill Deutschlander, Duane Brown, Sharon Mitchener, Jasmin Daniels, Jill Childers, and Lance Creech. Pat Becker, Rhonda Abke, Sharon Lindholm, and Mary Lou Taylor have contributed to the typing and retyping of various portions of the manuscript. Special thanks are given to Susan S. Tatterson and my two lovely children, Lisa and Benjamin.

Greensboro   Gary Tatterson

ABOUT THE AUTHOR

Gary B. Tatterson is a professor of chemical engineering at North Carolina A&T State University. He has published a number of journal articles, given many presentations, and organized professional society meetings dealing with mixing in agitated tanks. He has taught at Texas A&M University and the University of South Carolina. Dr. Tatterson has worked for E. I. Du Pont de Nemours & Company, the National Bureau of Standards, and the Morgantown Energy Technology Center. He holds a B.S. in chemical engineering from the University of Pittsburgh and M.S. and Ph.D. degrees from Ohio State University. He is a member of the American Institute of Chemical Engineers and is a past chair of Mixing, 3A. He is also an active member of the American Society of Mechanical Engineers and the Instrument Society of America.

Mixing Geometries

Introduction

Mixing in tanks is an important area when one considers the number of processes which are accomplished in tanks. Essentially, any physical or transport process can occur during mixing in tanks. As a portion of these, this book focuses on liquid mixing and gas dispersion in agitated tanks and will cover the following topics: power consumption in laminar and turbulent flow, fluid mechanics and flow fields in agitated tanks under conditions of laminar and turbulent flow, turbulence theory as applied to agitated tanks, and gas-liquid dispersion.

Typically, human beings will use a pot, a tank, or a vessel for processing liquids before most anything else, and they will have the desire to stir the material while processing. In a great percentage of these cases, they will find the processing results satisfactory and will end further investigations into other geometries or configurations for their processing unit. Hence, there is a need for this book. Other sources for technical literature and texts concerning mixing are available. Some of these are noted in the reference list at the end of this chapter.

There are basic tools which can be employed in the understanding of mixing. One tool is the general property balance written in all its forms, i.e., continuity, scalar balance, equations of motion, higher-level equations of motion, turbulent equations of motions, energy balances, population balances, etc. Dimensional analysis, flow analogies, computational fluid mechanics, experimental observations and data are also utilized extensively. Equations of state, constitutive models for fluids, and kinetic rate expressions are sometimes used. Models for mixing processes using these tools can either be lumped or distributed, depending upon the complexity of the processes.

Quite often, the mixing process of interest is not well understood. In these cases, qualitative and quantitative observations, experimental data, and flow regime identification are needed and should be emphasized in any experimental pilot studies in mixing. Considerable literature on mixing is devoted to the development of experimental techniques to obtain such information. There is a tendency to model mixing without first understanding the physics of the process. This tendency should be avoided.

Above all else, however, fluid mechanics and geometry are key to understanding mixing. The fluid mechanics transport the material about the tank, whereas the geometry determines, in part, the fluid mechanics. In fact, the geometry is so important that processes can be considered geometry specific. Solid suspension is very much dependent upon the shape of the tank bottom; gas-liquid and liquid-liquid dispersion depend upon the geometry of the impeller; blending, upon the relative size of the tank to the impeller; and power draw, upon the impeller geometry.

The Mixing Tank

Mixing tanks and impellers have come in all shapes and sizes. However, it was not until the late 1940s to early 1960s that a more or less standard geometry, shown in Fig. 1.1 for two different impellers, was established for single-phase mixing for the turbulent flow regime. The figure shows the tank diameter T, the impeller diameter D, the impeller blade width W, the off-bottom clearance C of the impeller, liquid height H, and wall baffles of width B. The dimensions are scaled upon the tank diameter for easy reference. This standard configuration evolved from power studies and should be viewed as a reference geometry and as a point of departure for studies in turbulent flow. The gross flow patterns using the disk style and pitched blade turbine impellers are also shown in the figure.

Any geometric configuration can always be criticized in mixing as not being the best or optimum geometry. However, the determination of the best geometry is very much a function of the process which is being carried out. The determination of true optimum geometry for any process is very difficult indeed.

The standard geometries in Fig. 1.1 are not the optimum geometry for all types of processing which can occur in a tank. A standard geometry for viscous or laminar mixing has not been established as yet. The geometries in Fig. 1.1 are far from adequate. Given the sophistication in laminar mixing, there may not be a need for a standard geometry in laminar mixing and, likewise, in other areas. In many respects, the standardization to a specific geometry has limited research

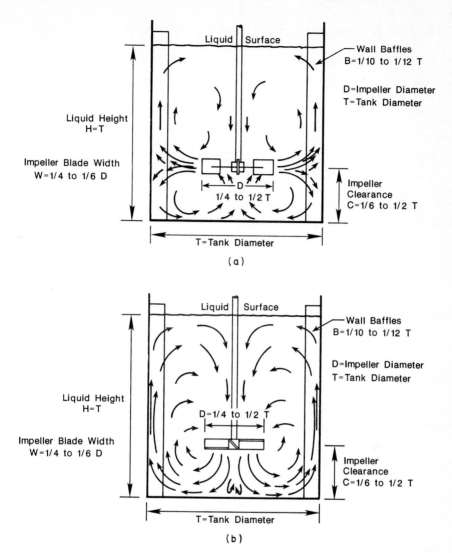

Figure 1.1 Turbulent mixing impellers. (*a*) Gross flow patterns for a radial flow impeller showing the standard tank geometry; (*b*) gross flow patterns for an axial flow impeller showing the standard tank geometry.

since it has been naturally assumed that the standard geometry is the optimum geometry.

The laminar to turbulent transition for an agitated tank occurs at an impeller Reynolds number $\rho ND^2/\mu$ from 1 to 10,000. The impeller Reynolds number is based upon the impeller tip speed πND, where N is the rotational speed in revolutions per time, not radians per time, and D is the impeller diameter. The thermodynamic viscosity μ used in the Reynolds number is the laminar viscosity for Newtonian fluids or the apparent viscosity μ_a for non-Newtonian fluids. Turbulent viscosities or eddy viscosities, μ_t or ν_t, are not used in defining the impeller Reynolds number.

Figure 1.1 shows tanks with standard wall baffles. Baffles are plates placed in the flow to disturb or redirect the flow. The most common types are wall baffles whose widths are usually expressed as a percentage, e.g., 10 percent, of tank diameter. There are also other baffle configurations: bottom baffles, surface baffles which float on the fluid surface, disk baffles placed on the impeller shaft, and baffles which are suspended from the surface at different radii. However, wall baffles have been well studied; they maximize power input to the fluid and minimize solid body rotation of the fluid in the tank, a situation which does not promote mixing.

In solid body rotation, Fig. 1.2, the fluid rotates as if it were a solid

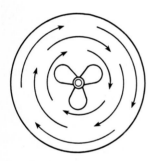

Solid Body Rotation

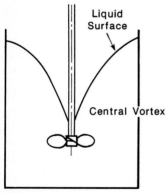

Liquid Surface

Central Vortex

Central Vortex

Figure 1.2 (*a*) Solid body rotation and (*b*) the central surface vortex.

mass; little mixing takes place as a result. Solid body rotation mostly occurs in the turbulent mixing regime. At high impeller rotational speeds, a large surface vortex, also shown in Fig. 1.2, forms at the center of the rotating fluid due to the centrifugal force of the impeller throwing the fluid out to the walls. In some cases, the surface vortex can reach the impeller, causing air entrainment. Solid body rotation can cause classification, stratification, and separation in multiphase systems and is undesirable for mixing. Surface and wall baffles are useful in hindering the formation of the central vortex, surface entrainment, and the detrimental effects of solid body rotation.

There are other types of vortices which are important in mixing: impeller-bound vortex systems and baffle-bound vortex systems, as well as toroidal vortices, free vortices, and vortices shed spatially throughout the mixing tank. These will be discussed in later chapters of this book.

Impeller Geometries

Impellers are typically classified according to mixing regime, laminar (i.e., creeping flow), or turbulent mixing. For laminar mixing, the impeller diameters approach the size of the tank, since the transport of momentum by laminar flow is poor. The impeller is required to bring fluid motion to the entire contents of the tank. For good mixing performance, the impellers should be three dimensional, moving throughout the tank during rotation. Due to mechanical stability problems and low batch volumes, the diameters of laminar impellers tend to be relatively small. In most laminar flow applications, baffles are not needed and can cause poor mixing behavior.

Typical laminar impellers are helical ribbons, screws, helical ribbon screws, and anchor impellers as shown in Fig. 1.3. Sigma, MIG, disks, paste rollers, high shear, and gate impellers are also used in laminar mixing applications. As is evident in Fig. 1.3, the geometries of laminar impellers are quite complex. Geometric variables include the tank diameter T, the impeller diameter D, the blade width W, the pitch p, the impeller wall clearance C, and the off-bottom clearance C_b. If a draft tube is used, as shown with the screw impeller in Fig. 1.3, the geometric variables include the draft tube diameter D_t, the length L_t, and the off-bottom clearance C_t of the draft tube.

Large-diameter impellers approaching the tank diameter are not used in turbulent mixing since they are not needed. Turbulent flow transports momentum well, and, hence, the impellers are typically one-fourth to one-half the tank diameter. Classification is further refined in turbulent flow regime into axial and radial flow impellers; the classification started as early as 1942 (Hixson and Baum, 1942). Axial

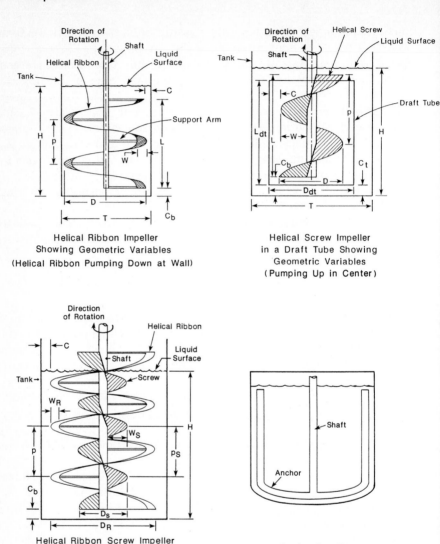

Figure 1.3 Laminar impellers.

flow impellers mainly discharge flow axially; radial flow impellers discharge the flow radially. Impeller geometry usually indicates the type of impeller, although the D/T ratio and clearance have significant effects.

Weetman and Oldshue (1988) offered another classification scheme for impellers in the turbulent mixing regime based upon whether the impeller was a flow, pressure, or a shear impeller. Impellers which have large pumping capacities are called flow impellers. Those which work against pressure gradients are pressure impellers, and those which produce shear are shear impellers.

Typical radial flow impellers are shown in Fig. 1.4 and include the disk style, flat blade, and curved blade impellers. The pitched blade turbine and the propeller are common axial flow impellers, also shown in Fig. 1.4.

Axial flow impellers can discharge material radially as well as axially. In particular, axial flow impellers discharge material radially at

Radial Flow Impellers

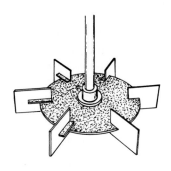

Disk Style Flat Blade Turbine
Commonly Referred to as
the Rushton Impeller

Sweptback or Curved Blade Turbine
(a Spiral Turbine)

Axial Flow Impellers

Propeller

45° Pitched Blade Turbine

Figure 1.4   Turbulent impellers.

low off-bottom clearances and large D/T ratios. At a D/T above 0.5, the centrifugal forces of the impeller blades dominate the flow, and the impeller discharges material radially (Zweitering, 1958). This phenomenon is not well studied, however, because of the high D/T ratio. At low clearances, there is no place for the material to flow in the axial direction.

Most studies in the published literature have used radial flow impellers. However, axial flow impellers are equally important. The major applications for axial flow impellers have been in blending and in solid suspension applications where the axial flow of the impeller sweeps solids off the tank bottom. The axial flow is usually discharged downward in this case. Axial flow impellers pumping upward tend to be viewed as a misapplication. However, in many applications, pumping upward should be considered and may be desirable, particularly in multiple impeller systems (Hixson and Baum, 1942; Chapman et al., 1983). Pumping up or pumping down is determined by the rotation and geometry of the impeller. The behavior of the central surface vortex has not been extensively studied for axial flow impellers pumping up.

The impeller types in Fig. 1.4 are the most commonly used industrial impellers for turbulent mixing. There are also hydrofoil or fluidfoil impellers, which have blades shaped as a hydrofoil at a low angle of attack. The geometry minimizes the formation of blade bound vortices and increases axial pumping efficiency of the impeller. Different specialty impellers include high shear impellers with pointed blades and disk style turbines with concave blades for gas dispersion.

Turbulent impellers have been used in laminar mixing, but their success is limited. Laminar impellers used in turbulent applications have not been extensively studied although they should perform well in such applications. Overall, the selection of impeller is dependent upon the application and the process objectives to be performed. There is no universal impeller type or tank geometry which is optimum for every application.

Additional Comments on Geometry

Impeller eccentricity and location.   The impeller shaft does not have to be placed along the centerline of the tank. The asymmetric placement of the impeller shaft from the tank centerline is referred to as eccentricity. Impeller eccentricity can be in all three directions. Various entry locations are possible; side-entering and bottom-entering impellers are common.

Tank geometry.   Tank bottoms can have any shape, from dished bottoms with cones placed on the bottom in the tank center, to dished bottoms, to flat bottoms, to conical bottoms, where the entire bottom is shaped as an inverted cone. Tanks also do not have to be open at the top to the atmosphere or a vapor space. In a full closed tank, the central vortex will not form, although solid body rotation will still occur under turbulent flow conditions if wall baffles are not present.

Clearance. In turbulent flow, clearance is the distance between the impeller and tank bottom. Clearance generally affects the overall flow patterns in an agitated tank for both radial and axial flow impellers. Axial flow impellers, placed close to the tank bottom, will discharge radially. The discharge flow angle from radial flow impellers is affected by clearance as well. At low clearances, the flow tends to be deflected upward. This deflection will be referred to as a clearance shift. Generally, flow patterns will change as a function of clearance.

Clearance in laminar mixing is defined as the distance between the impeller blade and tank wall or draft tube. The distance between the impeller and the tank bottom or the distance between the draft tube and tank wall is of lesser importance. The clearance or the distance of the draft tube from the tank bottom is important in draft tube studies.

Draft tubes. A draft tube is a tube in which the impeller is positioned to create strong axial flow. Its diameter is usually slightly larger than the impeller diameter. However, draft tubes can be as large as the tank diameter. The purpose of draft tubes is usually to promote more uniformity in the tank. Draft tubes are used in crystallization processes for crystal suspension and, in laminar mixing, usually with the screw impeller. The clearance of the draft tube from the tank bottom has a significant effect on performance in some applications.

Stator rings. Although not well studied, another geometric arrangement is the stator ring which is a stationary ring of vanes placed around the discharge perimeter of an impeller to redirect the impeller discharge flow. Stator rings are baffles placed very close to the impeller and have the same effect as baffles in hindering the formation of solid body rotation.

Pragmatic comments on geometry. Bissell et al. (1947) provided advice on the design and use of tank internals such as coils, baffles, draft tubes, and bearing supports while noting the need for vertical and horizontal fluid motion to accomplish mixing. Off-center and side-entering positioning of agitators were noted for their positive effects on general circulation patterns. The use of corner fillets was also mentioned. Vertical wall baffles were noted to cause strong vertical flow which promoted good circulation. In applications requiring bottom draw-off from cone bottom tanks, baffles were recommended to extend into the cone. In applications where wall baffles were not practical, a stator ring surrounding the impeller was recommended as a means to prevent solid body rotation. Steady bearings for the shaft were not recommended. Instead, impeller stabilizers were recommended up to shaft lengths of 4.5 m. For mixing of high-viscosity liquids, wall baffles were not needed. Comments on clearances of heating coils from the wall were given by Bissell et al. Baffle coils were also mentioned as an alternative to wall baffles. Draft tubes were not recommended except where there was a need to feed material directly into the impeller. For ordinary mixing problems, draft tubes were not necessary. Common-sense remarks were also made concerning feed locations. The use of false bottoms, dissolving baskets, and the general protection of the impeller were noted in solid suspension processes.

Lyons (1948) noted that the useful limits of actual mixer designs had to be determined using the full-scale equipment. Practical considerations in mixer design included the type of circulation patterns generated, position of inlets and outlets, the location of shaft and amount of overhang, and method of baffling. For the different impellers, Lyons noted that the pitched blade turbine produced both axial and radial flow and that propellers had a tapered blade tip which produced a large axial flow and minimized the effect of centrifugal forces and radial flow. An open cone impeller for dense fibrous slurries was mentioned which induced vertical flow through the cone. For gas dispersion applications, a stator ring with a hooded cover caused a downward flow of the impeller discharge which increased the residence time of the gas in the tank. Baffles and stator rings were noted generally to increase the efficiency of mixing, and the standard geometry in Fig. 1.1 was presented as being a basic geometric configuration to follow. Steady bearings were not recommended, and the overhung shaft without steady bearings was considered optimum engineering design.

Rushton (1954) discussed blending in large tanks in which mixing was accomplished by high-velocity streams entraining low-velocity fluid in the tank. Rushton noted that mixers produced equivalent flows in tanks for less energy than external recycle pumps. Rushton recommended large-diameter discharge turbulent jet streams from a properly positioned, side-entering agitator for blending in large-diameter tanks. Bates (1959) provided a discussion concerning bench-scale studies including advice on impellers, vessel size, flow patterns, multiple impellers, impeller location, impeller diameter, power measurements, and scaleup. Salzman et al. (1988) discussed a new composite impeller for abrasive flows.

Nongeometric Terms

Several nongeometric terms deserve mention.

Power draw and power consumption.   Power draw is the power transfer from the impeller to the fluid causing fluid motion. This power is eventually dissipated by the viscous heating of the fluid. Power dissipation is termed power consumption and is equal to the power draw at steady state by an energy balance.

Similarity.   There are three basic types of similarity which are of interest in mixing: geometric similarity, kinematic similarity, and dynamic similarity. Geometric similarity occurs when two different tank sizes have the same geometry and the dimensions scale by a multiplication factor. Kinematic similarity occurs when the respective velocities in two geometrically similar tanks differ by a multiplication factor. Dynamic similarity occurs where the respective forces differ by a multiplication factor. Generally, geometric similarity can be grossly obtained in practice but is rarely obtained exactly between two tank sizes. Relatively little experimental information is known about the requirements necessary to obtain kinematic and dynamic similarity although having the same impeller tip velocity in geometrically similar tanks is often stated as a condition needed to obtain kinematic similarity between the two tanks.

Scale.   Scale is an often-used term in mixing and indicates the relative size of a process. In a sense, scale is the multiplication factor in geometrical similarity. In turbulence theory, scale has a similar meaning. Scaleup or scaledown is the design process by which equipment and operating conditions are specified for a given size based on process information obtained at another size. This is also referred to as translation.

How processes change, going from one vessel size to another, is always difficult to determine and should be noted in reviewing mixing research. The mixing occurring at one scale is most likely quite different from the mixing at another scale.

Cavitation.   The impellers cited above are considered noncavitating impellers; the pressure drop occurring across the impeller blades is relatively small. However, cavitation is also a function of fluid temperature and the vapor pressure of the fluid. Obviously, if the fluid is near its boiling point, cavitation will occur with these impellers.

Typically, the static or ambient pressure is not a variable of study in mixing. Most of the work reported in this book is for low-pressure systems of 1 to 10 atm.

Some Older Literature

Some older literature deserves recognition. Wood et al. (1922) noted the lack of the "theory of stirring" in their mixing time study using a conductivity method. Wood et al. discussed the aspects of vertical and horizontal uniformity and noted how ineffective solid body rotation was in accomplishing mixing. Field (1922) mentioned that off-center positioning of the agitator promoted mixing in an unbaffled tank. The breakup of solid body rotation by wall baffles and the various process classifications in mixing were also mentioned.

Valentine and MacLean (1931) of the Turbo Mixer Corp. of New York noted the general process requirements of any mixer as: (1) to bring motion to the entire body of material in the tank by having both horizontal and vertical flows occurring in the tank, and (2) to move repeatedly the entire material into a zone of intense mixing. Agitator designs, as presented by Valentine and MacLean, were fairly well fixed in 1931 as they are today. Valentine and MacLean did note the use of hollow shaft designs and special baffles plates to increase gas residence times for gas dispersion processes.

Valentine and MacLean (1934) requested knowledge regarding the exact method by which momentum was transferred from the mixing element to the liquid and fundamental data with regard to mixing. MacLean and Lyons (1938) spoke of the problem of translation (i.e., scaleup) and the importance of liquid circulation. In their review, they mentioned some of the important translation factors such as: high circulation in the tank, high shear rates at one point in the tank, high velocity over the tank walls or tank bottom, high volume/low velocity circulation, etc. The major uses of agitation equipment in 1938, as noted by MacLean and Lyons, were chemical reactions requiring interfacial area, blending applications, and dissolution-solid suspension applications. These major areas have not changed greatly from the viewpoint of equipment usage.

Bissell (1938) noted the need for mixing criteria and recommended side-by-side comparisons of mixer designs for evaluating performance. Gunness and Baker (1938) noted the need for standards and tests in mixing. Testing involves two major areas: (1) the establishment of the optimum conditions for a working unit, and (2) the determination of the optimum design for a particular problem. They noted that testing using actual operating units eliminates the problems of translation. They discussed interfacial phenomena, the need for velocity measurements, rates of dissolution, particle-suspending capacity of agitators, agitation intensity, striae, high-speed motion pictures (1000 fps), limited use of optical methods, and specific mixing indices. Miller and Rushton (1944) suggested treating mixer performance based on three characteristics: power draw, discharge capacity, and discharge velocity. Lyons and Parker (1954) suggested the use of light streaks on photographs to determine the pumping capacity of impellers.

Valentine and MacLean (1935) reviewed the history of mixing equipment, briefly, noting the appearance of the rotary paddle mixer in 1600 to 1800. In Europe, the roller mixer for handling heavy pastes appeared in 1789, apparently designed by Ziborghi. A helical ribbon mixer appeared in 1829 designed by Lasgorseix. Boland in 1853 patented a screw-paddle mixer, also in Europe. The colloid mills and homogenizer appeared in the first part of the twentieth century. The date of the first patenting of a turbine mixer in the United States was 1890.

Lyons and Parker (1954) provided examples of impeller designs by leading equipment manufacturers and dates of equipment introduction in the United States. Examples include: pitched turbines made by New England Tank and Tower Co., before 1910; a shrouded turbine made by Johnson-Turbo-Mixer, 1913; a center disk turbine and a shrouded-covered gas type turbine made by Turbo-Mixer, 1934–35; and a vaned disk gas turbine by International Engineering Co., 1939. The last three designs were essentially the disk style turbine or the Rushton impeller as it is called now.

In these early articles on mixing, it is obvious that mixing was identified as an area for much needed research and that, although qualitative understanding was fairly high, quantitative understanding was very much needed. The early requests for an impeller-tank geometry which provided a specific desire, say, high velocity along the tank bottom, are essentially requests for solutions of the fundamental equations governing the process: equations of continuity, of motion, etc. Such solutions may be possible today, but experimental tests are still necessary.

Dimensional Analysis

Dimensional analysis is the basis from which mixing research first began. The following problem provides an example of dimensional analysis as applied to mixing.

Example Problem Power P in an agitated tank is thought to be a function of impeller diameter D; impeller blade width W; impeller blade number n; impeller rotational speed N; impeller blade angle α; tank diameter T; baffle number n_b; gravity g; density ρ; and viscosity μ of the fluid. Find the independent dimensionless groups which could be used to correlate power from these variables.

solution There are 11 variables: P, D, W, n, N, α, T, n_b, g, ρ, and μ. Three of these are dimensionless: blade number, baffle number, and angle. Choosing mass M, length L, and time t, as the fundamental constraining units, the dimensional matrix can be formed as:

Dimensions	Variables										
	P	D	W	n	N	α	T	n_b	g	ρ	μ
Mass	1	0	0	0	0	0	0	0	0	1	1
Length	2	1	1	0	0	0	1	0	1	−3	−1
Time	−3	0	0	0	−1	0	0	0	−2	0	−1

The rank of the largest submatrix of the dimensional matrix having a nonzero determinant is the number of constraints the units place on the variables. As a result, the number of independent dimensionless groups which can be formed is equal to the number of variables minus the rank of the dimensional matrix minus the number of variables which are dimensionless. Taking the last three columns as the first submatrix to test for a nonzero determinant:

$$\begin{vmatrix} 0 & 1 & 1 \\ 1 & -3 & -1 \\ -2 & 0 & -1 \end{vmatrix} = 0 - 1(-1 - 2) + 1(0 - 6) = -3$$

The rank of the matrix is equal to the order of its largest nonzero determinant. In this case, the order of the largest matrix having a nonzero determinant is 3. As a result, there are five [(11 variables) − (3 constraints) − (3 dimensionless variables)] dimensionless groups which can be formed using the variables P, D, W, N, T, g, ρ, and μ. These can become the impeller Reynolds number, the power number, the Froude number, and two length ratios W/D and D/T. A possible functional form which could be used to correlate power is then:

$$f\left(N_P, N_{Re}, N_{Fr}, \frac{W}{D}, \frac{D}{T}, n, n_b, \alpha\right) = 0$$

The geometric ratios and other dimensionless geometric quantities like number form the basis for geometric similarity. The Reynolds and Froude numbers are the bases for dynamic similarity.

A reference text for dimensional analysis procedures, as shown above, is Bober and Kenyon (1980).

Nomenclature

B	baffle width
C	impeller off-bottom clearance
C_b	bottom clearance
C_t	draft tube length
D	impeller diameter
D_t	draft tube diameter
g	acceleration of gravity
g_c	gravitational constant
H	liquid height
L	impeller length
L_t	draft tube length
M	mass

N_{Fr} Froude number, N^2D/g

N_P power number, $Pg_c/\rho N^3D^5$

N_{Re} impeller Reynolds number, $\rho ND^2/\mu$

N impeller rotational speed, revolutions per second

n blade number

n_b baffle number

P power

p impeller pitch

T tank diameter

t time

W blade width

Greek symbols

θ time

ρ density

μ viscosity

μ_a apparent viscosity

μ_t turbulent viscosity

ν_t turbulent kinematic viscosity

α blade angle

References

Reference books

Bober, W., and R. A. Kenyon, *Fluid Mechanics*, Wiley, New York, 1980.

Holland, F. A., and F. S. Chapman, *Liquid Mixing and Processing in Stirred Tanks*, Reinhold, New York, 1966.

Johnstone, R. E., and M. W. Thring, *Pilot Plants, Models and Scale-Up Methods in Chemical Engineering*, McGraw-Hill, New York, 1957.

Nagata, S., *Mixing: Principles and Applications*, Kodansha Ltd., Tokyo, and Wiley, New York, 1975.

Nauman, E. B., and B. A. Buffham, *Mixing in Continuous Flow Systems*, Wiley, New York, 1983.

Oldshue, J. Y., *Fluid Mixing Technology, Chemical Engineering*, McGraw-Hill, New York, 1983.

Patton, T. C., *Paint Flow and Pigment Dispersion*, 2d ed., Wiley-Interscience, Wiley, New York, 1979.

Sterbacek, Z., and P. Tausk, *Mixing in the Chemical Industry*, translated by K. Mayer, edited by J. R. Bourne, Pergamon, Oxford, England, 1965.

Contributed books and chapters in books

Brodkey, R. S., (ed.), *Turbulence in Mixing Operations*, Academic, New York, 1975.

Dickey, D. S., "Liquid Agitation," in *Handbook of Chemical Engineering Calculations*, edited by N. P. Chopey, McGraw-Hill, New York, 1984.

Harnby, N., M. F. Edwards, and A. W. Nienow (eds.), *Mixing in the Process Industries*, Butterworth, Stoneham, Mass., 1986.
Patterson, G. K., "Turbulent Mixing and Its Measurement," in *Handbook of Fluids in Motion*, edited by N. P. Cheremisinoff and R. Gupta, Ann Arbor Science, Ann Arbor, Mich., 1983, Chap. 5.
Skelland, A. H. P., "Mixing and Agitation of Non-Newtonian Fluids," in *Handbook of Fluids in Motion*, edited by N. P. Cheremisinoff and R. Gupta, Ann Arbor Science, Ann Arbor, Mich., 1983, Chap. 7.
Uhl, V. W., and J. B. Gray (eds.), *Mixing*: Vols. I and II, Academic, New York, 1966.
Uhl, V. W., and J. B. Gray (eds.), *Mixing*: Vol. III, Academic, Orlando, Fla., 1986.
Ulbrecht, J. J., and G. K. Patterson (eds.), *Mixing of Liquids by Mechanical Agitation*, Gordon Breach Science Publishers, New York, 1985.

Proceedings

Fluid Mixing, I. Chem. E. Sym., Series No. 64, Bradford, England, The Inst. of Chem. Engrs., Rugby, England, 1984.
Fluid Mixing II, I. Chem. E. Sym., Series No. 89, Bradford, England, The Inst. of Chem. Engrs., Rugby, England, 1984.
Mixing—Theory Related to Practice, AIChE—I. Chem. E. Sym., Series 10, London, England, AIChE and The Instit. of Chem. Engrs., London, England, 1965.
Proceedings of the First European Conference on Mixing, edited by N. G. Coles, Cambridge, England, BHRA Fluid Eng., Cranfield, England, 1974.
Proceedings of the Second European Conference on Mixing, edited by H. S. Stephens and J. A. Clarke, Cambridge, England, BHRA Fluid Eng., Cranfield, England, 1977.
Proceedings of the Third European Conference on Mixing, edited by H. S. Stephens and C. A. Stapleton, York, England, BHRA Fluid Eng., Cranfield, England, 1979.
Proceedings of the Fourth European Conference on Mixing, edited by H. S. Stephens and D. H. Goodes, Noordwijkerhout, Netherlands, BHRA Fluid Eng., Cranfield, England, 1982.
Proceedings of the Fifth European Conference on Mixing, edited by S. Stanbury, Wurzburg, Germany, BHRA Fluid Eng., Cranfield, England, 1985.
Proceedings of the Sixth European Conference on Mixing, published by AIDIC, Pavia, Italy, BHRA Fluid Eng., Cranfield, England, 1988.

Papers

Bates, R. L., *Ind. Eng. Chem.*, **51**, 1245, 1959.
Bissell, E. S., *Ind. Eng. Chem.*, **30**, 493, 1938.
Bissell, E. S., H. C. Hesse, H. J. Everett, and J. H. Rushton, *Chem. Eng. Prog.*, **43**, 649, 1947.
Chapman, C. M., A. W. Nienow, M. Cooke, and J. C. Middleton, *Chem. Eng. Res. Des.*, **61**, 71, 1983.
Field, C., *Trans. Am. Inst. Chem. Engrs.*, **14**, 444, 1922.
Gunness, R. C., and J. G. Baker, *Ind. Eng. Chem.*, **30**, 497, 1938.
Hixson, A. W., and S. J. Baum, *Ind. Eng. Chem.*, **34**, 194, 1942.
Lyons, E. J., *Chem. Eng. Prog.*, **44**, 341, 1948.
Lyons, E. J., and N. H. Parker, *Chem. Eng. Prog.*, **50**, 629, 1954.
MacLean, G., and E. J. Lyons, *Ind. Eng. Chem.*, **30**, 489, 1938.
Miller, F. D., and J. H. Rushton, *Ind. Eng. Chem.*, **36**, 499, 1944.
Rushton, J. H., *Petr. Refiner*, **33**, 107, 1954.
Salzman, R. N., W. Webster, G. Muratore, and P. K. Page, *Proc. 6th Eur. Conf. on Mixing*, Pavia, Italy, BHRA Fluid Eng., Cranfield, England, 79, 1988.
Valentine, K. S., and G. MacLean, *Chem. Met. Eng.*, **38**, 234, 1931.
Valentine, K. S., and G. MacLean, *Chem. Met. Eng.*, **41**, 237, 1934.
Valentine, K. S., and G. MacLean, *Chem. Met. Eng.*, **42**, 220, 1935.

Weetman, R. J., and J. Y. Oldshue, *Proc. 6th Eur. Conf. on Mixing*, Pavia, Italy, BHRA
    Fluid Eng., Cranfield, England, 43, 1988.
Wood, J. C., E. R. Whittemore, and W. L. Badger, *Chem. Met. Eng.*, **27**, 1176, 1922;
    Trans. Am. Inst. Chem. Engrs., **14**, 435, 1922.
Zwietering, T. N., *Chem. Eng. Sci.*, **8**, 224, 1958.

2

Power Consumption in Turbulent Mixing

Introduction

Power draw is the energy per time which is transferred from the impeller to the fluid. It is an integral quantity fundamental to mixing and dispersion processes since energy is needed to cause fluid motion necessary for mixing. The calculation of power draw can be performed in many ways depending upon the process, mixing regime, and fluid. However, for single-phase turbulent flow, power calculation has been mainly approached through dimensional analysis and experimental measurement of torque. Although not fully developed, drag analysis represents an alternate means for power calculations and provides the fundamental basis necessary to understand the results of dimensional analysis.

Dimensional Analysis

Dimensional analysis is based upon the fact that natural laws are independent of units or, in other words, the units constrain the variables. Power basically has the same fundamental units as the product of ρ, N^3, and D^5 or the product of μ, N^2, and D^3, and either dimensionless group $P/\rho N^3 D^5$ or $P/\mu N^2 D^3$ can be formed. The dimensionless group for turbulent flow is typically $Pg_c/\rho N^3 D^5$ and is called the power number. This dimensionless group will be a function of other geometric, kinematic, and dynamic dimensionless groups. For an agitated tank, a traditional relationship for power in the form of a series of assumed power laws can be written as:

$$\frac{Pg_c}{\rho N^3 D^5} = K\left(\frac{\rho N D^2}{\mu}\right)^a \left(\frac{N^2 D}{g}\right)^b \left(\frac{T}{D}\right)^c \left(\frac{C}{D}\right)^d \left(\frac{H}{D}\right)^e$$

$$\times \left(\frac{p}{D}\right)^f \left(\frac{W}{D}\right)^g \left(\frac{L}{D}\right)^h \left(\frac{n_1}{n_f}\right)^i \left(\frac{n_{b1}}{n_{bf}}\right)^j \left(\frac{b}{D}\right)^k \left(\frac{E}{D}\right)^l \left(\frac{C_t}{D}\right)^m \left(\frac{D_t}{D}\right)^n \cdots \quad (2.1)$$

The first dimensional group is the impeller Reynolds number, the second is the Froude number, and the rest account for the effects of geometry (e.g., impeller clearance and baffle width) which includes number effects (e.g., number of blades, baffles, and impellers). Geometric effects are typically expressed as length or number ratios for which some reference geometry has been defined.

Geometric similarity. If the geometry is fixed, a standard geometry is selected, and geometric similarity is maintained, Eq. (2.1) simply reduces to:

$$N_P = K(N_{\text{Re}})^a \quad (2.2)$$

for turbulent flow with no central vortex. Froude number has no effect on power number (Nagata et al., 1957; Novak et al., 1982) under these conditions. At high impeller Reynolds number, Eq. (2.1) reduces to:

$$N_P = K' \quad (2.3)$$

for baffled tanks, since power number depends mainly on the impeller geometry. For unbaffled tanks, there is a weak dependence upon impeller Reynolds number, and Eq. (2.2) applies.

K' is constant for a specific impeller-tank geometry. However, variations in K occur, and reasons for these variations will be discussed later. The justification for Eq. (2.3) rests upon geometric similarity and explains why geometric similarity is so popular. For geometrically similar systems with constant K', the study of the various geometric effects is no longer needed since geometry is no longer a variable. Experimentally it is easy to measure and correlate power data with such a simple equation. Scaleup and scaledown are easy as well since geometry is not a variable.

Negative aspects of the dimensional analysis are present, however. First, there is usually no attempt to establish the fundamental mechanism or mechanisms for power draw of the impeller when dimensional analysis is used. Second, experimental studies are needed to establish the exponents in Eq. (2.1). Third, there is no understanding, intuitive or otherwise, as to whether the experimental results and exponents are reasonable or why the variables form into the dimensionless groups as they do. Furthermore, the use of Eq. (2.1) presupposes that the phenomena will follow the various power laws as assumed in Eq. (2.1) which may not be the case. Phenomena do not have to follow a power law or a product of power laws. Equation (2.1) is also limitless due to the difficulty in the characterization of geometry. Any pre-selected standard geometry is suspect and can always be criticized as not being optimum.

The power number, as given above, is only correct with respect to geometrically similar systems but is not correct generally as will be shown later. As formulated above, the power number has been universally accepted and, as a result, has been misapplied in some cases. This is particularly true for those cases where geometric similarity has not been maintained or has been superficially maintained.

Drag-Velocity Analysis

The drag basis to power draw is more mechanistic and approaches the calculation of power draw using a drag coefficient. Under turbulent flow conditions, an impeller blade experiences the two forms of drag: skin or frictional drag and form or pressure drag. Typically, these mechanisms are considered in terms of a drag coefficient where the drag force becomes:

$$F = \tfrac{1}{2}C_D\rho U^2 A \qquad (2.4)$$

where A is the projected area or surface area of the body, ρ is the density of the fluid, and U is the relative velocity between the blade and fluid. If the drag coefficient is assumed to be a constant, then force is proportional to density, velocity squared, and projected area or:

$$F \propto \rho U^2 A \qquad (2.5)$$

Power is essentially force times velocity or:

$$P \propto FU \qquad (2.6)$$

By equating the relative velocity to the blade velocity:

$$P \propto \rho U^3 A \qquad (2.7)$$

For geometrically similar systems, the projected area of the impeller is proportional to the impeller diameter squared or:

$$A \propto D^2 \qquad (2.8)$$

since the impeller diameter D is typically the characteristic length upon which every other dimension is based for geometric similarity. The impeller tip speed U is the characteristic velocity and is πND using the impeller diameter and rotational speed. Equation (2.7) can now be written as:

$$P \propto \rho N^3 D^5 \qquad (2.9)$$

or establishing an equality using the power number:

$$N_P = \frac{Pg_c}{\rho N^3 D^5} \qquad (2.10)$$

As can be observed, N_P behaves similarly to a drag coefficient.

The same expression can be obtained from a differential drag force balance (Hixson et al., 1942) on a blade section:

$$dF = \tfrac{1}{2}\, C_D \rho U^2 W\, dr \qquad (2.11)$$

where U is the approach velocity of the fluid relative to the blade. W is the width of the blade, $W\, dr$ is the projected area of the blade section, and C_D is the drag coefficient of the blade section. The differential torque is then:

$$dT_0 = r\, dF = \tfrac{1}{2}\, C_D \rho U^2 W\, r\, dr \qquad (2.12)$$

Equating the relative velocity of the fluid to the blade with the impeller rotational speed, Eq. (2.12) becomes:

$$dT_0 = 2C_D N^2 \rho W r^3\, dr \qquad (2.13)$$

as shown in Fig. 2.1. Integrating Eq. (2.13) over the impeller radius:

$$T_0 = K \rho N^2 R^4 W \qquad (2.14)$$

For geometrically similar systems, the impeller radius R and blade width W are proportional to the impeller diameter D. Torque is therefore:

$$T_0 = K'' \rho N^2 D^5 \qquad (2.15)$$

and power:

$$P = K''' \rho N^3 D^5 \qquad (2.16)$$

since power is the product of torque and the impeller rotational speed. Power number can be obtained by rearrangement of Eq. (2.16) to show the independence of power number from the impeller Reynolds number for turbulent flow when K''' is constant.

Although the development of the power number appears to be straightforward, an average drag coefficient C_D has been used. A more accurate integration would require knowing C_D as a function of blade position, and such data are not readily available (Tay and Tatterson, 1985). The relative velocity between the blade and fluid may not equal

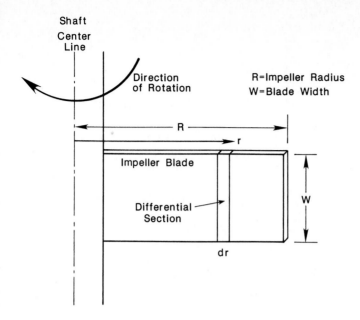

Drag Force per Differential Blade Section
$$\alpha \ \rho \ N^2 r^2 \ Wdr$$
Torque α (Drag Force) (Impeller Radius)

Figure 2.1   Drag force on an impeller blade section.

the impeller rotational speed. In fact, the solid body rotation in un-baffled tanks lowers the relative velocity between the fluid and the impeller blade. Equation (2.2) occurs for unbaffled tanks under turbulent flow conditions because of this effect. Furthermore, the flow velocities around the impeller are not actually constant with time. Power and torque fluctuate with time when the flow velocities fluctuate.

For laminar or creeping flow, inertial forces are not important, and viscous forces completely dominate the flow. When a solid body moves through a fluid, the opposing force of the fluid against the motion can be written as:

$$F \ \alpha \ \mu UD \qquad (2.17)$$

where μ is the fluid viscosity. If the same characteristic length D and same characteristic velocity πND are assumed as for turbulent conditions, Eq. (2.6) for laminar flow conditions becomes:

$$P \ \alpha \ \mu N^2 D^3 \qquad (2.18)$$

or:
$$\frac{Pg_c}{\mu N^2 D^3} = K \qquad (2.19)$$

The simple analysis also shows that:

$$N_P \propto N_{\text{Re}}^{-1} \qquad (2.20)$$

for low Reynolds number flows.

The mechanisms for power consumption are dependent upon drag regimes, that is, Stokes drag, skin drag, and form or pressure drag. Typically, for the impeller shapes which will be discussed, form or pressure drag dominates in turbulent flow regime and, hence, the dependence of the power number upon impeller geometry, i.e., its form.

Other factors which cause changes in form drag of the impeller will also affect power number. Blade width and blade number are examples. Given the crude form of the drag analysis, power draw of an impeller should be approximately proportional to blade number above some base value for the shaft and impeller hub since blade number is the number of projected blade areas of the impeller. Power draw should be proportional to blade width, since increasing blade width increases the projected areas of the blades. This assumes that the drag coefficient C_D remains constant with blade width. Data on how C_D varies with blade width are required. There are very little drag coefficient data of this type, partly because of the difficulty in establishing the relative velocity between the impeller blades and the surrounding flow for a rotating impeller.

For turbulent flow, actual power data show the approximate effects of blade number and blade width as reported by O'Connell and Mack (1950), Bates et al. (1963), Nagata et al. (1955; 1956), O'Kane (1974), and Rushton et al., (1950). In detail, power varies with the 1.2 to 1.45 power of blade width and the 0.75 to 0.85 power of blade number which shows that the simple drag analysis is only approximately correct.

The effects of fluid density and blade angle on power draw are apparent from the drag-velocity analysis.

The previous discussion assumed no blade-blade effects or interactions. It should be apparent that as blade width or blade number is increased drastically, the flow around the impeller blades, the drag coefficient, and relative velocity will change. In this situation, the drag-velocity concept is still applicable; however, the drag coefficient and velocity data are necessary. There is a limited amount of pressure distribution data for impeller blades in the mixing literature. Such data would be of interest for pressure or form drag calculations.

From the drag-velocity analysis, a Froude number effect upon power number is not present except through the effect of Froude number on the drag, relative velocity, and the pressure coefficients of the blade.

Other effects, if they are to have a substantial impact upon power draw, must affect the fluid mechanics around the impeller, the drag coefficients of the blades, and/or the projected area of the blades. The effects of minor changes in geometry have to be determined experimentally because of the complexity of turbulent flow. Other methods are not available.

Flow fields. Power draw of the impeller has to be transferred from the impeller to the surrounding fluid locally, as was shown in the drag-velocity approach to power draw. The power, thus transferred, is dissipated in fluid mixing and wall friction and into the heating of the fluid. The impeller and the fluid mechanical environment locally around the impeller are responsible for the transfer of the power draw. Except for viscoelastic fluids and perhaps highly shear thickening fluids, the fluid flows around the blades as the impeller blades move forward through the fluid. The fluid is not pushed by the impeller blades. Low pressure regions are created behind the blades, and the fluid reaches its highest accelerations as it flows over and behind the impeller blades. For the most part, the fluid flowing in front of the impeller blade undergoes very little acceleration as it passes through the impeller.

Viscoelastic fluids show flow reversals from that of Newtonian fluids, and shear thickening fluids solidify under shear in the impeller region.

Vertical wall baffles. For radial discharge impellers in an unbaffled tank, the fluid and impeller establish solid body rotation which is not conducive for mixing. Tank contents also tend to leave the tank more easily when the fluid is in solid body rotation, since the fluid level rises along the wall. Baffles disrupt the solid body rotation by stopping the tangential flow around the tank perimeter and by redirecting the flow vertically. Baffles also increase or maximize the power draw from the impeller to the fluid. The particular baffle width in the standard design shown in Chap. 1 was selected to maximize power input. The objective in practice has been to maximize mixing performance which has been equated directly to maximizing power input. Baffles also prevent the formation of the central vortex and also tend to minimize surface entrainment.

The central vortex is particularly undesirable in mixing. Uneven blade loading and tank wall vibrations can occur because of the presence of the vortex. The vortex causes complications in process scaleup or scaledown.

Power Studies

The following review of the research papers on power is given more or less in chronological order and provides the various data sources and studies which went into the power curve development. The information is presented in the form of a literature review since most of the studies were fairly independent of each other and a variety of systems were studied.

The objectives of many of the reported power studies were: (1) to establish the relationships necessary to calculate impeller power draw and the effects of geometry on power draw, (2) to establish a standard system geometry which maximized impeller power draw, and (3) to establish a scalable system (e.g., an impeller-tank geometry) where the flow and mixing behave similarly upon scaleup and scaledown.

Early studies

White and Brenner (1934), using an equation similar to Eq. (2.1), were the first to determine the various power law exponents using data of White, Brenner, Phillips, and Morrison (1934). The data for the correlation were obtained in an unbaffled tank, 1.3 m in diameter, with paddles approaching the tank diameter in shallow liquid depth (0.5 m). Their final power correlation was given as:

$$\frac{P}{D\mu N^2 T^{1.1} W^{0.3} H^{0.6}} = 0.000129 \left(\frac{\rho N D^2}{\mu}\right)^{0.86} \tag{2.21}$$

which has the same form as Eq. (2.16).

The equation fitted the experimental data for Reynolds number from 10^4 to 10^5 but diverged from the data below an impeller Reynolds number of 10. White et al. noted that the transition from turbulent to viscous flow occurred between Reynolds numbers of 2000 and 20,000. They further showed that viscosity played a minor role in the turbulent flow regime. White and Sumerford (1936) reported the graphical form of their power curve.

Hixson and Luedeke (1937), in a rather remarkable study for its time, examined wall drag occurring in an agitated tank. The tank diameter was used as the characteristic length in the Reynolds number. Effects of impeller pitch, clearance, and liquid depth were studied. The work was carried out in an unbaffled tank where $D/T = 0.33$ and the liquid depth was equal to the tank diameter. The laminar to turbulent transition Reynolds number for the wall flows was found to be approximately 5×10^5, the same as that for normal flat plate. Power, dissipated in overcoming wall drag, was not determined in this work since this required knowing the fluid velocity adjacent to the wall.

Hixson and Baum (1942) presented torque measurements, reduced to $T_0/\rho N^2 D^5$ versus the impeller Reynolds number, for unbaffled tank-pitched blade turbine arrangements using water, 40 and 60 percent sucrose solutions, glycerol, and two viscous oils. Their power law exponents, obtained from using Eq. (2.1), agreed with White et al. Their standard design had the pitched blade turbine pumping upward. Effects of reversed rotation (pumping down), impeller pitch, area, and T/D ratio were also studied and data provided. Substantial reductions in torque and power consumption occurred when the liquid level was less than the tank diameter because of the surface entrainment of air. When the liquid level was equal to the tank diameter, no surface entrainment was noted. Effects of baffles were also studied. Hixson and Baum apparently had little difficulty with the central vortex for the pitched blade turbine pumping upward.

Stoops and Lovell (1943) performed power studies on a marine type propeller with three blades in an unbaffled tank. They correlated their data as:

$$\frac{P}{\mu N^2 D^3} = 0.56\left(\frac{\rho N D^2}{\mu}\right)^{0.81}\left(\frac{T}{D}\right)^{0.93} \qquad (2.22)$$

Power consumption was independent of liquid depth, clearance, and condition of the surface. No abrupt changes occurred in their data due to laminar to turbulent transition.

Miller and Mann (1944) performed a two-phase power and performance study with oil and water, using seven impeller designs in an unbaffled tank. The agitators included: flat two- and four-bladed paddles, 45° pitched two- and four-bladed paddles (i.e., pitched blade turbines pumping upward), a 45° pitched four-bladed paddle (i.e., pitched blade turbine), a two-bladed propeller, and a closed spiral turbine. Power was correlated through the use of a geometric mean viscosity for the two-phase mixture. Individual power curves were given for all seven impeller types as a function of impeller Reynolds number and volume fraction of the oil phase. Differences in power curves for the impeller designs were given in terms of a shape factor based on a standard geometry. The shape factors were independent of impeller Reynolds number. In the two-phase systems, liquid depth had an effect on power which was handled in terms of a correction factor. Clearance had little effect on power except at the low and high clearances. Power was also found independent of the proximity of the initial static interface level. For a 50 percent kerosene-water mixture, there was a range of impeller rotational speeds in which it was possible to have two different power draws depending upon the direction of approach to the point of operation (i.e., increasing rotational speed or decreasing rotational speed). This was explained in terms of a shear dependent effective viscosity of the emulsions formed.

Martin (1946) presented a review of power curves for 13 impellers, noting several important observations concerning the curves: (1) viscosity had little effect on power consumption except at low impeller Reynolds numbers, (2) power numbers tended to a constant value at high Reynolds numbers and were higher for baffled tanks, and (3) the constant value was reached sooner for baffled tanks at lower Reynolds numbers than for unbaffled tanks. Martin further noted that baffling reduced the tendency toward solid body rotation and increased the relative velocity between the impeller and fluid. Impellers in baffled tanks operated at a higher effective angular velocity, higher effective Reynolds number, higher drag, and a higher power number than in unbaffled tanks. For unbaffled tanks, increasing the ratio of tank diameter to impeller diameter T/D caused the power number to become constant at lower Reynolds numbers. Tank-impeller geometries, having low tank-to-impeller-diameter ratios T/D, developed solid body rotation more readily which caused the impellers to operate at lower effective Reynolds numbers.

Rushton (1946) provided a comparison of data of White and Summerford and data taken by the Mixing Equipment Co. The comparison essentially justifies the initial work of White and his associates. Olney and Carlson (1947) studied an arrowhead impeller and a spiral turbine with a stator ring for single-phase liquids and two-phase liquids. They considered the effects of impeller speed and the properties and proportions of the liquids on power consumption while maintaining the equipment size and shape dimensions constant. Two-phase power data were correlated using average viscosities and densities for the mixture. Their power number group was arranged as $P/\mu N^2 D^3$ and was plotted versus with Reynolds number from 10^2 to 10^5 having the form:

$$\frac{P}{\mu N^2 D^5} = \frac{K \rho N D^2}{\mu} \tag{2.23}$$

The power data showed an increase in power draw as the volume percent of the heavier viscous phase increased. For the two-phase liquid mixtures, the density was calculated using a volume average and the viscosity from a geometric mean:

$$\mu_m = (\mu_x)^x(\mu_y)^y \qquad (2.24)$$

where x and y refer to the volume fractions of the two-phase mixture. Olney and Carlson also suggested the following geometry be used whenever possible: impeller-to-tank diameter ratio, D/T = 0.33; blade width ratio, W/D = 0.25; liquid height ratio, H/T = 1; clearance ratio, $0.2 < C/T < 0.5$; and baffle width ratio, B/T = 0.0833. This is the first appearance of geometric standardization in the mixing literature and is the standard geometry mentioned in Chap. 1. Olney and Carlson also studied different baffling arrangements including horizontal and surface baffles.

Mack and Kroll (1948) performed a study of the effects of baffles on agitator power consumption using a simple flat paddle in water. Baffle configurations which maximized power draw were described and termed fully baffled. Under this condition the power number became independent of liquid height, impeller position, and the dimensions of the vessel including the geometry of the baffles. For non-fully baffled geometries, the effects of various baffle sizes, angle and baffle position, impeller position, liquid depth, and tank diameter on power draw were also determined. The angle which the baffles made with the wall was found to be not important. A fully baffled condition was established for the baffle and impeller arrangements where D/T was varied between 1/4 to 1/2 with three or four baffles, B/T between 1/10 and 1/12. Mack and Kroll also showed that, for turbulent flow in a fully baffled tank, the characteristic length in the power number must be the impeller diameter and not just any characteristic length of the baffled tank such as the tank diameter.

O'Connell and Mack (1950) investigated the power consumption for flat blade turbines by varying the number of blades and the dimensions of the blades for the laminar and turbulent flow regimes under a fully baffled condition. Two, four, and six rectangular blades were studied. For turbulent flow, the typically accepted power correlation at this time was:

$$Pg_c = K\rho N^3 D^5 \qquad (2.25)$$

and for laminar flow:

$$Pg_c = K\mu N^2 D^3 \qquad (2.26)$$

However, for the various widths studied, Eqs. (2.25) and (2.26) were changed to include the effect of blade width. For turbulent flow:

$$\frac{Pg_c}{\rho N^3 D^5} = K\left(\frac{W}{D}\right)^b \qquad (2.27)$$

and for laminar flow:

$$\frac{Pg_c}{\rho N^3 D^5} = K\left(\frac{\mu}{\rho N D^2}\right)\left(\frac{W}{D}\right)^b \qquad (2.28)$$

where the constants K and b were to some degree dependent upon blade number as listed in Table 2.1. The exponents for the blade width effect for turbulent flow agree approximately with the form drag analysis given above.

TABLE 2.1   Effect of Blade Number on K and b in Eqs. (2.27) and (2.28)

Blade number	Turbulent flow		Creeping flow	
	K	b	K	b
2	13.8	1.23	113	0.52
4	19.4	1.15	141	0.45
6	23.7	1.09	161	0.38

Intermediate studies

At this point, the effects of the impeller and tank geometry, density, viscosity, rotational speed, baffling, and some secondary effects on power consumption had been established in the technical literature. This essentially set the stage for several classic studies in the mixing literature.

Rushton, Costich, and Everett (1950a; 1950b) summarized a research program by Mixing Equipment Corporation which was conducted to establish the power characteristics of agitators used commercially. Data were given on five impeller types, diameters from 60 mm to 1.2 m, tanks from 0.2 m to 2.5 m, baffled and unbaffled configurations, fluid viscosities from 0.001 to 40 Pa · s (1.0 to 40,000 cP), power measurements from 0.746 to 6000 W (0.001 to 8 hp) with the impeller Reynolds number from 1 to 10^6.

Part I (1950a) of the work provided data on propellers in unbaffled tanks including the effects of tank diameter, D/T ratio, propeller pitch, partial baffling, and Froude number effects. Detailed effects of the central surface vortex were discussed. The direction of rotation for the propeller was indicated to be pumping downward. No data were apparently given for a pumping upward configuration and the effect of this configuration on the nature of the central vortex. Data were also obtained for propellers in baffled tanks in the absence of the central vortex. Eccentricity was found to be equivalent to baffling where the eccentric positioning of the propeller maximized power consumption for the propeller in the absence of a central vortex.

Part II (1950*b*) dealt with other impeller types for which data curves were given. Besides the obvious effects of density, viscosity, impeller rotational speed, and impeller diameter, other effects on power consumption were noted, and the major effects have been summarized in Table 2.2. Effects of the blade length-width ratio for the disk style impeller were also discussed but have not been reported in Table 2.2.

At this point in the development of power draw correlations, further work was needed with regard to the effects of impeller clearance, blade pitch, liquid height, and blade width, and the length of the blade of the disk style turbine.

Nagata and his associates in 1955 to 1956 published a series of four papers which also have to be considered classics in the mixing literature. Nagata, Yoshioka, and Yokoyama (1955*a*) investigated the angular velocity distribution of liquid agitated by a paddle in an unbaffled tank. The flow pattern near the center of the vessel rotated at the same angular velocity as that of the agitator and was called a force vortex zone. (This zone is essentially solid body rotation which is established in unbaffled tanks mentioned earlier.) Beyond this region, the liquid velocity varied inversely proportional to distance from the

TABLE 2.2   Various Effects on Power Consumption

	Laminar regime $N_{Re} < 10$	Turbulent regime $N_{Re} > 10,000$	
		Baffled	Unbaffled
Tank diameter	—	No effect $2.0 < T/D < 7.0$	$(T/D)^{0.91}$ for propellers $2.7 < T/D < 4.5$
Liquid height	No effect	No effect	No effect
	$2.0 < H/D < 7.0$ but further data are needed.		
Off-bottom clearance	No effect	No effect	No effect
	$0.7 < C/D < 1.6$ but further data are needed.		
Number of baffles	No effect	No effect for fully baffled conditions	N/A
Blade pitch	Minor	$(p/D)^{1.7}$ for propellers $1.0 < (p/D) < 2.5$	Higher-power draw with higher pitch.
	Further data are needed.		
Number of blades	—	$(n/n_f)^{0.75}$	—

center of the tank. Under this condition, the flow was termed a free vortex zone. Near the wall, the velocity abruptly decreased to zero. The size of the force vortex zone was established as a function of liquid viscosity, paddle width, and paddle diameter. A model was presented to explain the effect of paddle width on power consumption.

Very few studies report the accuracy of power measurements. Nagata and Yokoyama (1955b) reviewed the various methods used to measure power summarized in Table 2.3.

Two major causes for error in power measurement were given by Nagata et al. (1955b) as being: (1) static friction which causes an initial force requirement as motion commences and (2) no load dynamic friction. In the measurement of small torque magnitudes, static friction is the primary source of error. Using a dynamometer having negligible static friction, Nagata found a negligible effect of Froude number on power draw. The effect of Froude number on power and power number, observed by Rushton et al. (1950a; 1950b), was attributed to static friction by Nagata and Yokoyama.

Nagata, Yokoyama, and Maeda (1956) used their dynamometers to establish the general power characteristics for laminar and turbulent regimes for both baffled and unbaffled tanks. This paper has to be considered a classic due to its depth and stated accuracy of the dynamometers involved.

Nagata's study covered a Reynolds number range from 0.1 to 10^6 in tanks from 0.15 to 1.17 m for two-bladed paddles having dimensions: $0.3 < D/T < 0.9$ and $0.05 < W/T < 0.9$. The power equation for the laminar region, $N_{Re} < 10$ to 100, was found to be:

$$\frac{Pg_c}{\mu N^2 D^3} = 310 \left(\frac{W}{T}\right)\left(\frac{D}{T}\right)^{1.15} \tag{2.29}$$

for $0.7 < D/T < 0.9$ and $0.4 < W/T < 0.9$. Power was directly proportional to blade width. Below an impeller diameter of $0.7T$, Nagata et al. found that the exponent 1.15 and the constant 310 in the equation changed to − 0.58 and 140.

TABLE 2.3 Power Measurement Methods

1. Measurement of net power input by subtracting no load power from load power

2. Torque measurements
   a. Torque measurement on the vessel while on a frictionless turntable
   b. Reaction torque upon the stator of a motor

3. Torsional dynamometers
   a. Coiled or spiral spring (spring elongation)
   b. Strain gauge to measure axial torsion of a rotating shaft
   c. Differential gear procedure

For the intermediate viscosity region, 10 to $100 < N_{\text{Re}} < 10^4$ to 10^5, Nagata et al. provided the expression:

$$\frac{Pg_c}{\rho N^3 D^5} = A * B(N_{\text{Re}})^p \tag{2.30}$$

where

$$\log_{10}(A) = \frac{W}{0.03T + 0.42W}$$

$$\log_{10}(B) = -\frac{WD}{0.01T^2 + 0.73W * T}$$

$$p = -\left[0.17\frac{T}{W} + 1.37\left(\frac{D}{T}\right) + 1.8\right]^{-1}$$

For the low viscosity range, $N_{\text{Re}} > 10^4$ to 10^6, for unbaffled tanks:

$$\frac{Pg_c}{\rho N^3 D^5} = \left(\frac{W/T}{0.05 + 1.05W/T}\right)\left(\frac{D}{T}\right)^{-1.23}(N_{\text{Re}})^{-0.113} \tag{2.31}$$

Power numbers for baffled tanks were essentially independent of the impeller Reynolds number for $N_{\text{Re}} > 10^4$ to 10^6. For paddles of $0.2T$ in width, power numbers were typically 3.5 and 6.0 for D/T of 0.3 and 0.5, respectively.

Various other effects were investigated. For low-viscosity liquids (high Reynolds number), power was proportional to $(H/T)^{0.4}$ to $(H/T)^{0.55}$; for viscous liquids (low Reynolds number), power was found independent of liquid height. For the intermediate-viscosity range (transitional Reynolds numbers), a complex relationship was given by Nagata. Rushton and others cited above found no effect of liquid height on power number. However, the paddles used by Nagata were fairly wide in comparison to those used by others. Nagata et al. also found no effect of paddle location on power consumption. Blade-number and blade-width effects were noted. With everything else being equal, impellers had the same power consumption if the product of blade width W and paddle number n_p was equal to the same constant K, or:

$$W(n_p) = K \tag{2.32}$$

This relationship comes from drag analysis as well. For four to six baffles, power consumption went through a maximum in the range of $0.08 < B/T < 0.15$. The relationship between baffle width B and baffle number n_b at maximum power input was roughly approximated by:

$$\left(\frac{B}{T}\right)n_b = 0.5 \tag{2.33}$$

The data are shown in Fig. 2.2.

For maximum power input under fully baffled conditions, Nagata et al. found:

$$\frac{P_{max}g_c}{\rho N^3 D^5} = 23 \left(\frac{W}{T}\right)^{1.27}\left(\frac{D}{T}\right)^{-1} \tag{2.34}$$

for $0.05 < W/T < 0.2$ and $0.3 < D/T < 0.8$, and:

$$\frac{P_{max}g_c}{\rho N^3 D^5} = 12 \left(\frac{W}{T}\right)^{0.85}\left(\frac{D}{T}\right)^{-1} \tag{2.35}$$

for $0.2 < W/T < 0.4$ and $0.3 < D/T < 0.7$.

These equations can be rearranged to approximately:

$$\frac{P_{max}g_c}{\rho N^3 D^4 W} = K \tag{2.36}$$

which agrees with the drag basis for power.

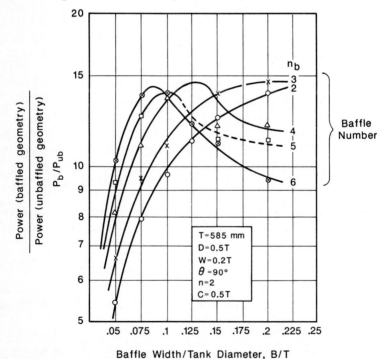

Figure 2.2 Power as a function of baffle width and baffle number. (*From S. Nagata, T. Yokoyama, and H. Maeda, Memoirs Fac. Eng., Kyoto Univ., 18, 13, 1956. By permission.*)

Nagata, Yamamoto, Yokyama, and Shiga (1957) combined the equations above to obtain one equation for power number for unbaffled tanks as:

$$N_P = \frac{A_1}{N_{Re}} + B_1 \left(\frac{10^3 + 1.2 N_{Re}^{0.66}}{10^3 + 3.2 N_{Re}^{0.66}} \right)^p \left(\frac{H}{T} \right)^{(0.35 + W/T)} (\sin \theta)^{1.2} \quad (2.37)$$

where A_1, B_1, and p were determined from:

$$A_1 = 14 + \left(\frac{W}{T} \right) \left[70 \left(\frac{D}{T} - 0.6 \right)^2 + 185 \right]$$

$$\log_{10}(B_1) = 1.3 - 4 \left(\frac{W}{T} - 0.5 \right)^2 - 1.14 \left(\frac{D}{T} \right)$$

$$p = 1.1 + 4 \left(\frac{W}{T} \right) - 2.5 \left(\frac{D}{T} - 0.5 \right)^2 - 7 \left(\frac{W}{T} \right)^4$$

For fully baffled tanks, Nagata et al. (1957) obtained an equation for the N_P at large Reynolds numbers:

$$N_P = \frac{A_1}{R_c} + B_1 \left(\frac{H}{T} \right)^{(0.35 + W/T)} \quad (2.38)$$

where

$$R_c = \frac{25(D/T - 0.4)^2}{W/T} + \left(0.11 - \frac{0.0048T}{W} \right)^{-1} \quad (2.39)$$

For this equation, A_1 and B_1 are the same as that given for Eq. (2.37). For impellers with inclined blades, R_θ replaced R_c but was calculated in terms of R_c as:

$$R_\theta = 10^{4(1 - \sin \theta)} R_c \quad (2.40)$$

R_c and R_θ are actually transition Reynolds numbers. The same ranges of W and D, given above, hold for these equations.

Most studies have been in open batch vessels with emphasis on baffling. Baffling prevents solid body rotation, improves mixing, dispersion, and transport generally, maximizes power input, and prevents the formation of the central vortex. However, if the air-liquid interface were eliminated, then the central vortex would not form and the need for baffling would not be as critical. Further, in some cases, baffling does not promote transport: for example, Overcashier, Kingsley, and Olney (1956).

Laity and Treybal (1957) studied the effects of no air-liquid interface. The objective of their study was liquid-liquid contacting in which the multiphase fluid properties of the dispersion, i.e., the density and the viscosity, had to be determined for use in the power and Reynolds numbers. For baffled and unbaffled tanks, power consumption was correlated with single curves using a weighted arithmetic mean density and mean viscosity. Several methods for the calculation of mixture properties, cited by Laity and Treybal, were:

For viscosity:

1. The use of a weighted geometric mean viscosity obtained from studies of Miller and Mann (1944):

$$\mu_m = (\mu_x)^x (\mu_y)^y \qquad (2.41)$$

2. The following viscosity relationship by Vermeulen et al. (1955) developed from studies in baffled vessels with an air-liquid interface:

$$\mu_m = \left(\frac{\mu_c}{x_c}\right)\left(1 + \frac{1.5x_d\mu_d}{\mu_c + \mu_d}\right) \qquad (2.42)$$

For density, the use of a weighted arithmetic mean density used by Miller and Mann (1944) and recommended by Vermeulen et al. (1955):

$$\rho_m = x\rho_x + y\rho_y \qquad (2.43)$$

where x and y refer to the volume fractions of the two-phase mixture. Neither viscosity equation was found to be adequate, particularly when a high-viscosity phase was continuous.

Instead, Laity and Treybal recommended:

For aqueous phase more than 40 percent by volume:

$$\mu_m = \left(\frac{\mu_a}{x_a}\right)\left(1 + \frac{6.0x_0\mu_0}{\mu_0 + \mu_a}\right) \qquad (2.44)$$

For aqueous phase less than 40 percent by volume:

$$\mu_m = \left(\frac{\mu_o}{x_o}\right)\left(1 - \frac{1.5x_a\mu_a}{\mu_0 + \mu_a}\right) \qquad (2.45)$$

The elimination of the air-liquid interface had little, if any, effect on the agitation. However, it was recommended that the impeller be placed in the low-viscosity phase whenever possible for better agitation (i.e., higher power draw under more turbulent flow conditions). It can also be recommended that the impeller be placed in the more dense phase.

Laity and Treybal found a measurable effect of flow rate on power consumption for continuous-flow systems, but, for most applications, the effect can be ignored. With an air-liquid interface present, the ef-

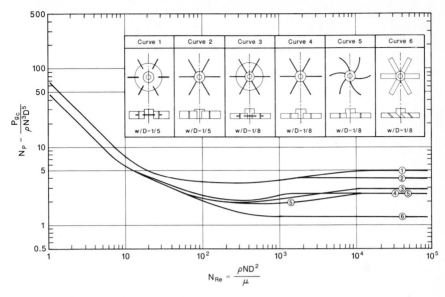

Figure 2.3 Power number-Reynolds number correlation in Newtonian fluids for various turbine impeller designs. *(Reprinted with the permission from R. L. Bates, P. L. Fondy, and R. R. Corpstein, I&EC Proc. Des. Dev., 2, 310, 1963. Copyright 1963 American Chemical Society.)*

fect of Froude number on power was also found to be negligible, and dynamic similarity was obtained in geometrically similar tanks with equal Reynolds numbers.

Bates, Fondy, and Corpstein (1963) examined the effects of geometry on impeller power draw. Their power draw curves for the various impeller designs are shown in Fig. 2.3. They reported a power number of 5.0 for the disk style impeller for 10 percent baffles differing from 6.3 as reported by Rushton et al. (1950). Data on the effect of blade angle on power were given as shown in Fig. 2.4, and the effects of blade width and blade number were also discussed. Bates et al. (1963) noted that it was not easy to separate the various geometric effects using a simple power law due to various interactions, particularly between blade width and blade number. The effects of tank-to-impeller diameter ratio, baffle width, and baffle number were studied and given as shown in Fig. 2.5. The effect of impeller spacing s is given in Fig. 2.6, where P_1 is referenced to a flat open style six-blade impeller. P_2 is the power of the two-impeller system. Effect of clearance on power number is shown in Fig. 2.7.

The data in the figures from Bates et al. (1963) show how the subtle effects of flow patterns around the impeller change as a function of impeller-tank geometry. The different geometries cause different flow

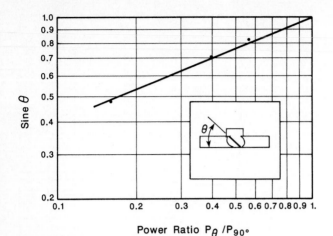

Figure 2.4   Effect of turbine blade angle on power.   (*Reprinted with the permission from R. L. Bates, P. L. Fondy, and R. R. Corpstein, I&EC Proc. Des. Dev., 2, 310, 1963. Copyright 1963 American Chemical Society.*)

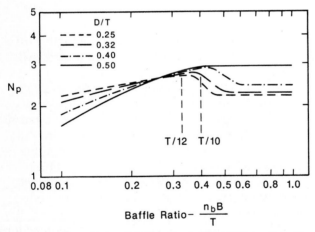

Figure 2.5   The effects of D/T ratio, baffle width, and baffle number on power number. (*Reprinted with the permission from R. L. Bates, P. L. Fondy, and R. R. Corpstein, I&EC Proc. Des. Dev., 2, 310, 1963. Copyright 1963 American Chemical Society.*)

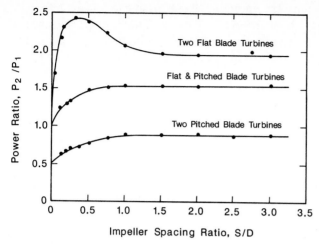

Figure 2.6   The effect of turbine spacing on power number. *(Reprinted with the permission from R. L. Bates, P. L. Fondy, and R. R. Corpstein, I&EC Proc. Des. Dev., 2, 310, 1963. Copyright 1963 American Chemical Society.)*

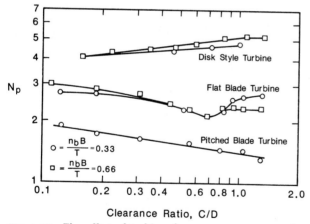

Figure 2.7   The effect of turbine clearance on power number. *(Reprinted with the permission from R. L. Bates, P. L. Fondy, and R. R. Corpstein, I&EC Proc. Des. Dev., 2, 310, 1963. Copyright 1963 American Chemical Society.)*

patterns around the impeller, and these flow patterns cause changes in the impeller-fluid relative velocity and in the local coefficients of drag which, in turn, appear in power data. For example, it is known that the discharge angle of flat blade turbines changes as a function of clearance as will be discussed in Chap. 4. This subtle change in the flow pattern changes the fluid-impeller relative velocities, the local

drag coefficients along the impeller blades, and the power draw. As a
result, power draw is a function of clearance. This interpretation holds
for any effect of the impeller-tank geometry which affects power draw
data. Another more obvious effect is that of baffling on power draw.

More recent studies

Following the period cited above, there was actually a 5- to 8-year period in the technical literature where very little work was accomplished in the area of power draw curves for Newtonian fluids under
turbulent flow conditions. However, following this, renewed interest
generated studies in various areas, including different geometries, off-
center positioning of the impeller, and investigations to resolve and
further support existing published data. Studies will be cited according to subject matter rather than chronological order.

Power studies.   Bertrand et al. (1980) studied five different impeller
configurations: the disk style impeller at clearances of $0.5H$ and
$0.33H$, two shrouded angle blade disk style impellers with different
blade widths and modified blade shapes, and an 8-angled blade disk
style turbine with the same modified blade shape. The power numbers
for these configurations were 5.1, 4.9, 4.2, 3.4, and 9.2, respectively. A
shrouded impeller has solid disks attached above and below the impeller blades as shown in Fig. 2.8. The flow to the impeller occurs near
the hub and exits radially from the impeller.

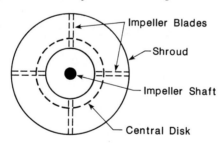

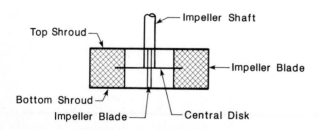

Figure 2.8   A four-blade shrouded disk style turbine.

The power number for the standard disk style turbine decreased as clearance decreased which agrees with data given by Bates et al. (1963) presented above. Shrouded turbines had much lower power numbers than the unshrouded impellers. Shrouds affected the impeller-fluid relative velocities and the drag coefficients of the blades which changed the power draw. All data were taken in the turbulent mixing regime, $1.5 \times 10^5 < N_{Re} < 6.7 \times 10^5$.

Shiue and Wong (1984) reported power number correlations for a six-blade disk style turbine, a six-blade disk style turbine with curved blades, an open curved four-blade turbine (no disk), a three-blade propeller, and two- and four-blade pitched blade turbines. At high impeller Reynolds numbers, the power numbers for the disk style turbine ranged from 4.3 to 5.5 and varied as a function of clearance and D/T ratio. The curved blade disk style turbine had a power number of 4.1; the open impeller, 1.56; propeller, 0.67; and the two- and four-blade pitched blade turbines, 1.20 and 1.74, respectively. The two- and four-pitched blade turbines with a low clearance draft tube $C_t = 0.15T$ had power numbers of 0.80 and 1.10, respectively. The data indicate the various effects of blade curvature, blade number, clearance, D/T ratio, blade angle and shape, and draft tubes on relative velocities and drag coefficients and hence on power draw. Landau and Prochazka (1963) found that high-clearance draft tubes did not affect power draw.

Sano and Usui (1985) reported power numbers for paddles and turbines in baffled tanks for the turbulent region for Reynolds numbers greater than 5×10^5. The paddles and turbines had various dimensions: D/T between 0.3 and 0.7, W/D between 0.05 and 0.4, and blade numbers between two and eight. The tanks were 0.2 and 0.4 m in diameter with four baffles, $B/T = 0.10$, and liquid height was equal to the tank diameter. Power draw was measured using a torque transducer mounted on the shaft. For paddles:

$$N_P = 7.3\left(\frac{D}{T}\right)^{-1.15}\left(\frac{W}{T}\right)^{1.15} n_p^{0.80} \tag{2.46}$$

and for turbines:

$$N_P = 3.6\left(\frac{D}{T}\right)^{-0.95}\left(\frac{W}{T}\right)^{0.75} n_p^{0.80} \tag{2.47}$$

Both of these correlations can be approximately rearranged to give:

$$\frac{Pg_c}{\rho N^3 D^4 W} = K n_p^{0.80} \tag{2.48}$$

Xanthopoulos and Stamatoudis (1986) studied impeller power numbers in closed baffled square tanks and found them to be similar to power numbers in closed baffled cylindrical tanks.

Weetman and Oldshue (1988) reported 0.30 as the power number for a fluidfoil impeller.

Papastefanos and Stamatoudis (1989) reported power number data for the disk style turbine, open flat blade, and 45° pitched blade turbines in baffled and unbaffled square and cylindrical closed tanks from the laminar-transition regime to fully turbulent flow conditions, i.e., $50 < N_{Re} < 10^5$. The power numbers for the six-blade disk style turbine varied between 2.5 and 6.5; flat six-blade open style impeller, 3.0 to 5.5; and the 45° pitched blade turbine, 3.5 in the laminar regime to a low of 1.1 in the transition region to 1.5 in the fully turbulent regime. The pitched blade turbines exhibited both axial and radial flow patterns. The radial flow patterns for the pitched blade turbine were suggested to be the cause of the increase in power number at the higher impeller Reynolds number.

Warmoeskerken and Smith (1989) reported power numbers for the hollow blade disk style turbine. The dimensions of the impeller are the same as the dimensions of the standard disk style turbine except for the blade shape. For the blade shape: Concave in the direction of rotation with the same projected blade area as a standard disk style impeller and used in the standard configuration, the power number was 3.8. The convex blade orientation had a power number of 4.3. The impeller blades were actually made from pipe sections, each symmetrically placed around the perimeter of the disk. The concave orientation was where the edges of the pipe sections were leading in the rotation.

Impeller geometry, _D/T_, closed vessels, clearance, and flow patterns effects.
Nienow and Miles (1971) studied power consumption in closed vessels and presented power numbers for six-blade disk style turbine, two-blade flat paddles, and four-blade, 45° pitched turbine in 0.15 and 0.30-m-diameter vessels with 10 percent baffles. Their major results can be summarized as: (1) Power number was a maximum when the flat blade impellers were placed midway in the tank at 50 percent clearance; (2) as the disk thickness increased for the disk style turbine, the power number dropped; (3) closed vessels caused higher N_P over open vessels by about 5 percent; (4) the pitched blade turbine gave minimum power number at 33 percent clearance; and (5) effects of D/T were noted for all impellers.

Nienow and Wisdom (1974) attributed the difference in power numbers between the flat blade and disk style impeller to low pressure vortices which formed behind the inside blade edge of the disk style turbine. As a result, the pressure or form drag for this impeller increased over that of the flat blade impeller whose blade extended to the hub. The same mechanism was used to explain the effect of the disk thickness on power for the disk style turbine.

Greaves and Kobbacy (1981) have also reported data on the effect of tank diameter and impeller clearance for a small-diameter tank. Significant variations in power number were observed. Gray et al. (1982) correlated the effect of clearance on power number for a six-bladed disk style turbine as:

$$N_P = 5.17\left(\frac{C}{D}\right)^{0.29} \qquad \text{for } \frac{C}{D} < 1.0 \qquad (2.49)$$

and $\qquad N_P = 5.17 \qquad\qquad \text{for } \frac{C}{D} > 1.0 \qquad (2.50)$

which shows a reduction in power number at lower clearances for the disk style turbine, similar to the results given above. Chapman et al. (1983) noted that minor differences in dimensions for the disk style turbine caused significant changes in power numbers between tank sizes of 0.29 and 1.83 m.

The effects of disk thickness X and scale on power were studied by Bujalski et al. (1987). The result indicated that:

$$N_P \propto \left(\frac{X}{D}\right)^{-0.2}\left(\frac{T}{1m}\right)^{0.065} \qquad (2.51)$$

for the disk style turbine.

Raghav Rao and Joshi (1988) obtained power numbers for three different six-blade impellers, $W/T = 0.2$, at the different clearances shown in Table 2.4. The lower power numbers at lower clearances for the disk turbine can be interpreted in terms of flow patterns. At lower clearances, the bottom circulation of the disk turbine was reduced, causing the lowered power draw. At lower clearances, the normal circulation pattern of the pitched blade turbine changed to that of a radial flow impeller. Higher-power numbers were a result of increased drag.

Papastefanos and Stamatoudis (1989) have also reported power data for baffled and unbaffled square and cylindrical closed tanks.

TABLE 2.4   Effect of Clearance on Power Numbers

C	Disk style turbine	Pitched blade turbine, up pumping	Pitched blade turbine, down pumping
$T/3$	5.18	1.29	1.29
$T/4$	4.70	1.24	1.35
$T/6$	4.40	1.81	1.61

Flow patterns in single-phase flow are typically not considered important in their effect on power draw. However, the effect on power draw can be significant. Unfortunately, the studies are rather limited in this area. Recirculation of the liquid back into the impeller also affects the local flow behavior around the impeller blades and the power draw to some degree. Local differences in liquid recirculation cause the drag coefficients to change, causing different power draw.

Baffling. Nishikawa et al. (1979) provided a minor modification to Nagata (1975) for the effect of baffle width and baffle number on power consumption:

$$\frac{N_{P\max} - N_{P0}}{N_{P\max} - N_{P_x}} = \left[1.0 - 2.9\left(\frac{B}{T}\right)^{1.2} n_b \right]^{4.0} \tag{2.52}$$

Hemrajani (1985) reported data on the effect of baffle width on power as a function of D/T ratio. Lehtola et al. (1988) discussed twisted or spiral element baffles.

Stator rings have not been studied as to their effect on power draw.

Blade width, blade number, and blade angle. O'Kane (1974) investigated the effects of blade width and number of blades for the flat blade turbine, the flat blade disk turbine, the 45° pitched blade turbine, and the curved blade open turbine. Power consumption was investigated independently for three, four, six, and eight blades and for four different blade widths. The results are in Tables 2.5, 2.6, and 2.7.

O'Kane cited Bates et al. (1963) as finding the relationship between blade angle and power number given as:

$$N_P = N_{P/90°} \, (\sin \theta)^{2.5} \tag{2.53}$$

for constant projected height conditions. However, O'Kane noted that, for constant blade width, the projected blade height changed with θ

TABLE 2.5   Effect of Blade Width

Blade width (mm)	W/D	Power number, N_P			
		Four-blade disk turbine	Four-blade open turbine	Four-blade 45° pitched turbine	Three-blade backswept turbine
25	0.10	1.17	1.53	0.88	1.21
38	0.15	2.40	2.63	0.96	2.16
51	0.20	3.50	3.62	1.29	3.18
76	0.30	6.15	5.93	1.69	4.77

TABLE 2.6   Effect of Number of Blades

Number of blades	Disk turbine	Power number, N_P		
		Open turbine	45° pitched turbine	Backswept turbine
3	—	3.09	1.04	3.18
4	3.5	3.62	1.29	3.67
6	5.05	5.20	1.52	5.06
8	6.35	—	1.82	5.78

TABLE 2.7   Summary of Results

	Exponent for blade width	Exponent for number of blades ($W/D = 0.2$)
Disk turbines	1.45	0.85
Open turbines	1.20	0.75
Pitched blade turbines	0.65	0.53
Backswept turbines	1.25	0.60

and that the effect of blade height on power followed the relationship:

$$N_P = N_{P/90°} (\sin \theta)^{3.1} \tag{2.54}$$

for pitched blade turbines. O'Kane also noted that a 2° error in blade angles led to an 11 percent change in power consumption. Medek (1980) also discussed power characteristics of agitators with angled blades.

Off-center position.   Rushton et al. (1950a) reported that off-centering propellers produced the same effect as baffling. However, studies of this effect are limited.

Nishikawa et al. (1979) studied the effects of off-centering and baffles on power consumption for flat blade impellers. The off-center ratio was based upon impeller diameter. The final correlation was given as:

$$\frac{N_{PE} - N_{P0}}{N_{P\max} - N_{P0}} = f \tag{2.55}$$

for the effect of off-centering on power. The various f functions are shown in Fig. 2.9 as a function of baffle number. As f approached 1, the power input was maximized, or, in other words, a fully baffled state was reached as shown in the figure. Under such conditions of off-centering, the flow patterns significantly affected the relative velocities and drag coefficients of the impeller, and baffling was not necessary.

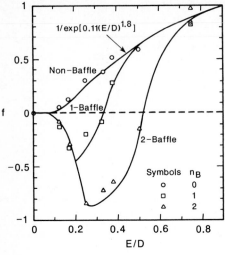

f vs. Off-Centered Distance

Figure 2.9 f versus off-centered distance. (*From M. Nishikawa, K. Ashiwake, N. Hashimoto, and S. Nagata, Intern. Chem. Eng., 19, 1, 153, 1979. By permission.*)

Novak et al. (1982) studied the influence of impeller position on power draw and surface entrainment for unbaffled vessels using a pitched three-blade turbine and a pitched six-blade turbine, both pumping downward in the study. The power numbers were 1.1 and 1.7, respectively. Cylindrical vessels having different bottom shapes were used: a flat-bottom vessel, a dished-bottom vessel, and a cone-bottom vessel. The liquid height equaled the tank diameter, and the ratio T/D was varied from 3 to 4. The impellers were placed at 0.3 and 1.0 impeller diameters from the bottom as measured from the impeller blade tip. For these geometries, a number of effects were noted. Off-centered impeller positioning increased the maximum power input to the fluid without surface entrainment almost up to the value reached in baffled tanks. Maximum power input occurred just before the central vortex reached the impeller; after this, power dropped due to surface entrainment. A correlation was provided for determining the critical impeller rotational speed for surface entrainment. At high impeller Reynolds numbers, the power number was found dependent upon off-center impeller position. No effect of Froude number upon power number was found which agrees with the results found by Nagata et al. (1955b, 1956) and Laity and Treybal (1957). Power input to dished- and cone-bottom tanks was less than for flat-bottom tanks as was noted by Martin (1946).

Power number was correlated as a function of Reynolds number and relative eccentricity E/T. The ratio of impeller power numbers for an unbaffled vessel with impeller placed in off-centered position, N_{PE},

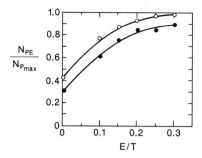

Figure 2.10 A plot of the ratio N_{PE}/N_{Pmax} versus the relative eccentricity E/T, in a flat-bottomed vessel, top curve: pitched three-blade turbine, bottom curve: pitched six-blade turbine. (*From V. Novak, P. Ditl, and F. Rieger, 4th European Conf. on Mixing, BHRA Fluid Eng., C1, 57, 1982. By permission.*)

and that of the standard vessel equipped with four 10 percent baffles, N_{Pmax}, is shown in Fig. 2.10. The best tank geometry for maximum power draw with no significant surface entrainment was at high eccentricities, $E/T = 0.3$, and large T/D ratios. Nishikawa et al. (1979) found maximum power draw at about the same eccentricity.

Pressure and drag coefficient measurements. Mochizuki and Takashima (1974) provided pressure and velocity information along the surface of flat blade impellers. The work is important in understanding how power is transferred from the blade to the fluid and for modeling purposes. Mochizuki and Takashima (1984a and 1984b) also discussed the effect of blade width on the pressure distributions and on power.

Nagase and Sawada (1977) determined drag coefficients of impeller blades for paddles and flat blade turbines by determining the relationship between the discharge flow rate and torque imposed on the impeller by the fluid. By measuring and modeling the discharge flow, a drag coefficient model was developed for torque. The discharge flow was modeled as:

$$Q_d = (1 + \alpha)nW' \int_0^{r_i} (2\pi Nr - V_e)\, dr \qquad (2.56)$$

Torque was modeled as:

$$T_0 g_c = \rho nW C_D \int_0^{r_i} r(2\pi Nr - V_e)^2\, dr \qquad (2.57)$$

where V_e is the rotational velocity of the fluid flowing into the region before the impeller blades and W is the impeller blade width. The parameter α in the model was approximately equal to 1, and W'/W was 0.85, i.e., the width of the discharge flow is typically less than the impeller blade width as noted in Chap. 4. The drag coefficient C_D was predicted to be 5.95. Using experimental data, C_D was found to be 4.6 and 5.2 for paddles and the disk style turbine. These drag coefficients are similar in magnitude to the power number for these impellers.

Tay and Tatterson (1985) successfully related power consumption for a pitched blade turbine to form drag. Drag coefficients and drag theory are the basis of power draw in an agitated tank, although drag coefficients are not typically measured for blades of a rotating impeller in mixing. The work by Tay and Tatterson provided pressure drop data across an impeller blade which resulted in pressure coefficient distributions at high impeller Reynolds number and estimates of the drag coefficients for the blades of a pitched blade turbine. Braginskii and Begachev (1972) reported drag coefficients for various impellers as well. Wall stress probes can also measure blade wall stress and skin or frictional drag although work in this area is still developing.

Effect of continuous flow on power.   Ito et al. (1979) studied the power consumption under continuous flow operation in an unbaffled agitated vessel using centrally located paddles covering a Reynolds number range from 10^3 to 2×10^5. Water and aqueous glycerin entered below the agitator or through the side wall of the tank at various heights as shown in Fig. 2.11. For the bottom-entering arrangement, the change in power number was correlated as a function of N_{Qf}, impeller and feed pipe radii, and the tangential flow velocity in the tank; the power data are shown in Fig. 2.12. The maximum change in power number was about 0.6. For the side-entering flows, power number first declined and then increased as a function of feed height and N_{Qf} as shown in Fig. 2.13. For $N_{Qf} < 0.04$, the side-entering feed did not disturb the tangential solid body rotation flow. However, above $N_{Qf} > 0.04$, the liquid, entering through the side pipe, acted as a sin-

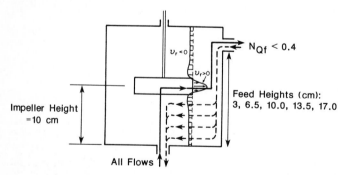

Figure 2.11   Flow patterns as a function of position.   (*From R. Ito, Y. Nirata, K. Sakata, and I. Nakahara, Intern. Chem. Eng., 19, 4, 605, 1979. By permission.*)

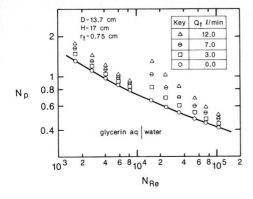

Figure 2.12 Relation between the power number N_P and Reynolds number N_{Re} in a case of the fluid entering through the pipe at the bottom center of the tank. (*From R. Ito, Y. Nirata, K. Sakata, and I. Nakahara, Intern. Chem. Eng., 19, 4, 605, 1979. By permission.*)

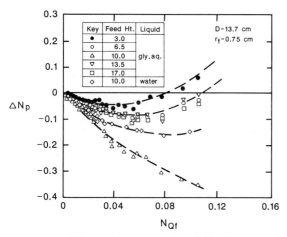

Figure 2.13 Relation between ΔN_P and N_{Qf} in a case of the flow entering through the pipe at the side wall. (*From R. Ito, Y. Nirata, K. Sakata, and I. Nakahara, Intern. Chem. Eng., 19, 4, 605, 1979. By permission.*)

gle baffle, causing secondary circulation and eccentricity of the main rotating flow. The baffle effect increased power consumption, and the power number was correlated taking this effect into account. The effect of a central feed position (feed height = 10 cm), where the side feed entered directly into the impeller stream, was not correlated.

Tank-diameter effect on power. Losev et al. (1967) investigated power consumption of propellers in unbaffled vessels using water and aqueous solutions of glycerin. The principal results of their study were data given on the effect of tank diameter on power number. Power number was given in the form:

$$N_P = A N_{\text{Re}}^{-m} \qquad (2.58)$$

Less common geometries. Ando et al. (1971) studied the power consumption in horizontally stirred vessels using two- and four-blade disk style turbines. Figure 2.14 shows the geometry. Gas-liquid interface was present, and four flow regimes were established. The change of regimes was related to the Froude number and liquid fraction in the tank. Regime 1 was characterized as a segregated liquid-gas regime with liquid sprayed against the wall, and regime 2, as an initial formation regime of a liquid annulus around the tank perimeter. The liquid annulus entrained gas in this regime. Regime 3 was a solid liquid annulus, and regime 4 was a solid liquid annulus which was unstable. A hysteresis loop was observed in the power number N_P as regime 2 collapsed to regime 3 and regime 4 collapsed to regime 1 as shown in Fig. 2.15. Power number was correlated with Froude number, liquid fraction and impeller diameter as shown in Fig. 2.16. Large gas-liquid contacting areas occurred at the transition point between regimes 4 and 1. Overall, the flow phenomena were quite complex and were actually a two-phase flow problem.

Sato (1980) interrelated impeller power draw of the disk style turbines to the turbulent flow fields which occurred in rectangular mixing tanks. The length of one side of the tank, eccentricity, and impeller off-bottom clearance were varied in the study. Power was measured with a torque meter, and power number data were reported as functions of the impeller diameter to tank width, clearance, length-to-width ratio of the tank, and eccentricity. The power numbers reported were significantly different from those typical of cylindrical tanks.

Tojo et al. (1979; 1981) discussed power draw in vibratory agitation

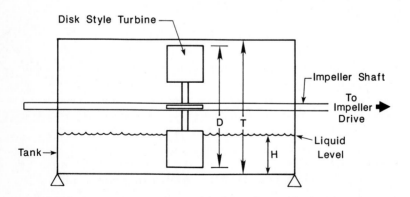

Figure 2.14 Horizontally stirred vessel. (*From K. Ando, H. Hara, and K. Endoh, Intern. Chem. Eng., 11, 4, 735, 1971. By permission.*)

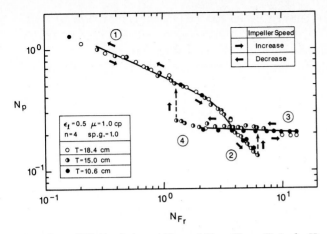

Figure 2.15   Relation between N_{Fr} and N_P.   (*From K. Ando, H. Hara, and K. Endoh, Intern. Chem. Eng., 11, 4, 735, 1971. By permission.*)

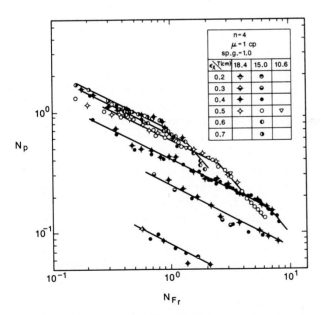

Figure 2.16   Correlation of N_P with N_{Fr} for regimes 1 and 2. (*From K. Ando, H. Hara, and K. Endoh, Intern. Chem. Eng., 11, 4, 735, 1971. By permission.*)

equipment. Nagase (1987) studied the torque on a new type of back-and-forth agitator with prism prongs. Nagase et al. (1988) reported power and torque data for reciprocating triangular beam impellers.

Hirato and Ito (1988) reported torque coefficients for rotating disk impellers as:

$$C_T = \frac{2(T_0)}{[\rho(2\pi N)^2(D/2)^5]}$$ (2.59)

where T_0 is the torque. The torque coefficient C_T was correlated as:

$$C_T = 7.4 \, \text{Re}^{-0.6} \qquad \text{for } 10^2 < \text{Re} < 6 \times 10^3$$ (2.60)

and $C_T = 3.1 \, \text{Re}^{-0.5} \qquad \text{for } 6 \times 10^3 < \text{Re} < 5 \times 10^5$ (2.61)

The Reynolds number was defined as $(\pi/2)ND^2/\nu$. The disk diameters were 0.08, 0.10, and 0.12 m used in a multistage configuration. The tank or column diameter was 0.14 m in the study.

Plant application.   For plant-size or production-scale agitators, it is often impossible to measure power input using laboratory devices. Uhl (1950) outlined a procedure to determine mixing power draw from shaft torque in a plant setting. Newitt et al. (1951) reviewed the various ways to measure power, noting that an electrical power meter was not accurate. Measurement of shaft torque was recommended. Brown (1977) and King et al. (1988) have outlined some procedures to estimate electrical and drive losses which would be useful in the measurement of on-site mixing power draw using wattmeters.

Multiphase considerations.   The various correlations as presented can be used for power calculations for processes involving liquid mixtures, liquid-liquid dispersions, and solid suspensions. Viscosities of liquid mixtures have been well studied in the literature and include McAllister (1960), Harada et al. (1975), Noda and Ishida (1977), and Przezdziecki and Sridhar (1985). Effective viscosities for liquid-liquid dispersions have also been studied, including Barnea and Mizrahl (1976). This is in addition to those methods already given by Laity and Treybal (1957), Miller and Mann (1944), and others.

Suspension of solids usually does not require any additional power considerations over that required for off-bottom suspension. At low percent solids, the impeller flow is unaffected by the solids and single-phase power correlations may be used. Power calculations for processes involving gas-liquid dispersion and solid slurries require other correlations. Power under gas dispersion conditions will be discussed in Chap. 6. Some power correlations for slurries having a high percent of solids are discussed in Chap. 3. Slurries tend to be shear thickening fluids. Mixing of slurries may be performed in media mills as discussed by Patton (1979). Quaraishi et al. (1976) and Ranade and Ulbrecht (1977) studied the effects of drag reducing materials in turbulent flow conditions.

Summary

Power draw from an impeller to the fluid under turbulent flow conditions is primarily determined by the geometry of the impeller. However, many secondary effects impact upon this power level. Variations in power draw can be expected because of variations in the impeller-tank geometry, the presence of tank internals, feed streams in the tank, etc.

Power number is not a constant quantity for a particular impeller at a specific Reynolds number, nor is it uniquely determined by the impeller Reynolds number and impeller geometry alone. In different geometric configurations, the flow patterns and the fluid mechanical environment of the impeller are different. The power number will change as a result. A fixed power number is an idealization and is questionable in practice. In the same manner, the distribution of the spatial energy dissipation rate for a specific impeller at a specific Reynolds number will also vary depending upon the geometry.

Nomenclature

A_1	constant
B	baffle width
B_1	constant
C	clearance of the impeller from the tank bottom
C_t	clearance of draft tube from the tank bottom
C_D	drag coefficient
C_T	torque coefficient
D	impeller diameter
D_t	draft tube diameter
E	eccentricity, impeller distance from tank center
F	drag force
f	f functions given by Nishikawa et al. (1979)
g	acceleration of gravity
g_c	gravitational constant
H	fluid height

L	blade length
K, K', K'', K'''	constants
N	impeller rotational speed
N_{Fr}	Froude number, $N^2 D/g$
N_{Re}	impeller Reynolds number, $\rho N D^2/\mu$
N_P	power number, $P g_c/\rho N^3 D^5$
$N_{P\mathrm{max}}$	power number, fully baffled
N_{P0}	power number, nonfully baffled
$N_{P\infty}$	power number, nonfully baffled state at $\mathrm{N_{Re}} = \infty$
N_{PE}	power number at an off-centered position
$N_{P/90°}$	power at 90° blade angle
ΔN_P	change in power number
N_{Qf}	feed flow number, Q_f/ND^3
n	number of impeller blades
n_b	number of baffles
n_f	number of impeller blades of reference geometry
n_p	number of paddles
n_{bf}	number of baffles of reference geometry
P	power
P_1	power of an open flat six-blade impeller
P_2	power of two-impeller system
P_b	power in a baffled geometry
P_{ub}	power in an unbaffled geometry
P_θ	power at an angle θ
$P_{90°}$	power at 90°
p	impeller blade pitch
Q_d	volumetric discharge rate
Q_f	volumetric feed rate
R	impeller radius
r	radius
r_i	impeller radius
r_f	feed pipe radius
R_1, R_2	transitional Reynolds numbers of Nagata et al.
S	impeller spacing
T	tank diameter
T_0	torque
V_e	rotational velocity of the fluid flowing into the impeller region

W	blade width
W'	discharge width
x	volume fraction x phase
X	disk thickness
y	volume fraction y phase

Greek symbols

θ	blade angle made with the horizontal shown in Fig. 2.4
ϵ	void fraction
ϵ_1	liquid void fraction
μ	viscosity
ρ	density

Subscripts

a	aqueous phase
c	continuous
d	discrete
m	mixture
o	organic phase
x	x phase
y	y phase
θ	θ direction

References

Ando, K., H. Hara, and K. Edoh, *International Chem. Eng.*, **11**, 4, 735, 1971.
Barnea, E., and J. Mizrahl, *I&EC, Funds.*, **15**, 120, 1976.
Bates, R. L., P. L. Fondy, and R. R. Corpstein, *I&EC Des. Dev.*, **2**, 4, 310, 1963.
Bertrand, J., J. P. Couderc, and H. Angelino, *Chem. Eng. Sci.*, **35**, 2157, 1980.
Braginskii, L. N., and V. I. Begachev, *Theo. Fund. Chem. Eng.*, **6**, 231, 1972.
Brown, D. E., *Chemistry and Industry*, **16**, August 20, 1977, p. 684.
Bujalski, W., A. W. Nienow, S. Chatwin, and M. Cooke, *Chem. Eng. Sci.*, **42**, 317, 1987.
Chapman, C. M., A. W. Nienow, M. Cooke, and J. C. Middleton, *Chem. Eng. Res. Des.*, **61**, 82, 1983.
Gray, D. J., R. E. Treybal, and S. M. Barnet, *AIChEJ.*, **28**, 195, 1982.
Greaves, M., and K. A. H. Kobbacy, *Fluid Mixing, I. Chem. E. Sym. Ser. No. 64*, H1, 1981.
Harada, M., M. Tanigaki, and W. Eguchi, *J. of Chem. Eng. Japan*, **8**, 1, 1975.
Hemrajani, R. R., *Proc. 5th Eur. Conf. on Mixing, Wurzburg, Germany*, BHRA Fluid Eng., Cranfield, England, 63, 1985.
Hirata, Y., and R. Ito, *Proc. 6th Eur. Conf. on Mixing, Pavia, Italy*, BHRA Fluid Eng., Cranfield, England, 109, 1988.
Hixson, A. W., and S. J. Baum, *Ind. Eng. Chem.*, **34**, 2, 194, 1942.
Hixson, A. W., and V. E. Luedeke, *Ind. Eng. Chem.*, **29**, 8, 927, 1937.

Ito, R., Y. Hirata, K. Sakata, and I. Nakahara, *International Chem. Eng.*, **19**, 4, 605, 1979.
King, R. L., R. A. Hiller, and G. B. Tatterson, *AIChEJ*, **34**, 506, 1988.
Laity, D. S., and R. E. Treybal, *AIChEJ.*, **3**, 2, 176, 1957.
Landau, J., and J. Prochazka, *Coll. Czech. Chem. Commun.*, **28**, 1866, 1963.
Lehtola, T., J. Soderman, and J. Laine, *Proc. 6th Eur. Conf. on Mixing, Pavia, Italy*, BHRA Fluid Eng., Cranfield, England, 85, 1988.
Losev, G. E., Yu. V. Tumanox, G. P. Solomakha, and P. I. Nikolaev, *Theoretical Found. Chem. Eng.*, **1**, 704, 1967.
Mack, D. E., and A. E. Kroll, *Chem. Eng. Progr.*, **44**, 3, 189, 1948.
Martin, J. J., *Trans. Amer. Inst. Chem. Engrs.*, **42**, 77, 1946.
McAllister, R. A., *AIChEJ.*, **6**, 427, 1960.
Medek, J., *Intern. Chem. Eng.*, **20**, 664, 1980.
Miller, S. A., and C. A. Mann, *Trans. Amer. Inst. Chem. Engrs.*, **40**, 709, 1944.
Mochizuki, M., and I. Takashima, *Kagaku Kogaku*, **38**, 249, 1974.
Mochizuki, M., and I. Takashima, *Kagaku Kogaku Ronbunshu*, **10**, 399, 1984*a*.
Mochizuki, M., and I. Takashima, *Kagaku Kogaku Ronbunshu*, **10**, 539, 1984*b*.
Nagase, Y., and S. Sawada, *J. Chem. Eng. Japan*, **10**, 229, 1977.
Nagase, Y., *Proc. SCEJ-CIChE Joint Seminar on Mixing Technology*, edited by W-L Lu, Taipei, Taiwan, p. 49, July 14–15, 1987.
Nagase, Y., Hirofuji, H., and Imiya, M., *Proc. 6th Eur. Conf. on Mixing, Pavia, Italy*, BHRA Fluid Eng., Cranfield, England, 97, 1988.
Nagata, S., N. Yoshioka, and T. Yokoyama, *Memoirs Fac. Eng.*, Kyoto Univ., **17**, 175, 1955*a*.
Nagata, S., and T. Yokoyama, *Memoirs Fac. Eng.*, Kyoto Univ., **17**, 253, 1955*b*.
Nagata, S., T. Yokoyama, and H. Maeda, *Memoirs Fac. Eng.*, Kyoto Univ., **18**, 13, 1956.
Nagata, S., K. Yamamoto, T. Yokoyama, and S. Shiga, *Memoirs Fac. Eng.*, Kyoto Univ., **19**, 274, 1957.
Nagata, S., *Mixing: Principles and Application*, Wiley, New York, 1975.
Newitt, D. M., G. C. Shipp, and C. R. Black, *Trans. Instn. Chem. Engrs.*, **29**, 279, 1951.
Nienow, A. W., and D. Miles, *I&EC Proc. Des. Dev.*, **10**, 1, 41, 1971.
Nienow, A. W., and D. J. Wisdom, *Chem. Eng. Sci.*, **29**, 1994, 1974.
Nishikawa, M., K. Ashiwake, N. Hashimoti, and S. Nagata, *International Chem. Eng.*, **19**, 1, 153, 1979.
Noda, K., and K. Ishida, *J. of Chem. Eng. Japan*, **10**, 479, 1977.
Novak, V., P. Ditl, and F. Rieger, *Proc. 4th Eur. Conf. on Mixing, Noordwijkerhout, Netherlands*, BHRA Fluid Eng., Cranfield, England, C1, 57, 1982.
O'Connell, F. P., and D. E. Mack, *Chem. Eng. Progr.*, **46**, 7, 358, 1950.
O'Kane, K., *Proc. 1st Eur. Conf. on Mixing and Cent. Sep., Cambridge, England*, BHRA Fluid Eng., Cranfield, England, A3, 23, 1974.
Olney, R. B., and G. J. Carlson, *Chem. Engr. Progr.*, **43**, 9, 473, 1947.
Overcashier, R. H., H. A. Kingsley, and R. B. Olney, *AIChEJ.*, **2**, 259, 1956.
Papastefanos, N., and M. Stamatoudis, *Chem. Eng. Res. Des.*, **67**, 169, 1989.
Patton, T. C., *Paint Flow and Pigment Dispersion*, 2d ed., Wiley-Interscience, Wiley, New York, 1979.
Przezkziecki, J. W., and T. Sridhar, *AIChEJ.*, **31**, 333, 1985.
Quaraishi, A. Q., R. A. Mashelkar, and J. J. Ulbrecht, *J. Non-Newtonian Fluid Mech.*, **1**, 233, 1976.
Raghav Rao, K. S. M. S., and J. B. Joshi, *Chem. Eng. Commun.*, **74**, 1, 1988.
Ranade, V. R., and J. J. Ulbrecht, *Proc. 2nd Europ. Conf. on Mixing, St. John's College, Cambridge, England*, BHRA Fluid Eng., Cranfield, England, F6-83, 1977.
Rushton, J. H., *Canad. Chem. Proc. Ind.*, 55, May 1946.
Rushton, J. H., C. W. Costich, and H. J. Everett, *Chem. Eng. Progr.*, **46**, 8, 395, 1950*a*.
Rushton, J. H., E. W. Costich, and H. J. Everett, *Chem. Eng. Progr.*, **46**, 9, 467, 1950*b*.
Sano, Y., and H. Usui, *J. Chem. Eng. Japan*, **18**, 47, 1985.
Sato, K., *Chem. Eng. Commun.*, **7**, 45, 1980.
Shiue, S. J., and C. W. Wong, *Canad. J. Chem. Eng.*, **62**, 602, 1984.
Stoops, C. E., and C. L. Lovell, *Ind. Eng. Chem.*, **35**, 8, 845, 1943.
Tay, M., and G. B. Tatterson, *AIChEJ.*, **31**, 11, 1915, 1985.

Tojo, K., K. Miyanami, I. Minami, and T. Yano, *Chem. Eng. J.*, **17**, 211, 1979.
Tojo, K., K. Miyanami, and H. Mitsui, *Chem. Eng. Sci.*, **36**, 279, 1981.
Uhl, V. W., *Paint Oil Chemical Review*, **113**, 24, 95, November 23, 1950.
Vermeulen, T., G. M. Williams, and G. E. Langlois, *Chem. Eng. Progr.*, **51**, 2, 85F, 1955.
Warmoeskerken, M. M. C. G., and J. M. Smith, *Chem. Eng. Res. Des.*, **67**, 193, 1989.
Weetman, R. J., and J. Y. Oldshue, *Proc. 6th Europ. Conf. on Mixing, Pavia, Italy*, BHRA Fluid Eng., Cranfield, England, 43, 1988.
Westerterp, K. R., *Chem. Eng. Sci.*, **18**, 495, 1963.
White, A. M., E. Brenner, G. A. Phillips, and M. S. Morrison, *Trans. Amer. Inst. Chem. Engrs.*, **30**, 570, 1934.
White, A. M., and E. Brenner, *Trans. Amer. Inst. Chem. Engrs.*, **30**, 585, 1934.
White, A. M., and S. D. Sumerford, *Chem. Met. Engrs.*, **43**, 7, 370, 1936.
Xanthopoulos, C., and M. Stamatoudis, *Chem. Eng. Commun.*, **46**, 123, 1986.

Power Consumption in Viscous Creeping Flow Mixing

Introduction

Rapid mixing and homogeneity cannot be easily obtained for highly viscous liquids. These difficulties cannot be eliminated by extending mixing time or increasing rotational speed of the agitator for many impeller types. This is in contrast to mixing in turbulent flow where mixing can be readily achieved by such methods. If an inappropriate impeller is used, considerable heterogeneity can occur in a mixing operation which cannot be eliminated without changing the impeller and possibly resizing the agitator drive. Generally, it is considered good practice to use full-tank, close-clearance impellers which sweep out the entire tank volume in laminar mixing operations.

Viscous liquids are typically high-value, high-technology materials. Power per unit volume and the relative cost of mixing equipment to vessel size are high compared to turbulent mixing. These factors impact upon the need for fairly well established power and torque correlations. Power draw and torque are integral dynamic-kinematic quantities which are required for proper design. Torque is an important quantity in impeller design but is more of a design choice rather than a quantity to be supplied to a process like power. The technical literature reports power correlations which will be emphasized in this chapter.

The power correlations in viscous mixing are more fundamentally based than similar correlations in turbulent mixing. Basic theory, flow analogies, and computer models are available from which basic understanding of flow and power draw phenomena can be obtained. There is considerably more basic research on power consumption in viscous creeping flow laminar mixing than for turbulent flow as well. However, power draw calculations are complicated by more complex agitator geometries and the characterization of viscosity for the non-Newtonian fluids.

The purpose of this chapter is to serve as a review and an index of the literature on power draw correlations for viscous mixing for all types of fluids. Correlations and their bases in theory for Newtonian and shear thinning fluids will be given first, followed by studies of more complex fluids. Each section of this chapter provides background theory followed by a review of the literature and experimental results.

Other reviews are also available which discuss power consumption for non-Newtonian fluids. These include Skelland (1983), Ulbrecht and Carreau (1985), and Godfrey (1985).

Newtonian and Shear Thinning Fluids

The classification scheme for power correlations for Newtonian and shear thinning fluids will be based upon the theory or fluid mechanics upon which the power correlations are based. Shear thinning fluids form, by far, the largest class of non-Newtonian fluids. Newtonian and shear thinning fluids have also been well studied in the area of power consumption in creeping flow laminar mixing. Shear thickening fluids will also be discussed in this section since they are the rheological counterpart to shear thinning fluids.

At the present time there are six discernible approaches into which the literature on power consumption can be divided for shear thinning, Newtonian, and shear thickening fluids:

1. Dimensional analysis

2. Phenomenological coaxial rotating cylinder analogy

3. Extended coaxial rotating cylinder analogy

4. Dimensional analysis based upon flow mechanism

5. Drag-based analysis

6. Computational analysis

These fall into the three major categories of dimensional analysis, flow analogies (i.e., 2, 3, 4, and 5) and computational analysis. The end results of the studies are power correlations. Dimensional analysis is usually used at the end of the development in order to treat the geometric effects which cannot be determined from theory.

To perform a power draw calculation, the characterization of the fluid rheology and the tank-agitator geometry is required.

Fluid rheology

Fluid rheology is a complex subject which cannot be dismissed easily. However, for shear thinning fluids, the characterization of the fluid rheology appears to be fairly simple. For these fluids, all other non-Newtonian phenomena are secondary. With very few exceptions, the power law, Eq. (3.1), has been assumed to describe the rheology of these fluids:

$$\tau = K\gamma^n \tag{3.1}$$

The exponent n is the fluid index and is less than 1 for shear thinning fluids, equal to 1 for Newtonian fluids, and greater than 1 for shear thickening fluids. The parameter K is called the consistency index. The apparent viscosity μ_a is also defined by Eq. (3.1) when Eq. (3.1) is written in terms of the apparent viscosity:

$$\tau = \mu_a \gamma \tag{3.2}$$

The apparent viscosity is then:

$$\mu_a = K\gamma^{n-1} \tag{3.3}$$

and is shear rate dependent. It is well known that the power law can represent shear stress-shear rate data only over a few decades in shear rate. However, if the power law fit is adequate over the important range of shear rates occurring in the mixing tank, then the power law will give useful results. This does require knowledge of the magnitudes of shear rates in the process and rheological data.

The first major work using the power law was performed by Metzner and Otto (1957). Godleski and Smith (1962) examined the general case of power law fluids having variable n and K. Bingham fluids have been analyzed using a power law by Nagata et al. (1970; 1971) with a plastic Reynolds number. Thixotropic liquids have been studied by Edwards et al. (1976) and Godfrey et al. (1974) using the power law. Studies of viscoelastic fluids have incorporated the power law for both shear and normal stresses in relation to the deformation rate as reviewed by Ulbrecht (1974).

An exception to the use of the power law in the literature concerning power consumption in agitated tanks is work by Mitsuishi and Hirai (1969), who used the Ellis and Sutterby models. However, the benefit derived from the use of the more sophisticated rheological models was not evident. Another exception is work by Walton et al. (1981), who found that the power law gave poor results in comparison to the Eyring model. Yap et al. (1979) and Bertrand and Couderc (1985) used a Carreau model for viscosity in their studies.

Despite the results of Walton et al. and others, power laws for both shear and normal stresses appear to be sufficient. Further, there appears to be no pressing need to use more sophisticated rheology models in the calculation of power draw which is an integral quantity.

Geometry

The tank-agitator geometry impacts directly upon the fluid mechanics which occur in the tank since the geometry determines the boundaries where the various flow conditions are set. The impeller geometries are also much more complicated in creeping flow laminar mixing than in turbulent flow mixing, since the impellers have to bring motion to the entire tank for good mixing.

There are two major types of agitators in creeping flow laminar mixing as shown in Fig. 3.1. The different geometric parameters have been reviewed in Chap. 1. The first is the helical ribbon agitator. Anchor and gate agitators are essentially helical ribbon agitators with infinite pitch, i.e., the impeller ribbon becoming vertical. The second type is the screw agitator, Fig. 3.1, also called a helical screw impeller. The combination of the ribbon with the screw, called the helical ribbon screw impeller, has not received much study but is also available. Numerous other impeller designs exist but typically have not been emphasized in the literature. These include the sigma blade impellers, intermeshing rotor mixers, planetary mixers, and wire or whip mixers.

The typical geometric arrangements for the helical ribbon and screw agitators are shown in Fig. 3.1. The screw impeller is often used with a draft tube to promote uniform flow of material through the impeller region and into the bulk of the tank.

The major geometric considerations for both the helical ribbon and screw agitators are: (1) pitch or the number of flights of the ribbon or screw per height and (2) the distance or clearance of the ribbon or screw from the wall of the tank or draft tube. The distance from the tank bottom and from the liquid surface is of lessor importance. Surface area of the screw or ribbon is an important geometric parameter, since this is the area over which the shear acts. The thickness of the ribbon or screw at the outer circumference is important due to the high shear rates which act in the clearance.

For the ribbon agitator, there is additional pitch which can be discussed. As stated above, the first type of pitch is the number of flights of the ribbon per height. The second type of pitch is the tilt or orientation of the ribbon about the first type of pitch. This second pitch adds further complications to the characterization of the impeller geometry and power draw correlations but is typically not studied in the literature. This second type of pitch can also be applied to screw impellers as well.

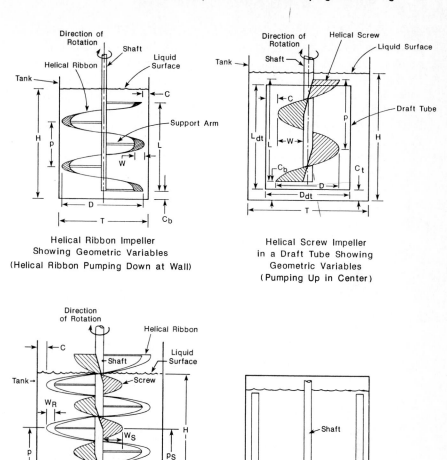

Helical Ribbon Impeller
Showing Geometric Variables
(Helical Ribbon Pumping Down at Wall)

Helical Screw Impeller
in a Draft Tube Showing
Geometric Variables
(Pumping Up in Center)

Helical Ribbon Screw Impeller
(Helical Ribbon Pumping Up at Wall
with Screw Pumping Down in Center)

Anchor Impeller

Figure 3.1 Helical ribbon, screw, and anchor impellers.

In creeping flow laminar mixing, Newtonian and shear thinning fluids are not pushed by the impeller blade. Instead, the fluid flows
around the impeller blade, and the shape, surface area, and orientation
of the impeller blade are important from drag considerations.

Dimensional analysis

Power basically has the same fundamental units as the product of μ,
N^2, and D^3. As a result, the dimensionless power number $Pg_c/\mu N^2 D^3$

can be formed. By dimensional analysis, this number may be written as a function of other dimensionless groups involving geometric variables and force ratios like the Reynolds number. However, it can be shown that the dimensionless number $Pg_c/\mu N^2 D^3$ is actually based upon a Stokes drag-velocity mechanism involving the pertinent fluid forces and velocities as discussed in Chap. 2. As a result, $Pg_c/\mu N^2 D^3$ becomes only a function of geometry and independent of impeller Reynolds number. Dividing by the impeller Reynolds number $\rho N D^2/\mu$, however, the traditional equation for the power number N_P is obtained:

$$\frac{Pg_c}{\rho N^3 D^5} = \frac{B}{\rho N D^2/\mu} \tag{3.4}$$

or as:

$$N_P = \frac{B}{N_{\text{Re}}} \tag{3.5}$$

where B is a function of the geometry. Typically B varies around 200 to 300, but individual values may span 30 to 4000.

Dimensional analysis can also be applied to the equations of motion, resulting in the impeller Reynolds number. If the power law is used to describe the rheology and the characteristic velocity and length are chosen as ND and D, respectively, the resulting impeller Reynolds number is $\rho N^{2-n} D^2/K$. This particular Reynolds number is specific to the use of the power law.

There are several other ways in which the power number can be obtained. One way is from Stokes drag analysis, and another is from energy dissipation equations. From Stokes drag, power can be calculated as:

$$P = \int \mathbf{V}^* \, \tau^* \, d\mathbf{A} \tag{3.6}$$

where τ is the total stress tensor and includes pressure and the normal and shear stresses which act on the impeller area. This power is the power draw which leaves the impeller and enters the fluid. The drag force is based upon the relative velocity of the impeller blade with respect to the surrounding fluid. The explicit velocity, appearing in Eq. (3.6), is fixed to laboratory coordinates.

From dimensional analysis of Eq. (3.6), assuming a power law for the shear stresses (and for normal stresses if necessary) and noting the dimensions on the pressure term are KN^n, the power number becomes $Pg_c/KN^{n+1} D^3$. When divided by the Reynolds number $N_{\text{Re}'} = \rho N^{2-n} D^2/K$ given above, the more common power number, given in Eq. (3.4), is obtained.

Power draw is energy added to the fluid per time by the impeller. By an energy balance, this power draw becomes power consumption in the fluid from which power and power number can also be obtained. Power consumption is the volume integral of the specific energy dissipation rate Φ, or:

$$P = \int \Phi \, dV \qquad (3.7)$$

This integration has never been accomplished using experimental information but can be obtained two-dimensionally by computational fluid mechanics as will be described below. The power number can be obtained as above.

As a result of the above analysis, two dimensionless parameters have been obtained: the power number and the impeller Reynolds number. These have been obtained from velocity or kinematic considerations and fluid force or dynamic considerations.

Geometric considerations are specific to the individual process and cannot be made general with only one dimensionless group. Changes in the geometry cause changes in the magnitudes of the velocities and forces involved. These changes or effects are written using power laws about some standard geometry or reference. In correlations, geometric effects appear as the constant B in Eq. (3.5) for a fixed geometry. However, for variable geometries, B will vary, and the determination of B as a function of geometry is a major objective of many power studies. The power correlation is thus established after the function B has been established.

Geometric effects do not have to be written as power laws, and their effects on power draw may not follow a simple power law relationship. Thus, care must be appropriately taken in establishing the function B.

Another parameter, which incorporates rheological effects and which is dimensionless, is the flow index n of the rheological power law used to model the fluid.

In general, then, power correlations appear as:

$$f\left(\frac{Pg_c}{KN^{n+1}D^3}, \frac{\rho N^{2-n}D^2}{K}, n, B\right) = 0 \qquad (3.8)$$

or as:

$$f\left(\frac{Pg_c}{\rho N^3 D^5}, \frac{\rho N^{2-n}D^2}{K}, n, B\right) = 0 \qquad (3.9)$$

Such correlations contain dynamic, kinematic, rheological, and geometric parameters. Traditionally, the power number is written as shown in Eq. (3.9). However, since μ and not ρ is the important quality in viscous mixing and since inertial forces and, hence, the Reynolds number are not important in creeping flow laminar mixing, Eq. (3.8) or Eq. (3.9) reduces to:

$$f\left(\frac{Pg_c}{KN^{n+1}D^3}, n, B\right) = 0 \qquad (3.10)$$

However, this is not the traditional form of power correlation which has been developed. The traditional form is Eq. (3.9).

Stokes drag is the important drag mechanism in laminar mixing and is typically written as the product of the fluid viscosity, a velocity, and a size measure. The viscosity is dependent upon the rheology, and the velocity can be based upon the impeller diameter and rotational speed in mixing. The object size measure should be based upon the blade size or a size measure of blade area. Impeller diameter alone does not contain information about area or size of the impeller blade. Care should be exercised when using a power draw correlation which uses only impeller diameter. In such situations, absolute geometric similarity should be maintained.

There are four characteristic lengths which characterize an impeller: diameter, length, blade width, and thickness. However, in the literature the impeller diameter is most often chosen as the size measure. As a result, impeller diameter may appear inappropriately in power draw correlations.

Literature based upon dimensional analysis. The following studies are solely based upon dimensional analysis and are arranged according to impeller type and date.

Turbulent impellers. Nagata et al. (1955a; 1955b; 1956; 1957), using highly accurate dynamometers, established power characteristics for creeping flow laminar and turbulent flow mixing for both baffled and unbaffled tanks. Their power correlation for the creeping flow laminar mixing was given as:

$$\frac{Pg_c}{\mu N^2 D^3} = 310 \left(\frac{W}{T}\right)\left(\frac{D}{T}\right)^{1.15} \qquad (3.11)$$

for two-bladed paddles having $0.3 < D/T < 0.9$ and $0.05 < W/D < 0.9$. As indicated by the correlation, power was found to vary proportionally with blade width and independent of liquid height so long as the paddles were beneath the liquid surface. Nagata et al. found no effect of Froude number upon power consumption.

Lee, Finch, and Wooledge (1957) performed an experimental study on power number-Reynolds number relationships of non-Newtonian fluids. Flat bladed turbines and shrouded and unshrouded impellers were used in the study. They found little effect of baffles on power; shrouded impellers drew 50 percent more power than unshrouded impellers (possibly due to higher surface area); the power draw for dual impellers was approximately twice that of single impellers. Spacing between impellers and impeller position had no effect on power consumption.

Anchor impellers. Foresti and Liu (1959) studied power consumption in non-Newtonian liquids for a six-bladed turbine, an anchor, and two cone impellers. Their correlation for all agitators and for shear thinning Newtonian fluids was:

$$N_P = 160B^{n-1}(N_{Re'})^{-1}\left(\frac{L}{H}\right)^n\left(\frac{D+T}{D}\right) \tag{3.12}$$

where B was an empirical constant with an approximate value of 50. Shear thickening liquids were also studied; however, the results were not correlated. Flow reversals were also observed for polyisobutylene and CMC solutions.

Uhl and Voznick (1960) studied the anchor agitator. Power data were correlated with Reynolds number and a clearance parameter. It was shown that power draw increased as blade clearance decreased. Anchor blade height was used in power calculations.

Sawinsky, Havas, and Deak (1976) performed some power measurements and reviewed a number of articles on power consumption for anchor and helical ribbon impellers. For Newtonian fluids, they provided:

$$P = B(\mu N^2 D^3)\left(\frac{L}{D}\right)\left(\frac{T}{C}\right)^x \tag{3.13}$$

where B and x were given as functions of impeller type and ranges of geometric parameters. Agreement with 15 previous studies for helical impellers and 9 previous studies for anchor impellers was noted. Effects of clearance on power in the range of $0.84 < D/T < 0.95$ were not observed. For shear thinning liquids, a similar correlation was given.

Helical ribbon impellers. Gray (1963), citing Nagata's work for helical ribbon impeller, noted that:

$$N_P N_{Re} = 300 \tag{3.14}$$

The relationship shows the order of magnitude expected for the product of the power number and the impeller Reynolds number.

Hall and Godfrey (1970) studied helical ribbon impellers for Newtonian fluids. Their final correlation was:

$$N_P N_{Re} = 66\left(\frac{p}{D}\right)^{-0.73}(N_R)\left(\frac{H}{D}\right)\left(\frac{W}{D}\right)^{0.5}\left(\frac{C}{D}\right)^{-0.6} \tag{3.15}$$

The correlation takes into account the effects of helical pitch, number of ribbons, impeller height, blade width, and wall clearance. The effect of scale was found insignificant. The surface area of the helical ribbon impeller was also important. For non-Newtonian fluids, an apparent viscosity was used in the impeller Reynolds number.

Kappel (1979) provided an excellent review of power correlations for the helical ribbon impeller. The results showed agreement in power number at large clearances; however, considerable differences in power number occurred at low clearances close to the wall. Kappel (1979) studied 12 different geometric arrangements for the helical ribbon impellers. The product of power number and Reynolds number was found to be:

$$N_P N_{Re} = 60(N_R)^{0.8} \left(\frac{p}{D}\right)^{-0.5} \left(\frac{C}{D}\right)^{-0.3} \tag{3.16}$$

where $C/D = (T/D - 1)/2$ for $H/D = 1.1$, $W/D = 1.1$, $W/D = 0.1$, $L/D = 1.0$, $N_R = 1$ or 2, $0.5 < p/D < 1.0$, $1.02 < T/D < 1.1$, and $1 < N_{Re} < 30$.

Bourne et al. (1979), in a literature summary, cited the following correlation for power number for helical ribbon impellers:

$$N_P N_{Re} = 134\left(\frac{L}{T}\right)\left(\frac{p}{T}\right)^{-0.3} \left(\frac{C}{T}\right)^{-0.3} \tag{3.17}$$

Blasinski and Rzyski (1980) gave a power correlation for helical ribbon impellers as:

$$N_P N_{Re} = 34.1\left(\frac{C}{D}\right)^{-0.53} \left(\frac{H}{D}\right)^{0.45} \left(\frac{p}{D}\right)^{-0.63} \left(\frac{L}{D}\right)^{1.01} \left(\frac{W}{D}\right)^{0.14} N_R^{0.79} \tag{3.18}$$

The ranges for the various quantities are: $0.01 < C/D < 0.095$; $1.02 < T/D < 1.19$; $0.357 < p/D < 1.28$; $1.02 < H/D < 1.64$; $0.862 < L/D < 1.11$; and $0.071 < W < 0.167$ for $N_R = 1$ or 2.

The correlation exponents for clearance, pitch, blade length, and blade number are listed in Table 3.1 for comparison purposes.

Helical screw impellers. Novak and Rieger (1969) studied the influence of geometry of screw agitator systems on power consumption. The following screw geometries are listed in order of decreasing power draw: screw centered in a draft tube, screw off-centered using no draft tube, screw centered with wall baffles, and screw centered without wall baffles. The first two geometries were found to be the most efficient geometries for mixing. The centered position was found to be the least efficient for mixing. Figure 3.2 shows the power correlation for centrally located screw impellers without a draft tube for Natrosol and PAA by (Chavan and Ulbrecht, 1973b).

TABLE 3.1   Correlation Exponents from the Various Dimensional Analysis Studies
for the Helical Ribbon Impeller

Variable	C	p	L	N_R	W
Nagata et al.	−0.5	—	—	—	—
Hall and Godfrey (1970)	−0.6	−0.73	1.0	1.0	0.5
Kappel (1979)	−0.3	−0.5	—	0.8	—
Bourne et al. (1979)	−0.3	−0.3	1.0	—	—
Blasinski and Rzyski (1980)	−0.53	−0.63	1.01	0.79	0.14

Seichter (1971) studied the efficiency of screw agitators. The most efficient mixers were found to be those which produced axial flow. Power measurements were carried out indirectly.

Chavan and Ulbrecht (1973b) studied power requirements of off-centered helical screw impellers in viscous Newtonian and elastic and inelastic shear thinning fluids. The approach taken was simply based upon dimensional analysis rather than their prior work using coaxial rotating cylinder theory, which will be discussed later. They noted that the agitator-tank geometry was very complicated, and changing geometry was not an easy task. At the time of their work, there was no generalized correlation for power consumption for screw impellers. They assumed that: (1) power was proportional to the surface area of

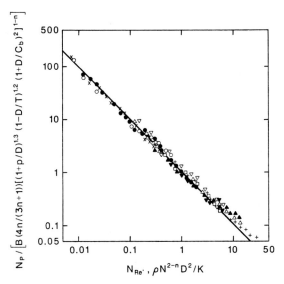

Figure 3.2   Power correlation for centrally located screw impellers without a draft tube for non-Newtonian Natrosol and PAA solutions in various geometries. (*From V. V. Chavan and J. J. Ulbrecht, Trans. Instn. Chem. Engrs., 51, 349, 1973. By permission.*)

the screw, (2) the effect of clearance of the impeller from the top liquid surface was negligible, and (3) the effect of the liquid height was incorporated into other quantities like impeller length and clearance. They found for Newtonian fluids:

$$N_P N_{Re} = 13.5a \left[\left(1 - \frac{D}{T} \right) \left(1 + \frac{D}{C_b} \right) \right]^{0.7} \tag{3.19}$$

where the dimensionless area a is a geometric parameter given as a complicated function of length, diameter, pitch, and blade width of the impeller. The correlation provides a finite power number as C_b/D and T/D approach large values. The above correlation also shows that power draw increased with tank diameter and that power draw increased as clearance from the bottom decreased. Their correlation for off-centered screw impellers in shear thinning liquids was given as:

$$N_P N_{Re'} = 13.5a \left(\frac{4n}{3n + 1} \right) \left[\left(\frac{T - D}{T - E} \right) \left(1 + \frac{D}{C_b} \right) \right]^{0.7}$$

$$\times \left[1.9 \left(1 + \frac{p}{D} \right)^{-1.40} \left(1 - \frac{D}{T} \right)^{0.94} \left(1 + \frac{D}{C_b} \right)^{1.32} \right]^{n-1} \tag{3.20}$$

The equation holds for centrally located screw impellers as well. Novak and Rieger (1975) reported power data on the screw impeller also, but their study was primarily an efficiency study.

Sawinsky, Deak, and Havas (1979) studied screw agitators in Newtonian liquids in centered and off-centered positions and in draft tubes. Although, following Chavan and Ulbrecht's analysis using screw surface area, they could not confirm any agreement with Chavan and Ulbrecht (1973a; 1973b). However, agreement with several other studies was noted.

Helical ribbon screw impellers.   Chowdhury and Tiwari (1979) studied power consumption for helical ribbon screw impellers and obtained a correlation similar to Eq. (3.5).

Merquiol and Choplin (1988) studied the power consumption for helical ribbon screw impellers in the laminar and transition regimes. Their power number correlation for Newtonian fluids was:

$$N_P N_{Re}^{-b} = A \left(\frac{D}{p} \right)^{0.7} \left(\frac{W}{D} \right)^{0.03} \tag{3.21}$$

where p is the pitch of the ribbon and where:

$$A = 330 \qquad b = -1 \qquad \text{for Re} < 100$$

$$A = 118 \qquad b = -0.8 \qquad \text{for } 100 < \text{Re} < 500$$

The impellers were single-ribbon impellers.

Comparison studies. Chapman and Holland (1965) performed a comparison between turbine and helical screw agitators. Power curves were obtained for a number of helical screw systems both with and without baffles. Helical screw agitators drew substantially more power than turbulent agitators. An off-centered position very close to the wall with no wall baffles was found to be the most efficient mixing position for the helical screw system.

Hoogendoorn and den Hartog (1967) gave a summary of power results concerning turbines, screw, propellers, anchors, helical impellers, and draft tubes. They found that draft tubes increased power consumption considerably.

Phenomenological coaxial rotating cylinder analogy

In this analogy, a helical ribbon or a screw impeller is considered to be similar to a rotating cylinder inside another cylinder, i.e., the tank. The system for this analogy is shown in Fig. 3.3 where the inner cylinder is rotating and the outer cylinder is stationary. From a force balance on the system or from the equations of motion (Brodkey, 1967):

$$\frac{\partial(r^2\tau_{r0})}{\partial r} = 0 \tag{3.22}$$

Integrating this equation, the constant of integration being $-T_0/2\pi$, the shear stress $\tau_{r\theta}$ is obtained as:

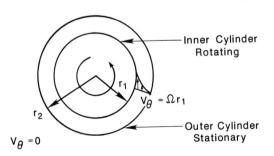

$$\Omega = 2\pi N$$

Figure 3.3 Coaxial rotating cylinders showing conditions.

$$\tau_{r\theta} = \frac{-T_0}{2\pi r^2} \tag{3.23}$$

where T_0 is an unknown torque. Assuming a power law relationship for $\tau_{r\theta}$, the equation becomes:

$$\frac{T_0}{2\pi r^2} = -K\left(r\frac{\partial(V_\theta/r)}{\partial r}\right)^n \tag{3.24}$$

or

$$r\frac{\partial(V_\theta/r)}{\partial r} = -r^{-2/n}\left(\frac{T_0}{2\pi K}\right)^{1/n} \tag{3.25}$$

Integrating again:

$$\frac{V_\theta}{r} = +\left(\frac{T_0}{2\pi K}\right)^{1/n}\left(\frac{n}{2}\right)r^{-2/n} + c \tag{3.26}$$

Evaluating the constant at $r = r_2$ where $V_\theta = 0$:

$$\frac{V_\theta}{r} = \left(\frac{T_0}{2\pi K}\right)^{1/n}\left(\frac{n}{2}\right)(r^{-2/n} - r_2^{-2/n}) \tag{3.27}$$

Evaluating at r_1:

$$\Omega = \left(\frac{T_0}{2\pi K}\right)^{1/n}\left(\frac{n}{2}\right)(r_1^{-2/n} - r_2^{-2/n}) \tag{3.28}$$

where Ω is $2\pi N$. Rearranging for the unknown torque T_0:

$$\left(\frac{T_0}{2\pi K}\right)^{1/n} = \frac{(2/n)\Omega}{r_1^{-2/n} - r_2^{-2/n}} \tag{3.29}$$

Thus, from Eq. (3.25), the shear rate at any radius becomes:

$$\gamma = \frac{-r\partial(V_\theta/r)}{\partial r} = \left(\frac{1}{r}\right)^{2/n}\frac{(2/n)\Omega}{r_1^{-2/n} - r_2^{-2/n}} \tag{3.30}$$

and at $r = r_1$, the radius of the inner cylinder:

$$\gamma_1 = \frac{(2/n)\Omega}{1 - (r_1/r_2)^{2/n}} \tag{3.31}$$

Two limits are of interest. The first occurs most often since the impeller shear rates are contained in a volume very close to the blade. In this case, r_1/r_2 essentially approaches zero and:

$$\gamma_1 = \frac{4\pi N}{n} \qquad (3.32)$$

When r_1/r_2 approaches 1, as in the case of high-friction film wipers:

$$\gamma_1 = \frac{2\Omega r_2^2}{r_2^2 - r_1^2} \qquad (3.33)$$

The important conclusion from these relationships is that there is a linear relationship, e.g., Eq. (3.32), between γ and N at fixed values of n, i.e., $\gamma = k\,N$, for specified geometries. The approximate order of magnitude of the constant k is 4π for situations where r_1/r_2 approaches zero. These situations occur often in creeping flow laminar mixing where the flow behavior, generated by the impeller, does not reach the tank wall. The impeller often behaves as if it is in an infinite fluid.

Metzner and Otto (1957), following work by Magnusson (1952), used the coaxial rotating cylinder relationship as an analogy to rotating impellers and postulated phenomenologically a relationship between the average shear rate, γ_{av}, around the impeller, and the impeller rotational speed N as:

$$\gamma_{av} = kN \qquad (3.34)$$

where k is a constant of proportionality. In the postulation, the flow around the impeller was considered to have one characteristic shear rate, the average shear rate γ_{av}. However, the constant of the proportionality k was specific to the particular agitator-tank geometry and had to be determined experimentally for each geometry of interest. The postulation has been applied to a number of impellers in laminar mixing, and k values have been reported for a number of impellers.

Equation (3.34) is used directly in the calculation of power draw. From the impeller rotational speed N, the average shear rate γ_{av} can be obtained as kN. From the average shear rate, the apparent viscosity can be obtained from the power law as:

$$\mu_a = K(\gamma_{av})^{n-1} \qquad (3.35)$$

Power is then $B\mu_a N^2 D^3$ where B is obtained from independent experimental studies of the effects of geometry on power.

The value of the constant k is established through the use of duplicate studies, one with a Newtonian fluid, the other with a non-Newtonian fluid, in the same agitator-tank geometry at the same impeller rotational speeds. From the respective power measurements, the apparent viscosity of the non-Newtonian fluid is calculated as:

$$\mu_{nN} = \mu_N \left(\frac{P_{nN}}{P_N} \right) \qquad (3.36)$$

From the apparent viscosity, the average shear is obtained by Eq. (3.35) which establishes the relationship between γ_{av} and N.

To perform power draw calculations, the reverse procedure is used. For the particular geometry of interest for which there is a relationship established between γ_{av} and N, the design rotational speed is selected which fixes the average shear rate. The average shear rate, in turn, determines the apparent viscosity and power.

Literature based upon coaxial rotating cylinders.   In their initial studies, Metzner and Otto (1957) found the constant of proportionality to be 13 for flat bladed turbines. They also found no effects of baffles or T/D ratio on power consumption. Their lowest T/D studied was 1.3. The impellers behaved as though in an infinite fluid, and there was very little wall movement of the fluid observed in the studies. Poor mixing was occurring. The impeller Reynolds numbers, given for the initial appearance of movement at the wall, were in the transition region. No effect of impeller position on power consumption was found except where a surface vortex was formed. Since shear thinning fluids become more viscous after leaving the impeller region than their respective Newtonian counterpart, multiple impellers and low T/D ratios were recommended for the mixing of these fluids.

A correlation for power was given by Metzner and Otto as:

$$P \doteq 71 \, \frac{\mu N^2 D^3}{g_c} \qquad (3.37)$$

for an impeller with six flat blades. Metzner and Otto also noted that, for shear thickening fluids, solidification of the fluid occurred near the impeller due to the high shear rates and the solidified core approached the wall as the rotational speed was increased.

Metzner and Otto's initial work is very important in creeping flow laminar mixing for three reasons. The first is the postulation of the relationship between γ_{av} and impeller rotational speed. The relationship has had considerable influence on power draw correlations. Second, their list of impeller Reynolds numbers for initial motion in the entire tank showed the turbulent type impellers are inadequate for mixing at impeller Reynolds numbers below the laminar to turbulent transition. Lastly, they recommended multiple impellers and low T/D ratios or, in other words, full tank impellers such as the helical ribbon and screw impellers.

Calderbank and MooYoung (1959) validated Metzner and Otto's

postulated relationship between average shear rate and impeller rotational speed for Bingham, shear thinning, and shear thickening fluids. Power and speed were measured for various non-Newtonian fluids from which the apparent viscosity was obtained from Newtonian power curves specific to the geometry used. They found:

$$\gamma = 10N \qquad (3.38)$$

for Bingham and shear thinning fluids with $0.05 < n < 0.6$ and:

$$\gamma = 38N\left(\frac{D}{T}\right)^{0.5} \qquad (3.39)$$

for shear thickening fluids with $1.28 < n < 1.68$. A power correlation for shear thickening fluids is shown in Fig. 3.4 for turbulent agitators.

Metzner et al. (1961) extended their work and again found excellent agreement with the proportionality postulated between rotational speed and shear rate. In this study, k was found to be 11.5 for flat bladed turbine impellers, 13 for pitched blade turbines, and 10 for marine propellers, Fig. 3.5, irrespective of shaft position, pitch, or direction of fluid pumping. Power curves were reported for flat bladed turbines in shear thinning fluids, Fig. 3.6, and for a single pitched blade turbine in shear thickening fluids.

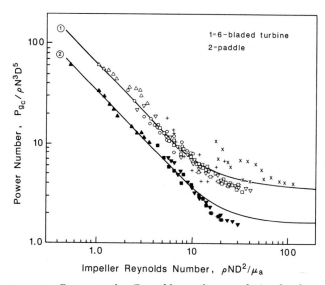

Figure 3.4   Power number-Reynolds number correlation for shear thickening fluids, $n > 1$. Fluids used included corn flour sugar solutions and gum arabic borax solutions. (*From P. H. Calderbank and M. B. Moo Young, Trans. Instn. Chem. Engrs., 37, 26, 1959. By permission.*)

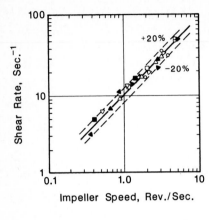

Figure 3.5 Dependence of mean shear rate upon impeller speed for marine propellers pumping either upward or downward in non-Newtonian fluids, $n < 1$. (*From A. B. Metzner, R. H. Feehs, H. L. Ramos, R. E. Otto, and J. D. Tuthill, AIChE Journal, 7, 1, 3, 1961. Reproduced by permission of the American Institute of Chemical Engineers.*)

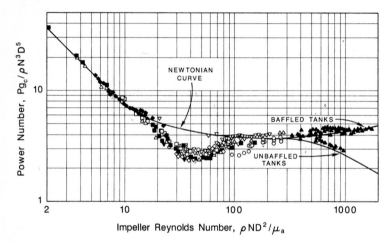

Figure 3.6 Power number-Reynolds number correlation for a single flat bladed turbine in non-Newtonian fluids. Fluids used included CMC, $n = 0.34$; Attasol, $n = 0.38$; carbopol, $0.18 < n < 0.54$; permagel, $n = 0.16$, 0.21; and pliovic, $n = 1.5$. (*From A. B. Metzner, R. H. Feehs, H. L. Ramos, R. E. Otto, and J. D. Tuthill, AIChE Journal, 7, 1, 3, 1961. Reproduced by permission of the American Institute of Chemical Engineers.*)

Metzner et al. noted that two impellers drew twice the power of a single impeller, providing the T/D ratio remained above about 1.25. At lower T/D ratios, wall effects caused additional increases in power consumption. Wall baffles were found to have no effect on power except in the laminar to turbulent transition region. In this transition region, power numbers, Fig. 3.6, for shear thinning fluids were substantially lower than Newtonian fluids. For the shear thickening fluids, data indicated that the shear rate varied with the square of the impeller rotational speed (except for data taken at $T/D = 1.06$), Fig.

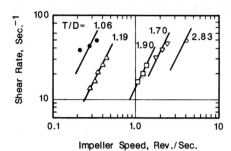

Figure 3.7 Mean shear rate-impeller speed for propellers in highly shear thickening fluid systems. (*From A. B. Metzner, R. H. Feehs, H. L. Ramos, R. E. Otto, and J. D. Tuthill, AIChE Journal, 7, 1, 3, 1961. Reproduced by permission of the American Institute of Chemical Engineers.*)

3.7. A considerable dependence upon T/D ratio was noted, but their results did not agree with the simple relationship of Calderbank and MooYoung given above for shear thickening fluids. As with the other impellers, a solidified core was observed around the propeller for shear thickening fluids.

For shear thickening fluids, local high power input from impellers is not desirable. A uniform distribution of low shear rates throughout the vessels is more appropriate for mixing.

Calderbank and MooYoung (1961) developed rather complicated correlations between a modified power number and modified Reynolds number. The correlation, however, has been reviewed by Penney and Bell (1967) and should be used with caution.

Godleski and Smith (1962) investigated shear thinning fluids whose power law parameters n and K varied as a function of shear rate. They found that the procedures developed by Metzner and Otto could be used for fluids with a variable fluid index n.

Bourne and Butler (1965) reported a value of 30 for the proportionality constant k for the helical ribbon impeller although they approached power draw in a different manner which will be discussed at length in the next section.

Beckner and Smith (1966) studied simple anchors using Newtonian and shear thinning fluids. They noted that as wall clearance decreased, power draw increased. Effects of breaker bars (i.e., stationary vertical baffles added to the tank inside the perimeter of the anchor) were studied. The first bar caused a 50 to 75 percent increase in power number; additional bars had little effect on power number. The effect of blade angle pitch (i.e., the second type of pitch) was correlated with a change in slope of power number versus Reynolds number relationship. The relationship between shear rate and rotational speed was not as well defined as that by Metzner et al. (1957; 1961) and Calderbank et al. (1959; 1961) given above. Beckner and Smith correlated k as:

$$k = c(1 - n) \tag{3.40}$$

where c was correlated with wall clearance and n is the exponent for the rheological power law. Separate lines were generated; one for the pitched anchor and one for the flat blade anchor. Their expression for apparent viscosity was:

$$\mu_a = K[c(1 - n)]^{n-1}N^{n-1} \tag{3.41}$$

and their Reynolds number was:

$$N_{Re}^* = \frac{\rho N^{2-n}D^2}{\{K[c(1 - n)]\}^{n-1}} \tag{3.42}$$

Power was found to vary with T/C to the 0.25 power. Their final power correlation was given as:

$$N_P = 82\,(N_{Re}^*)^{-0.93}\left(\frac{T}{C}\right)^{0.25} \tag{3.43}$$

for n between 0.27 and 0.77. The constant c was correlated as:

$$c = 37 - 120\left(\frac{C}{T}\right) \tag{3.44}$$

for flat anchor and pitched anchors and:

$$c = 106 - 1454\left(\frac{C}{T}\right) \tag{3.45}$$

for pitched anchors when $C/T < 0.06$. The wall clearance C is the distance from the anchor blade to the wall.

Hall and Godfrey (1970) suggested that their correlation, Eq. (3.15), given above for helical impellers could be used for non-Newtonian fluids using the apparent viscosity in the impeller Reynolds number. They suggested 27 as the constant of proportionality between average shear rate and impeller rotational speed.

Nagata et al. (1971) have summarized k values for different impellers from a number of different authors: for the helical ribbon impeller, 30, Fig. 3.8; for an anchor, 20 to 25; for paddles and the disk style turbine, 10 to 13, Fig. 3.8; for a curved blade impeller, 7.1; and for a propeller, 10.0.

Rieger and Novak (1973) related power number and Reynolds number using Eq. (3.5). Assuming the Metzner-Otto proportionality between shear rate and rotational speed and using the definition for apparent viscosity, the following was obtained:

$$N_P = \frac{Bk^{n-1}}{N_{Re'}} \tag{3.46}$$

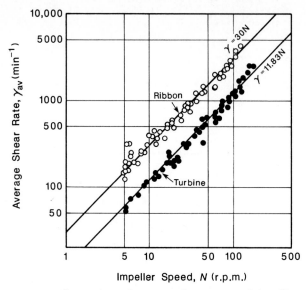

Figure 3.8 Comparison of average shear rate and impeller speed. (*From S. Nagata, M. Nishikawa, H. Tada, and S. Gotoh, J. Chem. Eng. Japan, 4, 72, 1971. By permission.*)

Rieger and Novak provided plots of Bk^{n-1} versus $1 - n$ for a helical screw agitator in a draft tube, helical screw agitator in the off-centered position, a helical ribbon agitator, a standard anchor agitator, and a pitched blade anchor agitator. Separate data were provided for k and B. The appeal of this method is its simplicity, but the method does not provide information of specific geometric effects.

Takahashi, Yokota, and Konno (1984) cited k values for helical ribbon impellers as:

$$k = 11.4\left(\frac{C}{T}\right)^{-0.411}\left(\frac{p}{T}\right)^{-0.361}\left(\frac{W}{T}\right)^{0.164} \tag{3.47}$$

The k values as a function of clearance, pitch, and blade width are shown in Fig. 3.9.

Shamlou and Edwards (1985) cited a relationship for k as:

$$k = 34 - 144\left(\frac{C}{D}\right) \tag{3.48}$$

for helical ribbon impellers which shows good agreement with Beckner et al. (1966) given above. The study by Shamlou et al. was performed at two different scales. Their power number correlation was:

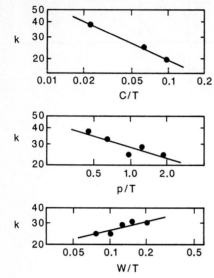

Figure 3.9 Relation between k and each of three geometrical ratios. (*From K. Takahashi, T. Yokota, and H. Konno, J. Chem. Eng. Japan, 17, 657, 1984. By permission.*)

$$N_P N_{\text{Re}} = 150\left(\frac{L}{D}\right)(N_R)^{0.5}\left(\frac{p}{D}\right)^{-0.5}\left(\frac{C}{W}\right)^{-0.33} \tag{3.49}$$

Merquiol and Choplin (1988) proposed the following correlation for k for helical ribbon screw impellers:

$$k = 60\left(\frac{p}{D}\right)^{0.65} \tag{3.50}$$

where p is the pitch of the ribbon.

Extended coaxial rotating cylinder analogy

The coaxial cylinder analogy can be extended to calculate power directly as the product of the shear stress $\tau_{r\theta}$, area A, and velocity πDN:

$$P = (\tau_{r\theta})A(\pi DN) \tag{3.51}$$

without any postulated relationship between average shear rate and impeller rotational speed. Assuming a power law for the shear stress, using Eq. (3.31) to define the shear rate and noting that the area of a cylinder is πDL and impeller tip speed is $\Omega D/2$, the equation for power can be written as:

$$P = K\left(\frac{2\Omega}{\{n[1 - (D/T)^{2/n}]\}}\right)^n (\pi DL)\left(\frac{\Omega D}{2}\right) \tag{3.52}$$

or as:
$$P = \frac{(\pi K/2)T^2D^2L\Omega}{(T^{2/n} - D^{2/n})^n}\left(\frac{2\Omega}{n}\right)^n \qquad (3.53)$$

Using apparent viscosity and shear rate as defined in Eq. (3.2) instead of the power law for the $\tau_{r\theta}$, power can also be calculated as:

$$P = \mu_a\left(\frac{4\pi N}{\{n[1 - (D/T)^{2/n}]\}}\right)(\pi DL)(\pi ND) \qquad (3.54)$$

Upon dividing by $\rho N^3 D^5$:

$$N_P = \mu_a\left(\frac{4\pi^3}{\{n[1 - (D/T)^{2/n}]\}}\right)\frac{N^2 D^2 L}{\rho N^3 D^5} \qquad (3.55)$$

Rearrangement of this expression with L/D incorporated into a constant by geometric similarity, the traditional form of Eq. (3.5) is obtained where:

$$B = \frac{4\pi^3}{\{n[1 - (D/T)^{2/n}]\}}\left(\frac{L}{D}\right) \qquad (3.56)$$

and the impeller Reynolds number is based upon the apparent viscosity.

Literature based upon extended coaxial cylinders. Bourne and Butler (1965) performed a study of helical ribbon impellers in an agitated tank. They found that the ribbon produced extensive swirling and a central vortex which was the deepest when the ribbon was pumping upward. The flow field between the central shaft and the ribbon rotated as a forced vortex. The shear rates were relatively low in the forced vortex or core, and the energy dissipation in the core was considered negligible. Power was consumed in overcoming the tangential viscous drag forces between the forced vortex and the tank wall. Bourne and Butler correlated their power data using Eqs. (3.5) and (3.54) in the manner above. They noted that the shear thinning nature of the fluid had a considerable effect on power and that power consumed when the impeller was pumping upward was 30 percent higher than when pumping downward.

In continued study of helical ribbon impellers, Bourne and Butler (1969) noted that: (1) the Froude number had no influence on power although a surface vortex was formed and (2) increasing n increased the wall shear stress and increased power draw. Power was proportional to liquid height except where the liquid exceeded the agitator height, in which case, power was proportional to agitator height. The off-bottom clearance had a minor effect on power. The direction of rotation had no influence on power consumption, and the effects of the shear thinning nature of the material or the pseudoplasticity was accounted for in the correlation by using n. The effect of scale was accounted for in the analogy. The transition from laminar to partially turbulent flow was determined by a combination of Reynolds number and n and not just Reynolds number alone. For shear thinning fluids, the laminar to turbulent transition occurred at higher impeller Reynolds number which caused low power numbers in the transition region. However, with the onset of turbulence, the power number curves returned to the power number levels appropriate for turbulent flow, power in turbulent flow being dependent upon fluid density and not viscosity.

Bourne and Butler also studied the influence of clearance and found that the coaxial analogy failed as D/T approached 1. This was due, in part, to larger variations in shear rates in the clearance volume as D/T approached 1 and the assumptions made in the coaxial analogy. The effect of pitch was also investigated. As pitch increased (i.e., the number of ribbon flights per height increased in this case), the coaxial theory was shown to be adequate; the ribbons approached a solid cylinder. As the number of flight decreased, the coaxial cylinder analogy became less adequate. This is shown in Fig. 3.10.

Chavan and Ulbrecht (1972), using the coaxial cylinder analog, obtained an equivalent diameter for the helical ribbon impeller. The shear rate at this diameter was equal to the average shear rate in the

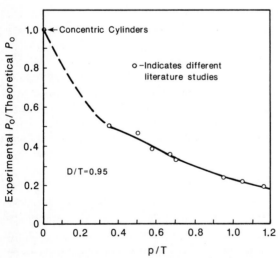

Figure 3.10   Influence of relative pitch on the accuracy of the coaxial cylinder theory. Data refer to Newtonian liquids.   (*From J. R. Bourne and H. Butler, Trans. Instn. Chem. Engrs., 47, T263, 1969. By permission.*)

gap between the tank wall and the impeller. Their expression for the equivalent diameter was:

$$\frac{D_e}{D} = \frac{T}{D} - \frac{2(W/D)}{\ln \dfrac{(T/D) - [1 - (2W/D)]}{(T/D) - 1}} \tag{3.57}$$

and their resulting power number correlation was:

$$N_P = 2.49\pi a \left(\frac{D_e}{D}\right)\left(\frac{T}{D_e}\right)^2 \left(\frac{4\pi}{\{n[(T/D_e)^{2/n} - 1]\}}\right)^n N_{Re'}^{-1} \tag{3.58}$$

where a was:

$$a = \frac{A}{D^2} \tag{3.59}$$

where A is the surface area of the impeller. The use of the concept of an equivalent diameter where the average shear can be considered to act and the inclusion of the area of the impeller into the correlation allowed for more of the geometric variables of the helical ribbon impeller to be incorporated into the power correlation.

Chavan, Jhaveri, and Ulbrecht (1972) studied helical screw impellers with and without draft tubes using their equivalent diameter approach. They noted that there were actually 10 geometric variables to be considered in power correlations, some of which were not totally independent of others. Their power number correlation for helical screw agitators with a draft tube was:

$$N_P = 0.11 \left(\frac{C_t}{L_t}\right)^{-0.9} \pi a \left(\frac{D_e}{D}\right)\left(\frac{T}{D_e}\right)^2 \left(\frac{4\pi}{\{n[(T/D_e)^{2/n} - 1]\}}\right)^n N_{Re'}^{-1} \tag{3.60}$$

When $C_b < C_t$, C_b should be used in the equation in replacement of C_t. For helical screw impellers without a draft tube, they obtained:

$$N_P = 5.3a \left(\frac{T}{D}\right)^{0.9}\left(\frac{C_b}{D}\right)^{-0.6} (k_s)^{n-1} N_{Re'}^{-1} \tag{3.61}$$

where k_s was a constant for a particular geometry and flow index. Chavan and Ulbrecht (1973b) noted that care should be exercised in using these correlations since, as T/D or C_b/D approach infinity, power consumption approaches infinity or zero, respectively. The correlations, however, can be used in the ranges of dimensionless variables which were studied in the work. Chavan and Ulbrecht (1973a; 1974) extended their results by taking into account clearance between the impeller and vessel bottom C_b/D; the spacing between the draft tube and tank $(T - D_t)/D$; the length of the draft tube L_t/D; and the clearance between the draft tube and vessel bottom (C_t/D). Their final correlation for screw impellers in draft tubes was:

$$N_P = 0.5\pi a \left(\frac{D_e}{D}\right) \left(\frac{4\pi}{n[(T/D_e)^{2/n} - 1]}\right)^n N_{\text{Re}}^{-1}$$

$$\times \left[1 + \left(\frac{D}{C_b}\right)\right]^{0.37} \left(\frac{T - D_t}{L_t}\right)^{-0.046} \left(\frac{C_t}{D}\right)^{-0.036} \quad (3.62)$$

The spacing between the draft tube and tank and the clearance between the draft tube and vessel bottom were geometric variables outside the draft tube and do not characterize the impeller. Hence, these geometric variables had relatively little effect on power.

For ribbon impellers, Chavan and Ulbrecht also tested their correlation, Eq. (3.58), given above. Power consumption was predicted satisfactorily for 28 different geometries having T/D varying between 1.05 and 1.58. The correlation was recommended when C_b/D was about 0.1. The correlation failed to correlate data for three studies reported in the literature, being 31.6 percent on average too low. Chavan et al. (1975b) and Chavan (1983) should also be consulted before using this correlation.

Chavan and Ulbrecht noted further that the ribbon screw impeller was by far the most geometrically complicated impeller. However, it appeared from their work that if the surface area of the ribbon was larger than the screw, the power consumption for the ribbon screw impeller can be predicted using the ribbon impeller geometry only because of the exponent of 3 for the impeller diameter. However, Merquiol and Choplin (1988) found that if the screw area was equal to or slightly lower than the ribbon area, the geometry of the screw would have to be included in power correlations.

Dimensional analysis based upon flow mechanism

Penney and Bell (1967) published an article which essentially forms a category for several reasons. First, the work is by an equipment user and is one of the first, as such, to appear in the literature on power consumption for laminar mixing. Second, Penney and Bell critically reviewed a number of literature correlations for power draw calculations, noting whether the correlations were physically realistic. Lastly, although not performing any experiments of their own but alluding to extruder screw theory, Penney and Bell provided a form for power correlations based upon two effects: (1) bulk effects and (2) shear effects.

Power consumed in bulk effects was correlated as:

$$P_b = \beta \mu D^2 N^2 L \left[f\left(\frac{C}{D}, \frac{p}{D}\right) \right] \tag{3.63}$$

which has a basis in drag theory. The Stokes drag term is μNDL, and the velocity is ND. Stokes drag is also affected by clearance C and pitch p. Clearance affects the flow field and drag around the impeller blade and a reduction in pitch p effectively increases the area of the impeller blade.

Power consumption also occurred due to the high shear rates between the impeller along its length and the wall of the vessel. To correlate this power consumption, Penney and Bell suggested:

$$P_c = 2\pi D^2 N^2 \left(\frac{\mu}{g_c}\right)\left(\frac{t}{C}\right) L \tag{3.64}$$

Although Penney and Bell did not show the basis of this equation in theory, it can be obtained from the coaxial cylinder analogy above. The two power expressions combined provided an overall correlation given as:

$$N_P = \frac{P_c + P_b}{\rho N^3 D^4 L}$$

or:
$$N_P = (\beta + 2\pi)N_{\mathrm{Re}}^{-1}\left[\frac{t}{C} + f\left(\frac{C}{D}, \frac{p}{D}\right)\right] \tag{3.65}$$

Penney and Bell had included an effect of shaft diameter, but this was dropped in the final equation since shaft effects are usually considered negligible.

Drag-based analysis

In the section on dimensional analysis above, the basis for the calculation of power from drag considerations was discussed. However, very few studies have actually used drag coefficients of the impeller blades for the calculation of power consumption. Very few studies in laminar mixing have discussed drag coefficients.

Takaisi (1958) studied the slow motion of an elliptic cylinder with an arbitrary angle of attack in a viscous liquid, bounded by a plane, based upon Oseen's linearized equations of motion. Lift and drag coefficients were obtained approximately for cylinder thicknesses of zero, 0.1, 0.5, and 1.0. General expressions for drag and lift coefficients were given.

Patterson, Carreau, and Yap (1979) studied power consumption for helical ribbon agitators based upon drag coefficients. Their final power correlation for Newtonian fluids was:

$$N_P = 24 N_R (N_{Re})^{-0.93} \left(\frac{D}{T}\right)^{0.91} \left(\frac{L}{D}\right)^{1.23} \qquad (3.66)$$

Yap, Patterson, and Carreau (1979) extended their correlation to include non-Newtonian fluids using a generalized Reynolds number and an effective viscosity based upon the Carreau model. The appearance of the correlation was the same as that given above except that the effective viscosity of the Carreau model was used in the impeller Reynolds number. This viscosity was calculated at an effective rate of deformation which was given as:

$$\gamma_e = 4^{1/(2s)} \left(\frac{D}{T}\right)^2 \left(\frac{L}{D}\right) N \qquad (3.67)$$

which is similar to the Metzner-Otto (1957) postulation. The k values calculated from this equation were in the range of values already reported above.

Takahashi, Arai, and Saito (1980), using Takaisi's work cited above, carried out power consumption measurements for anchor and helical ribbon agitators under laminar flow conditions in Newtonian liquids in flat bottom tanks. Takahaski et al. (1980) reduced Takaisi's drag approximation and obtained an expression for fluid drag on a vertical plate per unit height perpendicular to the wall as:

$$F = \left(\frac{8\pi\mu U}{g_c}\right) \left\{ \frac{1}{[2 \ln (4 + 8c^*/w^*) - 1]} \right\} \qquad (3.68)$$

where c^* and w^* are specially defined clearance and blade width, respectively. This equation can be used to define the drag coefficient C_D, which is defined in terms of the drag force per unit height, density, relative velocity, and blade width as:

$$F = (\tfrac{1}{2}) C_D \rho U^2 W \qquad (3.69)$$

Torque per unit height per blade was given as:

$$T_0 = F\left(\frac{D}{2}\right) \qquad (3.70)$$

or:

$$= \left(\frac{4\pi\mu ULD}{g_c}\right) \left\{ \frac{1}{[2 \ln (4 + 8c^*/w^*) - 1]} \right\} \qquad (3.71)$$

Power per blade was given as:

$$P = \Omega T_0 \tag{3.72}$$

or:

$$P = \left(\frac{8\pi^2 N\mu ULD}{g_c}\right)\left\{\frac{1}{[2 \ln (4 + 8c^*/w^*) - 1]}\right\} \tag{3.73}$$

For total torque and power draw for impellers with multiple blades, multiplication by blade number is required. This assumes no blade-blade interactions which could reduce the power consumption. Power for two blade lengths was then:

$$P = \left(\frac{16\pi^2\mu U^2 L}{g_c}\right)\left\{\frac{1}{[2 \ln (4 + 8c^*/w^*) - 1]}\right\} \tag{3.74}$$

Rearranging in terms of power number, the expression became:

$$N_P N_{Re} = \left\{\frac{16\pi^3}{[2 \ln (4 + 8c^*/w^*) - 1]}\right\}\left(\frac{L}{D}\right) \tag{3.75}$$

which was used to correlate power consumption for anchor impellers.

The correlation agreed with experimental data at large clearances. At low clearances (i.e., low values of C/D) the correlation under-predicted power and an empirical factor was added to correct for this effect. Their final correlation for anchor impellers was:

$$N_P N_{Re} = \left\{\frac{16\pi^3}{[2 \ln (4 + 8c^*/w^*) - 1]}\right\}\left(\frac{L}{D}\right)\left[1 + 0.00735\left(\frac{D}{C}\right)^{0.832}\right] \tag{3.76}$$

For helical ribbon impellers, Takahaski et al. obtained:

$$N_P N_{Re} = \left\{\frac{16\pi^3}{[2 \ln (4 + 8c^*/w^*) - 1]}\right\}\left(\frac{L}{D}\right)\left[1 + 0.00735\left(\frac{D}{C}\right)^{0.832}\right](\sin \theta_B)^{0.50} \tag{3.77}$$

where:

$$\sin \theta_B = p/[(\pi D)^2 + p^2]^{0.5} \tag{3.78}$$

An enclosed liquid surface at the top of the tank had little effect on power consumption.

An additional term was added to the correlation above to take into account the effect of the supporting arms for the helical ribbon (Takahashi, Arai, and Saito, 1982). Agreement with other studies was also noted as shown in Fig. 3.11.

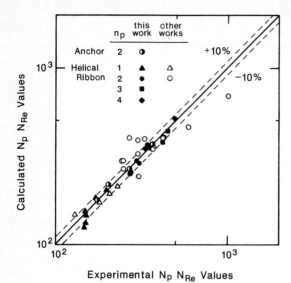

Figure 3.11 Power correlation for both anchor and helical ribbon impellers. (*From K. Takahashi, K. Arai, and S. Saito, J. Chem. Eng. Japan, 15, 77, 1982. By permission.*)

Computational analysis

A number of investigators have used the equations of motion to calculate the flow field in agitated tanks and have been able to calculate the two-dimensional viscous dissipation function to estimate power consumption. The calculations have been only two dimensional in the r and θ directions to date; however, three-dimensional calculations are likely in the near future. Sweeney and Patrick (1977) first initiated this type of work in mixing by their simulation of flow patterns around a perpendicular stirrer for power law fluids. For an agitated tank, specifically, work by Hiraoka, Yamada, and Mizoguchi (1978; 1979) is of particular interest. In this work the velocity components V_r and V_θ were given in the typical fashion as partials of a stream function Ψ:

$$V_0 = \frac{\partial \Psi}{\partial r} \tag{3.79}$$

and:

$$V_r = -\left(\frac{1}{r}\right)\frac{\partial \Psi}{\partial \theta} \tag{3.80}$$

The shear stresses of interest were given in usual manner as:

$$\tau_{rr} = -2\mu\frac{\partial V_r}{\partial r} \tag{3.81}$$

$$\tau_{\theta\theta} = -\mu\left\{2\left[\left(\frac{1}{r}\right)\frac{\partial V_\theta}{\partial r} + \frac{V_r}{r}\right]\right\} \tag{3.82}$$

$$\tau_{r\theta} = \tau_{\theta r} = -\mu\left[r\frac{\partial(V_\theta/r)}{\partial r} + \left(\frac{1}{r}\right)\left(\frac{\partial V_\theta}{\partial\theta}\right)\right] \tag{3.83}$$

where viscosity μ can be the apparent viscosity. In terms of the stream function, these became:

$$\tau_{rr} = +2\mu\left[\left(\frac{1}{r}\right)\frac{\partial^2\Psi}{\partial r\,\partial\theta} - \left(\frac{1}{r^2}\right)\frac{\partial\Psi}{\partial\theta}\right] \tag{3.84}$$

$$\tau_{\theta\theta} = -\tau_{rr} \tag{3.85}$$

$$\tau_{r\theta} = \tau_{\theta r} = -\mu\left[\frac{\partial^2\Psi}{\partial r^2} - \left(\frac{1}{r}\right)\frac{\partial\Psi}{\partial r} - \left(\frac{1}{r^2}\right)\frac{\partial^2\Psi}{\partial\theta^2}\right] \tag{3.86}$$

The equations of motion written in terms of the stream function do not have pressure as a variable. The stream function equation (Hiraoka et al., 1978) was written as:

$$\frac{(N_{Re}/r)\partial(\Psi, \nabla^2\Psi)}{\partial(r, \theta)} = -\left\{\left[\left(\frac{2}{r^2}\right)\frac{\partial\tau_{\theta\theta}}{\partial\theta} + \left(\frac{2}{r}\right)\frac{\partial^2\tau_{\theta\theta}}{\partial r\,\partial\theta}\right]\right.$$

$$\left. + \left[\frac{\partial^2\tau_{r\theta}}{\partial r^2} + \left(\frac{3}{r}\right)\frac{\partial\tau_{r\theta}}{\partial r} - \left(\frac{1}{r^2}\right)\frac{\partial^2\tau_{r\theta}}{\partial\theta^2}\right]\right\} \tag{3.87}$$

The vorticity equation was then:

$$\nabla^2\Psi = -\omega \tag{3.88}$$

where vorticity was:

$$\omega = \left(\frac{1}{r}\right)\frac{\partial V_r}{\partial\theta} - \frac{\partial(rV_{\theta_i})}{\partial r} \tag{3.89}$$

With various substitutions, the following equation was obtained:

$$\frac{(N_{Re}/r)\partial(\Psi, \omega)}{\partial(r, \theta)} = \mu\nabla^2\omega + F(\Psi, \mu) \tag{3.90}$$

$F(\Psi, \mu)$ was an additional term which arose due to the dependence of viscosity on shear rate and is zero for Newtonian fluids. Hiraoka et al. (1979) provided an equation for $F(\Psi, \mu)$. Ultimately, the equations can be solved two-dimensionally for the V_θ and V_r flow field in the tank.

Literature based upon computational analysis. For Newtonian fluids, numerical values of power input were obtained by Hiraoka et al. (1978) as a product of a friction factor and a modified Reynolds number:

$$f^*N_{Re''} = 2\epsilon\left[\left(\frac{T}{D}\right)^2 - 1\right](\omega_\omega + 2) \tag{3.91}$$

where $f = \tau_w/(\rho V_\theta^2/2)$ and $\epsilon = 1 + \exp\{-10[(T/D) - 1]\}$. ω_ω is the average vorticity at the vessel wall. The modified Reynolds number was defined as $N_{Re''} = rV_\theta(T/2)\epsilon \ln (T/D)/\mu$. The power number for a two-bladed paddle was given as:

$$N_P N_{Re} = \frac{\pi^3 L/D}{\{\epsilon[1 - (T/D)]^2\}} f^*N_{Re''} \tag{3.92}$$

The function $f^*N_{Re''}$ was found: (1) to be relatively independent of the D/T, (2) to have a value of 2 below $N_{Re''}$ of 10, and (3) to vary with the one-third power of paddle number. The function is shown in Fig. 3.12.

There are a number of advantages to the numerical calculations as outlined by Hiraoka et al., since the velocity distributions, streamlines, dissipation function, vorticity, and pressure distributions can be obtained. Stream and dissipation functions are shown in Fig. 3.13. Hiraoka et al. found that the energy dissipation near the impeller blade was very high.

Hiraoka, Yanada, and Mizoguchi (1979), using non-Newtonian fluids, developed a similar relationship for power and an expression for the apparent viscosity based upon vorticity. By doing so, they provided the numerical basis for the Metzner-Otto (1957) postulation. Their work will be discussed more fully in the laminar mixing chapter.

Bertrand and Couderc (1981), in a similar analysis as Hiraoka et al. (1978; 1979), cited the power dissipation function as a method to calculate power for Newtonian fluids. They provided detailed informa-

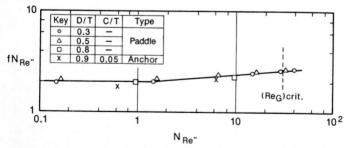

Figure 3.12 Relationship between power input and modified Reynolds number in laminar flow region. (*From S. Hiraoka, I. Yamada, and K. Mizoguchi, J. Chem. Eng. Japan, 11, 487, 1978. By permission.*)

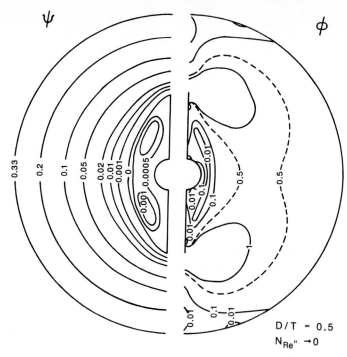

Figure 3.13 Profiles of the stream function and dissipation function for a paddle impeller of $D/T = 0.5$. (*From S. Hiraoka, I. Yamada, and K. Mizoguchi, J. Chem. Eng. Japan, 11, 487, 1978. By permission.*)

tion as to the distributions of velocity, shear stress, and streamlines as well as the distribution of the dissipation function. Their power draw correlation obtained from the calculations was:

$$N_P N_{\text{Re}}^{0.96} = 201 \qquad (3.93)$$

The main area of viscous dissipation was found to be near the impeller tip.

Walton et al. (1981) carried out similar calculations using the stream function-vorticity equations using the power law and the Eyring models for viscosity. The stream function contours for a four-bladed paddle were given. Couette type flow occurred near the wall with a recirculating cavity flow between the blades. Vorticity and dissipation distributions were also given. The numerical results agreed well with flow visualization experiments. Power requirements were obtained from: (1) integration of the dissipation function over the volume and (2) the integration of wall shear stress around the outer boundary. The integration of the dissipation function gave the best results. The power law was found to give unrealistically high viscosities for shear thinning fluids, and an Eyring model was used instead. The numerically calculated power numbers were obtained, and power consumption was also studied as a function of n and D/T.

Kuriyama et al. (1982) provided numerical solutions for power consumption of an anchor impeller in a viscous fluid in an agitated tank. Their results showed excellent agreement with Takahaski et al. (1980).

Bertrand and Couderc (1985) studied gate agitators for viscous Newtonian fluids and fluids whose viscosity was calculated by the Carreau model. The stream and vorticity functions, the radial and tangential velocity profiles, shear and normal stress distributions, and the dissipation function were presented. Their power correlation for Newtonian and shear thinning fluids was:

$$N_P N_{Re} = 169[1 + 3.05(\lambda)^2]^{(n-1)/2} \qquad (3.94)$$

The Reynolds number in this equation was based upon the viscosity of the fluid at rest, and λ was a characteristic time in the Carreau model.

Correlation recommendations

There are certain attributes which a correlation should have. First, a correlation should provide a reasonable prediction of actual power draw. Further, a particular correlation should have at least a basis in theory. All six general categories do; however, particular correlations may have limited theoretical support and poor development. Correlations should make sense on physical grounds. The article by Penney and Bell (1967) should be consulted concerning this matter. When using a particular correlation, the ranges of variables over which the correlation was developed should be approximately those involved in the power calculation. A correlation should not be expected to perform outside the ranges of variables from which it was developed. Correlations should include all relevant geometric variables. The correlation for a particular use is very much dependent upon the geometry at hand. The use of as many correlations as possible is recommended to identify agreement.

Table 3.2 lists suggested sources for power correlations for power calculations. Table 3.3 lists literature sources where correlations have been compared with other literature data as well as studies of efficiencies of agitator systems. Computational correlations have not been recommended since such work is two dimensional. It should also be recognized that quite a few studies, reviewed above, were pioneering studies, some were efficiency studies, and many studies were incomplete.

TABLE 3.2   Suggested Correlation Sources

Paddle Impellers
Nagata et al. (1955)
Mitsuishi and Hirai (1969)

Helical Ribbon Impellers
Hall and Godfrey (1970)
Chavan and Ulbrecht (1972)
Chavan and Ulbrecht (1973a)
Bourne et al. (1979)
Kappel (1979)
Patterson et al. (1979)
Yap et al. (1979)
Takahaski et al. (1980; 1982)
Blasinski and Rzyski (1980)
Takahashi et al. (1984)
Shamlow and Edwards (1985)

Helical Screw Impellers
Mitsuishi and Hirai (1969)
Chavan, Jhaveri, and Ulbrecht (1972)
Chavan and Ulbrecht (1973a; 1973b; 1974) while noting Chavan et al. (1975b) and Chavan (1983)

Anchors
Bechner and Smith (1966)
Takahaski et al. (1980; 1982)

Ribbon Screw Impeller
Chowdhury and Tiwari (1979)
Chavan and Ulbrecht (1973a)
Merquiol and Choplin (1988)

Viscoelastic Fluids

Good introductory articles concerning power draw of agitators in viscoelastic fluids are provided by Hocker et al. (1981) and Kelkar et al. (1972). Viscoelastic flow patterns are quite different from those for Newtonian fluids and shear thinning fluids. As a result, differences in power correlations can be expected. General references and references

TABLE 3.3

I. Comparisons of Correlations

General Comparison

Penney and Bell (1967)

Helical Ribbon Impellers

Chavan and Ulbrecht (1972)

Chavan and Ulbrecht (1973a)

Sawinsky, Deak, and Havas (1975)

Kappel (1979)

Patterson et al. (1979)

Yap et al. (1979)

Takahaski et al. (1980; 1982)

Blasinski and Rzyski (1980)

Takahaski et al. (1984)

Shamlow and Edwards (1985)

Helical Screw Impellers

Chavan, Jhaveri, and Ulbrecht (1972)

Chavan and Ulbrecht (1973a; 1973b)

Sawinsky, Deak, and Havas (1978)

Anchors

Takahaski et al. (1980; 1982)

Ribbon Screw Impeller

Chowdhury and Tiwari (1979)

Chavan and Ulbrecht (1973a)

II. Comparison of Equipment and Efficiency

General Comparisons of Different Impeller Types

Hoogendoorn and den Hartog (1967)

Rieger and Novak (1973)

Novak and Rieger (1975)

Helical Screw Geometry Comparisons

Novak and Rieger (1969)

Seichter (1971)

on flow patterns include: Ide et al. (1974), White et al. (1977), and Ulbrecht and Carreau (1985).

There needs to be further work on impeller power draw for viscoelastic fluids, although there is evidence which indicates that so long as the normal stresses are equal to or less than the shear stresses, elasticity of the fluid has no effect on power draw. Power correlations for shear thinning fluids can be used. However, there is still debate concerning this matter.

The following reviews the literature available to calculate power consumption for viscoelastic fluids.

Fluid rheology

The structural properties and the identification of elastic properties for viscoelastic fluids, particularly polymeric solutions, such as PAA (polyacrylamide) and CMC (carboxymethycellulose), are important in understanding the fluid mechanics of these fluids. In such fluids, the molecular structure of the long chain molecules have to be taken into account in the continuum mechanics.

PAA has been studied as a viscoelastic fluid and CMC as a shear thinning fluid. However, both exhibit a first normal stress difference which classifies these fluids as viscoelastic. For CMC solutions, the viscoelasticity degrades with time over a short time period (12 h), after which CMC solutions may not exhibit elastic properties (Nishikawa et al., 1981).

The classification of polymeric solutions and viscoelastic fluids can be done according to the kinetic conditions of the molecules. For dilute solutions, the molecules are independent of each other. Viscosity is then proportional to the concentration of the molecules, and viscoelastic effects are not observed. For higher concentration, strong interactions such as cross-linking can occur between molecules. When flow occurs, the strong interactions are disrupted, the molecules tend to align themselves with the flow, and energy is stored in the molecules as they align. When the motion stops, the molecules tend to relax back to their original orientation, and some of the stored energy is returned to the fluid motion.

In such flows, there are several different times which require discussion. There is a characteristic time for the molecules to align. There is also the time span of the flow and the time in which the fluid remains aligned as well as a time for the molecules to return back to their original position, known as the relaxation time of the molecules. These characteristic times may be associated with normal stresses and shear stresses differently. Normal stresses arise due to the elasticity of the molecules, and the elastic behavior is characterized by normal stress properties.

The flow index n of the power law can be considered a measure of the elastic properties for aqueous polymeric solutions. The lower the value of n, the more likely the fluid will be elastic also. The first normal stress difference and shear stresses are functions of shear rate, and either one can dominate. The flow can shear and elongate fluids. Shear flows have been well studied, but extensional or elongation flows have not. Elongational or extensional viscosity arise in such work although very little is known about such viscosities and how to model them in mixers.

There is a certain convention used when discussing normal stress differences which should be noted (Bird, Armstrong, and Hassager, 1977). When a fluid moves in one direction, this is referred to as the "1 direction." Under such motion there is typically a velocity variation in another direction, which is referred to as the "2 direction." The remaining direction is referred to as the "3 direction." Under this convention, then, the primary normal stress difference is $\tau_{11} - \tau_{22}$, and the secondary normal stress difference is $\tau_{22} - \tau_{33}$.

Ulbrecht (1974) provided an excellent review paper upon the non-Newtonian aspect of fluids, noting particularly non-Newtonian behavior for polymer solutions and melts. Flow anomalies exist for these fluids and include recoil and stress relaxation in transient experiments, normal stress differences under steady-state conditions, and shear dependent viscosity. The first normal stress difference can be measured using a cone and plate viscometer by measuring the force which tends to separate the rotating cone from a stationary plate. Reversing the direction of the cone rotation does not change the sign of the force. Shearing and stretching of liquids produce normal forces which always act perpendicular to the shearing surfaces.

The first normal stress difference is a function of shear rate and is written as:

$$\tau_{11} - \tau_{22} = F_1(\gamma) \tag{3.95}$$

and where typically:

$$F_1(\gamma) = \sigma_1(\gamma)\gamma^2 \tag{3.96}$$

The coefficient $\sigma_1(\gamma)$ is called the primary normal stress coefficient and is the counterpart to viscosity for normal stresses. The first normal stress coefficient $\sigma_1(\gamma)$ is typically a constant at low shear rates but decreases at higher shear rates as shown by typical data in Fig. 3.14. Shear stress data τ_{12} are also shown in Fig. 3.14. For polymer solutions, the first normal stress difference is of the same order of magni-

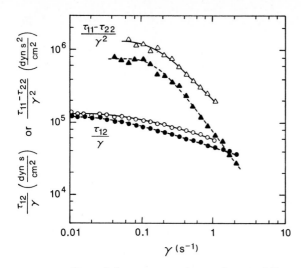

Figure 3.14 Data of shear stress and normal stress difference for high- and low-density polyethylene; top curve, low-density polyethylene. (*From J. J. Ulbrecht, Chem. Engr., 286, 1974. By permission.*)

tude as the shear stress but is much greater than the shear stress τ_{12} for polymer melts. Polymer solutions are of interest to mixing in agitated tanks.

There are other properties which can be discussed, for example, the second normal stress difference and the second normal stress coefficient. These, however, are not particularly important to mixing. There appears to be very little known about the second normal stress difference which is considered to be an order of magnitude smaller than the first normal stress difference. Thus, to characterize a viscoelastic fluid, the viscosity and the first normal stress coefficient are required. It should also be noted that both of these are no longer solely considered properties of the fluid and are also functions of the shear rates in the flow.

There are numerous ways to model rheological data. The viscosity for shear thinning fluids was modeled above using the power law, Eq. (3.1). The viscosity can also be modeled using an Ellis model as:

$$\mu = \mu_0 \left[1 + \left(\frac{\tau_{12}}{\tau_0} \right)^{i-1} \right]^{-1} \tag{3.97}$$

where μ_0 is the zero shear viscosity and τ_0 is the shear stress at $\mu = \mu_0/2$. The first normal stress coefficient can be modeled in a similar fashion through the use of a power law or an Ellis type model; the power law being:

$$\sigma_1(\gamma) = h(\gamma)^{m-2} \qquad (3.98)$$

which fits reasonably well except at low shear rates where σ is a constant. The parameter h is equivalent to K used in the power law for viscosity. The Ellis model is:

$$\sigma = \sigma_0 \left[1 + \left(\frac{F_1}{F_0} \right)^{q-1} \right]^{-1} \qquad (3.99)$$

where F_1 is the primary normal stress difference, σ_0 is the zero shear normal stress coefficient, and F_0 is the zero shear primary normal stress difference. The exponents i, m, and q in Eqs. (3.97), (3.98), and (3.99) are obtained from the data as a result of the curve fits. Data of this type are equilibrium data with time-dependent behavior excluded. Other models are also available in Brodkey (1967) and Bird et al. (1977). Time-dependent behavior, such as degradation, is quite common for viscoelastic fluids but has not been studied to any extent in studies on power consumption.

The various methods used to establish power correlations for Newtonian and shear thinning fluids have been used with success for viscoelastic fluids as well. These are reviewed below.

Geometry

The geometry typically reported in the study of viscoelastic fluids uses turbulent agitators as reported in Chaps. 1 and 2, and ribbon and screw impellers only to some degree. Disks and spheres have been used as impellers to gain insight into the fluid mechanics of viscoelastic fluids. It is also clear that the most optimum impeller geometry to mix viscoelastic fluids in a tank configuration has not yet been found.

Dimensional analysis

Pawlowski (1969), in a dimensional analysis study of Newtonian and non-Newtonian fluids, pointed to the need of including a time constant θ in addition to a characteristic viscosity in power correlations of non-Newtonian fluids. He suggested that correlations would take the following form:

$$f\left(\frac{P}{\rho N^3 D^5}, \frac{\rho N D^2}{H^*}, N\theta \right) = 0 \qquad (3.100)$$

where H^* is a rheological parameter. Additional comments were also made concerning temperature dependence of rheological properties.

Rieger and Novak (1974) in a short paper on viscoelastic liquids correlated power number with impeller Reynolds number as:

$$N_P = \frac{C(n)}{N_{\mathrm{Re}}} \tag{3.101}$$

Plots of $C(n)$ versus $(1 - n)$ were given for helical ribbon and anchor agitators as:

$$C(n) = Bk^{n-1} \tag{3.102}$$

The parameters k and B were 36.7 and 300, respectively, for the helical ribbon impeller. For the anchor impeller, k equaled 16.2 and B equaled 180. Power measurements were carried out with polyisobutylene in decalin. The first normal stress differences were of the same order as shear stress levels. However, the results showed no effect of liquid elasticity on agitator power draw.

Collias and Prud'homme (1985) employed an elastic fluid having a constant viscosity to the study of the effects of elasticity on both power consumption and mixing for a disk style turbine for a Reynolds number range up to 30. To produce elastic fluids with constant viscosity, a few parts per million of PAA were added to a Newtonian fluid which degraded quite rapidly under agitation conditions. After degradation, the normal stresses were constant with time. To correlate their results, they introduced an elasticity number as:

$$El = \frac{\sigma}{\rho D^2} \tag{3.103}$$

which represents a balance between inertial and elastic forces. A torque correlation was given as:

$$N_T = 13.12 N_{\mathrm{Re}} + 0.001167 N_{\mathrm{Re}}^3 + N_{\mathrm{Re}}^3 (71 El - 3200 El^2) \tag{3.104}$$

where N_T is defined as a torque number, $T_0 \rho / \mu^2 D$. The last term in this equation represented the additional torque required to agitate an elastic fluid above that necessary for a Newtonian fluid.

Collias and Prud'homme noted that the polymeric fluids increased the torque and power consumption for mixing because of the occurrence of sustained secondary flows and the increase in viscosity due to elongational flow (i.e., the elongational viscosity was greater than three times the shear viscosity). From the work, equal impeller Reynolds number and elasticity number upon scaleup cannot be obtained.

Merquiol and Choplin (1988) developed the following correlation for power number for helical ribbon screw impellers in viscoelastic fluids:

$$N_p = A\left(\frac{D}{p}\right)^{0.7}\left(\frac{W}{D}\right)^{0.03}\left(\frac{N_{\text{Re}'}}{k^{n-1}}\right)^{b}[1 + (CWi)^{x}]^{y} \qquad (3.105)$$

where the geometric variables are for the ribbon. The constants A and C and the exponents b, x, and y were flow regime dependent, i.e., laminar or transition regime. The constant C was dependent upon the impeller geometry as well.

Phenomenological coaxial rotating cylinder analogy

For the CMC and PAA solutions studied by Hocker et al. (1981), the first normal stress difference became important at high shear rates and, hence, the solutions became more elastic at higher impeller rotational speeds. For the CMC solutions, power draw of the turbulent type agitators was correlated by the Metzner-Otto (1957) method. The proportionality constant between shear rate and impeller rotational speed was 11 which implied that, for the CMC solutions studied, the viscoelasticity of the fluid could be ignored. The CMC solutions behaved as a shear thinning fluid at these concentrations (0.1 to 0.2 wt%) although the first normal stress difference was the same order of magnitude as the shear stress for the concentration studied.

For the PAA solutions studied, the first normal stress difference exceeded the shear stress by an order of magnitude. No well-defined laminar flow was achieved, and polymer agglomerates were noted to exist in some solutions. However, the Metzner-Otto method was still applied to determine the power number-Reynolds number curves shown in Figs. 3.15 and 3.16. At low impeller Reynolds number (i.e., low shear rates and, hence, low first normal stress differences), some of the power curves were the same as those for Newtonian and CMC solutions. However, most of the power data were above the Newtonian curves, particularly at higher impeller Reynolds numbers. The data curves exhibited substantial changes in slope over small changes in impeller Reynolds number. In the turbulent regime, the PAA solutions showed an increased drag and power draw over the Newtonian fluids. Hocker et al. stressed the need to establish the main flow directions concurrently with power measurements during studies of viscoelastic fluids.

Simple analogies for viscoelastic flows

Solution of the equations of motion under conditions of agitation is extremely difficult. This is due to the complicated geometry of the impeller and the three dimensionality of the flow. However, these problems can be minimized if the impeller is replaced by a simpler

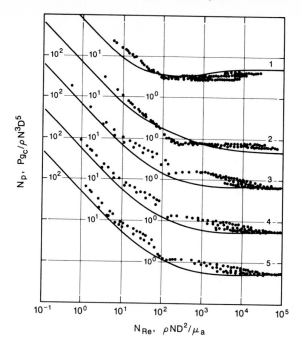

Figure 3.15 Power characteristics in PAA solutions. (1) Flat blade disk turbine; (2) flat disk; (3) MIG-4; (4) MIG-6; (5) MIG-7. (*From H. Hocker, G. Langer, and U. Werner, German Chem. Eng., 4, 113, 1981. By permission.*)

geometry. This was done earlier in the extended coaxial cylinder analogy where the impeller was replaced by a cylinder. For viscoelastic fluids, because of their complex flow fields, simple shapes have also been assumed for the impellers. With such geometries, solutions of the equations of motions are easier to obtain and the kinematics and the dynamics of the flow can be described. Such work provides results and guidelines for understanding flows around the complex geometries of impellers. The coaxial rotating cylinder analogy is an example of a flow analogy being successfully applied.

Kelkar, Mashelkar, and Ulbrecht (1972) suggested simulating impeller flow fields using simpler geometries and correlated power for viscoelastic flows based upon these. Selection of rheological models and the need for two time parameters, one to describe the natural time of a viscoelastic fluid t_n, another to describe the characteristic time of a viscoelastic fluid t_c, were discussed. They noted work by Giesekus (1963) and Schummer (1969) for torque on a rotating sphere generated by a third-order fluid. In dimensionless form, an equation for torque was given as:

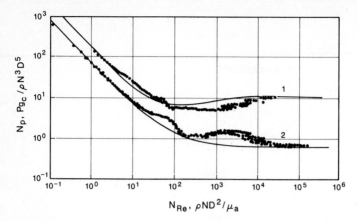

Figure 3.16 Power characteristics in PAA solutions. (1) Flat blade disk turbine; (2) MIG-6. (*From H. Hocker, G. Langer, and U. Werner, German Chem. Eng., 4, 113, 1981. By permission.*)

$$Y = C_M N_{\mathrm{Re}^0} = 8\pi \left[1 + \frac{(N_{\mathrm{Re}^0})^2}{600} - 0.6(\mathrm{Vi})^2 - \frac{(\mathrm{Wi})^2}{30} - \frac{9.0 \mathrm{Wi} N_{\mathrm{Re}^0}}{280} \right] \quad (3.106)$$

$$\quad\; \mathrm{I} \qquad\quad \mathrm{II} \qquad\quad \mathrm{III} \qquad\quad \mathrm{IV} \qquad\quad \mathrm{V}$$

The dimensionless groups were specifically defined as follows. The torque number is C_M or:

$$C_M = \frac{T_0}{\rho R^5 \omega^2} \quad (3.107)$$

The Reynolds number N_{Re^0} is based upon zero shear viscosity μ_0 or:

$$N_{\mathrm{Re}^0} = \frac{\rho \omega R^2}{\mu_0} \quad (3.108)$$

The Weissenberg number is Wi or:

$$\mathrm{Wi} = \omega t_n \quad (3.109)$$

and the viscosity number:

$$\mathrm{Vi} = \omega t_c \quad (3.110)$$

The various terms in the equation are contributed to: I—inertialess flow of a Newtonian liquid; II—inertial contribution; III—additional viscous effects; IV—additional elastic effects; and V—the interaction between inertial, viscous, and elastic effects.

For Reynolds numbers below 10, terms II and V were considered negligible. The resulting equation became:

$$Y = C_M N_{Re^0} = 8\pi\left(1 - 0.6Vi^2 - \frac{Wi^2}{30}\right) \qquad (3.111)$$

By analogy then, for any shaped agitator, the same general form of the equation can be written as:

$$Y = Y(Vi, Wi) \qquad (3.112)$$

where more general definitions of Vi and Wi can be used in which:

$$t_n = \frac{\sigma_0}{\mu_0} \qquad (3.113)$$

and

$$t_c = \frac{\mu_0}{\tau_0} \qquad (3.114)$$

where μ_0 is the zero shear viscosity, τ_0 is the shear stress at $\mu = \mu_0/2$, and σ_0 is the zero shear normal stress coefficient. Since dimensionless parameters may also appear such as i and q from models of rheological data, e.g., Eqs. (3.97) through (3.99), these may also be added to the correlation.

Simplifications can arise due to differences in flow behavior. For liquids having a shear dependent viscosity and negligible elasticity:

$$Y = f(Vi, i) \qquad (3.115)$$

where i is the exponent of the shear dependent viscosity using the Ellis model. For fluids showing elasticity only and no shear dependent viscosity:

$$Y = f(Wi, q) \qquad (3.116)$$

where q is the exponent to fit normal stress difference using the Ellis model.

Kelkar et al. (1972) tested the model for spheres, disks, and agitators using three types of fluids:

Case 1: Silicone oils: no shear dependent viscosity and normal stresses present; elastic phenomena only.

Case 2: Aqueous CMC solutions: shear dependent viscosity and no normal stresses present.

Case 3: Aqueous PAA solutions: shear dependent viscosity and normal stresses present.

The rheological data of shear and normal stress difference were corre-

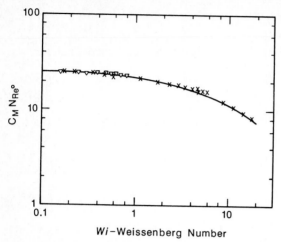

Figure 3.17   Effect of elasticity on the power consumption for spheres (R = 1.27 cm) rotating in silicone oils. (*From J. V. Kelkar, R. A. Mashelkar, and J. J. Ulbrecht, Trans. Instn. Chem. Engrs., 50, 343, 1972. By permission.*)

lated using the Ellis models and the magnitudes of the natural and characteristic times of the fluids were given. For the silicone fluids, Case 1, Y was fairly constant but did vary with the Weissenberg number as shown in Fig. 3.17 for spheres and Fig. 3.18 for a standard turbine. Other dimensionless groups, such as q, had an insignificant effect although q was not varied over a significant range.

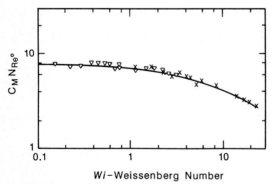

Figure 3.18   Effect of elasticity on the power consumption for standard turbines (R = 1.0 cm) rotating in silicone oils. (*From J. V. Kelkar, R. A. Mashelkar, and J. J. Ulbrecht, Trans. Instn. Chem. Engrs., 50, 343, 1972. By permission.*)

TABLE 3.4   Values of Constants for Eq. (3.117)

Geometry	A	B
Sphere	25.1	1.0
Disk	8.9	0.8
One-sided turbine	11.4	0.8
Standard turbine	6.9	0.8

For the CMC solutions, Case 2, power data were obtained for three spheres, three disks, and three agitators for five aqueous CMC solutions. The data were correlated using the equation:

$$Y = A\left(\frac{4}{3+i}\right)\text{Vi}^{B(1/i-1)} \tag{3.117}$$

where A and B are functions of geometry as listed in Table 3.4. The data are shown in Figs. 3.19, 3.20, and 3.21.

The figures also show the results obtained for the PAA solutions, Case 3. The PAA solutions exhibited shear dependent viscosities and large normal stresses. From the comparison of the data and correlations between the two types of fluids, it appears that elastic effects may be neglected in the calculation of power consumption for viscoelastic fluids. This is in contrast to the results of Hocker et al. (1981) and Collias and Prud'homme (1985). In the study by Hocker et al. (1981) with PAA solutions, polymer agglomerates were present, and no steady, well-defined laminar flow was established. Merquiol

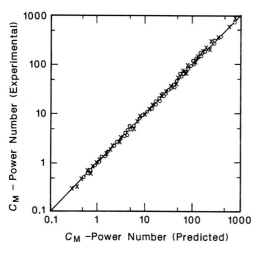

Figure 3.19 Comparison of the experimental values of C_M with those predicted for the case of spheres. (*From J. V. Kelkar, R. A. Mashelkar, and J. J. Ulbrecht, Trans. Instn. Chem. Engrs., 50, 343, 1972. By permission.*)

o: CMC solutions (inelastic liquids)

x: PAA solutions (viscoelastic liquids)

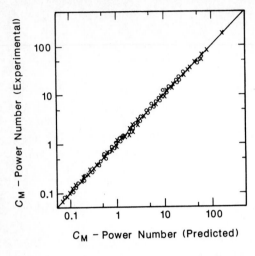

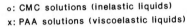

o: CMC solutions (inelastic liquids)

x: PAA solutions (viscoelastic liquids)

Figure 3.20 Comparison of the experimental values of C_M with those predicted for the case of disks. (*From J. V. Kelkar, R. A. Mashelkar, and J. J. Ulbrecht, Trans. Instn. Chem. Engrs., 50, 343, 1972. By permission.*)

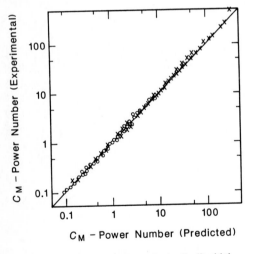

o: CMC solutions (inelastic liquids)

x: PAA solutions (viscoelastic liquids)

Figure 3.21 Comparison of the experimental values of C_M with those predicted for the case of agitators. (*From J. V. Kelkar, R. A. Mashelkar, and J. J. Ulbrecht, Trans. Instn. Chem. Engrs., 50, 343, 1972. By permission.*)

and Choplin (1988) found that elasticity increased the torque for helical ribbon screw impellers.

Elasticity plays a much more important role on the kinematics (velocity fields) around an agitator than on the dynamics. It should be noted, however, that most studies have been at large T/D ratios. Studies of viscoelastic fluids have not used ribbon or screw impellers extensively. Close clearance impellers may cause rather strange flow behavior and different power consumption levels due to the elastic behavior of the fluid. More research is needed in these areas.

Bingham Plastic Fluids

There have been very few studies involved in correlating power draw for Bingham plastic fluids. Nagata et al. (1970; 1971) modeled the power necessary to move a Bingham plastic fluid as the sum of the power to break the internal structure of the fluid and the power necessary to overcome the viscous resistance. The relationships can be developed as follows. The constitutive equation of state for a Bingham plastic fluid can be written as:

$$\tau = \tau_y + \eta\gamma \qquad (3.118)$$

where τ_y is the yield point shear stress and η is the plastic viscosity obtained from:

$$\eta = \frac{\tau - \tau_y}{\gamma} \qquad (3.119)$$

Rearranging this, the following relationship can be obtained:

$$\frac{\tau}{\gamma} = \eta + \frac{\tau_y}{\gamma} \qquad (3.120)$$

However, τ/γ is typically the apparent viscosity for any fluids including a Bingham plastic. As a result, the apparent viscosity for a Bingham plastic is the sum of a plastic viscosity plus a viscosity associated with the yield stress or:

$$\mu_a = \eta + \frac{\tau_y}{\gamma} \qquad (3.121)$$

Assuming the form of Eq. (3.5), the power becomes:

$$N_P = \frac{B'}{N_{\text{Re}}} \qquad (3.5)$$

$$N_P = \frac{B'\mu_a}{\rho ND^2} \qquad (3.122)$$

Making the substitution for apparent viscosity, the equation becomes:

$$N_P = B'\left(\frac{\eta}{\rho ND^2} + \frac{\tau_y/\gamma}{\rho ND^2}\right) \qquad (3.123)$$

$$= \frac{B'\eta}{\rho ND^2} + \left(\frac{B'N}{\gamma}\right)\left(\frac{\tau_y}{\rho N^2 D^2}\right) \qquad (3.124)$$

$$= B'\,(N_{\text{Rep}})^{-1} + \alpha N_y \qquad (3.125)$$

where the N_{Rep} is the impeller Reynolds number based upon the plastic viscosity and N_y is $\tau_y/\rho N^2 D^2$. α is based upon the Metzner-Otto postulation in that the shear rate γ is the average shear rate γ_{av} which is kN. B'/k is α.

The first term on the right in Eq. (3.123) is the power necessary to overcome viscous resistance. The second term is the power needed to break down the internal structure of the fluid. An additional term I was added to the equation by Nagata et al. to account for the N_P required for the turbulent flow regime but was not included here since it is zero in creeping flow laminar mixing. In the transition regime such a term may become important.

The Metzner-Otto postulation permits the separation of the two independent quantities, τ_y and shear rate, and weights τ_y to the impeller tip speed. At low rotational speeds, power draw depends mainly upon the yield stress.

To complete the correlation, Nagata et al. (1970; 1971) used the Hedstrom number defined as:

$$\text{He} = \frac{\tau_y \rho D^2}{\eta} \qquad (3.126)$$

The power draw correlation becomes:

$$N_P = \frac{B'}{N_{\text{Rep}}} + \frac{\alpha \text{He}}{N_{\text{Rep}}^2} \qquad (3.127)$$

In the correlation of the data, α was found to be constant for a particular agitator. B' was found to be a function of the Hedstrom number and was correlated as:

$$B' = B + c\text{He}^{0.333} \qquad (3.128)$$

where B is for a Newtonian fluid as used in Eq. (3.5). The parameters B and c were found to be functions of agitator geometry.

For very low impeller rotational speeds:

$$N_P = \alpha N_y \qquad (3.129)$$

Torque was found to be:

$$T_0 g_c = \left(\frac{n_b W D^2}{4}\right) F \qquad \text{for } n_b < 3 \qquad (3.130)$$

where F is the drag force per area acting on the blade. Substituting this into the previous equation:

$$\left(\frac{n_b W D^2}{4}\right)\frac{FN}{\rho N^3 D^5} = \alpha\left(\frac{\tau_y}{\rho N^2 D^2}\right) \qquad (3.131)$$

and rearranging for α:

$$\alpha = \left(\frac{F}{\tau_y}\right)\left(\frac{W}{D}\right)\frac{n_b}{4} \qquad (3.132)$$

The data showed that F/τ_y was fairly constant and α was proportional to blade number up to a blade number of 3, as shown in Fig. 3.22. At higher rotational speeds or increased blade number above $n_b = 3$, torque reached a constant level which was attributed to the inability of the fluid to rebuild the internal structure before the next blade passage.

Helical ribbon impellers, anchors, paddle, and disk style turbines were investigated in the study. Typically, the turbine and paddles had lower power numbers, Fig. 3.23, for the same plastic Reynolds numbers than the helical ribbon and anchor impellers, Fig. 3.24. The

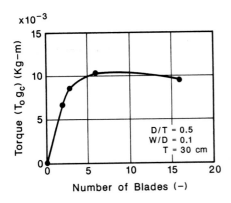

Figure 3.22 Effect of number of blades on torque for paddle impeller. (*From S. Nagata, M. Hishikawa, H. Tada, H. Hirabayashi, and S. Gotoh, J. Chem. Eng. Japan, 3, 237, 1970. By permission.*)

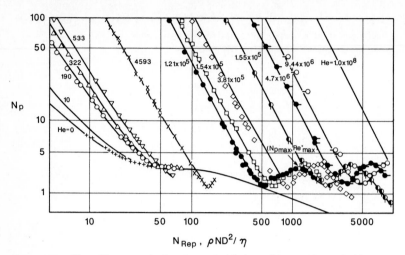

Figure 3.23 N_P – N_{Rep} correlation for a six-blade turbine. (*From S. Nagata, M. Hishikawa, H. Tada, H. Hirabayashi, and S. Gotoh, J. Chem. Eng. Japan, 3, 237, 1970. By permission.*)

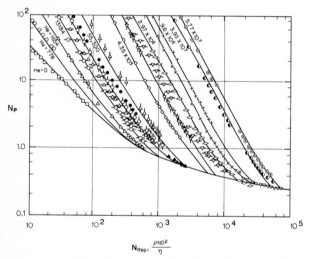

Figure 3.24 N_P – N_{Rep} correlation for a ribbon impeller. (*From S. Nagata, M. Hishikawa, H. Tada, H. Hirabayashi, and S. Gotoh, J. Chem. Eng. Japan, 3, 237, 1970. By permission.*)

laminar to turbulent transition for the turbine and paddle agitators was 200 to 5000. The transition was also rather abrupt and was dependent upon the Hedstrom number; the higher the Hedstrom number, the higher the transition Reynolds number. High Hedstrom number implied high yield stress and/or low viscosity. The power number at transition was equal to the power number in the turbulent regime for these agitators. Below the laminar-turbulent transition Reynolds number, a considerable portion of the tank was not being mixed by turbulent impellers. The plastic impeller Reynolds number for the laminar to turbulent transition for the helical ribbon and anchor agitators spanned from 100 to 10^5.

Nagata et al. (1971) showed that power correlations obtained for Bingham plastic fluids provided the necessary information to obtain the power correlations for shear thinning fluids and vice versa through an empirical factor. The fundamental reason why such a relationship exists needs further study. The effects of clearance and pitch of the ribbons on the power correlations for shear thinning fluids were also given by Nagata et al. (1971).

Thixotropic Liquids

Godfrey et al. (1974) and Edwards et al. (1976) proposed a method for the prediction of the power requirements of a thixotropic liquid. The rheological characterization of thixotropic behavior has been discussed by Chavan, Deysarker, and Ulbrecht (1975a). Such liquids show a reduction in viscosity as a function of time which is typically viewed as a breakdown of an internal fluid structure. Thixotropic fluids are also quite common. During the mixing process, the thixotropic fluid will experience high shear rates near the impeller and low shear rates in the tank bulk. As a result, the rheological state of the fluid is dependent upon the shear history which is difficult to model exactly.

For power consumption considerations, the Metzner-Otto postulation and the viscometer analogy of coaxial cylinders were extended by Godfrey et al. and Edwards et al. to thixotropic fluids in an agitated tank. The thixotropic nature of the fluids in an agitated tank was assumed equivalent to the thixotropic nature of the fluids in a coaxial cylinder viscometer. Following the time-independent studies, the power number and average shear rate were written as:

$$N_P = \frac{B}{N_{\mathrm{Re}}} \tag{3.6}$$

and
$$\gamma_{\mathrm{av}} = kN \tag{3.34}$$

Since the Metzner-Otto procedure performed well for variable n and K power law shear thinning fluids (Godleski and Smith, 1962), the procedure should also perform well for time-dependent thixotropic fluids. However, the procedure required: (1) the constant B for the various impellers, (2) the proportionality constant k for the impellers, and (3) the viscometric data for the thixotropic fluid.

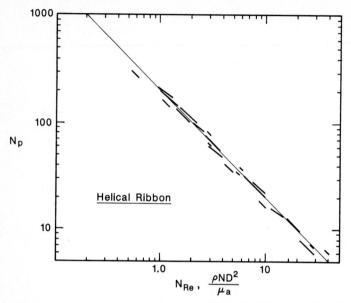

Figure 3.25 Power number data for a helical ribbon impeller for thixotropic liquids. (*From M. F. Edwards, J. Ç. Godfrey, and M. M. Kashani, J. of Non-Newtonian Fluid Mech., 1, 309, 1976. By permission.*)

Two effects were present in the studies: inertial effects during the startup of the agitator and viscosity breakdown due to the thixotropic phenomena. The apparent viscosity was a function of both time and the impeller rotational speed through the relationship between shear rate and impeller rotational speed or:

$$\mu_a = f(\theta, N) \tag{3.133}$$

The power number data were obtained from power measurements. The impeller Reynolds numbers were obtained using the apparent viscosity from the viscometer data collected at a shear rate equal to the average shear rate in the tank and at the same time. Figures 3.25 and 3.26 show the results for helical ribbon and screw impellers as data time lines which parallel the equilibrium power data of time-independent fluids. The length of the lines indicate the extent of the thixotropic breakdown and the lowering of the viscosity.

The work assumed that the rate of thixotropic breakdown in the viscometer was the same as that occurring in the tank which was the weakest assumption in the study. However, the study extended the average shear rate concept of Metzner and Otto to power-time variations during mixing of thixotropic fluids.

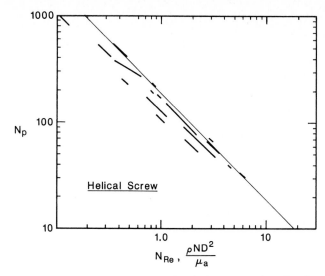

Figure 3.26 Power number data for a helical screw impeller for thixotropic liquids. (*From M. F. Edwards, J. C. Godfrey, and M. M. Kashani, J. of Non-Newtonian Fluid Mech., 1, 309, 1976. By permission.*)

Other Impeller Types

Cheng et al. (1974) discussed the power requirements of a two-rotor positive displacement mixer which trapped and squeezed material through a narrow gap. The unit was used for mixing pastes and other very high viscosity fluids. Power was correlated using a simple model and dimensional analysis similar to Penney and Bell. Essentially, pressure, generated in the trapped material, was a function of shear through the gap and varied proportionally with gap length and inversely with gap width. Torque on the rotors was obtained from the product of pressure, area of the rotor, and rotor radius.

Kappel (1979) provided power data on the MIG impeller and the sigma mixer. The MIG impeller is a large-diameter, multistage, pitched blade impeller with the blade pitch switched between the inner and outer portions of the blades. The sigma blade mixers have blades which resemble a high-pitch helical ribbon impeller with thick s- or z-shaped ribbons. Two of these are usually positioned horizontally in troughs inside a mixing container. The blades rotate in opposite directions at different speeds.

For the MIG impeller, Henzler and Hohnel (Kappel, 1979) provided $N_P N_{Re}$ of 110 and 170 for a four- and five-stage MIG impeller, respectively. Kappel (1979) provided for the effect of wall clearance for a five-stage MIG and reported:

$$N_P N_{\text{Re}} = 73.4\left(\frac{C}{D}\right)^{-0.2} \qquad (3.134)$$

The calculated results from this correlation was 131 for a C/D of 0.055 which was in agreement with Henzler and Hohnel.

For the sigma mixer, Kappel used Hall and Godfrey's data and reported a $N_P N_{\text{Re}}$ value of 415 for a one-arm and 1120 for a two-arm sigma mixer.

Murakami et al. (1980) have performed an extensive study of up-down impellers, considering a number of different blade geometries. These types of agitators swept out the entire volume of the tank in their motion. Mixing and power requirements were measured, and, generally, shorter mixing times at equal or less power requirements than those for helical ribbon impellers were found. Power numbers were on the order of 1 to 20 in the transition region for impeller Reynolds numbers of 10 to 100.

Studies are available on floating scraper blade impellers, for example: Toh and Murakami (1982a, 1982b); Leung, (1967); and Gluz and Pavlushenko (1969). Toh and Murakami (1988) presented power consumption relationships and velocity-pressure data for fluid-loaded and spring-loaded scraper blades as a function of blade geometry and scraping conditions.

Suggested Research Areas

Close clearance impellers need further study both experimentally and theoretically. The ribbon screw impeller needs further study. Combined efficiency and detailed power studies are rare. Multiphase power correlations for the various impellers cited have been relatively unexplored. For example, there is considerable work done on gas dispersion with turbulent type agitators (Nienow et al., 1983 and Ulbrecht and Ranade, 1977), but there appears to be nothing on power consumption during gas dispersion by helical ribbon or screw agitators.

Ribbon and screw impellers should be more extensively studied for the mixing of viscoelastic fluids. Shear thickening fluids warrant further investigations. Computational modeling needs to be extended to three dimensions. Substantial flow phenomena occur around impeller blades, and most of this phenomena are three dimensional in nature. Computational modeling has mainly focused upon Newtonian and shear thinning fluids. Computational modeling of the behavior of shear thickening, Bingham plastic, and viscoelastic fluids in agitated tanks should be undertaken. Drag-based and computational analyses should be emphasized in the future. Orbiting and planetary mixers have not been investigated as much as they should. The effects of different tank geometries have not been investigated.

Nomenclature

A	area of blade, cylinder or impeller, constant
a	dimensionless area of the impeller
B	constant, a function of geometry
B'	constant, a function of geometry
b	exponent
C	clearance of impeller from tank or draft tube wall, constant dependent upon geometry
C_D	drag coefficient
C_t	clearance of the draft tube from the vessel bottom
C_b	clearance of the impeller from the vessel bottom
C_M	torque number $T_0/(\rho R^5 \omega^2)$, also referred to as power number
$C(n)$	function used by Rieger and Novak (1974)
c	constant, a function of geometry
c^*	clearance as defined by Takahaski et al. (1980)
D	impeller diameter
D_e	equivalent diameter of impeller
D_t	diameter of the draft tube
E	eccentricity: distance from the center of the tank
El	elasticity number $\sigma/\rho D^2$
F	force, drag force, drag force per area
F_1	first normal stress difference
F_0	first normal stress difference at zero shear
f	friction factor
g	acceleration of gravity
g_c	gravitational constant
H	fluid height
He	Hedstrom number: $\tau_y \rho D^2/\eta$
H^*	rheological parameter used by Pawlowski
h	constant in power for primary normal stress coefficient
i	exponent in Ellis viscosity model
I	turbulent power number for Bingham plastic fluids
K	consistency index constant of the power law
k	constant of proportionality between γ_{av} and N
k_s	constant

L length or height of impeller

L_t length or height of draft tube

m exponent in power law for primary normal stress coefficient

N impeller rotational speed

N_{Re} impeller Reynolds number: $\rho ND^2/\mu$

$N_{Re'}$ impeller Reynolds number: $\rho N^{2-n}D^2/K$

$N_{Re''}$ impeller Reynolds number used by Hiraoka et al. (1979)

N_{Re^*} impeller Reynolds number defined by Beckner and Smith

N_{Re^0} impeller Reynolds number based upon zero shear viscosity

N_{Rep} impeller Reynolds number based upon plastic viscosity

N_P power number: $Pg_c/\rho N^3 D^5$

N_R number of ribbons

N_T torque number: $\tau \rho/\mu^2 D$

N_y yield stress power number: $\tau_y/\rho N^2 D^2$

n flow index of the power law

n_b blade number

P power

P_b power due to bulk effects or Stokes drag

P_c power due to clearance effects

P_N power draw of a Newtonian fluid

P_{nN} power draw of a non-Newtonian fluid

p pitch, either as height rise or height rise-impeller diameter

q exponent for Ellis model for primary normal stress coefficient

r radius in the coaxial cylinders

R radius of sphere, disk, or impeller

Re   impeller Reynolds number

s dimensionless parameter in the Carreau model

T tank diameter

t blade thickness

t_c characteristic time

t_n natural time

T_0 torque

U approach velocity

V tank volume

V velocity vector

V_0 velocity in the θ direction

W blade width

Wi	Weissenberg number
Vi	viscosity number
w^*	blade width as defined by Takahaski et al. (1980)
x	exponent
Y	$C_M N_{Re}{}^0$

Greek symbols

Ψ	stream function
ϕ	viscous dissipation function
η	plastic viscosity
σ, σ_1	primary normal stress coefficient
σ_0	primary normal stress coefficient at zero shear
α	constant
β	constant
θ	time
ρ	density
λ	characteristic time in the Carreau model
μ	viscosity, apparent or otherwise
μ_a	apparent viscosity
μ_0	zero shear viscosity
γ	shear rate, deformation rate
γ_{av}	average shear rate
γ_e	effective deformation rate
τ	stress tensor, total stress tensor, shear stress
τ_0	shear stress at $\mu = \mu_0/2$
τ_{ij}	the i, j component of τ
τ_w	wall shear stress
τ_y	yield stress
ω	angular velocity, vorticity
ω_ω	average vorticity at the wall
Ω	angular velocity, $2\pi N$, rad/s (radians per second)

References

Beckner, J. L., and J. M. Smith, *Trans. Instn. Chem. Engrs.*, **44**, T224, 1966.
Bertrand, J., and J. P. Couderc, *Proc. 5th Eur. Conf. on Mixing, Wurzburg, Germany*, BHRA Fluid Eng. Centre, Cranfield, England, 313, 1985.
Bertrand, J., and J. P. Couderc, *Fluid Mixing I. Chem. E. Sym. Series: No. 64*, Bradford, England, B1, 1981.

Bird, R. B., R. C. Armstrong, and O. Hassager, *Dynamics of Polymeric Liquids: Volume I Fluid Mechanics*, Wiley, New York, 1977.
Blasinski, H., and E. Rzyski, *Intern. Chem. Engrs.*, **16**, 4, 751, 1976.
Blasinski, H., and E. Rzyski, *Chem. Eng. J.*, **19**, 157, 1980.
Bourne, J. R., W. Knoepfli, and R. Riesen, *Proc. 3rd. Eur. Conf. on Mixing, Univ. of York, England*, BHRA Fluid Engineering, Cranfield, England, 1, 1979.
Bourne, J. R., and H. Butler, *AIChE-I. Chem. E. Symposium Series: No. 10*, **89**, 1965.
Bourne, J. R., and H. Butler, *Trans. Instn. Chem. Engrs.*, **47**, T263, 1969.
Brodkey, R. S., *The Phenomena of Fluid Motion*, Addison-Wesley, Reading, Mass., 1967, p. 423.
Calderbank, P. H., and M. B. MooYoung, *Trans. Instn. Chem. Engrs.*, **37**, 26, 1959.
Calderbank, P. H., and M. B. MooYoung, *Trans. Instn. Chem. Engrs.*, **39**, 337, 1961.
Chapman, F. S., and F. A. Holland, *Trans. Instn. Chem. Engrs.*, **43**, T131, 1965.
Chavan, V. V., A. S. Jhaveri, and J. J. Ulbrecht, *Trans. Instn. Chem. Engrs.*, **50**, 147, 1972.
Chavan, V. V., and J. J. Ulbrecht, *Chem. Eng. J.*, **3**, 308, 1972.
Chavan, V. V., and J. J. Ulbrecht, *I&EC Proc. Des. Dev.*, **12**, 472, 1973*a*.
Chavan, V. V., and J. J. Ulbrecht, *Trans. Instn. Chem. Engrs.*, **51**, 349, 1973*b*.
Chavan, V. V., and J. J. Ulbrecht, *I&EC Proc. Des. Dev.*, **13**, 309, 1974.
Chavan, V. V., A. K. Deysarkar, and J. Ulbrecht, *Chem. Eng. J.*, **10**, 205, 1975*a*.
Chavan, V. V., P. J. Diggory, and J. J. Ulbrecht, *Chemie. Eng. Techn.*, **47**, 74, 1975*b*.
Chavan, V. V., *AIChEJ.*, **9**, 181, 1983.
Cheng, D. C-H., C. Schofield, and R. J. Janes, *Proc. 1st Eur. Conf. on Mixing and Centrifugal Sep., Churchill College, Cambridge, England*, BHRA Fluid Eng., Cranfield, England, C2–15, 1974.
Chowdhury, R., and K. K. Tiwari, *I&EC Proc. Des. Dev.*, **18**, 2, 227, 1979.
Collias, D. J., and R. K. Prud'homme, *Chem. Eng. Sci.*, **40**, 8, 1495, 1985.
Edwards, M. F., J. C. Godfrey, and M. M. Kashani, *J. of Non-Newtonian Fluid Mech.*, **1**, 309, 1976.
Foresti, R., and T. Liu, *Ind. Eng. Chem.*, **51**, 7, 860, 1959.
Giesekus, H., *Rheol. Acta.*, **3**, 59, 1963.
Gluz, M. D., and I. S. Pavlushenko, *Theoretical Found. Chem. Eng.*, **3**, 626, 1969.
Godfrey, J. C., "Mixing of High-Viscosity Fluids," *Mixing in the Process Industries*, edited by N. Harnby, M. F. Edwards, and A. W. Nienow, Butterworth, Stoneham, Mass., 1985, Chap. 11.
Godfrey, J. C., T. H. Yuen, and M. F. Edwards, *Proc. 1st Eur. Conf. on Mixing and Cent. Sep.*, BHRA Fluid Eng., Cranfield, England, C3, 1974.
Godleski, E. S., and J. C. Smith, *AIChEJ.*, **8**, 5, 617, 1962.
Gray, J. B., *Chem. Eng. Prog.*, **59**, 3, 55, 1963.
Hall, K. R., and J. C. Godfrey, *Trans. Instn. Chem. Engrs.*, **48**, T201, 1970.
Hiraoka, S., I. Yamada, and K. Mizoguchi, *J. of Chem. Eng. Japan*, **11**, 6, 487, 1978.
Hiraoka, S., I. Yamada, and K. Mizoguchi, *J. of Chem. Eng. Japan*, **12**, 56, 1979.
Hocker, H., G. Langer, and U. Werner, *Ger. Chem. Eng.*, **4**, 133, 1981.
Hoogendoorn, C. J., and A. P. den Hartog, *Chem. Eng. Sci.*, **22**, 1689, 1967.
Ide, Y., and J. L. White, *J. of Applied Polymer Sci.*, **18**, 2997, 1974.
Kappel, M., *International Chem. Eng.*, **19**, 571, 1979.
Kelkar, J. V., R. A. Mashelkar, and J. J. Ulbrecht, *Trans. Instn. Chem. Engrs.*, **50**, 343, 1972.
Kelkar, J. V., R. A. Mashelkar, and J. J. Ulbrecht, *Chem. Eng. Sci.*, **28**, 664, 1973.
Kuriyama, M., H. Inomata, K. Arai, and S. Saito, *AIChEJ.*, **28**, 385, 1982.
Lee, R. E., C. R. Finch, and J. D. Wooledge, *Ind. Eng. Chem.*, **49**, 11, 1849, 1957.
Leung, L. S., *Trans. Instn. Chem. Engrs.*, **45**, T179, 1967.
Magnusson, K., *Iva*, **23**, 86, 1952.
Merquiol, T., and L. Choplin, *Proc. 6th Eur. Conf. on Mixing, Pavia, Italy*, BHRA Fluid Eng., Cranfield, England, 465, 1988.
Metzner, A. B., R. H. Feehs, H. L. Ramos, R. E. Otto, and J. D. Tuthill, *AIChEJ.*, **7**, 1, 3, 1961.
Metzner, A. B., and R. E. Otto, *AIChEJ.*, **3**, 1, 3, 1957.
Mitsuishi, N., and N. Hirai. *J. of Chem. Engrs. Japan*, **2**, 2, 217, 1969.

Murakami, Y., T. Hirose, M. Takao, T. Yamato, H. Fujiwara, and M. Ohshima, *J. Chem. Eng. Japan*, **14**, 488, 1981.
Nagata, S., N. Yoshioka, and T. Yokoyama, *Memoirs Fac. Eng.*, Kyoto Univ., **17**, 175, 1955*a*.
Nagata, S., and T. Yokoyama, *Memoirs Fac. Eng.*, Kyoto Univ., **17**, 253, 1955*b*.
Nagata, S., T. Yokoyama, and H. Maeda, *Memoirs Fac. Eng.*, Kyoto Univ., **18**, 13, 1956.
Nagata, S., K. Yamamoto, T. Yokoyama, and S. Shiga, *Memoirs Fac. Eng.*, Kyoto Univ., **19**, 274, 1957.
Nagata, S., M. Nishikawa, H. Tada, H. Hirabayashi, and S. Gotoh, *J. of Chem. Engr. Japan*, **3**, 2, 237, 1970.
Nagata, S., M. Nishikawa, H. Tada, and S. Gotoh, *J. of Chem. Engr. Japan*, **4**, 1, 72, 1971.
Nienow, A. W., D. J. Wisdom, J. Solomon, V. Machon, and J. Vlcek, *Chem. Eng. Commun.*, **19**, 273, 1983.
Nishikawa, M., M. Nakamura, and K. Hashimoto, *J. Chem. Eng. Japan*, **14**, 227, 1981.
Novak, V., and F. Rieger, *Trans. Instn. Chem. Engrs.*, **47**, T335, 1969.
Novak, V., and F. Rieger, *Chem. Eng. J.*, **9**, 63, 1975.
Patterson, W. I., P. J. Carreau, and C. Y. Yap, *AIChEJ.*, **25**, 508, 1979.
Pawlowski, J., *AIChEJ.*, **15**, 2, 303, 1969.
Penney, W. R., and K. J. Bell, *Ind. Eng. Chem.*, **59**, 4, 40, 1967.
Rieger, F., and V. Novak, *Chem. Eng. Sci.*, **27**, 39, 1972.
Rieger, F., and V. Novak, *Trans. Instn. Chem. Engrs.*, **51**, 105, 1973.
Rieger, F., and V. Novak, *Chem. Eng. Sci.*, **29**, 2229, 1974.
Rieger, F., and V. Novak, *Trans. Instn. Chem. Engrs.*, **52**, 285, 1974.
Sawinsky, J., A. Deak, and G. Havas, *Chem. Eng. Sci.*, **34**, 1160, 1979.
Sawinsky, J., G. Havas, and A. Deak, *Chem. Eng. Sci.*, **31**, 507, 1976.
Seichter, P., *Trans. Instn. Chem. Engrs.*, **49**, 117, 1971.
Schummer, P., *Chemie-Ingr-Tech.*, **41**, 1156, 1969.
Shamlou, P. A., and M. F. Edwards, *Chem. Eng. Sci.*, **40**, 1773, 1985.
Skelland, A. H. P., "Mixing and Agitation of Non-Newtonian Fluids," *Handbook of Fluids in Motion*, edited by N. P. Cheremisinoff and R. Gupta, Ann Arbor Science, Ann Arbor, Mich., 1983, Chap. 7.
Slattery, J. C., and R. A. Gaggiolo, *Chem. Eng. Sci.*, **17**, 893, 1962.
Solomon, J., T. P. Elson, and A. W. Nienow, *Chem. Eng. Commun.*, **11**, 143, 1981.
Sweeney, E. T., and M. A. Patrick, *Proc. 2nd Eur. Conf. on Mixing, St. John's College, Cambridge, England*, BHRA Fluid Eng., Cranfield, England, A4–43, 1977.
Takahashi, K., K. Arai, and S. Saito, *J. of Chem. Eng. Japan*, **13**, 2, 147, 1980.
Takahashi, K., K. Arai, and S. J. Saito, *J. of Chem. Eng. Japan*, **13**, 2, 77, 1982.
Takahashi, K., T. Yokota, and H. Konno, *J. of Chem. Eng. Japan*, **17**, 657, 1984.
Takaisi, Y., *J. of Phy. Soc. Japan*, **13**, 5, 496, 1958.
Toh, M., and Y. Murakami, *J. Chem. Eng. Japan*, **15**, 242, 1982.
Toh, M., and Y. Murakami, *J. Chem. Eng. Japan*, **15**, 493, 1982.
Toh, M., and Y. Murakami, *Proc. 6th Eur. Conf. on Mixing, Pavia, Italy*, BHRA Fluid Eng., Cranfield, England, 501, 1988.
Uhl, V. W., and H. P. Voznick, *Chem. Eng. Progr.*, **56**, 72, 1960.
Ulbrecht, J. J., *Chem. Engr.*, **286**, 347, 1974.
Ulbrecht, J. J., and V. R. Ranade, *Proc. 2nd Eur. Conf. on Mixing, St. John's College, Cambridge, England*, BHRA Fluid Eng., Cranfield, England, F6, 1977.
Ulbrecht, J. J., and P. Carreau, "Non-Newtonian Fluids," *Mixing of Liquids by Mechanical Agitation*, edited by J. J. Ulbrecht and G. K. Patterson, Gordon Breach Science, New York, 1985.
Walton, A. C., S. J. Maskell, and M. A. Patrick, *Fluid Mixing, I. Chem. E. Sym. Series: No. 64*, Bradford, England, C1, 1981.
White, J. L., S. Chankraiphon, and Y. Ide, *Trans. Soc. Rheo.*, **21**, 1, 1, 1977.
Yap, C. Y., W. Patterson, and P. J. Carreau, *AIChEJ.*, **25**, 516, 1979.

Chapter

4

Mixing in Turbulent Agitated Tanks

Introduction

Turbulent mixing processes occurring in a tank are spatially hetero-geneous. Fundamental to the understanding of any spatially distrib-uted flow process is the quantification of how the flow moves about material and properties. In the study of turbulent flow fields in agi-tated tanks, the emphasis has been upon experimental velocity and mixing measurements since the equations of motion cannot be solved for the turbulent flows occurring in an agitated tank. Such work is ba-sically an attempt to find the important mechanisms to mixing in tanks. Velocity information can be expressed as velocity profiles, pumping rate, and circulation data. Velocity flow fields are also more or less a prerequisite to the understanding of mixing and higher-level quantities such as shear, elongation, and vorticity fields.

Velocity alone does not indicate the amount of mixing actually tak-ing place in the flow field, and the matter becomes quite complicated. Some important questions are:

1. How is mixing accomplished?

2. What are the basic mechanisms important in mixing?

3. What information is available about the flow fields in an agitated tank?

4. Where do the various studies agree, and where do they disagree?

5. How much mixing is performed by shear, elongation, and vorticity?

6. Are there higher-level flow properties which perform mixing?

Velocity indicates the amount of these higher-level quantities in the flow or the amount of these quantities which could develop in the flow.

Turbulent flow fields in agitated tanks. The flow fields occurring in an agitated tank under turbulent flow conditions are quite complex. In one single tank, there are at least the following time-dependent flow systems:

1. Impeller flows which include high-speed discharge flows, impeller blade boundary layers, blade wake regions, boundary layer separation regions on blades and baffles, trailing vortex systems (multiple numbers per blade), blade bound vortex systems
2. Wall flows which include impinging jets originating from the impeller, boundary layers, corner flows, baffle bound vortex systems, shed vortex systems from the impeller and baffles
3. Bulk tank flows including large recirculation zones, toroidal vortices, and decaying vortex systems

Some of these flow systems are shown pictorially in Figs. 4.1 and 4.2. Most of these flows have an interior structure, making the entire tank rather a mess. However, that is the point.

The flow behavior in an agitated tank is quite reproducible in an average fashion, and there are temporal and spatial sequences which repeat. A systems approach to the fluid mechanics occurring in the tank is necessary, and the fact that such systems exist in the tank and have interior structures and levels of scale separates turbulent mixing from laminar mixing behavior. For industrial problems, most of the complexity need not be studied but only that portion which is relevant to the mixing process and to the level which provides basic understanding of the controlling process mechanisms in the mixing.

General literature categories. The literature on the turbulent flow fields in an agitated tank can be divided into three categories: overall or gross-scale studies, statistical turbulence studies, and structural turbulence studies.

Gross-scale studies are concerned with the overall flow patterns throughout the tank, gross integral flow measurements, and agitator pumping capacity. The techniques used in such work appear to be rather crude, and the results are rather simplistic. However, the results are quite useful.

Gross flow studies should not be overlooked in an analysis of the flow field in an agitated tank since they can provide a basis for understanding the complex nature of the velocity field in a gross sense. Qualitative and quantitative results can be inexpensively obtained, and initial impeller selections can often be based upon such studies. Using side-by-side comparison of alternative designs, gross flow stud-

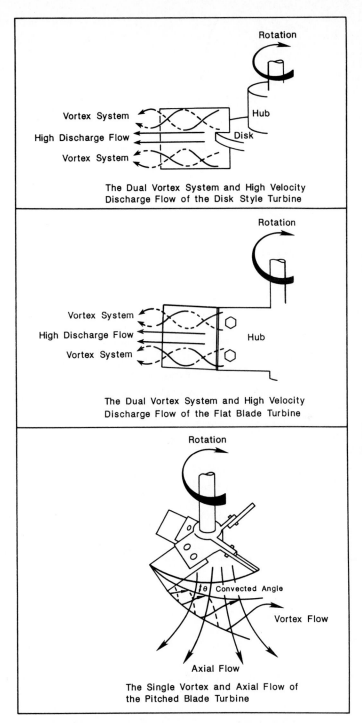

Figure 4.1   The various flow systems around turbulent agitators.

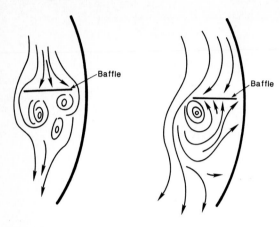

Baffle Off Wall   Baffle On Wall

Figure 4.2 Baffle bound vortices.

ies can often focus upon the most advantageous designs. This is also
true for multiphase studies.

The statistical turbulence studies are performed in support of the
application of the statistical turbulence theory to flows occurring in an
agitated tank. The approach is that the flows are highly complex and
random in nature, requiring statistical treatment. Statistical turbulence studies involve, in part, the measurement of average velocity
profiles, fluctuating velocities, turbulence intensities, and pumping
capacities. Experimental work relies heavily upon hot-wire or film
anemometry, pitot tubes, and laser doppler anemometry. The data are
point measurements, usually referenced to laboratory coordinates,
and the concept of eddies, particularly those of the smallest scale, and
the energy casade are emphasized. Indiscriminate averaging is typically used.

Structural turbulence studies involve the identification of the mean
flow fields at various scales and includes the other two categories.
However, discriminant averaging or conditional averaging is done.
Such work seeks a higher level of understanding of the flow than possible through gross or statistical studies. Attention is paid to the
large-scale flows which are treated in a deterministic fashion. Fluid
velocities are measured relative to the fluid itself and to laboratory
coordinates. The identification of flow systems, measurement of their
internal nature, the determination of how such systems interact and
the integration of the major events into overall flow schemes are technical issues of interest. Time is included in such studies as the flow
evolves. Large-scale flows and high-velocity flows of all sizes are emphasized.

The two approaches, statistical and structural turbulence, will now be reviewed. Gross flow studies will be discussed in later sections on integral circulation and mixing times.

General Comments on Statistical Turbulence Theory

Turbulence in this approach is generally considered to be highly irregular and random, having high diffusivities for momentum, mass, and heat, but is still considered a continuum phenomenon independent of any molecular structure. Turbulence is rotational and three dimensional with high levels of fluctuating angular momentum or vorticity. Turbulence is dissipative and requires a continuous supply of energy from the large mean flows. The mechanical energy associated with the flow velocities is dissipated by viscosity at the smallest eddies or scales into increases in the internal energy of the fluid.

Scales. The term scale is used in the discussion of turbulence to represent a size measure of the motions, whether it is a time scale, size scale, or velocity scale. In statistical turbulence, a continuous variation of scales, i.e., cascade of scales, is considered to exist and used to characterize the various size levels of the motion. There are the large scales involved in the large mean flows, intermediate scales involved with intermediate motions, and the smallest scales involved with the dissipation of the energy into internal energy by molecular viscosity. On the smallest scale, the molecular thermodynamic viscosity μ is important. On the largest and intermediate scales, the turbulent eddy viscosity μ_t is important and is a mechanical flow property generated by the flow.

The dissipation of turbulence into internal energy occurs at the smallest scales of the motion. The thermodynamic kinematic viscosity ν and viscous energy dissipation ϵ determine the size of the smallest scales and are the parameters which govern the small-scale motions. By dimensional analysis first performed by Kolmogorov, the dissipation rate and kinematic viscosity can be arranged to give a length scale, a time scale, and a velocity scale as shown:

Length scale: $$\eta = \left(\frac{\nu^3}{\epsilon}\right)^{1/4} \tag{4.1}$$

Time scale: $$t_K = \left(\frac{\nu}{\epsilon}\right)^{1/2} \tag{4.2}$$

Velocity scale: $$\upsilon = (\nu\epsilon)^{1/4} \tag{4.3}$$

These have been named the Kolmogorov microscales. Since the viscous dissipation process occurs at a very small length scales, the Reynolds number at the smallest scales is taken as 1, and turbulence is considered to be quite viscous. Large scales cause relatively little viscous dissipation.

At very high Reynolds number (i.e., high energy levels received from the large-scale motions), small length scales are very much smaller than the large-scale motions and are considered independent of the large-scale behavior. At low Reynolds numbers (i.e., low energy levels received from the large scale), the small length scales are coarse and may be within an order of magnitude of the large scales. This difference in scales is very important to recognize in application and can be used to separate highly turbulent flow from other turbulence. If present, the large scale can be sized by using the magnitude of the Reynolds number which is based upon the thermodynamic viscosity. A highly turbulent flow contains all sizes of scales; the large scale with the thermodynamic viscosity used to define the Reynolds number and the small scale with the thermodynamic viscosity used to calculate the viscous dissipation.

There are ranges in the scales of viscosity, velocity, and time in turbulence. Viscosity ranges from the eddy viscosity to the molecular or thermodynamic viscosity; velocities range from the large-scale mean velocities to the dissipation velocities of the energy dissipation; and time scales range from the times of the large-scale motions to the times involved with the energy dissipation. Such scales should be noted in any detailed analysis of turbulence.

Eddy viscosity. The ways in which momentum is transported about in turbulent flow are not clearly understood. The eddy viscosity concept is an attempt at understanding this transport phenomenologically and is the parameter relating turbulent stresses to mean deformation rate. The defining equation for the eddy viscosity μ_t is:

$$\tau = (\mu + \mu_t)\frac{d\overline{U}}{dy} \tag{4.4}$$

where $d\overline{U}/dy$ is the gradient of the average velocity. The term μ_t/ρ is the turbulent eddy kinematic viscosity, or ν_t. The overbar indicates an average quantity. In terms of the Reynolds stresses, $\overline{u_i u_j}$, the kinematic eddy viscosity is defined as:

$$\rho\overline{u_i u_j} = \mu_t\left(\frac{\partial \overline{U_i}}{\partial x_j} + \frac{\partial \overline{U_j}}{\partial x_i}\right) \tag{4.5}$$

The Reynolds stresses arise from time averaging the instantaneous equations of motion and are considered to be the terms which account for the turbulent transport of momentum.

There are a number of other concepts which have been developed to describe momentum transport as reviewed by Brodkey (1967). These include Prandtl's mixing length theory, Taylor's vorticity-transport theory, Von Karman's similarity hypothesis and dimensional analysis. However, in mixing, the eddy viscosity is used almost exclusively.

Methods of analysis. There are a variety of methods used to analyze mixing and turbulence in an agitated tank. A general approach is based upon the use of the fundamental conservation equations, performing order-of-magnitude analyses on the governing equations from which models of the flow can be developed. This approach is infrequently used in mixing because of the variety and complexity of the flows in an agitated tank. However, computational methods and results from the equations of motion are more and more prevalent and will become progressively more useful and trusted.

Experimentation and simplistic models, based upon experimental data, are used heavily in mixing and are useful in establishing basic scaling relationships. Dimensional analysis plays an important role in the scaling.

Often, mixing phenomena are dependent upon a few variables, and similar flow and mixing behavior can be expected if there is similarity in variables. Dimensional analysis can be applied in such cases, resulting in basic length, time, and velocity scale relationships. The objective is to obtain general relationships which are applicable to a number of different systems. Such relationships are very useful in mixing and are important results of gross flow studies. Mixing and circulation time correlations are examples of such relationships.

Equations of the theory of statistical turbulence

Turbulence in the context of the statistical turbulence theory (Brodkey, 1967) is viewed as a sea of eddies of different scales or sizes which can be treated in a statistical manner. The instantaneous Navier Stokes equations should be valid for turbulent flow at the smallest scales. However, little information can be gained by direct application of the basic instantaneous equations of motion since the variables in the equations vary considerably and are impossible to treat analytically.

Time averaging rules. Typically, turbulence is presented as a waveform, as shown in Fig. 4.3. To treat this waveform, time averaging is used where the instantaneous velocity U is divided into an av-

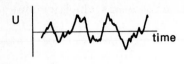

U

time

Velocity Waveform

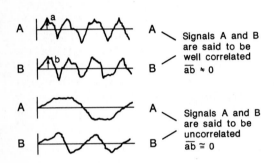

A   A   Signals A and B
            are said to be
            well correlated
B   B $\overline{ab} \neq 0$

A   A   Signals A and B
            are said to be
            uncorrelated
B   B $\overline{ab} \cong 0$ **Figure 4.3** Velocity waveform
                                    and correlations.

Correlations

erage velocity $\overline{U}$ and a fluctuating component u:

$$U_i = \overline{U}_i + u_i \tag{4.6}$$

where

$$\overline{U}_i = \frac{1}{t_2 - t_1} \int_{t_1}^{t_2} U_i \, dt \tag{4.7}$$

Pressure can be treated in the same manner:

$$P = \overline{P} + p$$

and

$$\overline{P} = \frac{1}{t_2 - t_1} \int_{t_1}^{t_2} P \, dt \tag{4.8}$$

The root mean square, or rms, of the fluctuating quantity is denoted
with a prime as:

$$u'_i = \left[\frac{1}{t_2 - t_1} \int_{t_1}^{t_2} (U_i - \overline{U}_i)^2 \, dt \right]^{0.5} \tag{4.9}$$

As Eq. (4.9) indicates, u_i' is dependent upon the average velocity and how it is calculated. Turbulence data are reported as either the root mean square, i.e., the rms, or as a relative intensity or a turbulent intensity, i.e., $u_i'/\overline{U}$ or $u_i'/\overline{U}_i$, where an average velocity or average component velocity is used.

The time span over which the average is taken and the length of time of the flow events are important in such averaging. Depending upon the research objectives, the time span can be much greater than the time of the flow events or much smaller than the time scale of the flow events. In the former, the details of the flow are lost and, in the latter, too much flow detail may be included. Hopefully, a proper balance is obtained in the time averaging process. Unfortunately, the time span over which the averaging is done is typically not reported in studies.

Any variable can be written in terms of an average, plus a fluctuating component, including pressure and correlations. However, in using Eq. (4.6), one unknown instantaneous quantity becomes two unknown quantities: an average quantity and a fluctuating component.

The term correlation needs some definition. Turbulence, as measured at a fixed point, can appear as a fluctuating waveform as shown in Fig. 4.3. If two instantaneous waveforms show a correspondence or appear to have some corresponding behavior in some fashion, they are said to be correlated. A measure of this correspondence is referred to as a correlation.

The purpose of developing correlations is to understand turbulence by studying the corresponding behavior in different waveforms. Correlation may involve any number of flow properties, not just two, and some can actually be measured. As will be shown later, correlations appear in the equations which attempt to describe turbulence. The Reynolds stresses are correlations between fluctuating velocity components.

Averages in mixing are usually defined relative to laboratory coordinates or coordinates fixed to a rotating geometry. Averging tends to be very arbitrary, and care must be taken in performing the actual averages. If the average is taken over the time span which is much larger than that of interest, then time averaging can be detrimental to the understanding of the physics of the process under study.

To obtain averages of waveforms as shown in Fig. 4.3, heuristic rules of averaging were developed by Reynolds (Hinze, 1975; Brodkey, 1967) and can briefly be stated as follows. The overbar indicates averaging and lowercase indicates the fluctuating component. A and B can be any quantity, e.g., velocity, pressure, or partial derivatives of these quantities.

1. Quantities which have been averaged may be considered constant in subsequent averaging:

$$\bar{\bar{A}} = \bar{A} \quad \text{or} \quad \overline{\bar{A}\,\bar{B}} = \bar{A}\,\bar{B} \tag{4.10}$$

2. Averaging obeys the distributive law. For example, an average of a sum of instantaneous quantities is the sum of their averages:

$$\overline{A + B} = \bar{A} + \bar{B} \tag{4.11}$$

This rule is important in treating the subtraction involved in partial derivatives, for example:

$$\overline{\frac{\partial A}{\partial x}} = \frac{\partial \bar{A}}{\partial x} \tag{4.12}$$

and, in the limit, for a time derivative:

$$\overline{\frac{\partial A}{\partial t}} = \overline{\lim_{\Delta t \to 0} \left(\frac{A_{t+\Delta t} - A_t}{\Delta t}\right)} = \lim_{\Delta t \to 0} \left(\overline{\frac{A_{t+\Delta t} - A_t}{\Delta t}}\right)$$

$$= \lim_{\Delta t \to 0} \left(\frac{\bar{A}_{t+\Delta t} - \bar{A}_t}{\Delta t}\right) = \frac{\partial \bar{A}}{\partial t} \tag{4.13}$$

Here it is assumed that the average of the limit process is equal to the limit of averages. The average of an instantaneous quantity and its fluctuating component is given as:

$$\bar{A} = \overline{\bar{A} + a} = \bar{\bar{A}} + \bar{a} = \bar{A} \tag{4.14}$$

and, therefore:
$$\bar{a} = 0 \tag{4.15}$$

The average of a fluctuating component is zero.

3. The product of two instantaneous terms is:

$$\overline{AB} = \overline{(\bar{A} + a)(\bar{B} + b)} = \overline{\bar{A}\bar{B} + a\bar{B} + b\bar{A} + ab} = \overline{AB} + \overline{ab} \tag{4.16}$$

since:
$$\overline{A b} = \bar{A}\,\bar{b} = 0 \quad \text{since } \bar{b} = 0$$

and:
$$\overline{B a} = \bar{B}\,\bar{a} = 0 \quad \text{since } \bar{a} = 0$$

but:
$$\overline{ab} \neq 0 \tag{4.17}$$

even though $\bar{a} = 0$ and $\bar{b} = 0$. These relationships can be observed more clearly in Fig. 4.3.

Products of instantaneous terms appear in the Navier Stokes equations which subsequently, upon averaging, give rise to unknown correlations, like $\overline{ab}$. An important $\overline{ab}$ type correlation is the Reynolds stresses. These correlation quantities are nonzero when the time signals have similar appearance or systematic relationships. The quantities a and b are said to be correlated if $\overline{ab}$ is nonzero. Higher-level equations can be written for terms like $\overline{ab}$, but, in the averaging process, terms like $\overline{abc}, \overline{abb}, \overline{aab},$ or $\overline{Abb}$ appear which are again unknown quantities. These correlations give rise to the need for experimental data or other higher-level equations which describe their behavior.

It is very important to recognize what is being done when an average is being taken. First, an average has to be considered an approximation. Second, information is lost. Basically, an average is the result of a filtering process. Whatever the waveform was, it has been replaced with a constant value and a fluctuating component. Simplicity is gained in the averaging process. Unfortunately, in some cases, the physics of the process is dependent upon the information which has been averaged away. The averaging rules, given above, are heuristic and, in some cases, are probably unrealistic.

Example Problem Obtain the time average of the following product of instantaneous quantities $U_i U_j U_j$.

solution

$$\overline{U_i U_j U_j} = \overline{(\overline{U_i} + u_i)(\overline{U_j} + u_j)(\overline{U_j} + u_j)}$$

$$= \overline{\overline{U_i}\,\overline{U_j}\,\overline{U_j}} + \overline{2u_j\overline{U_j}\,\overline{U_i}} + \overline{\overline{U_i}\,u_j^2} + \overline{u_i\,\overline{U_j}\,\overline{U_j}} + \overline{2u_i u_j\,\overline{U_j}} + \overline{u_i\,u_j^2}$$

Upon time averaging:

$$= \overline{U_i}\,\overline{U_j}\,\overline{U_j} + 2\overline{u_i u_j}\,\overline{U_j} + \overline{U_i}\,\overline{u_j^2}$$

since $\overline{u_i}$ and $\overline{u_j}$ are zero.

It should be noted that, only upon time averaging are the three terms, having an independent fluctuating component, set to zero. In the actual instantaneous flow, these quantities are not zero and participate in the fluid mechanics as part of the instantaneous quantities.

Example Problem A velocity record has the waveform as shown. Perform a time average of the time derivative of the velocity over the time span shown in Fig. 4.P1.

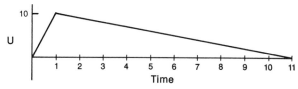

Figure 4.P1

solution   The time average is zero. The time derivative can be plotted as shown in Fig. 4.P2

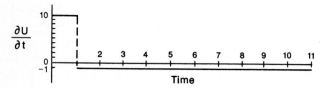

Figure 4.P2

In the equations to follow, average mean quantities appear with an overbar. Fluctuating quantities can only appear in products of fluctuating quantities and cannot appear alone or as a part of product with a mean value because of averaging rules cited above.

Continuity.   The general equation of continuity is:

$$\frac{\partial \rho}{\partial t} + \nabla \cdot (\rho \mathbf{U}) = 0 \qquad (4.18)$$

The equation of continuity for a constant density fluid is:

$$\nabla \cdot \mathbf{U} = 0 \qquad (4.19)$$

or

$$\frac{\partial U_i}{\partial x_i} + \frac{\partial U_j}{\partial x_j} + \frac{\partial U_k}{\partial x_k} = 0 \qquad (4.20)$$

Continuity is obeyed instantaneously by velocity, and if averaging is done correctly, the average velocity obeys continuity since:

$$\nabla \cdot \overline{\mathbf{U}} = 0 \qquad (4.21)$$

and

$$\overline{\nabla \cdot (\overline{\mathbf{U}} + \mathbf{u})} = \nabla \cdot \overline{\mathbf{U}} + \overline{\nabla \cdot \mathbf{u}} = 0 \qquad (4.22)$$

From this, one can observe that velocity fluctuations obey continuity as well:

$$\overline{\nabla \cdot \mathbf{u}} = 0 \qquad (4.23)$$

In blending liquids of different densities, density effects are present and sometimes have to be taken into account. Time averaging the general continuity equation with fluctuating density results in:

$$\frac{\partial \hat{\rho}}{\partial t} + \frac{\partial}{\partial x_i}(\overline{\rho U_i} + \overline{\hat{\rho} u_i}) = 0 \qquad (4.24)$$

$\hat{\rho}$ denotes the fluctuations in density, and the subscript i is repeated for all three directions. For the equations which follow, fluctuations in density and molecular viscosity will be ignored.

The Reynolds time-averaged equations of motion. The instantaneous equations of motion and the Navier Stokes equations for liquids are force balances or momentum rate balances for liquids which take into account pressure, viscous, and body (e.g., gravity) forces, in addition to any other forces F_s which might be present in the flow. Generally, the equations can be written as:

$$\rho \frac{DU}{Dt} = \rho \frac{\partial U}{\partial t} + \rho[\nabla \cdot (UU)] = -\nabla p + \mu \nabla^2 U + \Sigma \rho_s F_s \quad (4.25)$$

This is a vector equation containing a force balance in each of the three directions. If each term is replaced by its average and a fluctuating component and then time averaged, the resulting equation appears as:

$$\rho \frac{D\overline{U}}{Dt} = \rho \frac{\partial \overline{U}}{\partial t} + \rho(\nabla \cdot \overline{U})\overline{U} = -\nabla \overline{p} + \mu \nabla^2 \overline{U} - \nabla \cdot \rho \overline{uu} + \Sigma \rho_s F_s \quad (4.26)$$

where $\nabla \cdot (UU) = (U \cdot \nabla)U + U(\nabla \cdot U)$. This is also a vector equation. The term $\rho \overline{uu}$ arises from the left-half side of the equation from the **UU** term in the previous equation. This term has been named the Reynolds stresses and represents turbulent accelerations. The equations can also be written out showing the appropriate subscripts as:

$$\rho \left(\frac{\partial \overline{U_i}}{\partial t} + \overline{U_j} \frac{\partial \overline{U_i}}{\partial x_j} \right) = -\frac{\partial \overline{P}}{\partial x_i} + \frac{\partial}{\partial x_j} \left(\mu \frac{\partial \overline{U_i}}{\partial x_j} - \rho \overline{u_i u_j} \right) + \Sigma \rho_s \overline{F_{si}} \quad (4.27)$$

where these are individual component equations. The subscript i designates the direction for which the force balance is made and j is a counter over the three directions.

The averaged equations have the same form as the original instantaneous equations except that average properties have replaced instantaneous properties and an additional set of terms has appeared which are the unknown Reynolds stresses. Other equations must be written for these or experimental data must be obtained to describe their behavior for closure. The Reynolds stresses are unknown correlations which are not well understood in general.

It should also be noted that the instantaneous equations of motion were solvable in principle. However, because of time averaging and the generation of the higher-order unknown correlations, the time-averaged equations are no longer solvable in principle. Various attempts are presented in the literature to provide data and methods to provide the closure to these equations. The objective in using the equations as presented is to calculate the average velocity $\overline{U}_i$ for the purposes of describing mixing.

The turbulent Reynolds stresses can be written as the tensor:

$$
\begin{bmatrix}
\overline{u_i u_i} & \overline{u_i u_j} & \overline{u_i u_k} \\
\overline{u_j u_i} & \overline{u_j u_j} & \overline{u_j u_k} \\
\overline{u_k u_i} & \overline{u_k u_j} & \overline{u_k u_k}
\end{bmatrix}
$$

u_i equation of motion.   The u_i equation of motion can be obtained by subtracting the time-averaged equations of motion from the instantaneous equation of motion. The u_i equation is then:

$$
\frac{\partial u_i}{\partial t} + (\overline{U}_j + u_j)\frac{\partial u_i}{\partial x_j} = -\frac{1}{\rho}\frac{\partial p}{\partial x_i} + \nu\frac{\partial^2 u_i}{\partial x_j^2} \tag{4.28}
$$

where p is the fluctuating pressure. The time average of this equation is zero.

Turbulent kinetic energy equation.   If a vector dot product is performed between the instantaneous flow velocity and the instantaneous equations of motion, the equation for the instantaneous kinetic energy equation is obtained as:

$$
\frac{1}{2}\frac{\partial U_i U_i}{\partial t} = -\frac{\partial}{\partial x_j}U_j\left(\frac{P}{\rho} + \frac{U_i U_i}{2}\right) + \nu\frac{\partial}{\partial x_j}U_i\left(\frac{\partial U_i}{\partial x_j} + \frac{\partial U_j}{\partial x_i}\right)
$$

$$
\quad\text{I}\qquad\qquad\qquad\text{II}\qquad\qquad\qquad\qquad\text{III}
$$

$$
-\nu\left(\frac{\partial U_i}{\partial x_j} + \frac{\partial U_j}{\partial x_i}\right)\frac{\partial U_i}{\partial x_j} \tag{4.29}
$$

$$
\qquad\qquad\qquad\text{IV}
$$

By inspection, the various terms have the following meanings:

I.   The local change of the total kinetic energy in the differential control volume

II.   The convective transport of the total kinetic energy in or out of the control volume and the work done by the dynamic pressure

III. The work done by viscous stresses

IV.   Viscous dissipation

The time-averaged kinetic energy equation becomes:

$$\frac{1}{2}\frac{\partial \overline{U_i U_i}}{\partial t} + \frac{\partial \overline{k}}{\partial t} = -\frac{\partial}{\partial x_j}\overline{U_j\left(\frac{\overline{P}}{\rho} + \frac{\overline{U_i U_i}}{2}\right)} + \nu\frac{\partial}{\partial x_j}\overline{U_i\left(\frac{\partial \overline{U_i}}{\partial x_j} + \frac{\partial \overline{U_j}}{\partial x_i}\right)}$$

$$-\nu\overline{\left(\frac{\partial \overline{U_i}}{\partial x_j} + \frac{\partial \overline{U_j}}{\partial x_i}\right)\frac{\partial \overline{U_i}}{\partial x_j}} - \frac{\partial}{\partial x_j}\overline{u_j\left(\frac{p}{\rho} + k\right)} - \frac{\partial}{\partial x_j}(\overline{U_i u_i u_j}) - \frac{\partial(\overline{U_j}\,\overline{k})}{\partial x_j}$$

$$+ \nu\frac{\partial}{\partial x_j}\overline{u_i\left(\frac{\partial u_i}{\partial x_j} + \frac{\partial u_j}{\partial x_i}\right)} - \nu\overline{\left(\frac{\partial u_i}{\partial x_j} + \frac{\partial u_j}{\partial x_i}\right)\frac{\partial u_i}{\partial x_j}} \quad (4.30)$$

where k is the turbulent kinetic energy defined as:

$$k = 1/2\,(u_i^2 + u_j^2 + u_k^2) \quad (4.31)$$

Kinetic energy equation of the turbulent velocity fluctuations. If a vector dot product is performed between the time-averaged velocity and the time-averaged equations of motion, the following equation is obtained as:

$$\frac{1}{2}\frac{\partial \overline{U_i U_i}}{\partial t} = -\frac{\partial}{\partial x_j}\overline{U_j}\left(\frac{\overline{P}}{\rho} + \frac{\overline{U_i U_i}}{2}\right) + \nu\overline{U_j}\frac{\partial}{\partial x_j}\left(\frac{\partial \overline{U_i}}{\partial x_j} + \frac{\partial \overline{U_j}}{\partial x_i}\right) - \overline{U_j}\frac{\partial}{\partial x_j}(\overline{u_i u_j})$$

$$(4.32)$$

If this equation is subtracted from the time-averaged total kinetic equation above, the equation for the turbulent kinetic energy of the velocity fluctuations can be obtained as:

$$\frac{D\overline{k}}{Dt} = -\frac{\partial}{\partial x_j}\overline{u_j\left(\frac{p}{\rho} + k\right)} - \overline{u_i u_j}\frac{\partial}{\partial x_j}(\overline{U_i}) + \nu\frac{\partial}{\partial x_j}\overline{u_i\left(\frac{\partial u_i}{\partial x_j} + \frac{\partial u_j}{\partial x_i}\right)}$$

$$\text{I} \qquad\qquad \text{II} \qquad\qquad \text{III} \qquad\qquad\qquad \text{IV}$$

$$- \nu\overline{\left(\frac{\partial u_i}{\partial x_j} + \frac{\partial u_j}{\partial x_i}\right)\frac{\partial u_i}{\partial x_j}} \quad (4.33)$$

$$\text{V}$$

The substantial derivative, term I, in this equation uses the average velocity. The various terms in the equation have the following interpretations:

I.   The change of the kinetic energy of the turbulence with time plus the convective transport of the turbulent kinetic energy by the mean velocity.

II.   Convective-diffusional transport of turbulent kinetic energy by the turbulent fluctuations u_j and pressure fluctuations p.

III.   Energy transferred from the mean motion by the Reynolds stresses. This is typically called turbulent production and represents the source term of turbulent kinetic energy.

IV.   Work done by viscous shear stresses of the velocity fluctuations.

V.   Viscous dissipation of the turbulent kinetic energy by the turbulent motion.

For very high Reynolds number flows, terms II and IV are considered unimportant.

The last two terms in the last equation can be rewritten as:

$$\nu \frac{\partial^2 \overline{k}}{\partial x_j^2} - \nu \overline{\left(\frac{\partial u_i}{\partial x_j} \frac{\partial u_i}{\partial x_j} \right)}$$

and as a result, the equation may appear as:

$$\frac{D\overline{k}}{Dt} = -\frac{\partial}{\partial x_j} \overline{u_j \left(\frac{p}{\rho} + k \right)} - \overline{u_i u_j} \frac{\partial}{\partial x_j}(\overline{U_i}) + \nu \frac{\partial^2 \overline{k}}{\partial x_j^2} - \nu \overline{\left(\frac{\partial u_i}{\partial x_j} \frac{\partial u_i}{\partial x_j} \right)} \qquad (4.34)$$

The $\overline{u_i u_j}$ equations. The next level of complexity in the hierarchy of equations is to write the equations for $\overline{u_i u_j}$ since they are unknown quantities in the time-averaged equations of motion. To obtain these equations, the time-averaged equations of motion are subtracted from the time-dependent equations to obtain equations for u_i as was done above. The resulting equation for u_i is multiplied by the fluctuating component u_j of interest. The u_j equation is obtained in the same manner and multiplied by the fluctuating component u_i. The two equations are then added and time averaged to obtain the $\overline{u_i u_j}$ equation. These equations appear in general form as:

$$\underbrace{\frac{\partial \overline{u_i u_j}}{\partial t}}_{\text{I}} + \underbrace{\frac{\overline{U_l} \partial}{\partial x_l}(\overline{u_i u_j})}_{\text{II}} = \underbrace{-\frac{\partial}{\partial x_l}(\overline{u_l u_i u_j})}_{\text{III}} - \underbrace{\frac{1}{\rho}\left(\frac{\partial}{\partial x_i}\overline{u_j p} + \frac{\partial}{\partial x_j}\overline{u_i p} \right)}_{\text{IV}}$$

$$\underbrace{-\overline{u_i u_l}\frac{\partial(\overline{U_j})}{\partial x_l} - \overline{u_j u_l}\frac{\partial(\overline{U_i})}{\partial x_l}}_{\text{V}} + \underbrace{\frac{p}{\rho}\overline{\left(\frac{\partial u_i}{\partial x_j} + \frac{\partial u_j}{\partial x_i} \right)}}_{\text{VI}}$$

$$\underbrace{+\nu \frac{\partial^2 \overline{u_i u_j}}{\partial x_l^2}}_{\text{VII}} - \underbrace{2\nu \overline{\left(\frac{\partial u_i}{\partial x_l} \frac{\partial u_j}{\partial x_l} \right)}}_{\text{VIII}} \qquad (4.35)$$

In these equations, the subscripts i and j are specific to the correlation of interest and the subscript l is the direction counter. There are six equations in all; one for each $\overline{u_i u_j}$ term. The meanings to the various terms are:

I.   The rate of change of $\overline{u_i u_j}$ in the control volume
II.   Convection of $\overline{u_i u_j}$ by the mean flow velocities
III.   Convective transport of the $\overline{u_i u_j}$ by the turbulent fluctuations
IV.   Pressure-fluctuating velocity correlation (work or energy term)
V.   Mean shear production
VI.   Pressure-fluctuating strain
VII.   Diffusional transport of $\overline{u_i u_j}$
VIII.   Viscous dissipation

These equations are just the preliminary beginning to classical statistical turbulence theory. A more complete description of their development can be found in Hinze (1975).

Similar equations, which parallel those above, can be written in terms of vorticity. Taking the curl of the instantaneous Navier Stokes equation, the instantaneous vorticity equation is obtained. This equation is time averaged to obtain the mean vorticity equation which has unknown correlations about which other equations can be written. The equation of the mean square vorticity fluctuation can also be written. The development of these equations is available in Tennekes and Lumley (1972) and need not be repeated here. The use of vorticity does provide some insight into how energy is transferred to the smaller eddies. Vortices and their axes are thought to align with the mean flow. During this alignment process, the vortices are stretched. Their cross section is reduced so these vortices are now a different eddy of a smaller scale and, due to the conservation of angular momentum, their rotational speed increases, which causes energy dissipation. The analogy can be drawn to an ice skater drawing in his arms during a spin. This obviously is a crude analogy but a useful one.

There has been considerable work done in wave number space by taking the Fourier transform of the equations and developing the energy spectra. These concepts have been developed more thoroughly in Hinze (1975), Tennekes and Lumley (1972), and Brodkey (1967).

The number of equations, written to describe turbulence, is actually limitless due to the closure problem. As more equations are written, higher levels of complexity occur which make the equations difficult to use in any practical sense.

Scalar transport.   The equation used to model the transport of a scalar quantity can be written as:

$$\frac{\partial \Gamma}{\partial t} + U_i \frac{\partial \Gamma}{\partial x_i} = \frac{\partial}{\partial x_i}\left(\alpha \frac{\partial \Gamma}{\partial x_i}\right) + S_v \qquad (4.36)$$

where α is the molecular diffusivity for the diffusion of the scalar quantity and S_v is a source term. If the diffusivity is assumed constant, Γ and U are replaced by an average and fluctuating components and, when time averaging of the various terms is done, the following equation is obtained:

$$\frac{\partial \overline{\Gamma}}{\partial t} + \overline{U_i}\frac{\partial \overline{\Gamma}}{\partial x_i} = \frac{\partial}{\partial x_i}\left(\alpha \frac{\partial \overline{\Gamma}}{\partial x_i} + \overline{u_i\gamma}\right) + \overline{S}_v \qquad (4.37)$$

The distribution of $\overline{\Gamma}$ in the flow is affected by molecular diffusion and by turbulent convection, i.e., the $\overline{u_i\gamma}$ term. The $\overline{u_i\gamma}$ term is the turbulent flux of the scalar Γ and can be considered to be transport by turbulent acceleration.

Correlations. Correlations between the various velocity fluctuations have been developed and used extensively to gain some understanding of turbulence. Correlations should contain information about how velocities and other flow properties are related in turbulent flow. From these correlations, various scales and the turbulent energy spectrum can be obtained. Turbulent velocity fluctuations at two points can be statistically correlated if the distance between the two points are small.

The individual ij correlation is written as:

$$Q_{ij}(r) = \overline{u_i(x)u_j(x + r)} \qquad (4.38)$$

where the overbar indicates the time-averaged value. Since there are three velocity fluctuations, the individual correlations form a tensor having nine components. Typically, within an eddy, the correlation can be quite high. If the separation distance r is zero, the correlations become the Reynolds stresses. The entire tensor is also called the energy tensor since the units of $u_i u_j$ are energy units.

$Q_{ij}(r)$ can be normalized using the rms velocity fluctuations to form the Eulerian correlation function where the components are:

$$R_{ij}(r) = \frac{\overline{u_i(x)u_j(x + r)}}{u_i'(x)u_j'(x + r)} \qquad (4.39)$$

Other higher-level correlations of the velocity fluctuations can also be defined. For example, the components of the triple-velocity correlation are:

$$S_{ijk}(r) = \overline{u_i(x)u_j(x)u_k(x + r)} \tag{4.40}$$

This particular correlation can be considered as a measure of the energy of the velocity fluctuations which is transported by the turbulent velocity fluctuations.

For isotropic turbulence, the rms velocity fluctuations are all equal, and $R_{ij}(r)$ simply becomes:

$$R_{ii}(r) = \frac{\overline{u_i(x)u_i(x + r)}}{u_i'(x)u_i'(x + r)} \tag{4.41}$$

For isotropic turbulence, all $i \neq j$ terms are zero. In addition, to specify $R_{ii}(r)$, only one scalar function is needed, called $f(r)$ or the Eulerian correlation function for isotropic turbulence, where the velocity fluctuations have been measured in the same longitudinal direction:

$$f(r) = \frac{\overline{u_r(x)u_r(x + r)}}{(u')^2} \tag{4.42}$$

Another correlation $g(r)$ exists in a direction normal to r and is defined as:

$$g(r) = \frac{\overline{u_n(x)u_n(x + r)}}{(u')^2} \tag{4.43}$$

$R_{ij}(r)$, $f(r)$, and $g(r)$ are dimensionless and have been obtained experimentally in many studies of the flow field in an agitated tank.

There are also Lagrangian correlations which can be defined.

Length scales.   Since $R_{ij}(r)$, $f(r)$, and $g(r)$ are dimensionless, integration over the distance from 0 to ∞ results in the appearance of length units. One such length is obtained from the integral of $f(r)$ and is referred to as the integral length scale or macroscale:

$$L_f = \int_0^\infty f(r)\, dr \tag{4.44}$$

$f(r)$ contains information on how velocity fluctuations correspond over distance, and L_f is a length measure of the correspondence. In studies of an agitated tank, $f(r)$ and L_f have been determined; L_f is typically referred to as the turbulence macroscale Λ. The macroscale can also be calculated from:

$$\Lambda = \frac{a(u')^3}{\epsilon} \tag{4.45}$$

where a is a constant and ϵ is the energy dissipation. Individual length scales are also possible and are obtained directly from the individual $R_{ij}(r)$ correlations:

$$L_{ij} = \int_0^\infty R_{ij}(r)\,dr \qquad (4.46)$$

Data concerning $g(r)$ and its integral are not typically reported in studies in agitated tanks although such data can be obtained as well.

A turbulent microscale λ_f can also be defined from $f(r)$ as:

$$\lambda_f = \left[-\frac{1}{\partial^2 f(r)/\partial r^2|_{r=0}} \right]^{0.5} \qquad (4.47)$$

The microscale can also be calculated from:

$$\frac{1}{\lambda_f^2} = \frac{1}{2(u_1')^2} \int_0^\infty k^2 E_1(k)\,dk \qquad (4.48)$$

where $E_1(k)$ is the energy spectrum of u_1', the subscript indicates the direction of an important velocity fluctuation. For isotropic turbulence, the energy dissipation rate and the microscales are related as:

$$\epsilon = 30\nu\left(\frac{u_1'^2}{\lambda_f^2}\right) = 15\nu\left(\frac{u_2'^2}{\lambda_g^2}\right) \qquad (4.49)$$

where λ_g is lateral microscale obtained from the integration of $g(r)$.

Lagrangian correlations can also be integrated over time to obtain Lagrangian time scales which, when multiplied by velocity, can result in Lagrangian length scales.

In the discussions on mixing to follow, no distinction will be drawn between the Eulerian and Lagrangian length scales. Instead, these will be considered as simply macro and micro length scales.

Energy spectrum. The description of a turbulent flow field can involve double- and triple-velocity correlations between two points. Equations can be written for these correlations just as was done for the $\overline{u_i u_j}$ Reynolds stresses. For example, the equation for $Q_{ij}(r)$ (Brodkey, 1967) can be developed as follows. Assuming the average velocity components are constant, an equation for the velocity fluctuations are written about a point A in the flow. This equation is then multiplied by the velocity fluctuations occurring at point B. The reverse is done for point B where the separation distance between A and B is r. The two equations are then added to obtain a single equation for $Q_{ij}(r)$ which appears as:

$$\frac{\partial(\overline{u_A u_B})}{\partial t} = -\nabla_r \cdot (\overline{u_A u_B u_B} - \overline{u_A u_A u_B}) - \frac{\nabla_r(\overline{p_B u_A} - \overline{p_A u_B})}{\rho} + 2\nu\nabla_r^2\overline{u_A u_B}$$

(4.50)

where the ∇_r operators are defined in the nomenclature (Brodkey, 1967). In terms of correlations, the equation becomes:

$$\frac{\partial Q_{ij}(r)}{\partial t} = T_{ij}(r) - P_{ij}(r) + 2\nu\left[\frac{\partial^2 Q_{ij}(r)}{\partial r_k^2}\right]$$

(4.51)

where $T_{ij}(r)$ and $P_{ij}(r)$ are the triple-velocity correlation and pressure velocity correlation, respectively. This equation collapses to the Reynolds equation when r is zero. Ignoring the separation of points A and B, the equation contains overtly energy units. However, the separation distance between A and B is present which means the units of the equation is not exactly energy. However, the equation contains information to describe the energy behavior or energy "flow" between two points A and B which is brought about by the velocity fluctuations. In a sense, the $Q_{ij}(r)$ equation models how the turbulence energy is spread over different eddy sizes and the transfer mechanisms of the energy exchange through the energy cascade. The last term in both equations represents the energy dissipation rate ϵ, which is where the energy flow eventually is dissipated.

Taking the Fourier transform of the $Q_{ij}(r)$ equation to obtain the equation in the frequency domain or the wave number space, the equation for the energy spectrum $E(k)$ is obtained:

$$\frac{\partial E(k)}{\partial t} = W(k) - 2\nu k^2 E(k)$$

(4.52)

where $W(k)$ is the Fourier transform of $[T_{ij}(r) - P_{ij}(r)]$. The energy spectrum is actually a function of time but is shown here to be independent of time for simplicity. In wave number space, the wave number k can be considered a frequency or reciprocal length. The turbulent velocity fluctuations can then be considered to be distributed over various spectral frequencies. In wave number space, the equation also describes the energy cascade. $W(k)$ is a transfer term which describes the transfer of energy from the large-scale eddies to the small-scale eddies and $2\nu k^2 E(k)$ is the dissipation of this energy by the small eddies.

The energy spectra $E(k)$ and the dissipation spectra $k^2 E(k)$, as shown in Fig. 4.4, are divided into different regions according to wave

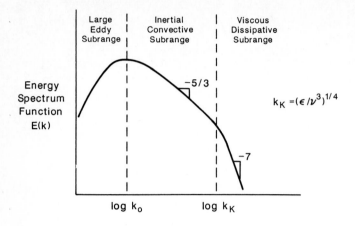

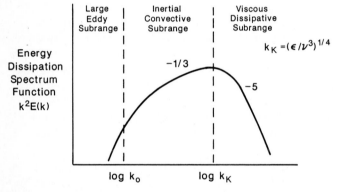

Figure 4.4   Energy and energy dissipation spectra in wave number space.

number k. The large eddy region occurs below k_0 where $E(k)$ is considered a constant or falls to zero as k goes to zero. The inertial convective subrange occurs from k_0 to the wave number associated with the Kolmogorov microscale η. In this region, mean kinetic energy is convectively transferred to the small-scale eddies. $E(k)$ varies with $k^{-5/3}$ under equilibrium conditions. The viscous dissipative region is considered to be where the mechanical energy of the eddies is converted into internal energy of the fluid. The flow is very viscous with a Reynolds number of 1 and $E(k)$ varies with k^{-7} for equilibrium conditions. The energy dissipation spectrum, $k^2 E(k)$, varies with $k^{1/3}$ and k^{-5} for the same subranges.

If the energy spectrum is measured, the energy dissipation rate can then be calculated from the energy spectrum by:

$$\bar{\epsilon} = 2\nu \int_0^\infty k^2 E(k)\, dk \tag{4.53}$$

where the energy dissipation spectrum is $k^2 E(k)$. Further, the velocity fluctuations can be obtained from the energy spectrum as well:

$$\overline{u_A u_B} = \int_0^\infty E(k)\, dk \tag{4.54}$$

The last two relationships come from a property of Fourier transforms in which the original quantity in the time domain is recovered by integration of the Fourier transform. The energy spectrum $E(k)$ is often used in agitated tank studies to calculate the velocity fluctuations if points A and B are assumed to be at the same location and isotropic turbulence is assumed. The energy dissipation spectrum $k^2 E(k)$ is used to calculate the energy dissipation rate in mixing literature studies.

Concentration spectrum.   Concentration fluctuations are considered to be distributed over various frequencies in a concentration spectrum just as the velocity fluctuations are considered to be distributed over various frequencies in the energy spectrum. These fluctuations give rise to concentration spectra and the integral and microscales of concentration. The function $f_c(r)$, similar to $f(r)$, can be defined:

$$f_c(r) = \frac{\overline{c(x)c(x + r)}}{c'(x)c'(x + r)} \tag{4.55}$$

where c' is the root mean squared of the concentration fluctuations and is similar to u'. The concentration fluctuation c' is dependent upon the average concentration and the time of integration to establish the average concentration just as the velocity fluctuation u' was. The integral scale is:

$$L_c = \int_0^\infty f_c(r)\, dr \tag{4.56}$$

The concentration microscale is:

$$\lambda_c = \left[-\frac{1}{\partial^2 f_c(r)/\partial r^2 \big|_{r=0}} \right]^{0.5} \tag{4.57}$$

These equations and the concentration spectra parallel the equations developed for velocity but are less complicated because concentration is a scalar. The purpose of the concentration equation also parallels the purpose of the velocity equations as well. The mean concentration equation is used to calculate the distribution of the mean concentration in the flow. The concentration correlations contain information about how concentration at one point correlates with concentration at another point.

The equation for the concentration spectrum is obtained in the same manner as the equation for the energy spectrum. An equation is written for the concentration fluctuations at point A in the flow. This equation is multiplied by a concentration fluctuation at point B. The reverse is done for point B. The two equations are added and time averaged which results in an equation for the correlation for $\overline{c_A c_B}$. Taking the Fourier transform of this equation results in:

$$\frac{\partial E_c(k)}{\partial t} = W_c(k) - 2D_L k^2 E_c(k) \tag{4.58}$$

where D_L is the molecular mass diffusivity. $E_c(k)$ is the concentration spectrum and $k^2 E_c(k)$ is the concentration dissipation spectrum. These spectra are divided into different regions according to wave number k, as shown in Fig. 4.5 in much the same manner as the energy spectrum given above. The viscous convective, diffusive subrange occurs at the highest wave numbers in the concentration spectrum to the Batchelor wave number of $(\epsilon/\nu D_L^2)^{1/4}$. Owing to the large differences between ν and D_L, the concentration fluctuations extend to much higher wave numbers or to smaller length scales than the velocity fluctuations. This concept encompasses the final diffusional stages of mixing and has important implications to micromixing.

The concentration spectrum can be integrated to obtain the concentration fluctuation as:

$$\overline{c_A c_B} = \int_0^\infty E_c(k)\, dk \tag{4.59}$$

Brodkey (1967) contains a complete discussion of these concepts.

It should also be observed that mixed-mode correlations, e.g., $\overline{u_i c_A}$, are possible. The correlation $\overline{u_i c_A}$ should contain information on turbulent mixing. Unfortunately, for mixing in agitated tanks, $\overline{u_i c_A}$ has not been measured experimentally.

The review of statistical turbulence theory above forms the basis for experimental measurements of turbulent quantities in an agitated tank. Many methods exist which can be used to calculate various statistical turbulence properties as outlined in Brodkey (1967) and Hinze

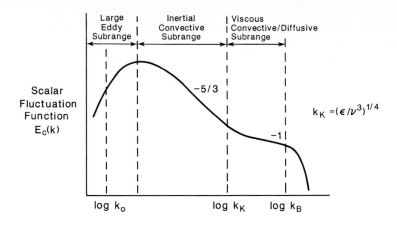

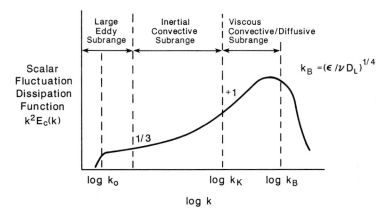

Figure 4.5   Scalar fluctuations and scalar fluctuation dissipation spectra in wave number space.

(1975). Brodkey (1967) and Hinze (1975) should be consulted for a more in-depth review of statistical turbulence theory as well.

Other discussions of turbulence and mixing can be found in Brodkey (1975). Articles dealing with turbulence and scalar mixing available for review at a fairly elementary level are Rushton and Oldshue (1953) and Hughes (1957).

Cylindrical coordinates.   Since the cylindrical geometry is very common to mixing, continuity and various equations of motion are listed here for a constant density Newtonian fluid for the cylindrical coordinate system (r, θ, z) (Hinze, 1975). Ju et al. (1987) also provided these equations in the development of the k-ϵ model described later.

Continuity:

$$\frac{\partial U_r}{\partial r} + \frac{U_r}{r} + \frac{1}{r}\frac{\partial U_\theta}{\partial \theta} + \frac{\partial U_z}{\partial z} = 0 \qquad (4.60)$$

Time-averaged continuity:

$$\frac{\partial \overline{U}_r}{\partial r} + \frac{\overline{U}_r}{r} + \frac{1}{r}\frac{\partial \overline{U}_\theta}{\partial \theta} + \frac{\partial \overline{U}_z}{\partial z} = 0 \qquad (4.61)$$

The instantaneous equations of motion:

The r component:

$$\rho\left(\frac{DU_r}{Dt} - \frac{U_\theta^2}{r}\right) = -\frac{\partial P}{\partial r} + \mu\left(\nabla^2 U_r - \frac{U_r}{r^2} - \frac{2}{r^2}\frac{\partial U_\theta}{\partial \theta}\right) + \Sigma\rho F_r \qquad (4.62)$$

The θ component:

$$\rho\left(\frac{DU_\theta}{Dt} + \frac{U_r U_\theta}{r}\right) = -\frac{1}{r}\frac{\partial P}{\partial \theta} + \mu\left(\nabla^2 U_\theta - \frac{U_\theta}{r^2} + \frac{2}{r^2}\frac{\partial U_\theta}{\partial \theta}\right) + \Sigma\rho F_\theta \qquad (4.63)$$

The z component:

$$\rho\left(\frac{DU_z}{Dt}\right) = -\frac{\partial P}{\partial z} + \mu(\nabla^2 U_z) + \Sigma\rho F_z \qquad (4.64)$$

Time-averaged equations of motion:

The r component:

$$\rho\left(\frac{D\overline{U}_r}{Dt} - \frac{\overline{U}_\theta^2}{r}\right) = -\frac{\partial \overline{P}}{\partial r} + \mu\left(\nabla^2\overline{U}_r - \frac{\overline{U}_r}{r^2} - \frac{2}{r^2}\frac{\partial \overline{U}_\theta}{\partial \theta}\right) - \frac{\rho}{r}\frac{\partial r\overline{u_r^2}}{\partial r} - \rho\frac{\partial \overline{u_r u_\theta}}{r\partial \theta}$$
$$- \rho\frac{\partial \overline{u_r u_z}}{\partial z} + \rho\frac{\overline{u_\theta^2}}{r} + \Sigma\rho\overline{F}_r, \qquad (4.65)$$

The θ component:

$$\rho\left(\frac{D\overline{U}_\theta}{Dt} + \frac{\overline{U}_r\overline{U}_\theta}{r}\right) = -\frac{1}{r}\frac{\partial \overline{P}}{\partial \theta} + \mu\left(\nabla^2\overline{U}_\theta - \frac{\overline{U}_\theta}{r^2} + \frac{2}{r^2}\frac{\partial \overline{U}_r}{\partial \theta}\right) - \rho\frac{\partial \overline{u_r u_\theta}}{\partial r}$$
$$- \frac{\rho}{r}\frac{\partial \overline{u_\theta^2}}{\partial \theta} - \rho\frac{\partial \overline{u_z u_\theta}}{\partial z} - 2\rho\frac{\overline{u_r u_\theta}}{r} + \Sigma\rho\overline{F}_\theta \qquad (4.66)$$

The z component:

$$\rho \frac{D\overline{U}_z}{Dt} = - \frac{\partial \overline{P}}{\partial z} + \mu \nabla^2 \overline{U}_z - \frac{\rho}{r} \frac{\partial \overline{ru_r u_z}}{\partial r} - \frac{\rho}{r} \frac{\partial \overline{u_z u_\theta}}{\partial \theta} - \rho \frac{\partial \overline{u_z^2}}{\partial z} + \Sigma \rho \overline{F}_z \qquad (4.67)$$

For the above equations, the following operators are defined as:

The substantial derivative:

$$\frac{D}{Dt} = \frac{\partial}{\partial t} + U_r \frac{\partial}{\partial r} + \frac{U_\theta}{r} \frac{\partial}{\partial \theta} + U_z \frac{\partial}{\partial z} \qquad (4.68)$$

The Laplacian operator:

$$\nabla^2 = \frac{\partial^2}{\partial r^2} + \frac{\partial}{r \, \partial r} + \frac{1}{r^2} \frac{\partial^2}{\partial \theta^2} + \frac{\partial^2}{\partial z^2} \qquad (4.69)$$

The substantial derivative contains either instantaneous or average velocities in the derivative depending upon the equation.

The total kinetic energy equation, the turbulent kinetic energy equation, and scalar equations in cylindrical coordinates can also be obtained.

The k-ϵ model

Turbulent motions are extremely difficult to predict theoretically, and experimentation is increasingly more expensive. As a result, computational methods to estimate turbulent quantities are increasing in popularity. However, the exact calculations which describe turbulent motion cannot be performed at the present time primarily because of the presence of the large range of time-dependent length scales in the flow which require description for absolute accuracy.

Generally, turbulence models relate the turbulent stresses $\overline{u_i u_j}$ and the average velocity gradient $d\overline{U}/dy$ by simple relationships as reviewed by Brodkey (1967) and Rodi (1984). Of the more recent models, the k-ϵ model has received acceptance in engineering calculations and in mixing. The model is simply an eddy viscosity model based upon the turbulent kinetic energy and a length-scale equation. However, validation of any model in application has to be accomplished before any trust can be placed upon the modeling results. The k-ϵ model has limitations, some of which have been documented, some of which have not.

In most cases of engineering relevance, the description of all turbulent quantities is not necessary. In some cases, only the mean velocity and scalar fields are of interest. Such information can be obtained from the time-averaged equations of motion if a model is used to calculate the Reynolds stresses $\overline{u_i u_j}$ and turbulent scalar fluxes $\overline{u_i \gamma}$. For modeling flows in a turbulent agitated tank, the turbulent Reynolds stresses and scalar fluxes are important.

Eddy viscosity is the oldest form of a relationship between turbulent stresses and the mean deformation rate. The relationship (Rodi, 1984) can be written as:

$$- \overline{u_i u_j} = v_t \left(\frac{\partial \overline{U_i}}{\partial x_j} + \frac{\partial \overline{U_j}}{\partial x_i} \right) - \frac{2}{3} k \delta_{ij} \qquad (4.70)$$

where v_t is the turbulent kinematic eddy viscosity or μ_t/ρ and the $2/3 k \delta_{ij}$ term is added so that continuity and the definition of turbulent kinetic energy are satisfied. The eddy viscosity varies throughout the flow field as the Reynolds stresses, the mean velocity gradient, and the turbulent kinetic energy k vary. If the eddy viscosity is known as a result, then closure to the equations of motion and continuity can be obtained.

The units of kinematic eddy viscosity are length squared per time which indicates a possible relationship between eddy viscosity and a product of a velocity scale and a length scale or:

$$v_t \propto \hat{V} L \qquad (4.71)$$

If such a relationship holds, then the eddy viscosity distribution can be calculated from the distributions of the length and velocity scales.

The problem is one of determining the eddy viscosity distribution in the flow. A reasonable candidate for the velocity scale is the square root of the turbulent kinetic energy $k^{1/2}$. The length scale can be obtained from the energy dissipation rate ϵ and the turbulent kinetic energy k as:

$$L \propto \frac{k^{3/2}}{\epsilon} \qquad (4.72)$$

As a result, the k-ϵ model involves two equations, one to model the transport of turbulent kinetic energy and the other to model the transport of the energy dissipation.

A transport equation to determine k can be written in the form as given above:

$$\frac{D\overline{k}}{Dt} = - \frac{\partial}{\partial x_j} \overline{u_j \left(\frac{p}{\rho} + k \right)} - \overline{u_i u_j} \frac{\partial}{\partial x_j} (\overline{U_i}) + v \frac{\partial}{\partial x_j} \overline{u_i \left(\frac{\partial u_i}{\partial x_j} + \frac{\partial u_j}{\partial x_i} \right)}$$

$$\text{I} \qquad\qquad \text{II} \qquad\qquad \text{III} \qquad\qquad \text{IV}$$

$$- v \overline{\left(\frac{\partial u_i}{\partial x_j} + \frac{\partial u_j}{\partial x_i} \right) \frac{\partial u_i}{\partial x_j}} \qquad (4.73)$$

$$\text{V}$$

The rate of change of k in term I is determined by convective transport due to the mean motion in term I, its diffusive transport due to velocity and pressure fluctuations in term II, the rate of its production by the Reynolds stresses and the mean velocity gradients in term III, the work by viscous shear stresses of the turbulent motions in term IV, and its rate of dissipation in term V. The production describes the transfer of the kinetic energy from the mean flow to the turbulent motion. The viscous dissipation term describes the transfer of kinetic energy into internal energy.

The terms in the k equation are quite complex involving a number of unknown correlations. To be useful, simplifications and further models are necessary. Term IV is typically dropped as being unimportant in comparison with the other terms. The correlations in term II are modeled by a diffusion mechanism using the gradient of k and the eddy viscosity.

$$\overline{u_j\left(\frac{p}{\rho} + k\right)} = \left(\frac{\nu_t}{\sigma_k}\right)\left(\frac{\partial k}{\partial x_j}\right) \tag{4.74}$$

The Reynolds stresses in term III are replaced by the eddy viscosity and the mean velocity gradients as shown in Eq. (4.70). Term V is simply modeled as energy dissipation ϵ:

$$\overline{\nu\left(\frac{\partial u_i}{\partial x_j} + \frac{\partial u_j}{\partial x_i}\right)\frac{\partial u_i}{\partial x_j}} = \epsilon \tag{4.75}$$

Using these models, the final k equation becomes:

$$\frac{Dk}{Dt} = \frac{\partial}{\partial x_j}\left(\frac{\nu_t}{\sigma_k}\frac{\partial k}{\partial x_j}\right) + \left[\nu_t\left(\frac{\partial \overline{U_j}}{\partial x_i} + \frac{\partial \overline{U_i}}{\partial x_j}\right) - \frac{2}{3}k\delta_{ij}\right]\frac{\partial \overline{U_j}}{\partial x_i} - \epsilon \tag{4.76}$$

The transport equation of length scale can be written in the usual form of a transport equation or a general property balance. The terms include the change of the length scale with time, the convection of the length scale, the diffusion of the length scale, and production and dissipation of the length scale. This may appear as an abstraction; however, length scales are produced from larger scales and length scales are dissipated as eddies are dissipated. Vortex stretching produces smaller length scales, and length scales are convected and diffused in the same sense as eddies are convected and diffused.

Many length-scale equations exist in the literature as noted by Rodi (1984). However, we should note that a length-scale equation does not necessarily have to have length scale as the dependent variable. For example, with ϵ and k, the length scale L can be calculated from $k^{3/2}/\epsilon$, as in Eq. (4.72).

The energy dissipation equation for ϵ is essentially a length-scale equation as a result and makes up the second equation for the eddy viscosity. The ϵ equation, again with many models being assumed, can appear as:

$$\frac{D\epsilon}{Dt} = \frac{\partial}{\partial x_j}\left(\frac{\nu_t}{\sigma_\epsilon}\frac{\partial \epsilon}{\partial x_j}\right) + c_{1\epsilon}\frac{\epsilon}{k}\left[\nu_t\left(\frac{\partial \overline{U}_i}{\partial x_j} + \frac{\partial \overline{U}_j}{\partial x_i}\right) - \frac{2}{3}k\delta_{ij}\right]\frac{\partial \overline{U}_j}{\partial x_i} - c_{2\epsilon}\frac{\epsilon^2}{k} \quad (4.77)$$

From the calculations of k and ϵ using the above equations, the eddy viscosity (Rodi, 1984) is calculated as:

$$\nu_t = \frac{c_\mu(k^{1/2})(k^{3/2})}{\epsilon} = \frac{c_\mu k^2}{\epsilon} \quad (4.78)$$

The various constants in the equations have been assigned the following values by Rodi (1984):

$$c_\mu = 0.09, c_{1\epsilon} = 1.44, c_{2\epsilon} = 1.92, \sigma_k = 1.0, \sigma_\epsilon = 1.3$$

The k-ϵ model provides closure of the time-averaged equations of motion. The eddy viscosity is determined for a particular flow by solving the equations for the turbulent kinetic energy k and the energy dissipation ϵ. k and ϵ are then used to calculate the eddy viscosity which is simultaneously used in the mean flow equations to obtain the spatial mean flow fields. The time-averaged equations of motion are closed in the calculation, and the eddy viscosity is determined as a function of spatial location in the flow.

The mean velocity flow fields are important in mixing since the velocity fields determine the displacement of material necessary to model mixing.

Eddy diffusivity can also be introduced for scalar transport modeling by assuming:

$$- \overline{u_i\gamma} = \frac{\vartheta_t\partial \overline{\Gamma}}{\partial x_i} \quad (4.79)$$

where ϑ_t is the eddy diffusivity of the scalar. Eddy diffusivities can be related back to eddy viscosity by a simple relationship such as:

$$\vartheta_t = \frac{\nu_t}{\sigma_t} \quad (4.80)$$

where σ_t is the Prandtl or Schmidt number for heat or mass transport. If the eddy viscosity is calculated, then the eddy diffusivities can be obtained and scalar transport can be modeled.

Overall, the k-ϵ model in its simplest form calculates a turbulent velocity scale in the form of k, a turbulent length scale, and one value for eddy viscosity independent of direction. The k-ϵ model is simple and fairly inexpensive to use. It has been successful in treating shear and recirculating flows. The time-averaged equations of motion with the k-ϵ model can provide only the mean velocity profiles and the distribution of the turbulent kinetic energy k and the energy dissipation rate ϵ spatially in an agitated tank. Such models do not provide accurate calculation of the Reynolds stresses.

The application of the k-ϵ models for agitated tanks is relatively new. Platzer (1981), Harvey and Greaves (1982a; 1982b), Middleton et al. (1986), Placek et al. (1986), Chen and Wood (1986), and Ju et al. (1987) have calculated the flow fields in agitated tanks using the k-ϵ type models. Of these, Ju et al. (1987) appears to be the most successful using a nonisotropic, three-dimensional k-ϵ model to simulate the flow in an agitated tank.

For the most part the k-ϵ models have attempted to duplicate the flow in a batch mixing tank. Little if any modeling of a continuous flow system has been attempted. The treatment of the impeller and the three dimensionality of the flow, boundary, and initial conditions, and inlet and outlet conditions for continuous flow systems are major difficulties in the application of k-ϵ modeling of turbulent agitated tanks.

Structural Turbulence

Basis of structural turbulence

Time-averaged equations of motion are approximations to the actual physics of the flow phenomena. As such, the approach may not provide an accurate representation of the mixing process. Coherent structures are present in the flow, which, in the time-averaging process, may be averaged away. In the averaging process, too, the important physics may not be properly represented. An example of a flow sequence showing the presence of coherent structures (Praturi and Brodkey, 1978) is given in Fig. 4.6. Such a sequence may actually perform most of the mixing.

An approach which characterizes the coherent structures is necessary if the physics of the mixing is to be accurately characterized. This requires the three-dimensional, time-dependent computations of the turbulent flow field. Unfortunately, the range of turbulent scales which exist in the flow is very large, 10^3, which would involve computations involving 10^9 points for three-dimensional computational studies. Analytical solutions are not possible.

Large-scale motions are present in the flow field and are very de-

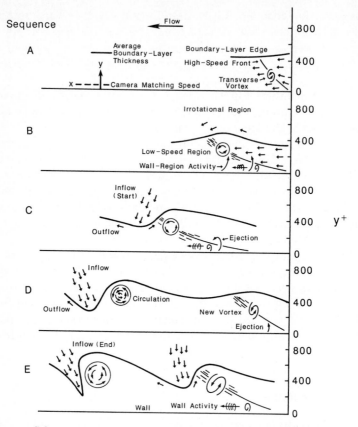

Figure 4.6   Coherent structures in turbulent shear flow.   *(From A. K. Praturi and R. S. Brodkey, J. of Fluid Mech., 89, Part 2, 251, 1978. By permission.)*

pendent upon the general broad nature of flow and the overall flow geometry. The largest-scale eddies dominate the general physics of turbulent flow and interact strongly with mean flow. However, the smallest scales at high Reynolds number are considered to be independent of the gross flow behavior and independent of the large-scale coherent structures. For example, the smallest eddies have much smaller length and time scales than the large-scale eddies. As such, the smallest eddies are considered universal and perform most of the energy dissipation. Due to these different dependencies, a dual approach is possible in which the large-scale structures are actually computed and the small-scale turbulence is modeled.

Computation of time-dependent,
three-dimensional turbulence

The approach requires computations on the computer in which a spatial grid is used which is much smaller than k-ϵ model grids. The time-dependent behavior of the large-scale eddies is calculated as the structures move across the spatial grid. The scale of resolution for the eddies is determined by the size of the spatial averaging operator. Typically, this is approximately one grid space since the size of the spatial operator is selected to be the same size as the grid spacing. Any turbulent behavior occurring below this grid length is modeled. The approach has been named large eddy simulation (LES) with subgrid scale (SGS) modeling of the small-scale turbulence.

The advantage of such modeling is that the actual transport of properties is computed. Such modeling is less sensitive to model inaccuracies and offers a more universal approach to turbulent flow modeling. Disadvantages include higher computational costs as a result of the large arrays needed in the calculations.

The equations of motions are not time averaged but, instead, are averaged over each discrete grid volume. Spatially averaged equations are written which are a result of a spatial averaging operator. This operator is essentially a low-pass filter which removes or filters out the small-scale behavior. The small-scale turbulence is the only portion of the flow that is modeled. The upper bound of the small scale is the size of the spatial averaging operator. This limits the importance and the effects of the averaging process.

The objective of LES is to calculate the time-dependent, three-dimensional flow behavior above the size of the filter. Typically, the filter width or size Δ is selected such that the corresponding wave number k, i.e., $k = 2\pi/2\Delta$, lies in the inertial convective subrange of the energy spectrum. In this range, energy transfer occurs between the large eddies and small eddies with no turbulence production or dissipation occurring. The turbulence above this wave number is modeled whereas the turbulence below this wave number is computed. The requirement for closure still exists, however, and is accomplished in the modeling of the turbulence above the filter wave number, $2\pi/2\Delta$.

Turbulent flows which do not have an inertial convective subrange are typically low Reynolds number flows where the difference between length scales of the largest eddies and the smallest eddies is small. For such flow, it may be possible to compute the flow field directly using the instantaneous Navier Stokes equations (Kawamura and Kuwahara, 1985).

The spatial averaging process (Leonard, 1974) depends upon the filter used to average the equations of motion. Leonard defined a flow variable $A(x)$, e.g., flow velocity component, which can contain large- and small-scale components. The filtered or large-scale components of $A(x)$ can be defined as:

$$\overline{A(x)} = \int G(x - x')A(x') \, dx' \qquad (4.81)$$

where $G(x - x')$ is the spatial filter and $\overline{A(x)}$ is the space-averaged component of $A(x)$ which contains all scales above the size of the filter. The overbar indicates spatial averaging. $A(x)$ is then:

$$A(x) = \overline{A(x)} + a' \qquad (4.82)$$

where a' is referred to as the residual component or the unresolved component of $A(x)$ which requires modeling. Spatially filtered velocities and their residuals are shown in Fig. 4.7. A particularly useful filter is the gaussian filter, defined as:

$$G(x - x') = \frac{\gamma^{3/2}}{\pi \Delta^3} e^{-\gamma(x-x')^2/\Delta^2} \qquad (4.83)$$

where γ is a constant, typically selected as 6, and Δ is the size of the filter. The gaussian filter is shown in Fig. 4.8.

The gaussian filter has several advantages over other types of filters (Kwak et al., 1975) and is most often used. The gaussian filter misses only a small fraction of the large-scale behavior and places most of the small-scale motions in the so-called residual field for modeling. The size of the filter is typically selected to be the size of the grid increment.

Kwak et al. (1975) noted several filtering rules useful in reducing the equations of motion to space-averaged equations. First, a filtered partial derivative of a variable is equal to the partial derivative of the filtered variable. For spatial derivatives:

$$\overline{\frac{\partial A}{\partial x}} = \frac{\partial \overline{A}}{\partial x} \qquad (4.84)$$

and for time derivatives:

$$\overline{\frac{\partial A}{\partial t}} = \frac{\partial \overline{A}}{\partial t} \qquad (4.85)$$

Upon application of this rule, continuity and the Navier Stokes equations become:

$$\frac{\partial \overline{U_i}}{\partial x_i} = 0 \qquad (4.86)$$

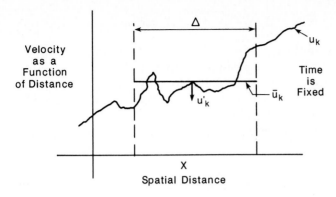

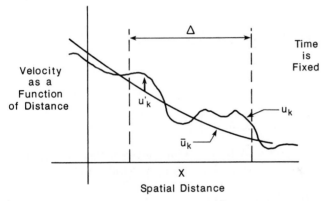

Figure 4.7 Spatially filtered velocities $\bar{u}_k$ and subgrid scale residuals u_k' as a result of filtering.

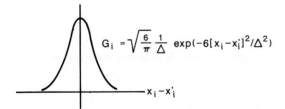

Figure 4.8 The gaussian filter.

$$\frac{\partial \overline{U}_i}{\partial t} + \frac{\partial \overline{U_i U_j}}{\partial x_j} = -\frac{1}{\rho}\frac{\partial \bar{p}}{\partial x_i} + \nu \nabla^2 \overline{U}_i \qquad (4.87)$$

Another filtering rule is that a filtered product of variables results in:

$$\overline{A_iA_j} = \overline{\overline{A_i}\,\overline{A_j}} + \overline{\overline{A_i}a_j'} + \overline{a_i'\overline{A_j}} + \overline{a_i'a_j'} \tag{4.88}$$

None of the terms in the equation can be considered zero, and only the first term on the right half side can be computed. The remaining three terms involve the small-scale unresolvable components of the A_i field. The a_i' terms must be modeled using subgrid scale models, i.e., models for turbulence below the size of the filter.

For the U_iU_j terms in the equations of motion, the filtered product results in:

$$\overline{U_iU_j} = \overline{\overline{U_i}\,\overline{U_j}} + \overline{\overline{U_i}u_j'} + \overline{u_i'\overline{U_j}} + \overline{u_i'u_j'} \tag{4.89}$$

where the last three terms on the right half side are called the residual stresses R_{ij} when multiplied by density. For scalar equations, a similar relationship is:

$$\overline{U_iC_j} = \overline{\overline{U_i}\,\overline{C_j}} + \overline{\overline{U_i}c_j'} + \overline{u_i'\overline{C_j}} + \overline{u_i'c_j'} \tag{4.90}$$

where some model is used to account for the last three terms.

Generally, however:

$$\overline{\overline{A_i}\,\overline{A_j}} \neq \overline{A_i}\,\overline{A_j} \tag{4.91}$$

In order to treat the term $\overline{A}\,\overline{B}$ a Taylor series expansion with a gaussian filter can result in the following approximation:

$$\overline{\overline{U_i}\,\overline{U_j}} = \overline{U_i}\overline{U_j} + \frac{\Delta^2}{4\gamma}\nabla^2(\overline{U_i}\overline{U_j}) + 0(\Delta^4) \tag{4.92}$$

where the difference between $\overline{U_i}\,\overline{U_j}$ and $\overline{U_i}\,\overline{U_j}$ is referred to as the Leonard stresses.

The residual stress terms R_{ij} can be treated in several ways. For example, an equation can be written for these in much the same way as was done for the Reynolds stresses except that space averaging is performed in the procedure rather than time averaging. Models for the residual stress terms R_{ij} have also been developed. An obvious model is an eddy viscosity model where R_{ij} (Smagorinsky, 1963) is:

$$R_{ij} = -\nu_t\left(\frac{\partial \overline{U_i}}{\partial x_j} + \frac{\partial \overline{U_j}}{\partial x_i}\right) + \frac{1}{3}R_{kk}\delta_{ij} \tag{4.93}$$

Using an eddy viscosity for the model and dropping all higher-order terms greater than Δ^2, the final space-averaged equation of motion is then:

$$\frac{\partial \overline{U}_i}{\partial t} + \frac{\partial}{\partial x_j}\left[\overline{U}_i\,\overline{U}_j + \frac{\Delta^2}{24}\,\nabla^2(\overline{U}_i\,\overline{U}_j) - \nu_t\left(\frac{\partial \overline{U}_i}{\partial x_j} + \frac{\partial \overline{U}_j}{\partial x_i}\right)\right] = -\frac{1}{\rho}\frac{\partial P}{\partial x_i} + \nu\nabla^2\overline{U}_i$$

(4.94)

where γ was selected as 6 and P is $\overline{p}/\rho + 1/3R_{ii}$.

The space-averaged equations of motion explicitly contain the filter size and the eddy viscosity which must still require a model.

Smagorinsky (1963), Deardorff (1970), Clark et al. (1979), and Ferziger (1983) provide additional reading in this area. More advanced articles include Leslie and Quarini (1979) and Moin and Kim (1982).

Spaced-averaged equations of motion have not as yet been applied to model flows in an agitated tank.

Flow Fields in Agitated Tanks

The different types of flow pattern and velocity studies in mixing become apparent during classification. Gross-scale studies provide a general picture of the flow behavior whereas other studies provide a more detailed analysis of the flow behavior. The following will review these detailed studies. Gross-scale studies will be reviewed in a later section under circulation and mixing times.

Detailed studies, for the most part, have relied heavily upon the time-averaged statistical turbulence concepts. Statistical quantities such as velocity profiles, turbulence intensities, energy spectra, various length scales, and energy dissipation rates are reported. Turbulence intensities are reported as a percent of the average velocity or the average velocity component, and all fluctuations are root mean square data. Structural studies are also becoming more prevalent in the literature as periodicity and trailing vortex systems have been identified. Velocity profiles, eddy viscosity, and shear rates in such structures have been determined to some degree although work is continuing. Disagreements between the various studies are very obvious.

Nomenclature is given at the end of this chapter; however, the following symbols are often used in this section and are repeated here for clarity: D is the impeller diameter, T is the tank diameter, N is impeller rotational speed, W is the blade width, C is the off-bottom clearance of the impeller, B is baffle width, and H is the liquid height. In addition, ϵ is the energy dissipation, and V_t is the impeller tip velocity.

Standard geometry.   In the following, reference to the standard configuration is made. This geometry has a cylindrical tank with: (1) the impeller diameter being about ¼ to ½ of the tank diameter, (2) the impeller positioned from ¼ to ½ of a tank diameter from the bottom or one impeller diameter from the tank bottom, (3) the liquid height in the tank equal to one tank diameter, (4) four symmetrically placed wall baffles, $B = T/10$ or $T/12$, and (5) impeller blade width to impeller diameter W/D equal to ⅙. The typical impeller is the disk style turbine with six blades, commonly referred to as the Rushton turbine. The standard geometry is discussed in Chap. 1.

Primary flow patterns depend upon the impeller geometry and the presence of baffles. Impeller clearance, liquid height, blade pitch, and baffle length are also important factors. However, the effects of liquid height and baffle length have not been studied extensively. Variations in impeller blade number and width, baffle width, and the tank geometry, e.g., dished or flat bottom, are most often not significant and do not affect gross flow patterns. Usually studies do not mention such effects. Generally, too, flow patterns can be considered independent of viscosity in turbulent flow.

Impeller discharge flows in baffled tanks

Knowledge of the discharge flow is important in the modeling of the flows in an agitated tank. The different types of impellers have been discussed in Chap. 1. Without a doubt, though, the disk style turbine or the Rushton impeller is the most studied impeller. Some studies have been performed on the discharge flow of the pitched blade turbine, propellers, and open flat blade impeller. However, these impellers have not yet received the same level of study as the Rushton impeller. The reasons for this are unclear.

The following discussion is divided into a review of the published literature on the discharge flows of the various impellers. The studies are reviewed in chronological order in most cases. Experimental techniques are also noted. Data are usually reported according to fixed laboratory coordinates and were obtained in a standard baffled tank geometry unless otherwise stated.

The disk style impeller. The important flow systems in the discharge of a disk style turbine originate from behind the blade in the blade's wake region. The specific flow systems near the blade in this region are shown in Fig. 4.1 and consist of a high-velocity flow along the centerline of the blade and the two vortex systems, one along the top of the blade, the other along the bottom of the blade. Since these systems are discharged continuously and since the flows from in front of the blade are much less intense, a stationary measurement probe records a periodicity. Velocity data depend upon the periodicity. Further away from the blade, the organized vortices and high-speed centerline flows are disrupted, causing a significant increase in turbulent velocity fluctuations. The vortex systems and the centerline flow are discharged from the blade at different angles in the $r\theta$ plane. Mixing can be considered to occur in two areas: (1) the flow passing through the vortex systems of the disk style impeller and (2) the area where the discharge flow is disrupted by the flow in the bulk of the tank. In a sense, the discharge flow slams into the relatively stationary flow in the bulk of the tank. For a considerable portion of the literature, the flow systems, as described above, were ignored or not recognized at the time of the study.

Among the earliest studies, Sachs and Rushton (1954) used streaks recorded on photographic film formed by neutrally bouyant immiscible liquid tracer drops to study the flow field of a four-blade turbine. From the length of the streaks and the time of exposure, velocity profiles were obtained in the discharge flow for this turbine. Unfortunately the data collection rate for the experimental technique was low and the reported velocity profile data did not appear to obey continuity. However, the data obtained in such a manner are spatial data, and such experimentation is difficult, given the tediousness of the data collection.

Nagata et al. (1960) found that the product of the radial position r and the radial velocity at that position was equal to a constant for the impeller stream of the disk style turbine. This follows from continuity with U_θ and U_z being considered zero.

Cutter (1966) considered power per unit volume a useful criterion in judging mixing performance. To obtain local energy dissipation rates, i.e., local power per unit volume, mean velocities were determined from streak photographs of lycopodium particles in the flow field of a disk style turbine in the standard geometry. A Kiel impact probe, which measures dynamic pressure, was used to determine velocities as well. The radial and tangential velocities, taken at different rotational speeds, were found proportional to impeller tip speed which demonstrated the similarity of velocity flow field when scaled on impeller tip speed. Near the impeller, both the radial and tangential velocities were of the same magnitude, but the tangential velocities fell off much more rapidly with radial distance. The turbulence fluctuations were on the order of 13 to 22 percent. The radial correlation coefficient $f(x)$ was obtained and found independent of impeller speed. Radial and tangential integral length scales were found to be on the order of the blade width. The radial scale increased with radial distance from the impeller. Angular momentum of the discharge flow was also calculated from the velocity data.

Cutter considered that the flow visualization provided information only in the lower frequency range of the energy spectrum, to eddies containing the bulk of the turbulent kinetic energy. Cutter found that it was not possible to measure turbulent velocity gradients to obtain reliable estimates of the energy dissipation rate. Such data were obtained but were found to provide estimates much smaller than the actual dissipation rates. Two indirect methods were employed to estimate the energy dissipation rate instead.

The energy dissipated in the impeller was found to be 20 to 25 percent of the total input. Of the energy dissipation, 35 to 50 percent occurred in the impeller stream, and 25 to 40 percent in the bulk of the tank. The local dissipation rates at the tip of the impeller were found to be 70 times the average dissipation rate; at the wall in the impeller stream, 3.5 times the average and, in the bulk of the tank, 0.26 times the average.

Mujumdar et al. (1970) reported turbulence parameters in an agitated tank using a disk style turbine in more or less the standard geometry, D/T = 0.33 and 0.4, H/T = 1.33, and C/T = 0.47. The process fluid was air. Mean and turbulent flow velocities were measured in the horizontal plane at the center of the impeller discharge using hot-wire anemometry. The mean velocities decreased inversely proportional to the distance from the impeller tip, and the ratio of the mean velocities to impeller tip speed was found to be independent of impeller rotational speed. A local maximum was found in the rms velocity fluctuations at some distance from the impeller; beyond this location the fluctuations decreased as the wall was approached. The energy spectra showed a sharp peak at the blade passing frequency, and the −5/3 region was not observed in the spectra taken near the impeller as shown in Fig. 4.9. The energy dissipation spectra exhibited a peak at the blade passing frequency as shown in Fig. 4.10. The rate of energy dissipation increased with distance from the impeller and then decreased as the wall was approached.

Cutter had observed a periodic velocity component in the impeller discharge. Mujumdar et al. also noted such a periodicity in the flow as well and considered the presence of this periodic component as giving rise to the peaks in the energy and dissipation spectra. Mujumdar et al. corrected the rms velocity fluctuations for this periodicity to obtain true turbulence as shown in Fig. 4.11. However, the periodic component was found to make up 80 to 90 percent of the turbulence intensity close to the impeller blade but decayed quickly to a negligible amount by about 0.66 of the tank radius. The turbulent intensities increased from about 5 to 10 percent near the impeller to about 35 percent away from the impeller. Macroscales were also determined to be on the order of the blade width, and dissipation scales were on the order of 1/5 to 1/10 the size of the macroscales.

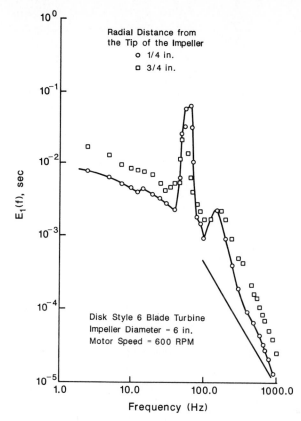

Figure 4.9   One-dimensional frequency spectrum of the radial velocity fluctuations. *(From A. S. Mujumdar, B. Huang, D. Wolf, M. E. Weber, and W. J. M. Douglas, Canad. J. Chem. Eng., 48, 475, 1970. By permission.)*

DeSouza and Pike (1972) modeled the impeller discharge as a free jet and an impinging wall jet. The impeller was a disk style turbine in roughly the standard configuration. The mean velocity profile for the jet as a function of radial and axial position (r, z) was given as:

$$\overline{V}(z, r) = \frac{A}{2}\left(\frac{\sigma}{r}\right)^{1/2} \frac{1}{(r^2 - a^2)^{0.25}}\left[1 - \tanh^2\left(\sigma\frac{z - z_0}{r}\right)\right] \quad (4.95)$$

where z_0 is the distance from the impeller centerline to the centerline of the jet. $\overline{V}(z, r)$ is the resultant velocity of V_r and V_θ. Typically, z_0 was found to be very small. The other parameters are a volumetric flow parameter A, the jet half width parameter σ, and the radius of the source parameter a, which was found independent of impeller size.

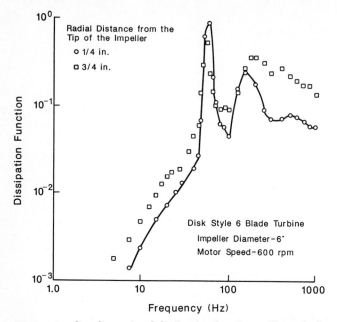

Figure 4.10 One-dimensional dissipation function. *(From A. S. Mujumdar, B. Huang, D. Wolf, M. E. Weber, and W. J. M. Douglas, Canad. J. Chem. Eng., 48, 475, 1970. By permission.)*

Correlations for these parameters were given in work by DeSouza and Pike (1972) and by Ju et al. (1987) as:

$$\sigma = 12.62 \tag{4.96}$$

$$a = 0.08354 \left(\frac{T - D}{T} \right)^{-1.7281} \tag{4.97}$$

and
$$A = 1.1436 \left\{ \frac{ND^3}{[(D/2)^2 - a^2]^{0.25}} \right\}^{0.8337} \tag{4.98}$$

Equations for eddy viscosity and the radial and tangential velocities were also given by DeSouza and Pike.

Fort et al. (1972) and Kratky et al. (1974) modeled the discharge flow profiles from the disk style turbine with a parabolic equation and determined pumping capacities from the curve fit. Velocity data were obtained using a five-hole Pitot tube. The discharge flow was found to be maximum at the center of the blade and decreased rapidly at the top and bottom edges of the blade. The width of the discharge flow was much less than the width of the impeller blade (Fig. 4.1). The maximum discharge velocity was equal to about the impeller tip velocity V_t. The velocity profiles, normalized using impeller tip velocity, were

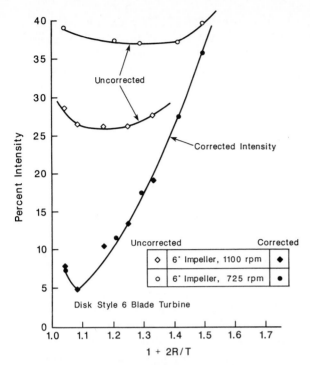

Figure 4.11 Comparison of relative turbulence intensity corrected for periodicity with the uncorrected intensities. R is the radial distance from the tip of the impeller. *(From A. S. Mujumdar, B. Huang, D. Wolf, M. E. Weber, and W. J. M. Douglas, Canad. J. Chem. Eng., 48, 475, 1970. By permission.)*

found independent of impeller rotational speed but were dependent upon the size of the impeller. Asymmetry was observed in the profiles due to the impeller clearance used in the study. The angle of the discharge flow in the $r\theta$ plane was low along the centerline of the blade being roughly 40° from the radial direction. The major portion of the discharge flow went in this direction. Along the top and bottom of the blade, the flow was not as strong and was more tangential in its direction, approaching angles of 80° from the radial direction. DeSouza and Pike (1972) noted a similar angle dependency in the discharge flow as well. The axial velocities were found to be on the order of $0.2V_t$, and the axial spread angle of the discharge flow at the top and bottom of the blade was about 20° from the horizontal. The results indicated nonuniform behavior across the discharge flow. Fort et al. also took exception to the use of air as the working fluid to study liquid mixing behavior in agitated tanks.

Gunkel and Weber (1975) investigated the discharge flow of a disk style turbine using air as the working fluid. They considered the impeller discharge stream to be a jet with a decreasing centerline velocity and expanding entrainment. Close to the impeller, the centerline velocity was on the order of 0.8 times the impeller tip velocity. Turbulence fluctuations rose to about 30 percent of the mean. A substantial portion of these fluctuations was periodic. Far from the impeller, the periodic contribution decreased to zero leaving only random turbulence. Close to the impeller, a peak was observed in the energy spectra at the blade passing frequency which was attributed to vortex shedding. The energy spectra showed the $-5/3$ and -7 regions. Gunkel and Weber considered that the energy dissipation was small in the impeller volume and impeller discharge flows, being about 38 percent of the total, and that 62 percent of the dissipation occurred in the bulk of the tank.

Komasawa et al. (1974a; 1974b), using high-speed cine photography, found turbulence intensities on the order of 30 to 50 percent in the impeller discharge stream of the disk style turbine. The turbulence intensities tended to increase with radial distance, but the energy dissipation was found to decrease with radial position. The flow had to be considered three dimensional.

Ito et al. (1974; 1975) described a multiple electrode spherical probe used for the measurement of three-dimensional, unsteady velocities. Using a tank in the standard configuration with a disk style turbine, they obtained mean velocity data, turbulence intensities, distributions of the Reynolds stresses shown in Fig. 4.12 and triple-point velocity correlations shown in Fig. 4.13. The $\overline{u_r u_\theta}$ Reynolds stress was much larger than the other stresses. The $\overline{u_r u_\theta}$ and $\overline{u_r u_z}$ stresses changed signs vertically from the blade centerline as shown in Fig. 4.12. This sign change is indicative of the trailing vortex systems shed from the impeller. A turbulent energy budget, following the turbulent energy equation given above, was also performed on the impeller discharge stream which showed the convection and production in the impeller stream were much higher than the diffusion and dissipation of the turbulent kinetic energy as shown in Fig. 4.14. The pressure diffusion term was found by differencing to be large. The maximum rate of turbulent energy transfer occurred at the boundary between the impeller stream and surrounding flow. Overall, the discharge flow was shown to be highly nonisotropic and heterogeneous.

Reed et al. (1977), using laser Doppler anemometry, obtained a set of fairly complete data of the average and rms turbulent velocities for an agitated tank in the standard configuration using a disk style turbine with a clearance ratio of 0.33. Additional flow visualization supported the results. In the impeller discharge stream, the radial profiles were not found symmetric along the radial centerline of the

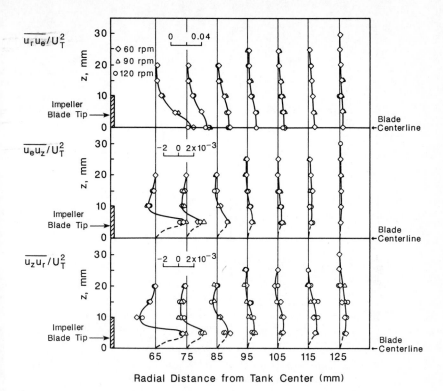

Figure 4.12   One-point double correlation as a function of position.   *(From S. Ito, K. Ogawa, and N. Yoshida, J. Chem. Eng. Japan, 8, 206, 1975. By permission.)*

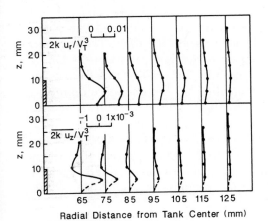

Figure 4.13   One-point triple correlation as a function of position. *(From S. Ito, K. Ogawa, and N. Yoshida, J. Chem. Eng. Japan, 8, 206, 1975. By permission.)*

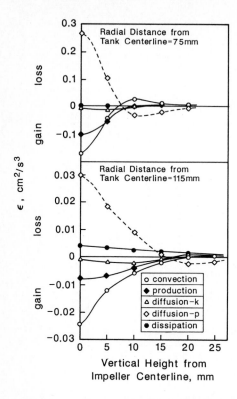

Figure 4.14 Turbulent energy balance. *(From S. Ito, K. Ogawa, and N. Yoshida, J. Chem. Eng. Japan, 8, 206, 1975. By permission.)*

impeller. The tip speed of the impeller was 3.35 m/s, and the maximum observed radial velocity was 2.6 m/s near the impeller. No angular dependence was noted in the radial profiles. Maximum tangential velocity was about 0.4 m/s. The axial velocity profiles showed little or no axial velocities occurring near the impeller, but, close to the wall, the axial velocities increased substantially and had a large angular dependency because of the presence of baffles. The turbulent velocity fluctuations were considered by Reed et al. as being the source of turbulent diffusion in mixing. The radial rms fluctuations were also asymmetric near the impeller but became more uniform as the wall was approached. Both the radial and tangential turbulent intensities typically reached 100 percent of the mean velocity values and 200 to 300 percent behind or in front of the baffles. Intensities of 1000 percent were also noted. Such high values indicate that the assumption of isotropic turbulence cannot be assumed for agitated tanks and that a mean velocity is either meaningless or has not been defined appropriately.

In a study of real and pseudoturbulence (i.e., turbulence with periodicity) in the discharge stream, van't Riet and Smith (1976) noted that the strong periodicity in the velocity near the blade was caused by trailing vortices being shed by the impeller. Typically, the energy spectra of such flows were not consistent with that proposed from isotropic turbulence theories. For a stationary probe, pseudoturbulent signals were generated by the convective velocity of the vortices and circumferential and axial velocities inside the trailing vortices. The circumferential velocity was on the order of the stirrer tip speed and gave rise to fluctuations of this magnitude. The location of the trailing vortices from blade passage to blade passage was not fixed, which caused further difficulty in the interpretation of the turbulent signal. The vortices were observed to exist all the way to the wall. The presence of trailing vortices gave rise to peaks in the energy spectra at the blade passing frequency and its harmonics and contributed to the level of spectra in the frequency range below about 60 N. It was apparent that the actual contribution of the trailing vortices to the energy spectra could not be calculated. The macroscale levels reported in the literature were on the order of the vortex diameter, which provides a physical interpretation of the macroscale.

van der Molen and van Maanen (1978) performed detailed measurements of the average velocity, the properties of the trailing vortices, and turbulence power spectra close to the blade in three different tank sizes. Close to the blade, the discharge flow was very periodic and caused a sharp peak in the energy spectra at the blade passing frequency as well as several sub and higher harmonics. The total average velocity closed to the tip of the blade was $1.2V_t$. The radial velocity was the most significant component and was correlated as:

$$\frac{V_r}{V_t} = 0.85\left(\frac{r}{r_0}\right)^{-7/6} \tag{4.99}$$

where r_0 is the radius of the impeller. The exponent $-7/6$ includes the effect of entrainment. The random turbulence was averaged out in the signal, leaving only the periodic component of discharge. The periodic component was caused by vortices in the discharge flow which decayed rapidly with radial distance. The vortices had no significant effect beyond $r/r_0 = 1.5$. Close to the blade, the axial velocity indicated the presence of the dual vortex system of the disk style turbine. Maximum values of the axial velocity equaled V_t as expected in vortices, and the diameter of the vortices was about 0.5 the blade width. The circulation velocity in the vortices was assumed to scale with $T^{1/6}$. Effects of blade thickness and impeller mounting collar also appeared in the data and affected the circulation velocity in the trailing vortex systems.

The turbulent intensities, obtained from the integration of the turbulence energy spectra without the harmonics of the periodic component, were found to be on the order of 30 to 45 percent. The turbulence intensities were lower nearer to the impeller blade, i.e., $r/r_0 < 1.5$. The turbulence with the periodic component removed was considered isotropic; an increase in intensity with scale was noted and correlated with $T^{1/6}$.

Although recognizing that the impeller discharge actually pulsed, Drbohlav et al. (1978a; 1978b; 1979) described the discharge flow from a disk style turbine as a tangential cylindrical jet. Following De Souza and Pike (1972), Drbohlav et al. developed the following expression for the radial velocity profile:

$$\overline{V}_r = A_1\{1 - \tanh^2 [A_2(Z - A_3)]\} \tag{4.100}$$

where Z is $2z/W$, A_1 is the maximum radial velocity component in the profile, A_2 is associated with the width of the discharge flow normalized by impeller blade width, and A_3 is the dimensionless axial coordinate of the maximum in the velocity profile. The equations for A_1 and A_2 were:

$$A_1 = \frac{A}{2\pi DN}\left(\frac{\sigma}{r^3}\right)^{1/2} (r^2 - a^2)^{1/4} \tag{4.101}$$

$$A_2 = \frac{W\sigma}{4r} \tag{4.102}$$

The parameters a, A, and σ were determined from the velocity profiles. The radius a of the cylindrical tangential jet was a function of the impeller Reynolds number. A and σ were considered universal constants: $A/2\pi ND^2$ ranged from 0.104 to 0.117; σ, from 11.86 to 14.81. The model for the velocity profiles also permitted the development of equations for the eddy viscosity ν_t and the turbulent stresses τ_{rz} and $\tau_{\theta z}$:

$$\nu_t = \frac{A}{2} \frac{1}{(\sigma^3 r)^{1/2}} \frac{2r^2 - a^2}{(r^2 - a^2)^{3/4}} \tag{4.103}$$

$$\tau_{rz} = -\frac{2\pi DN}{W} \rho\nu_t\left\{8A_1A_2 \frac{\exp[-2A_2(Z - A_3)]}{\{1 + \exp[-2A_2(Z - A_3)]\}^2}\right.$$

$$\left. \times \tanh[A_2(Z - A_3)]\right\} \tag{4.104}$$

The defining equations for the eddy viscosity ν_t and the shear stress τ_{rz} were:

$$\tau_{rz} = -\rho\overline{u_r u_z} \tag{4.105}$$

and
$$\tau_{rz} = -\frac{\rho\nu_t\partial\overline{V}_r}{\partial z} \tag{4.106}$$

Drbohlav et al. used a five-hole Pitot tube to determine the various velocity measurements and the Reynolds stress terms for a disk style turbine in the standard geometry. The results were used to justify the phenomenological models developed above. Data were obtained showing A_1, A_2, and A_3 as functions of radial position. The radius of the jet a as well as the other parameters of the velocity profiles were found independent of D/T.

The discharge jet widened, and the mean velocity profiles flattened with increasing distance from the impeller. A marked asymmetric shift in radial velocity profile upward was observed in the data as a result of the $0.33H$ clearance. The axial mean velocity was found to be small in comparison with the radial and tangential velocities. The overall turbulence intensities showed significant inhomogeneity and minimums in the centerline of the jet. The turbulence intensity profiles flattened as the wall was approached, and typically the intensities varied between 20 and 100 percent for the radial component and 200 to 800 percent for the axial component. The eddy viscosity values were found to be on the order of 1000 times the molecular viscosity, typically taking on values between 0.00485 to 0.0067 m^2/s for an impeller diameter of 0.33 m and rotational speed of 1.5 s^{-1}. The eddy viscosity was not constant in the discharge stream and varied in both the axial and radial directions.

The approach taken by Drbohlav et al. (1978) ignored the existence of the trailing vortex systems and the discharge pulsing from the impeller blades. The model failed to describe the flow field near the impeller. However, the pulsing dampened with distance from the impeller, and the models provided by Drbohlav et al. became more realistic as the distance from the impeller increased.

Costes and Couderc (1982), using hot-film anemometry, established velocity profiles and turbulence intensities for the disk style turbine centrally positioned in a tank in the standard configuration. The velocity profiles were similar when divided by impeller tip speed. Both the radial velocity and tangential velocity reached about $0.7V_t$. Turbulence intensities varied between 30 and 70 percent.

Kolar et al. (1984; 1985) have provided velocity profiles to model the discharge flow from a disk style turbine similar to work by DeSouza and Pike (1972) and Drbohlav et al. (1978a; 1978b; 1979). Wong and Huang (1988) also found that the De Souza-Pike model described the velocity profiles for four different impellers in the standard baffled geometry. The model was:

$$V_r = \frac{A}{2\pi D^2 N}\left(\frac{\sigma}{r^3}\right)^{0.5}(r^2 - a^2)^{0.25}\{1 - \tanh^2[A_2(z' - A_3)]\} \qquad (4.107)$$

where $a = r \sin \theta$, $A_2 = \sigma W/4r$, and $z' = 2z/W$. Values for the various parameters were given. Measurements were made in three geometrically similar flat bottom tanks, $T = 0.3$ to 0.6 m, using a disk style turbine, two curved blade disk style turbines, and a vane disk turbine. The profiles of $V_r/\pi ND$ were independent of impeller rotational speed N, tank size, and clearance. Maximum values of $V_r/\pi ND$ ranged from 0.62 to 0.72. The vaned disk had the highest radial velocity of $0.72V_t$. The pumping numbers for the different impellers decreased significantly in the transition regime, $10 < \text{Re} < 10{,}000$, a reduction of 50 percent in most cases. The disk style impeller had the highest pumping, and the curved blade turbine was considered the most efficient as measured by the ratio of pumping capacity and power number N_Q/N_P. The values were 0.19, 0.24, and 0.36 for the disk style turbine, vaned disk turbine, and the curved blade turbine, respectively.

Mahouast et al. (1988) have reported periodic and turbulent fluctuations as well as Reynolds stress terms and energy spectra for the disk style turbine. Weetman and Oldshue (1988) have reported spectra of $\overline{u_r u_z}$ for the disk style turbine and the pitched blade turbine.

Armstrong and Ruszkowski (1988) measured maximum radial velocities of $0.72V_t$ for a small hub disk style turbine and $0.60V_t$ for a large hub disk style turbine using LDA measurements synchronized with the impeller rotation. The velocity profiles were significantly different between the two geometries. As would be expected from the mean velocity data, the rms values for the smaller hub impeller were larger than the larger hub impeller. Armstrong and Ruszkowski also decomposed the velocity fluctuations into a periodic component and a random component. van der Molen and van Maanen (1978) reported hub effects as well.

Wu and Patterson (1989) made laser Doppler velocity measurements in a baffled agitated tank, $T = 0.27$ m, using a six-blade disk style turbine in the standard configuration. Mean velocity profiles, rms fluctuations, the Eulerian autocorrelation function, and the turbulence energy spectrum were obtained for the impeller discharge stream. Corrections were made to remove the periodic velocity fluctuations, i.e., the pseudoturbulence. The mean velocity profiles were similar to those reported in the literature and showed the clearance shift typical of data obtained at $C = 0.33H$. The accuracy of the velocity measurements were checked by mass balances. The maximum ratios of V_r/V_t and V_θ/V_t were 0.75 and 0.68, respectively. Total and random turbulence intensity profiles were obtained at the impeller tip. These profiles directly indicated the presence of a centerline jet and a dual vortex system shown in Fig. 4.1. Periodic pseudoturbulence, obtained from the difference, showed a double-peak distribution for u_z' and single-peak distributions for u_r' and u_θ' centered around the impeller blade centerline. The periodic pseudoturbulence disappeared at $r/R = 1.9$, which agrees with other work cited above. Energy spectra of total turbulence intensity showed peaks associated with pseudoturbulence which did not appear in the energy spectra of the random turbulence. The energy spectra of the random turbulence were typical of turbulence spectra shown in Fig. 4.4 except that the − 7 region was not observed.

Wu and Patterson obtained length macroscale distributions of the turbulence which showed minimums at the top and bottom of the blade and a local maximum at the blade centerline. The magnitudes of the macroscales at the minimums were roughly one-tenth of the blade width, i.e., $0.1W$, and at the maximum, $0.4W$. Such data indicate the formation of the systems for the disk style turbine shown in Fig. 4.1. Outside of the impeller discharge the length macroscales were much larger. Micro time and length scales were on the order of 1 to 3 ms and 0.3 to 1.0 mm, respectively. The turbulence Reynolds number based upon the velocity fluctuations and micro length scale varied between 50 and 190. Distributions of local energy dissipation rates were given as a function of position in the impeller discharge. Maximum energy dissipation rates were roughly 20 times the mean dissipation rate for the tank and occurred at r/R = 1.3 to 1.4 and not at the impeller blade tip. The local energy dissipation rate, the turbulent kinetic energy, and the resultant macroscale were found to be related as:

$$\epsilon = \frac{0.85k^{3/2}}{L_f} \tag{4.108}$$

which is a useful relationship for the k-ϵ model.

Open flat blade impeller. The open flat blade impeller has flow systems, shown in Fig. 4.1, similar to that of the flat blade disk style turbine. However, the flow patterns are not as stable as those of the disk style impeller due to the absence of the disk. It is suspected that on occasion flow passes vertically through the open impeller that would be stopped by a disk. The open flat blade turbine is far less studied than the disk style turbine.

Rao and Brodkey (1972) studied turbulence parameters in a continuous flow stirred-tank system using an open flat blade impeller in more or less the standard configuration with H/T = 1.5 with overflow at the top of the tank. The continuous flow was uniformly distributed from the bottom of the tank and had little influence on the flow patterns. Mean velocities, turbulent fluctuations, and one-dimensional energy spectra were reported using hot-film anemometer data. The discharge flow field was not symmetric along the centerline of the blade and was shifted because of clearance. Relative intensities were found to be 50 to 60 percent of the mean.

The work also emphasized the importance of determining the direc-

tion of the mean velocity vector and the presence of intermittent flow
in the discharge flow of the impeller. The velocity autocorrelation $f(r)$
showed a periodic oscillation close to the impeller which indicated
nonrandom flow. A peak was also observed in the energy spectra at
the blade passing frequency, and the energy spectra had the $-5/3$
and -7 regions. All spectra had the same general appearance.
Microscales, energy dissipation, and velocity probability density distributions were given. Skewed distributions indicated the existence of
nonturbulent flow in the impeller discharge and percent inter-
mittancy ranged from 60 percent near the impeller to 20 percent away
from the impeller.

Propellers and pitched blade turbines. The flow patterns of propellers
and pitched blade turbines are shown in Fig. 4.1 and are similar in
nature to each other. The flow patterns consist of a large axial flow
through the impeller. This flow, for the most part, is not mixed by the
impeller. Vortex systems are formed from the blade tips for both the
propeller and pitched blade turbine. Substantial mixing takes place in
these systems. The vortices can be considered to be similar to wing tip
vortices formed by the blade and similar to mixing layer vortex systems stabilized between the high-speed discharge flow and the recirculation flow in the bulk of the tank. The flow of the pitched blade
turbine tends to spread radially outward more so than the propeller
and contains both radial and axial velocity components. Propellers
have a more axial direction to their discharge.

The propeller and pitched blade turbine can pump in either direction, upward or downward. Pumping downward is the common design.
The flow patterns near the impeller in either pumping direction are
the same. Unfortunately, the up pumping mode has not been studied
extensively and has the stigma of being an improper design.

Generally, the flow fields of propellers and pitched blade turbines
are far less studied than the flow fields of the disk style impeller. An
elementary discussion of the discharge flow of the pitched blade turbine can be found in Fort (1986). This work need not be repeated here.
Unfortunately, the work does not mention the presence of trailing vortex systems in the flow and is mainly based upon potential flow theory. A good qualitative discussion of the flow fields of the pitched
blade turbine was given by Tatterson et al. (1980).

Plion et al. (1985) performed LDA measurements of the flow field
produced by a three-blade propeller in an agitated tank in the standard configuration with and without baffles in the downward pumping mode. The unbaffled case will be discussed below. For the baffled
tank, the discharge flow was parallel to the shaft and the pumping ca-

pacity was high. The maximum reported mean discharge velocity was about 0.5 m/s for an impeller tip speed of 2 m/s. The spatial turbulence intensities ranged from 31 to 1800 percent and were more homogeneously distributed in the tank than the unbaffled case.

The flow around hydrofoil or fluidfoil impellers have not been studied extensively.

Other impeller types. Bertrand et al. (1980) reported data on turbulence intensities for the disk style turbine at two different off-bottom clearances, two shrouded disk impellers with angled blades, and a disk style turbine with a modified blade shape and blade angle in a baffled tank with $H = T$. The turbulence intensities for the standard disk style turbine were found to vary between 10 and 40 percent across the discharge stream and were a function of clearance. The turbulence intensities at $C = 0.5H$ were lower than those at $C = 0.33H$. Shrouded (i.e., covered) blades of the disk impeller inhibited the formation of trailing vortices and caused a substantial reduction in discharge velocity and turbulence intensities. The turbulence intensities remained fairly uniform across the discharge stream at about 10 to 15 percent of the mean. Generally, the unshrouded impellers provided higher turbulence intensities than shrouded impellers. The modified blade shape of the disk style impeller had reduced turbulence intensities as well.

Obeid et al. (1983) have provided velocity profiles to model the discharge flow from the various impellers studied by Bertrand et al. (1980) following similar work by DeSouza and Pike (1972) and Drbohlav et al. (1978a; 1978b; 1979). Nagase et al. (1988) reported mean and rms velocity fluctuations for a reciprocating triangular beam impeller. The flow patterns around the blade contained vortices similar to those of flat blade impellers.

Specific flow patterns near and around impellers

Newitt et al. (1951) stated that the actual mechanisms which accomplished mixing were not known. Their article suggested that vortex systems being shed from the impeller blades were important in mixing. Newitt et al. also cited work by Proudman which showed that pure rotating liquids were stable in two dimensions and that any three-dimensional flows would dampen out. Rotation, like solid body rotation in an unbaffled tank, stabilizes the flow and dampens the turbulence. One can also assume that the vortex systems shed by an impeller stabilize turbulence because of their rotation. However, such systems leave the impeller and are shed into the discharge stream and the bulk of the tank, causing mixing.

Mean velocity flow fields must be established first in any study of flows in an agitated tank before any other measurements. The mean velocity is determined by the major direction of motion for the fluid which cannot be taken as simply the radial or tangential direction. It is only after the establishment of the mean velocity flow fields that turbulence can be defined and measured. This applies to other flow quantities as well.

The following studies approach the velocity flow fields near impellers from a structural turbulence viewpoint. Unless otherwise stated, the impeller and agitated tank will be in the standard configuration.

Flat blade impellers. Keey (1967) discussed: (1) the existence of vortex sheets which form in the wake region of the impeller blade and rapidly decay with distance, (2) regular fluctuations which decay with distance, and (3) entrainment. The regular fluctuations are due to alternating strong drafts behind the flat blades of the impeller which follow the blade and do not precede it.

Takeda and Hoshino (1966) and Takashima and Mochizuki (1971) were some of the earliest work which attempted to understand the flow phenomena around rotating impellers. Takeda and Hoshino (1966) obtained streak photographs of tracer particles in the impeller region for flat blade, pitched blade, and marine impellers with a camera rotating with the impeller. The work clearly showed the presence of trailing vortex systems and high-speed discharge flows along the centerline for flat blade impellers.

Takashima and Mochizuki (1971) performed a flow visualization study of the flat blade impeller noting that, at that time, the flow patterns of the various types of impellers had not been studied. Using neutrally buoyant tracer particles to mark the flow field, they photographed planes at different heights in the impeller region with a camera rotating with the impeller. The different flow patterns were observed at different heights. Roll vortices were observed to form along the top and bottom edges of the blade, and a strong exit flow occurred along the centerline of the blade as shown in Fig. 4.15. The flow along the back of the blade was higher than in front of the blade, and a sawtooth discharge flow pattern was formed on the outer perimeter of the impeller, i.e., high velocities behind the blade, low velocities in front of the blade. The exit flow along the centerline followed the blade angle for sweptback impeller blades. Unsteady cyclic phenomena were observed to occur between the blades.

van't Riet and Smith (1974) emphasized the importance of the impeller region and the trailing vortex systems generated by the disk

Direction of
Rotation

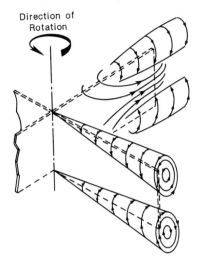

Figure 4.15   Double helical flow
model for agitator blade. *(From
I. Takashima and M. Mochizuki,
J. Chem. Eng. Japan, 4, 66, 1971.
By permission.)*

style turbine, as shown in Fig. 4.16, in understanding power draw, single-phase flow and multiphase dispersion mechanisms. van't Riet and Smith discussed the vortex behavior just as it left the blade to a point approximately 1.5 blade lengths away. Velocities were obtained using neutrally buoyant tracer particles photographed with a camera rotating with the impeller. The vortex axis was established visually. Pressure coefficient data were calculated from the measured circumferential velocity distributions and from direct measurements using a Pitot tube. They found that the angular velocity distribution did not vary along the core of the vortex and was only a function of the vortex radius and impeller Reynolds number. However, above an impeller Reynolds number of 1.5×10^4, the dependency on Reynolds number was negligible. At small radii, the vortex tended to a rigid solid body rotation; the size of the rigid core was larger at lower impeller Reynolds number. The circumferential velocity of the vortex, being the product of the angular velocity and vortex radius, was also obtained. At a high Reynolds number of 1.5×10^4, the circumferential velocity showed a flat maximum approximately equal to the impeller tip speed. At lower Reynolds numbers, the maximum value decreased to 0.5 of the impeller tip speed at an impeller Reynolds number of 300. The axial velocity along the vortex axis was independent of radius but increased linearly with distance along the axis.

Pressure coefficient data (i.e., $2\Delta P/\rho V_t^2$) were reported with maximum values of 3.4 to 4 behind the blade and a value of 2.1 in the region where the vortex left the blade. The centrifugal accelerations in the vortices were found to be extremely high, reaching values of 18 times the centrifugal force, $N^2 D/2$, of the impeller blades. At lower im-

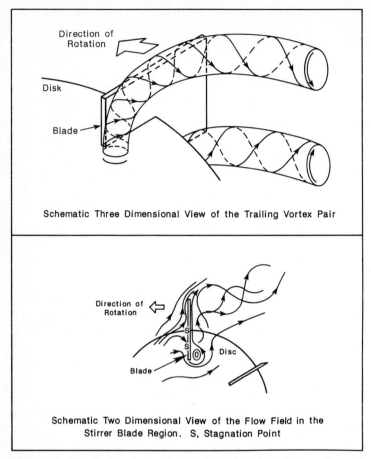

Schematic Three Dimensional View of the Trailing Vortex Pair

Schematic Two Dimensional View of the Flow Field in the
Stirrer Blade Region. S, Stagnation Point

Figure 4.16   The dual vortex system of the disk style turbine.   *(From K. van't Riet and J. M. Smith, Chem. Eng. Sci., 30, 1093, 1975. By permission.)*

peller Reynolds numbers, the value decreased rapidly. It was also noted that the presence of the vortex in the impeller discharge stream caused substantial error in turbulence measurements if they were ignored in such measurements.

van't Riet and Smith (1975) reported velocities, pressure distributions, and the spatial location of the trailing vortices as well as simplified equations of motion in understanding the flow behavior. In the study, the pressure distributions were obtained using an impact Pitot tube mounted on the rotating impeller. The pressure data were corrected for centrifugal acceleration and dynamic pressure. Most of the measurements were obtained away from the impeller blade tip where the vortex axis was linear. However, the vortex axes were not stationary from blade passage to blade passage, and its position had to be determined each time to perform the various measurements. The movement of the vortex axis varied between $\frac{1}{7}$ to $\frac{3}{7}$ of the blade width.

The trailing vortex systems formed between $150 < Re < 250$, and its average position and velocities were found to be dependent upon the impeller Reynolds number. The angular velocity in the vortex ω, as shown in Fig. 4.17, was measured by determining the time needed for a tracer particle to complete one cycle around the vortex axis and was found to be independent of the position along the vortex axis and only a function of impeller Reynolds number up to 1.5×10^4. Above this Reynolds number, the angular velocities were found to be constant with Reynolds number. At small vortex radii, a rigid core formed with the angular velocity being constant. At large vortex radii, ωr^2 was a constant. The circumferential velocity of the vortex was obtained from the angular velocities and showed the same dependency on impeller Reynolds number. At $Re = 1.5 \times 10^4$ and above, circumferential velocity approached the tip speed of the impeller and was modeled as:

$$\frac{V_c}{V_t} = \left(\frac{\Gamma_x}{V_t D}\right)\left[\frac{1}{2\pi(r/D)}\right]\left\{1 - \exp\left[-\frac{bD}{V_t}\left(\frac{r}{D}\right)^2\frac{V_t D}{2\nu_t}\right]\right\} \qquad (4.109)$$

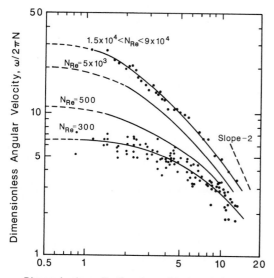

Figure 4.17   The dimensionless angular velocity distribution in a vortex of the disk style turbine.   *(From K. van't Riet and J. M. Smith, Chem. Eng. Sci., 30, 1093, 1975. By permission.)*

where Γ_x/V_t and b/V_t were constants evaluated to be 0.1 m and 4.0 m^{-1}, respectively. Γ_x is the circulation of the vortex at infinity, and $V_tD/2\nu_t$ is an impeller Reynolds number where ν_t is the eddy viscosity. The axial velocities in the vortices were found to increase along the vortex axis and were independent of Reynolds number when divided by the impeller tip speed. The turbulent nature of the flow was accounted for by using an eddy viscosity, which was a function of the impeller Reynolds number, varying between $2.4\nu_L$ at Re = 300 to above $35\nu_L$ above Re = 1.5×10^4. These were representative values since eddy viscosity varied across the vortex.

Gunkel and Weber (1975) used a shielded hot-wire anemometer mounted on a disk style impeller in a tank having the standard geometry to study the flow field in the impeller. Their data showed a large radial discharge flow occurring from behind the impeller blade with lower radial velocities in front of the blade. The radial discharge velocity was the same magnitude as the tip speed of the impeller. The turbulence levels were on the order of 4 to 16 percent.

Chang et al. (1981), using viscous oil as a tracer, found that the discharge flow from a disk style turbine consisted of: (1) a central high-speed jet along the centerline of the blade which had speeds up to $1.34V_t$ and (2) two roll vortices, one forming along the top of the blade and the other along the bottom edge of the blade. A pictorial representation of the flow of an oil drop in the flow field of a disk style turbine is shown in Fig. 4.18. The jet was formed because of a mass deficit generated by the centrifugal force of the blade throwing the fluid out as the blade moved forward. The mass deficit caused the formation of the vortices as well. The trailing vortices were not considered the major flow system of the disk style impeller in terms of mass flow rate. Furthermore, there were substantial elongational stresses which occurred in the high-speed jet.

Mochizuki and Takashima (1982) extended their earlier work and provided velocity, angular momentum profiles per unit area at the impeller tip, and velocity profiles in the vortical flows. They estimated that the power consumed in the impeller region was 16 to 36 percent of the power input and the energy dissipation per unit volume in this region was 10 to 20 times the dissipation in the bulk of the tank.

Popiolek et al. (1984) studied the trailing vortex systems of the disk style turbine in the standard geometry using laser Doppler anemometry synchronized with the rotation of the impeller. Six velocity fluctuation cycles were observed which indicated the presence of trailing vortices. The vortices were continually generated behind the blade but, upon leaving the blade, had a short duration time and broke up erratically. The axis of the vortex was not stationary from blade pas-

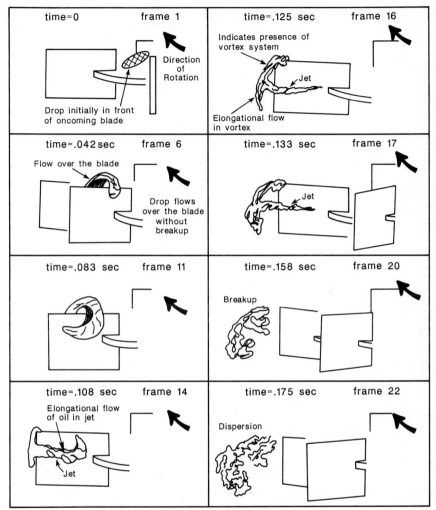

Figure 4.18 Ligament stretching mechanism for the disk style turbine. Frames taken from a film to show how oil is dispersed in water. *(From T. P. K. Chang, Y. H. E. Sheu, G. B. Tatterson, and D. S. Dickey, Chem. Eng. Commun., 10, 215, 1981. By permission.)*

sage to blade passage and moved about in erratic behavior similar to that reported by van't Riet and Smith (1975). The circumferential velocities over the blade were as much as 2 times the impeller tip speed indicating very strong trailing vortices and high shear rates over the blade. Turbulent kinetic energy contours were also obtained; the maximum value of k was $0.17V_t^2$. The flow was highly anisotropic and the flow from the impeller radius to 1.5 times the impeller radius was dominated by the trailing vortex systems with circumferential velocities of $0.5V_t$. Without synchronization with impeller rotation, Popiolek et al. noted that turbulence intensities could be overestimated by 400 percent.

Placek and Tavlarides (1985) viewed the discharge flow of the disk style turbine as consisting of trailing vortices which, upon exiting the impeller, turned tangentially due to strong Coriolis forces. The longitudinal velocity V_L along the axis of the trailing vortex varied linearly with impeller tip speed and was modeled as:

$$V_L = 1.386V_t \left(\frac{l}{D}\right) \quad \text{for } 0.3 < \frac{l}{D} < 0.9 \quad (4.110)$$

where l was the distance along the vortex axis. The effective kinematic viscosity was related to impeller tip speed using data provided by van't Riet and Smith as:

$$\nu_{\text{eff}} = 0.00013V_t \quad (4.111)$$

using SI units. Placek and Tavlarides also found discrepancies between the equation for circumferential velocity cited above by van't Riet and Smith and the actual data. They provided an equation for the radial coordinate of the vortex axis and developed a complicated set of equations for the discharge velocity of the disk style turbine, assuming a uniform outflow from the impeller. The mean discharge velocity was also found and, using this, Placek and Tavlarides evaluated the intensity of the periodic velocity fluctuations. Their model was limited to the presence of trailing vortices and cannot be used where the trailing vortices are not present in the flow.

Yianneskis et al. (1987) obtained ensemble averaged mean flow results for the trailing vortex systems formed by a disk style turbine in the tank of standard configuration. The impeller diameter, impeller rotational speed, and impeller tip velocity for the measurements were 0.098 m, 6 s^{-1}, and 1.847 m/s, respectively. The radial, axial, and tangential discharge velocities at the impeller in the discharge stream were found to be on the order of 0.7, 0.25, and $0.85V_t$, respectively, indicating a total velocity of $1.13V_t$. Mean flow data were obtained for the trailing vortex systems. The location of the center of the vortex was estimated assuming zero axial velocity at the vortex center. The center of the vortex was not established visually as was done by van't Riet and Smith (1975). The technique was based upon the assumption that the trailing vortex did not meander about the center of the impeller discharge. An equation was provided for the vortex axis as a function of radial and tangential position. Velocities in the trailing vortices away from the blade were on the order of 0.25 to $0.42V_t$ and were found to decay significantly with distance from the impeller. The vortex systems were considered to break up at a tank radius equal to two times the impeller radius, and the turbulent fluctuations were considered to increase with distance from the impeller up to a point. The maximum turbulent kinetic energy was about $0.4 \text{ m}^2/\text{s}^2$ in the impeller discharge flow, which was more than an order of magnitude larger than in the bulk of the tank. The energy dissipation was on the order of $25 \text{ m}^2/\text{s}^3$ in the impeller discharge stream for the impeller tip velocity of 1.847 m/s. A contour plot of turbulent kinetic energy was given for the impeller region.

Yianneskis et al. (1987) compared the instantaneous velocity data with their mean velocity data and found significant differences between the two. At times, the magnitude of the instantaneous axial velocity data were as much as two times the magnitude of the mean axial velocity data and significant differences occurred in the sign of the two types of velocities.

Stoots (1989) used a two-component laser Doppler anemometer to measure the average velocity, the rms turbulent velocity fluctuations, and Reynolds stresses in the flow field of a disk style turbine in a baffled tank in the standard configuration. The data acquisition was synchronized with impeller rotation into 1 degree increments over a fine spatial grid to obtain the flow field relative to the impeller blade. Vector plots of the mean velocity data were constructed to determine the size and shape of the trailing vortex systems. The mean and rms turbulent velocities were used to estimate the mean deformation rate, the turbulent kinetic energy, and the energy dissipation rate. The Reynolds stress terms were used to calculate the turbulent energy production rate. Spatial contour plots of these quantities were also obtained. The measurements were made at three impeller rotational speeds to determine how the various quantities scaled with rotational speed. To date, this study is the most extensive available in the literature.

Pitched blade impellers. Tatterson et al. (1980) used a stereoscopic flow visualization technique and neutrally bouyant tracer particles to study the flow in tanks agitated by pitched blade turbines. They found that each blade of the bench-scale pitched blade turbine (0.1 m diameter) acted as a hydrofoil. There was an accelerated high-speed flow over the top and backside of the blade and a low-speed flow below and in front of an advancing blade, as shown in Fig. 4.19. The high-speed flow over the back of the blade developed into a high-speed jet below the turbine. At the bottom outside tip of the blade, trailing vortices were formed. The main intake flow to the turbine came from above and moved smoothly toward the oncoming blade. Below the impeller, high-speed jets were observed to receive flow from different blades and had roughly a constant velocity. The discharge flow was substantially lower than the impeller tip speed.

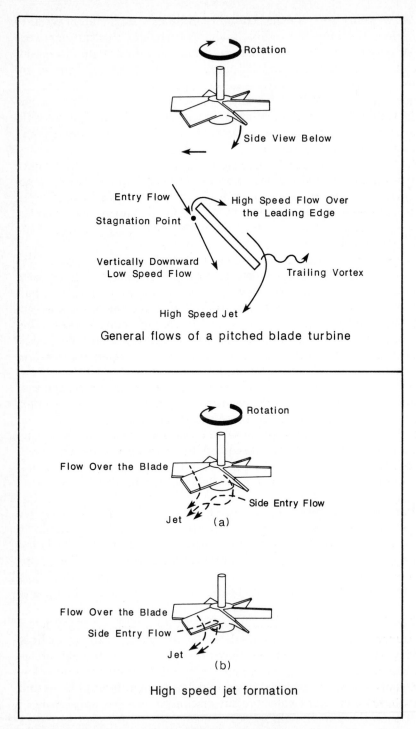

Figure 4.19 Flow patterns of a bench-scale pitched blade turbine. *(From G. B. Tatterson, H.-H. S. Yuan, and R. S. Brodkey, Chem. Eng. Sci., 35, 1369, 1980.)*

For a larger-scale pitched blade turbine, 0.3 m diameter, the flow patterns were different. The flow contained considerably less heterogeneity and is shown pictorially in Fig. 4.1. There was little acceleration of the fluid on the backside of the blade, and the velocities on the front and backside of the blade were approximately the same. A strong vortex formed at the blade tip. The vortices for the bench-scale turbine were not very coherent and exited the impeller region in a chaotic fashion below or directly into the path of the oncoming blade. The vortex of the large-scale pitched blade turbine was very coherent in the discharge flow. Overall, there were substantial differences in flow behavior between the small-scale and large-scale impellers. Propellers and hydrofoil impellers have a similar flow behavior as pitched blade turbines but have not been studied extensively.

Summary. Much of the work was done with flow visualization using dilute particle concentrations or with measurement techniques at stationary points or synchronized with the blade. Overall, the work indicates a region near the impeller dominated by high-speed jets and trailing vortices. Beyond this, the vortices are thought to break up and are not present in the flow upon reaching the wall. However, vortex remnants reach the wall and circulate into the bulk of the tank. Flow visualization using dilute particle concentrations, probes, or LDA synchronized with the blade are not capable of studing such behavior without considerable inaccuracies. Vortex systems have been visually observed to meander substantially in the axial location even in the region close to the impeller. Yianneskis et al. (1987), as cited above, made a comparison of instantaneous and mean velocity data which indicated significant differences between the two types of data.

Ensemble averaging of flow data according to location requires supporting documentation before the technique and the results can be accepted. Measurement of the location of the vortex axis for every velocity data set is required before an accurate understanding of the flow can be obtained.

General references on vortices include Lord Rayleigh (1916), Newman (1959), Dosanjh et al. (1962), Hoffman and Joubert (1963), Squire (1965), Saffman (1973), and Brasseur and Chang (1980).

Flow patterns in the bulk of the tank

Baffled tanks. Flow patterns in the bulk of a baffled tank are poorly understood and nondescript. Circulation zones are present and are referred to by various terms, e.g., toroidal vortices, ring vortices, or compartments. In baffled tanks, the discharge stream of impeller dissipates in the bulk of the tank, and the axial velocity is generally the most important velocity. The tank volume is also important in its effects on flow properties. Typically this effect is noted using the D/T ratio. Baffling is mainly at the wall but can also be located around the perimeter of the impeller as a stator ring or as finger baffles. Flow behavior in the bulk of the tank for stator rings has not been studied.

Bowers (1965) noted that baffles substantially increased axial circulation velocities, reducing tangential velocities which occur in unbaffled tanks. The flow velocities and turbulence intensities, measured using hot-wire anemometry in baffled tanks, were proportional to impeller tip speed at Reynolds numbers above 10^5. Reith (1965) presented the concept of universal velocity and turbulence intensity profiles for agitated tanks where velocities and turbulence intensities, when reduced by the impeller tip velocity, were expressed as functions of spatial location only.

Schwartzberg and Treybal (1968) measured the circulation velocity and the fluctuating component of velocity in a representative volume of the bulk of the tank. The impeller was a disk style turbine in roughly the standard configuration. The average velocity in the representative region was correlated as:

$$\overline{U} = \frac{1.586ND^2}{(T^2H)^{1/3}} \qquad (4.112)$$

and the fluctuating velocity as:

$$u' = \frac{0.581ND^2}{(T^2H)^{1/3}} \qquad (4.113)$$

which held for large ranges in operating parameters. As indicated by these correlations, both the average and fluctuating velocities can be considered proportional to the impeller speed ND and a size measure $D/(T^2H)^{1/3}$ or an effective D/T ratio. The flow was distinctly nonisotropic, and most of the velocity fluctuations were mainly due to changes in velocity direction rather than magnitude. The fluctuations were approximately 35 percent that of the mean. The flow velocities in the main discharge flow was much higher than in the bulk flow of the tank.

Levins and Glastonbury (1972), similar to Schwartzberg and Treybal (1968), studied particle liquid hydrodynamics using pitched blade turbines in the standard geometry. The turbulence intensities in the bulk of the tank were as high as 75 percent of the total velocity. The individual turbulence intensities $u_i'/\overline{U}_i$ were on the order of 200 percent. Such high turbulence intensities indicate poorly defined mean velocities in the bulk of the tank. Integral length scales were found to be on the order of the impeller blade width and increased as impeller speed decreased and the impeller diameter increased.

DeSouza and Pike (1972) modeled the circulation in the bulk of the tank as two potential flow corners and circular jet feed to the impeller. They incorporated "dead water" regions to treat the toroidal vortices. The impeller was a disk style turbine in roughly the standard configuration. The velocity flow field in the bulk of the tank was found to be three dimensional with a significant tangential velocity component. The flow field could only be qualitatively described using a two-dimensional model. Experimentally, DeSouza and Pike found a narrow high-speed flow along the tank wall with a thickness of about 16 percent of the tank diameter. This flow formed as the discharge from the impeller was split by the wall. This indicates that the discharge flow of the impeller does not decay away upon reaching the wall but continues along the tank wall.

Gunkel and Weber (1975) measured the axial, radial, and tangential fluctuating velocities in a large number of positions in the tank agitated by a disk style turbine using air as the fluid. The flow in the bulk of the tank was primarily in the axial direction. The velocity fluctuations in the axial direction were somewhat larger than in the radial and tangential directions. Overall, however, the flow was considered isotropic. One-dimensional energy spectra showed the $-5/3$ and -7 power regions and were very similar for the radial, tangential, and axial velocities despite differences in the operating conditions. The Eulerian velocity autocorrelations and integral scale of turbulence were also obtained. The integral scales were about one-half the impeller blade width. The dissipation function, similar to that in Fig. 4.4, showed that most of the dissipation occurred in the frequency range of 10 to 10^3 Hz. There were no peaks in the spectra at the impeller rotational speed. The scale of the energy containing eddies was found to remain fairly constant despite differences in operating conditions.

Reed et al. (1977) provided an excellent description of the flow patterns occurring in an agitated tank using a disk style turbine in the standard configuration at $C/H = 0.33$. Compartmentalization of the flow, the three dimensionality of the flow in the bulk of the tank, and strong recirculation below the impeller were noted. Compartmentalization is the tendency for the flow to form circulation cells or compartments. For the disk style turbine, two compartments were formed, one above the impeller and one below the impeller. Mixing between these cells occurred in the impeller discharge stream. Vortices were formed in front of and behind the baffles. Baffles were observed to stop the tangential flow of the impeller discharge stream and to cause the formation of strong axial drafts in front of the baffles. A circular, three-dimensional, proceeding recirculation zone was also present in the bottom compartment. The motions of the individual particles were reported as to how the particles moved between the baffles and in the recirculation loops. More flow was returned to the impeller from the vortices of the baffles than from the volume between the baffles. The axial velocities and recirculation in the lower compartment were much higher than in the top compartment. The strong recirculation flow in the lower compartment apparently caused the radial discharge velocity profile to be asymmetric. The profile was displaced upward as the flow approached the wall. This displacement of the discharge stream has also been reported by others including Rao and Brodkey (1972) and Drbohlav et al. (1978; 1979).

Tatterson et al. (1980), in an application of a cross-beam laser correlation technique, measured three-dimensional particle velocities in the discharge flows of a disk style turbine and pitched blade turbine as well as in the bulk of the tank. In one case in the discharge flow of a disk style turbine close to the wall, turbulence intensities in the axial velocity component were 800 percent. This essentially implied a poorly defined mean axial velocity. Organized flow structures were found to exist in the bulk of the tank. For the pitched blade turbine, Fort (1986) provided a discussion of the flow field in the bulk of the tank. This work need not be reviewed here.

Yuanneskis et al. (1987) investigated the bulk flow patterns in a baffled agitated tank in the standard configuration using a disk style turbine. Using flow visualization, ring vortices in the bulk of the tank were observed above and below the impeller. For a centrally located turbine, the ring vortices were symmetric about the impeller discharge stream. For lower clearances, $C = 0.25T$, the impeller discharge flow was displaced upward away from the bottom of the tank. This effect of clearance has been reported by others as cited above. The top ring vortex was more ordered, and the impeller discharge flow did not spread as much as for $C = 0.5T$. At higher impeller rotational speeds, the inclination of the discharge stream was less. The radial and axial flow velocities in these circulation systems were on the order of 0.1 to $0.2V_t$. At low clearances, $C = 0.25T$ and $D = 0.5T$, a second ring vortex was observed to form in the top quarter proportion of the tank and rotated in the same direction as the bottom ring vortex. Velocity measurements using LDA supported these conclusions. The turbulence level and turbulent kinetic energy were on the order of 0.1 to 0.2 m/s and 0.01 to 0.04 m^2/s^2, respectively, for an impeller diameter of 0.098 m and rotational speed of 6 s^{-1}; the impeller tip velocity was 1.85 m/s. Several surface vortices were also observed.

Laufhutte and Mersmann (1987) investigated local energy dissipation in agitated tanks using the disk style turbine, a three-blade propeller, and an Intermig impeller in the standard configuration. Fairly uniform distributions of the velocity fluctuations in the bulk of the tank were found. Larger-diameter impellers were found to have a more even distribution of fluctuating velocities. Local energy dissipation rates were correlated as:

$$\epsilon = \frac{C(u')^3}{D} \tag{4.114}$$

where C was found to be constant varying between 5.5 and 7.2 for the different impellers. A larger part of the energy dissipation was found to occur in the impeller discharge flow for smaller D/T ratios. Laufhutte and Mersmann recommended this configuration for obtaining the best micromixing with feed injection near the impeller in the highest energy dissipation zones. The most even distribution of energy dissipation was found using multiple large-diameter impellers.

Magni et al. (1988) also found the flow in the bulk of the tank to be approximately homogeneous. All three velocity components were the same order of magnitude.

Flow patterns of multiple impellers. Nienow and Kuboi (1984) investigated the flow patterns of a two-impeller system in the transition region, 70 < Re < 140 with D/T = 2: a disk style impeller on the bottom and a pitched blade turbine on top in either pumping-up mode and pumping-down mode. With the pitched blade turbine in the pumping-up mode, four recirculation zones were formed with the pitched blade turbine behaving much like a radial flow impeller. With the pitched blade turbine in the pumping-down mode, three recirculation zones were formed; two by the disk style turbine and one by the pitched blade turbine. The axial flow through the center of the pitched blade turbine formed the circulation loop. A small additional recirculation zone was formed between the top zone of the disk style turbine and the zone of the pitched blade turbine. For each configuration, flow followers were used to establish the mean circulation time, the pumping rate in each zone, the mean residence time for a fluid particle in each zone, and the transfer rate of fluid particles between each zone. These circulation loops are examples of compartmentalization mentioned earlier.

Scaleup of flow patterns. Scaleup rarely maintains full geometric similarity, and differences arise in the flow patterns produced. Unfortunately, little information is available on the scaleup of flow patterns. Typically it is assumed that the flow patterns will remain the same on scaleup.

Tatterson et al. (1980) made comparisons between impeller flow patterns of a small-scale six-blade pitched blade turbine, 0.10-m diameter, and a larger four-blade pitched blade turbine, 0.305 m in diameter. Substantial differences in flow behavior occurred between the two scales as cited above.

For design and scaleup, von Essen (1983) noted that it was advantageous to select the desired flow pattern which best suited the process and then to match the pattern with the impeller that has a similar flow pattern. The location of the impeller should be where their resulting flow patterns match the desired flow pattern. von Essen illustrated this design approach for a solid suspension problem using a hydrofoil impeller, a pitched blade turbine, and a flat blade turbine to obtain the same flow pattern. The tank was contoured to match the desired flow pattern. Cone bottoms, corner fillets, and draft tubes are examples of possible changes in tank geometry.

Unbaffled tanks

Unbaffled tanks are used for mixing for a number of reasons, and studies of the flow in unbaffled tanks have been performed. Flow in unbaffled tanks exhibit solid body rotation as the primary flow pattern. Solid body rotation tends to be independent of the geometry of the impeller. The central surface vortex is also an important flow pattern of unbaffled tanks.

Ironically, most studies of unbaffled tanks have been of the flow field in the bulk of the tank. The flow near the impeller has not been studied extensively. It is most likely that trailing vortices and high-velocity jets are present in the impeller region in much the same manner as in baffled tanks but with a lower level of coherence and intensity.

Given the tangential nature of the solid body rotation, the tangential velocity is the highest velocity. The axial velocity tends to be unimportant. This situation causes poorer circulation and mixing in unbaffled tanks in comparison with baffled tanks.

Martin (1946) noted that the relative velocity difference between the impeller and fluid in the bulk of the tank was much less in unbaffled tanks than baffled tanks, giving rise to less power draw for impellers in unbaffled tanks. Rushton et al. (1950a) reported on an extensive study of the flow patterns of a propeller in an unbaffled tank. Rushton et al. (1950b) found that the flow pattern for turbines was the same as that of a propeller in an unbaffled tank. Maximum power input for the turbine occurred as the central vortex reached the impeller.

Nagata et al. (1955) discussed the flow patterns in an unbaffled tank using paddles. In the turbulent regime, the liquid near the center of the vessel rotated with the same angular velocity as the agitator in a forced vortex motion. Velocity varied proportionately with radius.

Outside this region, the flow rotation varied inversely with distance. Close to the wall, the velocity dropped rapidly to zero.

Laity and Treybal (1957) suggested that the central surface vortex can be eliminated in enclosed tanks. However, solid body rotation still occurs. In an open unbaffled tank, a central surface vortex forms which makes scaleup of open unbaffled systems difficult. Proper scaleup requires parameters which describe the vortex, e.g., the Froude number.

Laity and Treybal found that the Froude number was unnecessary in scaleup for enclosed tanks since the central surface vortex did not form in such a configuration. Laity and Treybal also found that, for unbaffled tanks, the effect of continuous flow on agitation dynamics was small for the most part although Ito et al. (1979) documented the effect of feed streams on the flow patterns. Depending upon the orientation and the magnitude of the feed, the feed stream acted as a baffle.

Bowers (1965) measured flow velocities in unbaffled tanks for radial and axial flow impellers. The tangential velocity was found proportional to the impeller tip speed, and the ratio of tangential velocity to tip speed was approximately the same in geometrically similar systems. No effect of scale was observed.

Hiraoka et al. (1975) provided velocity profiles of the flow in unbaffled tanks, which will be reviewed in the section on shear rates below.

Nagata et al. (1975) studied turbulence in an unbaffled tank using a hot-film anemometer. Measurements were done at 43 locations in a 0.3-m-diameter tank agitated by a disk style turbine, 0.15 m diameter. The tangential velocity was the controlling velocity; the tangential turbulence intensities were 25 percent near the impeller tip and below 10 percent at the liquid surface. Radial and axial turbulence intensities were about $\frac{1}{3}$ to $\frac{1}{4}$ those of the tangential velocity, indicating nonisotropic flow. Mixing was generally poor outside the cylindrical rotating zone (i.e., CRZ) established near the impeller. Mixing in the axial and radial directions was much slower than in the tangential direction. Energy spectra were obtained and were used to calculate the turbulent microscale which was on the order of 0.8 to 1.5 mm or about one-tenth the impeller blade width. No periodicity in the impeller discharge flow was observed in the energy spectra. The flow was considered isotropic only on the small scale. The macroscale of the turbulence was found to be 2 to 20 mm or about the same order of magnitude as the impeller blade width. The macroscale Λ was used to calculate the turbulent diffusivities as:

$$D_{LT} = u'\Lambda \qquad (4.115)$$

which were found to be on the order of 0.1 to 50 cm^2/s and were direction dependent because of the dependency on the velocity fluctuations.

Yuu and Oda (1980) and Graichen (1980) studied turbulence parameters in unbaffled enclosed tanks at high impeller rotational speeds using air as the fluid. In such systems, no central surface vortex forms. Yuu and Oda used a standard disk style turbine centrally positioned with $D = 0.38T$ and $H = 1.05T$. Mean velocity distributions were obtained for the impeller stream and were found proportional to the impeller speed; being $0.6V_t$ at the impeller tip and $0.3V_t$ at $r = 0.4T$ close to the wall. The turbulence intensities varied in the same manner; varying between 10 to 20 percent at the impeller tip and 5 to 10 percent at $r = 0.4T$. The highest turbulence intensities were found for the tangential velocities. The velocity distributions outside the impeller stream did not fall on the same curve; however, the ratio of local velocity to impeller tip speed increased with decreasing rotational speed. The turbulence intensities outside the impeller stream varied between 2 and 7 percent. Large periodic fluctuations were also noted in the impeller discharge, indicating a nonuniform flow and the possible presence of trailing vortex systems. Eulerian autocorrelations, integral and dissipative scales of turbulence, and energy spectra were obtained. The dissipation scales were on the order of 1 to 5 mm and were about ⅕ the integral scales. Most energy spectra showed a $-5/3$ and -7 regions and were very similar. Energy spectra obtained near the impeller were different from the rest of the spectra.

Graichen (1980) studied the turbulence microstructure using spectral analysis of the turbulent velocity fluctuations generated by a four-blade paddle, $D = 0.5T$ and $W = 0.5D$. The distribution of turbulent energy dissipation for the impeller discharge flow showed a maximum away from the impeller but then decreased rapidly with increasing distance from the impeller; $\epsilon \propto r^{-3.55}$ in this region. The impeller discharge flow contained a considerable portion of the energy dissipation. Similar behavior has been noted in baffled tanks with trailing vortices in the discharge flow.

Plion et al. (1985) reported LDA measurements of the flow field produced by a three-blade propeller, pumping downward, in an unbaffled tank. The characteristic central vortex was formed in this configuration, the depth of which increased with impeller rotational speed. The fluid near the impeller rotated with the same angular velocity as the impeller, and the flow away from the impeller approached a free vortex. The discharge flow was not axial but moved away from the propeller with a substantial radial component. The maximum reported discharge velocity was 0.27 m/s for a tip speed of 2 m/s. Upon reaching the bottom, the flow either rose along the wall or recirculated back underneath the impeller to the central axis of the tank. Overall, the results agreed with those given by Nagata (1975). The spatial radial and axial turbulence intensities were high near the impeller and

ranged from 65 to 2920 percent. Such high values indicate poorly defined time-averaged velocities. The highest value, an axial turbulence
intensity of 2920 percent, was close to the impeller and indicated the
presence of trailing vortices in the flow.

Rieger et al. (1979) performed an extensive study of the central vortex in unbaffled tanks. Novak et al. (1982) have studied surface aeration in unbaffled tanks and the effect of eccentricity on the surface
vortex. Other studies on the central surface vortex and surface aeration are discussed in Chap. 6 on gas dispersion.

Feed locations in unbaffled tanks.   Ito et al. (1979) studied the effect of
feed location on flow patterns in an unbaffled closed tank for paddle
impellers, D/T = 0.5 to 0.75, and W/T = 0.1. The feed number, N_{QF} or
Q_F/ND^3, was varied up to 0.2. Given the radial flow of the impeller,
feed from below the impellers almost always entered the impeller flow
and was discharged outward radially, regardless of flow rate, feed pipe
diameter, rotational speed, and height of feed pipe into the tank. For
side-entry feed locations, the feed participated in the tangential flow
without any disruption of the flow pattern so long as N_{QF} was less
than 0.04. Above 0.04, the side feed acted as a baffle and established a
secondary recirculation flow. At side entrance locations, the feed apparently did not enter into the impeller discharge.

For a feed stream or any other internal flow arrangement to affect
the flow field in an agitated tank, the mass and momentum flows of
the stream have to be at least the same order of magnitude as that
generated by the impeller. In a properly designed agitated tank, the
impeller generates significant circulation and momentum.

One experimental difficulty occurs with probe measurements in
unbaffled tanks. Given the poor mixing which occurs radially, there
can be substantial wake effect from the probe which can recirculate
back to the probe, tangentially. This wake effect causes erroneous flow
data and indicates poor mixing.

Pressure fields in agitated tanks

Vortex pressure fields.   van't Riet and Smith (1974; 1975) reported
pressure distributions for the trailing vortex systems as contour plots
of the pressure coefficient C_p defined as:

$$C_p = \frac{2\Delta P}{\rho V_t^2} \qquad (4.116)$$

The data, as shown in Fig. 4.20, confirmed the general shape of the
vortex systems obtained from flow visualization and velocity data
cited above. The data indicated that the vortex systems were low-

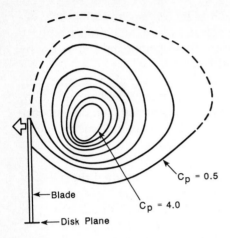

Figure 4.20 Pressure coefficient map, $C_P = \Delta P/(1/2\rho V_t^2)$, $D = 48$ cm, $N_{Re} = 3 \times 10^5$. *(From K. van't Riet and J. M. Smith, Chem. Eng. Sci., 30, 1093, 1975. By permission.)*

pressure regions. The pressure coefficient data increased linearly (i.e., to lower pressures) with the impeller Reynolds number as shown in Fig. 4.21. The position of minimum pressure $C_p = 5.0$ was located at about one-fourth of a blade length from the inner side of the blade. The pressure data also indicated that the flow in the vortex was not fully turbulent even at high impeller Reynolds numbers.

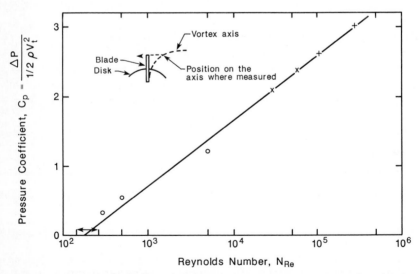

Figure 4.21 Pressure coefficient as a function of impeller Reynolds number showing the formation region of the vortex. *(From K. van't Riet and J. M. Smith, Chem. Eng. Sci., 30, 1093, 1975. By permission.)*

Spatial pressure fields.   Wolf and Manning (1966) used an impact tube
to determine the dynamic pressure distributions throughout the tank
and obtained estimates of velocities, impeller pumping capacities, and
flow entrainment rates. The work was performed in baffled tanks for
various turbines, propellers, and paddle impellers. The results on
pumping capacities are reported in a later section. The pulsation in
the pressure fields near the impeller was noted, but the dynamics of
the impact probe was not fast enough for use in determining instantaneous velocities.

Impeller blade pressure fields.   Pressure distributions and drag coefficients on rotating impeller blades have been obtained for various impellers (Mochizuki and Takashima, 1974, 1984a, 1984b; Tay and
Tatterson, 1985). The primary interest in such work is to identify how
power is transferred from the impeller blade to the fluid. Tay and
Tatterson attributed the major power transfer mechanism to a form or
pressure drag mechanism. Braginskii and Begachev (1972) reported
drag coefficients for the various impellers as well.

Wall pressure fields.   Nagase and Kikuchi (1983) measured wall pressure distributions for flat blade and disk turbines, propellers, and
pitched blade turbines as shown in Fig. 4.22. These wall pressure distributions were related to the circulation flow in an agitated tank using an overall vorticity balance.

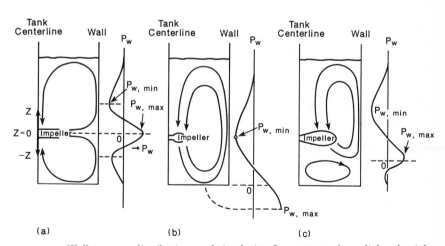

Figure 4.22   Wall pressure distributions and circulation flow patterns for radial and axial
flow impellers. (a) Radial discharge impeller (flat blade and disk turbine); (b) axial discharge impeller with low D/T ratio (pitched blade turbine and propeller); and (c) axial
discharge impeller with high D/T ratio (pitched blade turbine and propeller). (From Y.
Nagase and M. Kikuchi, J. Chem. Eng. Japan, 10, 164, 1977. By permission.)

Shear rates in agitated tanks

Shear rates have been studied by various methods and are important in processing, particularly multiphase processing. Generally, shear rates are considered either as an average and a maximum shear rate in the flow or an average and maximum shear on a boundary which is usually an impeller blade or tank wall.

Shear rates in the flow of baffled tanks. Metzner and Taylor (1960) discussed flow patterns in agitated vessels for the transition and turbulent flow regions and presented data on the level and distribution of shear rates and power dissipation rates resulting from velocity gradients within the tank. Most of the work was done in a baffled agitated tank in the standard configuration using a disk style turbine. Data were obtained from streak photographs of neutrally buoyant tracer particles. Local shear rates, as shown in Fig. 4.23, were obtained by differentiating the velocity data, and local power dissipation was calculated as the product of shear rates and shear stress obtained from

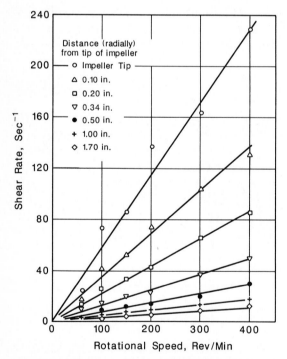

Figure 4.23   Shear rate distributions in Karo Syrup in the horizontal plane, 4-in impeller diameter. *(From A. B. Metzner and J. S. Taylor, AIChEJ., 6, 109, 1960. Reproduced by permission of the American Institute of Chemical Engineers.)*

viscometric data. The local shear rates were found to be proportional to impeller rotational speed:

$$\gamma = KN \qquad (4.117)$$

where K decreased rapidly with distance from the impeller blade tip. The constant K ranged from 0 to 90 and depended upon the viscosity of the fluid. Power dissipation occurred close to the impeller and decreased rapidly with distance from the impeller blade, covering a 1000 to 1 change in magnitude in some laminar-like flow cases. A simple phenomenological model of the process divided the tank into a well-mixed region confined to the impeller zone with the rest of the tank serving as a feed source. In early transition, mixing was performed by transport of fluid and not by any noticeable turbulence. Any local turbulence, particularly for shear thinning fluid, dampened out rapidly.

Midler and Finn (1966) considered both maximum shear rate and shear stress important parameters in agitated tanks. Using cells as a shear stress indicator, increasing the liquid viscosity at constant shear rate caused a substantial increase in cell disruption in a constant shear device, indicating the importance of shear stress. The effects of shear were dependent upon agitator diameter which implied the energy dissipation rate was important as well. The equivalent shear stresses in their study were 10 to 100 times the average shear stresses reported in laminar mixing power studies.

van't Riet and Smith (1975) obtained shear rates occurring in the trailing vortex system of the disk style turbine from their angular velocity data as shown in Fig. 4.24. The shear rates also varied along the axis of the vortex and average shear rates were 50 to 100 times the impeller rotational speed N in the vortex systems for $1.5 \times 10^4 < \text{Re} < 9 \times 10^4$. It was also estimated that 10 percent of the flow through the impeller experienced shear rates on the order of $50N$ for $1.5 \times 10^4 < \text{Re} < 9 \times 10^4$. However, no quantitative conclusions were made concerning the shear rates occurring immediately behind the blade. Centrifugal accelerations were also found to be very high in the vortex systems reaching as much as $700g$ at $\text{Re} = 2.2 \times 10^5$.

Shear rates on impeller blades. Mitschka and Ulbrecht (1966) in a study of power law fluids on a rotating disk found that shear rates varied as:

$$\frac{\gamma}{N} = (1 + 5.3n)^{1/n} \text{Re}^{1/(1+n)} \qquad (4.118)$$

Wichterle et al. (1984) considered that the maximum shear rates and stresses resided on the impeller blade. Direct measurement of

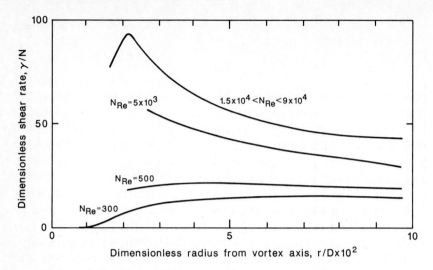

Figure 4.24 The shear rate distribution, obtained by angular velocity curves, as a function of dimensionless vortex radius and impeller Reynolds number. *(From K. van't Riet and J. M. Smith, Chem. Eng. Sci., 30, 1093, 1975. By permission.)*

these cannot be provided by velocity measurements. Using an electrochemical technique with flush mounted electrodes on the impeller and theory based on a simplified convective diffusion equation, shear rate was determined from measurements of a limiting current to the electrode. The electrode was mounted on the front face of a disk style turbine blade. Shear rates were found to vary between 1 to 2000 times the impeller rotational speed. The data were plotted versus the electrode Reynolds number and showed a remarkable correspondence with results of a study using a rotating disk by Mitschka and Ulbrecht (1966). For the calculation of shear rates in shear thinning liquids, Wichterle et al. recommended the same correlation as that given by Mitschka and Ulbrecht above. The Reynolds number in the correlation is the power law Reynolds number defined as:

$$\mathrm{Re} = \frac{\rho N^{2-n} D^2}{K} \qquad (4.119)$$

The fluid index is n, and K is the consistency coefficient of the power law fluid. Wichterle et al. (1985; 1988) also measured the shear stresses at the tank wall and found them to be about 30 percent of those occurring on the impeller as discussed below.

Robertson and Ulbrecht (1986) measured shear rate at the tip of a Rushton turbine using an electrochemical technique developed by Mitchell and Hanratty (1966). The electrode was imbedded in the forward-facing surface of an impeller blade near the tip. The shear rate was determined by measuring the transport of ions to the small electrode. Under suitable operating conditions, the current was limited by the rate of transport. Robertson and Ulbrecht correlated their results for shear rate as:

$$\frac{\gamma}{N} = 3.3 \left(\frac{ND^2}{\nu_z}\right)^{0.5} \tag{4.120}$$

where ν_z is the zero shear kinematic viscosity. The correlation was confirmed for a Reynolds number range of 100 to 29,000 for water and 0.5 and 1 percent polyox solutions.

Shear rates on the tank bottom and tank walls. The levels of shear stresses and shear rates on tank walls can have a substantial effect on heat transfer, particle abrasion, and crystal and cell growth. Boundary layers are present, and the level of the shear stresses and shear rates are of interest.

Hiraoka et al. (1975) presented wall shear stress data and velocity profiles in a fully turbulent unbaffled tank agitated by four blade paddles, $0.47 < D/T < 0.61$ and $0.1 < W/T < 0.30$. The shear stress, averaged over the side wall, was correlated as:

$$\frac{f}{2} = \frac{\tau_w}{\rho V_\theta^2} = 0.121 \left(\frac{LV_\theta}{\nu}\right)^{-0.333} \tag{4.121}$$

where V_θ and L are a characteristic velocity and a characteristic length, respectively, and were defined as:

$$V_\theta = \left(\frac{\pi}{2}\right) ND\beta \tag{4.122}$$

and
$$L = \left(\frac{T}{2}\right) \ln \left(\frac{T}{D}\right) \tag{4.123}$$

The parameter, β, was defined as:

$$\beta = \frac{2 \ln (T/D)}{(T/D) - (D/T)} \tag{4.124}$$

As with traditional pipe and flat plate flow, dimensionless velocity and distance were expressed as $v^+ = v/u_w^*$ and $\xi^+ = \xi u_w^*/\nu$ where $u_w^* = (\tau_w/\rho)^{0.5}$ and $\xi = (T/4)[(T/2r) - (2r/T)]$ and where r was the radius of consideration. For the range of $2r/D > 1$, a graphical velocity correlation was given which was essentially a velocity defect law for fully turbulent unbaffled agitated tanks. The velocity defect law was also written in terms of v^{2+} as:

$$v^{2+} = \frac{26}{2r/D} \qquad (4.125)$$

$0.5 < 2r/T < 1.0$. Close to the wall, a law of the wall was developed:

$$v^{2+} = 5.8 \log \xi^{2+} + 15.5 \qquad (4.126)$$

where $v^{2+} = (T\beta/D)v^+$ and $\xi^{2+} = (T\beta/D)\xi^+$. In the viscous sublayer of the wall, $v^{2+} = \xi^{2+}$. Shear stress distributions for $D/T = 0.66$ and $0.2 < W/T < 0.8$ have also been measured and reported (Mizushina et al., 1969).

Wichterle et al. (1985) used the electrochemical method cited in their previous work (Wichterle et al., 1984) to obtain shear rate distributions on the walls of a baffled agitated tank using the disk style turbine. From the measurement of the limiting diffusion current I_D and assuming a power law relationship for the fluid, wall shear rate γ_w and wall shear stress τ_w were obtained from the limiting current as:

$$\gamma_w = \alpha I_D^3 \qquad (4.127)$$

and

$$\tau_w = K\alpha^n I_D^{3n} \qquad (4.128)$$

where α is a proportionality constant dependent upon the fluid. Bottom and wall profiles of $f_w \mathrm{Re}^{0.5}$ were given for Newtonian fluids as a function of impeller Reynolds number where f_w was a friction factor defined as:

$$f_w = \frac{\tau_w T^2}{\rho N^2 D^4} = \frac{\tau_w}{\rho N^2 D^2}\left(\frac{T}{D}\right)^2 \qquad (4.129)$$

which shows the dependency of friction factor on the T/D ratio. The impeller Reynolds number was defined as:

$$\mathrm{Re} = \frac{\rho N^{2-n} D^2}{K} \qquad (4.130)$$

For the impingement region of the flow at the wall, the maximum friction factor for Newtonian fluids was found to vary as:

$$f_{\max} = 16 \mathrm{Re}^{-0.5} \qquad (4.131)$$

Bottom and wall distributions of the friction factor were given for various non-Newtonian fluids as well. The maximum friction factor in these data was correlated as:

$$f_{\max} = (1 + 15n)\mathrm{Re}^{-1/(1+n)}\left(\frac{T}{D}\right)^{2(1-n)/(1+n)} \qquad (4.132)$$

The wall shear rates were very high, being as much as 30 percent of the maximum shear rates obtained on impeller blades (Wichterle et al., 1984).

Wichterle et al. (1988) used the same electrochemical method to measure local mass transfer coefficients from which shear stress and shear rates were obtained. The relationship between shear stress τ and the electrode current and the definition of the wall friction factor were given above. The most important wall shear stress was considered to be caused by the impinging jet of the impeller discharge flow. Plots of f_w were given as a function of wall location, impeller Reynolds number, D/T ratio, and impeller-height-to-tank-diameter ratio. Typically $f_w(\mathrm{Re})^{1/2}$ varied from 1 to 40 for a pitched blade turbine pumping down in the standard configuration. The data for a down-pumping propeller and the disk style impeller varied in the same range. For the pitched blade turbine, the maximum value of wall shear stress was correlated by:

$$\tau_{max} = 10(\rho\mu)^{1/2}N^{3/2}D^3(D + C)^{-2} \tag{4.133}$$

for $0.2 < D/T < 0.4$, $0.2 < C/T < 0.4$, and $6000 < \mathrm{Re} < 110{,}000$. The data for $f_w\mathrm{Re}^{1/2}$ of an up-pumping propeller was quite different from the down-pumping data for the same impeller, so pumping direction can significantly affect wall stresses.

Shear rates for other impellers have not been reported. Elongational deformation rates have not been reported for agitated tanks generally and for the various vortex flow systems originating from the impeller.

Specific studies of energy spectra

The energy spectra and supporting theory were developed by a number of investigators including: von Karman and Howarth (1938), Kolmogorov (1941), Obukhoff (1941), Onsager (1945), Lin (1947; 1948), von Karman (1948), Kovasznay (1948), von Weizsacker (1948), Heisenburg (1948), Chandrasekhar (1949), von Karman and Lin (1949), and Tchen (1953). A clear explanation of the energy spectra can be found in Brodkey (1967).

The energy spectra describes how energy is transferred in the turbulent flow field from the large eddies to the smaller eddies. Spectra can be considered to be divided into three major regions: a large eddy range, an energy containing region, and a universal equilibrium region which is divided into an inertial subrange and a viscous dissipation subrange.

Kim and Manning (1964) measured the turbulence energy spectra in a baffled tank of standard disk style turbine geometry and studied

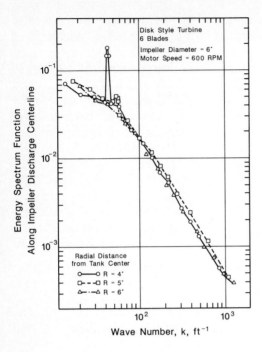

Figure 4.25   Radial energy spectrum as a function of wave number and radial distance from the impeller along the blade centerline. *(From W. J. Kim and F. S. Manning, AIChEJ., 10, 747, 1964. Reproduced by permission of the American Institute of Chemical Engineers.)*

the effects of impeller size and liquid viscosity on the spectra. They obtained the radial one-dimensional energy spectra, shown in Fig. 4.25, for the disk style turbine but essentially found the spectra insensitive to impeller size, impeller rotational speed, and probe position in the impeller stream. Close to the impeller, a single peak appeared in the spectra at the blade passing frequency, indicative of the pulsing nature of the discharge flow from the impeller. These spectra peaks decreased rapidly with distance from the impeller. Kim and Manning concluded that spectra data could probably not be used to determine the effects of design and operating variables on mixing and chemical reactions. Manning and Wilhelm (1963) also found that concentration spectra appeared similar to each other and roughly independent of conditions as well. The work will be reviewed below.

Fort et al. (1974) considered that the large vortex systems shed from the disk style impeller carried the largest fraction of energy supplied by the impeller. In the energy spectra taken close to the impeller, peaks were observed primarily at the blade passing frequency with harmonics also present at 0.5 and 1.5 times the blade passing frequency. As others have done, Fort et al. also noted the similarity in the energy spectra and independence from the impeller size and rotational speed. The peaks in the spectra occurring at the blade passing frequency decreased rapidly with distance from the impeller. The $-5/3$

inertial region and -7 dissipation region were also noted in the spectra.

Nishikawa et al. (1976) measured the turbulent energy spectra in baffled tanks, 0.15 to 0.6 m in diameter, agitated by disk style turbines. The impeller stream was nonisotropic, whereas in the bulk of the tank, the flow was isotropic. Although appearing roughly the same, the energy spectra did vary with conditions in this work. The spectra did not show the blade passing frequency as a dominant frequency. Spectra were collapsed to one spectrum, $E_1/u_1'^2$, using the mean square velocity fluctuation. This energy spectrum and the microscale of the turbulence showed little variation with impeller rotational speed. However, the microscale of turbulence varied with position and the size of the tank: 0.05 to 1.0 mm in a 0.15-m-diameter tank to 0.2 to 4 mm in a 0.6-m tank.

van der Molen and van Maanen (1978) obtained a fairly complete set of energy spectra as a function of radial position for a disk style turbine. Their spectra had a frequency resolution of 5 percent which permitted the recording of harmonics in the spectra at multiples of the blade passing frequency, particularly in the spectra of the axial component. Subharmonics were also noted. The basal spectra without the harmonics was modeled as a constant, $E_0(T)$, for $k < k_0(T)$:

$$E(k) = E_0(T) \qquad\qquad (4.134)$$

and
$$E(k) = E_0(T)\left(\frac{k}{k_0}\right)^{-5/2} \qquad\qquad (4.135)$$

for $k > k_0(T)$ where k_0 is the cutoff wave number of the spectra. k_0 varied with T^{-1}, and $E_0(T)$ varied with D^2. Generally, the spectra were not in an equilibrium state as represented by a $-5/3$ region. The high wave number portion of the spectra varied with $D^{-1/2}$ which indicated decreasing energy in the small eddies with increasing vessel size. The $-5/3$ region appeared in the spectra taken at the wall indicating a state of equilibrium in the energy transfer. The energy in the small eddies varied with T^{-1}, which indicated that the micromixing in large agitated tanks was not as intense as in small agitated tanks. This agrees with observations made by Tatterson et al. (1980) concerning the pitched blade turbine.

The flow behavior which established the spectra was summarized as follows. The turbulent kinetic energy was considered to be contained in the trailing vortex systems close to the impeller. These vortices decayed to low wave number eddies which transferred their energy to smaller eddies further away from the blade in the region between $1.5 < r/r_0 < 2$.

Most studies of the energy spectra have shown that spectra are not useful in discerning differences in the fluid mechanics occurring in the flow under different conditions at the same scale. The usefulness of the energy spectra in mixing remains unclear from these studies. Spectra may be more useful in comparing flow behavior at different scales.

Specific studies on energy dissipation rates

Energy dissipation rates are distributed throughout a mixing vessel, and differences of several orders of magnitude can be expected. An understanding of how this quantity is distributed in the tank provides information as to the state of mixing in the tank.

Metzner and Taylor (1960) reported local energy dissipation rates for the disk style turbine as shown in Fig. 4.26. The data show large changes in dissipation rates with distance from the impeller and with rotational speed.

Okamoto et al. (1981) reported distributions of energy dissipation rates for baffled and unbaffled tanks. In their study, the local energy dissipation rate was found by integrating the dissipation of the energy spectra:

$$\epsilon = 15\nu \int_0^\varkappa k^2 E(k) \, dk \qquad (4.136)$$

Dimensionless energy dissipation rates (i.e., actual dissipation rates divided by power input per volume) were reported spatially. For geo-

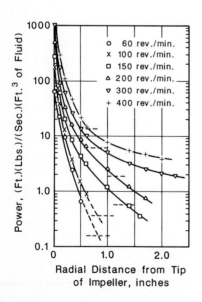

Figure 4.26 Local power or energy dissipation rates as a function of radial distance from the impeller, 1.2 percent CMC, 4-in impeller diameter. *(From A. B. Metzner and J. S. Taylor, AIChEJ., 6, 109, 1960. Reproduced by permission of the American Institute of Chemical Engineers.)*

metrically similar tanks, the distributions of energy dissipation rates in baffled and unbaffled tanks were found to be very similar and independent of impeller Reynolds number. As the D/T and W/T ratio increased, the dissipations became more uniform. For baffled tanks with flat blade impellers, the actual distributions were crudely represented using two levels of energy dissipation; one for near the impeller ϵ_i and another ϵ_b for the bulk of the tank. These were given as:

$$\epsilon_i = C_1 \left(\frac{P}{\rho V}\right)\left(\frac{W}{T}\right)^{-1.38} \exp\left(\frac{-2.46D}{T}\right) \tag{4.137}$$

and $\qquad \epsilon_b = C_2 \left(\frac{P}{\rho V}\right)\left(\frac{W}{T}\right)^{0.32}\left(\frac{D}{T}\right)^{0.78}$ $\qquad\qquad$ (4.138)

where $0.05 < W/T < 0.3$, $0.25 < D/T < 0.70$, and C_1 and C_2 are constants. For a six-blade disk style turbine impeller with a blade width of $0.2D$, the equations for the two dissipation levels were:

$$\epsilon_i = 7.8 \left(\frac{P}{\rho V}\right)\left(\frac{D}{T}\right)^{-1.38} \exp\left(\frac{-2.46D}{T}\right) \tag{4.139}$$

and $\qquad \epsilon_b = 0.90 \left(\frac{P}{\rho V}\right)\left(\frac{D}{T}\right)^{1.10}$ $\qquad\qquad$ (4.140)

Okamoto et al. found that the two-region model was a good representation of the actual energy dissipation distribution.

Laufhutte and Mersmann (1985) performed a study of power dissipation in an agitated tank using a disk style turbine. Using an analogy to pipe flow, the major contribution to the power number N_P for the transition region $10 < \text{Re} < 1000$ was modeled as:

$$N_{PL} = k_1 \left(\frac{T}{D}\right)^2\left(\frac{U_z}{\pi ND}\right)^3 \text{Re}^{-1} \tag{4.141}$$

where N_{PL} is a laminar power number, k_1 is a friction factor, and U_z is the mean axial velocity for the entire tank. Above 1000, the turbulent contribution to the power number was modeled as:

$$N_{PT} = k_2 \left(\frac{T}{D}\right)^3\left(\frac{u'}{\pi ND}\right)^3 \tag{4.142}$$

where u' is the mean value of the fluctuating velocity for the whole vessel and k_2 is another friction factor. The total power number was:

$$N_P = N_{PL} + N_{PT} \tag{4.143}$$

Above an impeller Reynolds number of 10,000, N_{PL} can be neglected.

In the turbulent regime, Laufhutte and Mersmann provided turbulent velocity fluctuations for the outflow of the disk style turbine as well as the axial velocity fluctuation distributions throughout the tank. For the discharge outflow, they removed the periodic component from the velocity fluctuations. For a fixed T/D ratio and using the equation for N_{PT}, the energy dissipation rates were calculated from the velocity fluctuations using:

$$\epsilon = A\left(\frac{u'^3}{D}\right) \tag{4.144}$$

where A was considered a constant between 6.0 and 6.5 and independent of tank geometry and impeller. The turbulent velocity fluctuations u' were either the turbulent portion of the velocity fluctuations without the periodic component or the total velocity fluctuations including the periodic component. Isoenergetic lines of $\epsilon/\bar{\epsilon}$ ($\bar{\epsilon}$ being the mean energy dissipation rate) were provided which showed the energy dissipation spatially throughout the tank. $\epsilon/\bar{\epsilon}$ values ranged from between 0.08 in the tank bulk to 30 in the impeller discharge flow. The periodic contribution to the energy dissipation was 60 to 70 percent of the total in the discharge flow near the impeller but decreased to a negligible contribution at $2r/T = 0.6$ where $\epsilon/\bar{\epsilon}$ equaled about 13.

In order to calculate the decay of concentration fluctuations, length scales and energy dissipation rates are needed. Patterson and Wu (1985) calculated the integral length scale of the turbulence L_s and the energy dissipation rate ϵ spatially in the tank. The integral length scale L_s was the sum of the individual length scales L_{si} of the three directions obtained from:

$$L_{si} = \left(\frac{U_i}{\overline{u_i u_i}}\right) \int_0^{\infty} \overline{u_i(t)u_i(t + \tau)}\, d\tau \tag{4.145}$$

The energy dissipation rate was calculated from:

$$\epsilon = A k^{3/2} L_s \tag{4.146}$$

where k was the turbulent kinetic energy obtained from:

$$k = \tfrac{1}{2} \sum_{i=1}^{3} \overline{u_i^2} \tag{4.147}$$

Patterson and Wu obtained the turbulent velocity fluctuations in the impeller discharge stream and in the bulk of the tank to calculate L_s and ϵ from these equations. The value of $k^{3/2}/L$ was integrated over the entire tank volume, and the constant A was found to be 0.4 which agreed with literature values for plane jets. Normalized energy dissipation rates in the impeller discharge stream were given as a function of radial position. No periodic component was noted in the turbulent energy dissipation distributions.

The differences in A values between Patterson and Wu (1985) and Laufhutte and Mersmann (1965) can be attributed to the differences between L_s and D. Typically, L_s is on the order of one-half the blade width; the blade width of a disk style turbine is typically $\frac{1}{6}$ the diameter. There is agreement between the two studies.

Experimental techniques

Experimental techniques and the development of experimental techniques are important in the understanding of the complex phenomena of mixing. Often the desired information cannot be obtained without first developing the experimental technique to acquire data.

Hot-wire or film and laser Doppler anemometry are very common experimental techniques which have been covered appropriately in the literature and thus will not be covered here. Hinze (1975) provided an excellent summary of hot-wire and film anemometry. Laser Doppler anemometry has been presented by many, including: Reed et al. (1977), Klaboch et al. (1988), and other authors whose work is cited above. Speckle and pulse laser velocimetry techniques are also available but have not been applied to agitated tanks. References for this technique include Adrian and Yao (1984; 1985) and Adrian (1984; 1985). Some other experimental techniques are included here.

Aiba (1958) measured velocities in the flow field of propellers, paddles, and turbines by measuring the deflection of a steel ball hung overhead by a thread. The velocity was determined from a force balance between the apparent weight of the ball and drag force.

Hjelmfelt and Mockros (1966) examined the various approximations to simplify the equation for the velocity of a particle in turbulent flow. Their results indicated that a particle follows the fluid motions in both phase and amplitude if the density ratio is unity. The particle was assumed to be entrained totally in an eddy. Substantial differences in amplitude and phase response occurred at various Stokes numbers when other density ratios were considered. Typically, when density matching is not done, large deviations between particle and fluid motions can be expected.

Gunkel et al. (1971) in an investigation of hot-wire anemometers found that these devices were inappropriate for use in flows of high turbulence intensity and in reversing flows. In the bulk of the agitated tank, in particular, they mentioned that the output of an anemometer was constant no matter how the probe was oriented. A single wire cannot detect the direction of the flow. To be able to detect reversing flow directions, two wires were used. A disk shield was also added to protect the probe from lateral velocity components. Although there are many difficulties with the probe as presented, such as flow disruption by the shield and the questions of lateral flow effects, the probe was shown to perform adequately in certain reversing flows.

Tracer particles have been used to follow the fluid motions in many visualization studies as cited above. To investigate this experimental technique, a specific study using high-speed cine photography (10,000 pps) was performed by Komasawa et al. (1974a; 1974b) in which instantaneous velocities of the tracer particles were determined from the films. Measurements were performed in two kinds of experimental systems: turbulent pipe flow and agitated tanks in the standard geometry. The turbulent microscales were on the order of 50 to 250 μm. The neutrally buoyant tracer particles comparable or smaller than the turbulent microscale were used. They found that large neutrally buoyant tracer particles, 900 μm, did not fully follow the turbulent eddies. Density differences attributed further to differences between eddy and particle motions. The presence of large particles in low concentrations had no effect on the energy spectra for the liquid, however. The energy spectra for both the pipe flow and impeller discharge flow were very similar indicating an independency of how the turbulence was generated.

Tatterson et al. (1980) developed a cross-beam laser correlation technique which parallels laser Doppler anemometry to some degree. The technique tracked particle shadows across arrays of photoelectric sensors. Two expanded and collimated laser beams were placed orthogonally in the same horizontal plane, and each beam was projected on a detection array. When a particle passed through the intersection of the two beams, particle shadows were cast upon the arrays. Time data from the arrays permitted the tracking of the particles as they passed through the intersection. Three-dimensional particle velocities were determined from the data and the geometry of the arrays. The work was also extended to much larger arrays by Yuan and Tatterson (1980) to obtain three-dimensional particle velocities and accelerations.

Sheu et al. (1982) for an agitated tank and Kent and Eaton (1982) for a cylinder developed a three-dimensional stereoscopic measurement technique to obtain particle paths spatially as a function of time in an attempt to obtain the entire flow field in such systems as a function of space and time. Some of the data is shown in Fig. 4.27. Sheu et al. specifically noted that idealized vortex models were conceptually reasonable in the initial understanding of the flow field near the impeller but were still highly qualitative. Much more data were needed for the characterization of flows in agitated tanks. Chang et al. (1985a; 1985b) extended the work by using image processing and au-

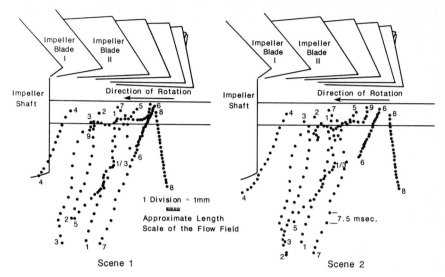

Figure 4.27   Tracer particle motions in the flow field of a pitched blade turbine. *(From Y. H. E. Sheu, T. P. K. Chang, G. B. Tatterson, and D. S. Dickey, Chem. Eng. Commun., 17, 67, 1982. By permission.)*

tomated tracking of particles to the point where the data needed for characterization could be obtained. However, such work needs to be extended to high particle concentrations to obtain structural information concerning the various flows. Studies using low particle concentrations cannot provide the spatial resolution necessary to determine structural information.

Faraday (1988) used a television camera linked to a computer to track neutrally buoyant flow followers in an agitated tank. Various types of data about the kinematics of the particle motions could be obtained with such a system. Tily et al. (1988) used acoustic emissions to monitor solid-solid mixing processes. The same techniques can be applied to other mixing processes such as chemical reactions.

Among the more exotic experimental techniques, Ali and Whittington (1979) mentioned the use of floating candles for the study of flow patterns in city reservoirs. Castellana et al. (1984) developed an approach to characterize the flow patterns and spatial dilution data in a container using a gamma ray camera.

Integral Circulation and Mixing Times

Ideally one would like to calculate the transport of scalars in a mixing tank. Unfortunately the flow field is turbulent which leaves significant uncertainty in the method of calculation and in any calculation results. It is also difficult to define exactly the flow patterns and the detailed flow behavior.

Circulation and mixing times are integral approaches which help in the understanding of scalar transport. To a great degree, the work has been successful and useful. Experimental data, correlations, and supporting phenomenological theories are available for use. Processes which occur in an agitated tank are distributed processes but are modeled as lumped systems by the use of circulation and mixing times.

Circulation times and pumping capacities

Circulation times are measures of the average velocity, bulk motion, or convective transport generated by the liquid pumping of the impeller in the stirred tank. Typically, only one circulation time is discussed for an impeller which is the controlling or overall circulation time. However, there are actually three circulation times present in an agitated tank, one for each direction.

Circulation times are associated with the tank's total volumetric flow rate, entrainment flow, and pumping capacity of the impeller. Turnover time is the time required for the tank contents to pass through the impeller or "turn over." Circulation times are associated with the direction of discharge for the different impellers and typically follow the classification of impeller type, i.e., axial impeller and radial impellers. However, it should be recognized that axial and radial flow impellers have substantial axial, radial, and tangential components in their circulation patterns and circulation times are weighted accordingly:

$$\theta_c = \frac{1}{1/\theta_r + 1/\theta_0 + 1/\theta_z} \tag{4.148}$$

However, the different circulation times are rarely reported or discussed and are implicitly incorporated in the measurement of an overall circulation time.

The total volumetric turnover rate in the tank is the total circulation rate occurring in the tank. Pumping capacity is the volumetric flow rate passing through the planes established by the impeller rotation. The entrained circulation rate is the difference between these. Entrained flow rate can be larger or smaller than the pumping capacity of the impeller depending upon geometry. Data are reported either in terms of pumping number N_Q or Q/ND^3 or in terms of the discharge coefficient K. The two report the same information since:

$$Q = KND^3 \tag{4.149}$$

Circulation time is also reported in such data since the total tank volume divided by the impeller pumping capacity Q is considered to be the circulation time.

The major variables affecting pumping capacity, circulation, and mixing times are the impeller rotational speed, the tank and impeller diameters, and the presence of baffles. Other geometry variables, e.g., blade number, blade width, impeller clearance, eccentricity, and shaft angle, also affect pumping capacity, circulation, and mixing times to some degree.

Axial flow impellers are used extensively in blending and mixing of miscible liquids, and studies on blending have been done in unbaffled tanks. For axial flow impellers, the need for baffling is not as great as for radial flow impellers. So long as the central surface vortex does not adversely affect operation, axial flow impellers in unbaffled tanks perform adequately especially if the feed is added directly into the impeller stream. Axial flow impellers also consume much less power than radial impellers. In the transition region between laminar and turbulent mixing, baffles are undesirable and may actually increase mixing times.

Pumping capacity is an important scaleup parameter and useful in correlating data of experiments. In many categories of mixing, e.g., heat transfer and power consumption, pumping capacities can be much more important than the impeller Reynolds number.

A considerable number of studies on circulation times rests on Danckwerts's study (1953) in which the average residence time of material in a volume is considered to be the volume of the container divided by the volumetic flow rate. The difficulty comes in the determination of the reactor volume. If there are unknown absolute dead zones in the tank, then the residence time of the material in the reactor volume is different from that expected.

In the discussion to follow, Q is used to indicate the impeller pumping capacity, Q_e is used to indicate entrainment flow, and Q_T is used to indicate total circulation. Unless otherwise stated, the following review of pumping capacities and circulation time data is for baffled tanks in the standard configuration.

Radial flow impellers.   Norwood and Metzner (1960) studied total volumetric flow rate in tanks having the standard configuration using a probe to measure velocity profiles around various turbines, e.g., the disk style turbine. The volumetric flow rate of the impellers was correlated by:

$$Q = 9.0 \times 10^{-4}ND^2W\left(\frac{D^{0.4}\rho}{\mu}\right)^{0.5} \tag{4.150}$$

where W is blade width.

Holmes et al. (1964) modeled the pumping capacity of disk style impellers located centrally in different size tanks having the standard geometric configuration using Eq. (4.149). The discharge coefficient K was found to be equal to 1.3 and independent of the impeller Reynolds number. The pumping capacities were obtained by measuring radial discharge velocities and integrating over the profiles to obtained volumetric flow rates. Circulation capacity was also determined which included the flow discharged through the impeller (i.e., the pumping capacity) and the surrounding flow entrained in the discharge flow. The ratio cf the circulation capacity to pumping capacity was found to be 1.8. Circulation times were obtained using a pulse tracer conductivity method and correlated with the impeller rotation speed and the impeller and tank diameters as:

$$ N\theta_c \left(\frac{D}{T}\right)^2 = K_1 \tag{4.151} $$

where the constant K_1 was 0.85 for impeller Reynolds numbers greater than 10^3 as shown in Fig. 4.28. Below an impeller Reynolds number of 100, the fluid near the wall was stagnant and circulation times were poorly defined. The circulation time was considered to be equal to the average circulation loop length divided by the average velocity where the loop length was determined mainly by liquid height and tank diameter. An estimate for the circulation velocity can be obtained by rearrangement of the equation above to:

$$ V_c \cong \frac{T}{\theta_c} \cong (ND)\left(\frac{D}{T}\right) \tag{4.152} $$

which shows that the circulation velocity is proportional to both the impeller tip speed ND and the D/T ratio.

For one of the smaller impellers, Holmes et al. also found the circu-

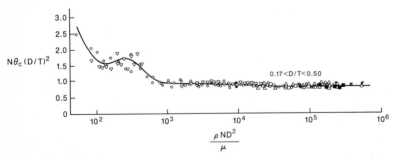

Figure 4.28 Circulation data as a function of impeller Reynolds number. *(From D. B. Holmes, R. M. Voncken, and J. A. Dekker, Chem. Eng. Sci., 19, 201, 1964. By permission.)*

lation times were 10 to 30 percent larger (i.e., poorer circulation) than expected because of an effect of the size of the shaft and coupling collar of the impeller on the discharge velocities. Holmes et al. found that the mixing times were reduced when the impeller was placed one-third tank diameter from the bottom instead of being centrally located. The circulation data in the transition region were also obtained but were difficult to characterize.

Cooper and Wolf (1967) developed an equation for pumping capacity of a disk style turbine that, when simplified, reduced to Eq. (4.149). The discharge coefficient K was a function of the W/D ratio (i.e., a blade-width effect) and the conditions of the flow field assumed in the development of the equation. The discharge flow from one of the turbines was not purely radial, and the discharge angle was found to be dependent upon the impeller blade geometry. Cooper and Wolf also showed the interrelationship between power and pumping:

Power = velocity head × pumping capacity × density

or: $$P = \left[\frac{(\pi DN)^2}{2}\right](0.75ND^3)\rho g_c \qquad (4.153)$$

where the discharge coefficient was assumed to be 0.75. Rearranging:

$$N_P = \frac{Pg_c}{\rho N^3 D^5} = 3.6 \qquad (4.154)$$

Assuming a power number of 5.5 for the impeller, 70 percent of the power input went into driving the circulation flow. The remaining 30 percent of the power was dissipated in the impeller region.

For the pumping capacity of the disk style turbine, DeSouza and Pike (1972) also used Eq. (4.149). The discharge coefficient K was found to be 0.95 ± 0.28. They further noted that the physical properties of the fluid had no effect on circulation times. Scaling according to blade width was appropriate for changes in blade width. Gunkel and Weber (1975) obtained a K value of 1.0 and cited Cooper and Wolf (1967) as having obtained a K value of 0.85 for the same impeller.

Fort et al. (1972) and Kratky et al. (1974) reported discharge coefficients of 0.70 and 0.76 for D/T ratios of ¼ and ⅓, respectively, for disk style turbines. They also noted substantial increases in pumping number over relatively short distances from the impeller, indicating the effect of entrainment flow on the measurement of pumping capacities.

Middleton (1979) described the use of a nearly neutrally buoyant radio pill to obtain circulation times in industrial-scale tanks. The technique used an antenna to detect the radio pill as it passed through the impeller region. The results indicated that circulation in large tanks was much longer than expected from traditional correlations which implied that mixing in industrial-scale tanks was much worse than expected. The circulation times for the disk style turbine were correlated using a dimensional correlation given as:

$$N\theta_c = 0.5V^{0.3}\left(\frac{T}{D}\right)^3 \qquad (4.155)$$

where θ_c is in seconds and V is the liquid volume in cubic meters. There is some question whether the slightly buoyant radio pill could have been lost in long circulation loops in low-velocity regions of the industrial-scale tanks, hence giving rise to the long circulation loops.

Sasakura et al. (1980) took a somewhat different view of liquid circulation in their study by considering that each point in an agitated tank had its own characteristic circulation pattern. General flow patterns were considered unimportant; however, the individual circulation paths were a characteristic of the impeller shape. Their study was performed in a baffled tank using a flat blade impeller. The number of blades and blade width were varied: two to eight blades and blade width of between 0.05 and 0.2 of the tank diameter. The time response of a tracer and the mixing time were inversely proportional to the impeller rotational speed. For different geometries of impellers studied, circulation time was correlated as:

$$N\theta_c = 4.10\left(\frac{T}{D}\right)^{1.4}\left(\frac{T}{W}\right)^{0.5} n_p^{-0.55} \qquad (4.156)$$

where n_p is the number of impeller blades and W is the impeller blade width. The circulation and mixing times were found to be independent of position.

Bertrand et al. (1980) investigated pumping capacities for four differently shaped radial flow impellers: a normal disk style turbine at two different clearances, two different shrouded and angled blade disk style impellers and an eight-angled blade, disk style turbine having a blade shape modification. Shrouded impellers are impellers with covered blades. The pumping capacity for a normal disk style turbine was a function of clearance: at $C = 0.5T$, $Q/ND^3 = 1.19$ and, at $C = 0.33T$, $Q/ND^3 = 1.61$. The two shrouded impellers had lower pumping numbers, 1.11 and 0.65, respectively; the pumping data showed a blade-width effect as well. Shrouded blade impellers had lower pumping because the shrouds, covering the blades, changed the nature of the flow. The eight-blade disk turbine had a pumping number of 1.31 despite the increase in the number of blades. This was attributed to the shape of the impeller blades.

Sato (1980) interrelated impeller pumping capacity of the disk style turbines to the tank width and impeller off-bottom clearance in rectangular mixing tanks. K varied between 0.5 to 1.0, depending upon the particular geometry and impeller location. The impeller discharge efficiencies N_Q/N_P were slightly higher for the rectangular tanks over that of cylindrical tanks.

Revill (1982) extensively reviewed pumping capacities for the standard disk style turbine. Literature values were noted to vary from 0.6 to 3.0 for geometrically similar tanks. The major reasons for such a large variation were due to differences in the calculation or the measurement of the discharge flow rates. Revill recommended:

$$N_Q = 0.75 \pm 0.15 \qquad (4.157)$$

for a D/T ratio between 0.2 and 0.5 and a clearance ratio C/T between 0.3 and 0.5.

Costes and Couderc (1982), using hot-film anemometry to establish velocity profiles, determined the pumping capacity for a disk style turbine centrally positioned in a tank in the standard configuration. They found the pumping number Q/ND^3 to vary between 0.91 and 1.05.

Keller (1985) reported pumping capacities for a disk style impeller in the standard configuration and noted a decrease in pumping number from 1.2 at an impeller Reynolds number of 4.2×10^4 to 1.0 at 6.7×10^4. This drop in the pumping rate was attributed to surface entrainment. The width of the impeller blade was large at $T/5$, which may be the reason for the large pumping number.

Sano and Usui (1985) reported correlations for circulation flow for turbines and paddles of various geometries. For paddles:

$$N_Q = 1.3\left(\frac{D}{T}\right)^{-0.86}\left(\frac{W}{T}\right)^{0.82} n_p^{0.60} \qquad (4.158)$$

and for turbines:

$$N_Q = 0.80\left(\frac{D}{T}\right)^{-0.70}\left(\frac{W}{T}\right)^{0.65} n_p^{0.60} \qquad (4.159)$$

These correlations agree roughly with Sasakura et al. (1980), cited above, in the effects of blade width W and blade number, n_p.

Magni et al. (1988) studied mean circulation rates in flat- and dished-bottom tanks using integrated LDA data for the disk style turbine. For the flat-bottom tank, the D/T and C/T ratios were 0.33 and 0.5, respectively; for the two rounded-bottom tanks, the D/T ratio was 0.45, and the C/T ratios were 0.33 and 0.5. The discharge profiles of the impeller just at the impeller exit were identical. At the wall, the discharge flow divided into two vertical wall flows, creating two large circulation loops in the upper and lower parts of the tank. Except in the discharge region, the turbulence was approximately homogeneous. For C/T = 0.33 geometries, the rms values were 1.25 higher in the lower part of the vessel however.

A number of different circulation rates were defined and were ratioed with impeller pumping capacity as listed in Table 4.1. As the data indicate, several conclusions can be drawn. The vertical position of the turbine affected the ratio of the upper and lower circulation rates. Larger D/T ratios caused lower relative circulation ratios in the tank, e.g., 1.99 as opposed to 1.34; 2.23 as opposed to 1.28; and 4.64 to 2.40. At large D/T ratios, the tank circulation was more dominated by the impeller. Although not listed, the angular circulation rates were 1.25 times the impeller pumping capacity for D/T = 0.45. In comparison with the tank circulation rates, the impeller circulation rates were low.

Armstrong and Ruszkowski (1988) reported pumping capacities in the range of 0.75 to 0.80 for the disk style impeller.

Wu and Patterson (1989) reported pumping capacities of the disk style turbine as a function of radial position. Q/ND^3 varied from 0.82 at the impeller tip to 2.0 at r = $2R$ away from the impeller, indicating considerable entrainment of fluid by the discharge flow.

The distribution of circulation times is also of interest. Bryant and Sadeghzaheh (1979), also using a radio pill technique, obtained mean values, standard deviations, and distributions of circulation times for flat blade impellers at low clearance in a baffled tank. A log-normal distribution was found to be a good approximation for the actual circulation time distributions. Mukataka et al. (1980; 1981), using a single impeller, found that the distributions of circulation times were log normal. For multiple impellers, the distributions of circulation times were log normal for each impeller. Funahashi et al. (1987) found bimodal distributions at low rotational speeds. Kudrna et al. (1989) found the distributions of circulation times for multiple impellers, e.g., flat blade and pitched blade turbines, much more complex than the log-normal distribution, particularly in the upper regions of the tank.

Propellers. Porcelli and Marr (1962) determined total circulation rate in the tank, impeller pumping rate, and entrainment circulation for propellers pumping downward in the center of the tank. Data were obtained experimentally using a neutrally bouyant tracer particle. The pumping capacity for propellers was found to follow Eq. (4.149), where the discharge coefficient K increased slightly with increasing impeller Reynolds number to a constant of 0.55 as shown in Fig. 4.29. The discharge coefficient was found to be independent of tank geometry and D/T ratio in the range studied. The total circulation rates in the tank were obtained and provided data on the entrainment flow Q_e, as

TABLE 4.1 Ratios of Vertical Circulation Rates to Impeller Pumping Capacities

D/T	C/T	Bottom shape	Upper wall flow	Upper tank return	Lower wall flow	Lower tank return	Total return	Upper/lower circulation
0.33	0.5	Flat	1.99	2.23	1.95	2.41	4.64	0.93
0.45	0.33	Dished	1.52	1.63	—	0.5	2.18	2.96
0.45	0.5	Dished	1.34	1.28	0.91	1.13	2.40	1.13

SOURCE: F. Magni, J. Costes, J. Bertrand, and J. P. Couderc, *Proc. 6th Eur. Conf. on Mixing, Pavia, Italy,* BHRA Fluid Eng., Cranfield, England, 7, 1988.

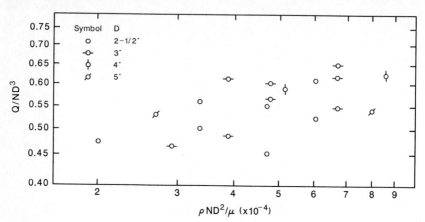

Figure 4.29 Correlation of propeller pumping capacity N_Q versus impeller Reynolds number. *(Reprinted with permission from J. V. Porcelli and G. R. Marr, I&EC Funds., 1, 172, 1962. Copyright 1962 American Chemical Society.)*

shown in Fig. 4.30. The ratio of entrainment flow to impeller pumping capacity ranged from 0.86 to 3.73. The entrainment flow Q_e was modeled as:

$$Q_e = K_2 Q \left[\left(\frac{T}{D} \right)^2 - 1 \right] \qquad (4.160)$$

where K_2 was equal to 0.16. The total circulation occurring in the tank for propellers was obtained by adding Q_e and Q. Although the pumping capacity was not dependent upon tank geometry, the en-

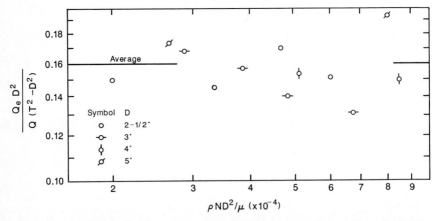

Figure 4.30 Correlation of entrainment flow data versus impeller Reynolds number. *(Reprinted with permission from J. V. Porcelli and G. R. Marr, I&EC Funds., 1, 172, 1962. Copyright 1962 American Chemical Society.)*

trained and total circulation flows were found to be dependent upon tank size.

Marr and Johnson (1963) reconsidered the work on the flow follower technique and assumed that the probability of a fluid particle entering a volume was proportional to the flow through the volume. They divided the tank into three zones: the impeller zone and zones above and below the impeller. The average circulation time was the sum of the times required to travel through each of these zones. Their final correlation for the impeller pumping capacity for a marine propeller was given as Eq. (4.149) where K was found to equal 0.61. They assumed that the marine propeller accelerated the flow from zero to an exit velocity of $4Q/\pi D^2$ and estimated the kinetic energy added to the flow by the impeller as $8K^2\rho N^2 D^2/g_c\pi^2$ and the power to pump the flow as $8K^3\rho N^3 D^5/g_c\pi^2$. Using K of 0.61, a pumping power number was estimated to be 0.18. Typical power numbers for the propeller are on the order of 0.3.

Sykes (1965) repeated the technique of Marr and Porcelli and obtained a similar correlation for the pumping capacity of a propeller. He found that the pumping capacity was independent of viscosity and that circulation time varied linearly with the liquid volume in the tank. The discharge coefficient was also less for smaller-diameter impellers. The average value of K was found to be 0.4.

Wolf and Manning (1966) obtained estimates of impeller pumping capacities and flow entrainment rates in baffled tanks for various turbines, propellers, and paddle impellers. The pumping capacity for propellers was found to follow Eq. (4.149) where K was 0.54. The entrainment was shown to be a complex function of distance from the impeller which roughly increased linearly with distance from the impeller.

Plion et al. (1985) provided pumping numbers for propellers in baffled and unbaffled tanks. For baffled tanks, N_Q varied between 0.43 and 0.47; for unbaffled tanks, N_Q was 0.19.

Merzouk et al. (1988) discussed the circulation capabilities of a submersible agitator. The impeller used in the study was a 0.2-m diameter, three-bladed propeller in a ring propector. The circulation experiments were carried out in a dished-bottom tank 1.6 m diameter. Values for $N\theta_c$ for different positions were found to vary around 110. The high $N\theta_c$ values were because of the large T/D ratio. This type of impeller does not require tank wall seals.

Pitched blade turbine. Medek and Fort (1979) noted the importance of convective circulation in mixing as being the source for the generation of turbulent diffusion or micromixing in an agitated tank. A measure of the ability of an agitator to produce turbulent diffusion is the pumping capacity of the impeller. Medek and Fort found that the pumping number for a six-blade, 45° pitched blade turbine was 0.92, and for three-blade turbines, 0.76. These pumping numbers were found to be independent of the D/T ratio for $D/T < 0.4$. For $D/T > 0.4$, the flow patterns changed drastically. The induced flow and total flow rate were found to increase linearly with increasing T/D ratios above 2. Essentially, the amount of induced and total circulation flow increased as the size of the circulation volume increased. The work, as presented by Medek and Fort (1979), permitted the calculation of pumping capacities for impeller with inclined blades at any angle. It is important to note that the pitched blade turbine can become a radial flow impeller at low clearances or at high D/T ratios, i.e., greater than 0.4 (Zwietering, 1958).

Weetman and Oldshue (1988) reported 0.56 as the pumping number N_Q for a fluidfoil impeller.

The interrelationship between pumping capacities and power number. Power input to mixing can be divided into power necessary to circulate the flow in the tank and power dissipated in the impeller. As a result, there exist relationships between impeller pumping capacities and power numbers.

Hiraoka and Ito (1975) developed a correlation between power input and impeller discharge flow using an angular momentum balance. Their final expression between the pumping number N_Q, defined as Q/ND^3, and power number N_P was:

$$N_Q = 0.26 \left(\gamma \frac{n_p W}{T} \right)^{0.33} \left(\frac{T^2 H}{D^3} \right)^{0.5} \left(\frac{N_P}{1 + \alpha} \right)^{0.5} \qquad (4.161)$$

In such correlations, the type of impeller need not be specified. Hiraoka and Ito also developed a relationship between the pumping number and the total circulation number N_{QC}, or Q_T/ND^3, as:

$$N_{QC} = 0.78(1 + \alpha)^{0.5} \left(\gamma \frac{n_p W}{T} \right)^{-0.33} \left(\frac{D}{H} \right)^{0.5} N_Q \qquad (4.162)$$

where α was the ratio of torque at the tank bottom to that at the tank side and was assumed to be 0.2. The parameter γ was a complicated function equal to $[(D/T)^5 \ln (T/D)/\beta^5]$ where $\beta = 2 \ln [(T/D)/(T/D - D/T)]$. The group $\gamma n_p W/T$ ranged from 0.08 to 1.0. Nagase et al. (1974), as cited by Hiraoka and Ito, provided a similar development which interrelated circulation flow and power draw using a vorticity balance and dimensional analysis. The results indicated that the impeller discharge flow and the total circulation flow were proportional to the square root of the power input.

Nagase et al. (1977a; 1977b; 1983) developed a relationship between

wall pressure and circulation flow in an agitated tank using an overall vorticity balance between the impeller blade and the circulation
flow. The expression was applied to the flow generated by flat blade
and disk turbines, pitched blade turbines, and propellers for various
ranges of D/T, W/D, and n_p in baffled tanks. By measuring the wall
pressure distributions, estimates of the discharge and circulation
flows were obtained. Higher circulation flows caused sharper drops in
wall pressure distributions. Their final correlations, for circulation
Reynolds numbers $Q_T/T\nu$ greater than 500 (Q_T is the total circulation), were expressed in terms of a circulation constant C_R which was
defined as $N_P/(N_Q D/T)^2$. The circulation constant C_R is a measure of
power input to pumping. Upon rearrangement, the expressions
became:

$$N_P = 2.3\left(\frac{N_Q D}{T}\right)^2 \tag{4.163}$$

for propellers and:

$$N_P = 5.7\left(\frac{N_Q D}{T}\right)^2 \tag{4.164}$$

for flat blade and pitched blade turbines. Nagase and Kikuchi cited
Nagata as providing:

$$N_P = 8.1\left(\frac{N_Q D}{T}\right)^2 \tag{4.165}$$

for disk style turbines. Below the circulation Reynolds number of 500,
the circulation constant C_R approached the laminar flow region where
C_R varied inversely with the circulation Reynolds number $Q_T/T\nu$.

Sano and Usui (1985) provided correlations between power number
and the discharge flow number for the two impellers:

For paddles:

$$N_P = 4.3 N_Q^{1.34} \tag{4.166}$$

and for turbines:

$$N_P = 6.6 N_Q^{1.34} \tag{4.167}$$

for Re > 5000. The tank-impeller geometry varied between $0.3 <
D/T < 0.7$, $0.05 < W/D < 0.4$, $2 < n_p < 8$, $H = T$, with four baffles,
$B/T = 0.1$. The number of circulation times for mixing m was corre-
lated as:

$$m = 3.8\left(\frac{D}{T}\right)^{0.5}\left(\frac{n_p W}{T}\right)^{0.1} \tag{4.168}$$

Power draw was measured using a torque transducer mounted on the shaft, and the discharge flow number was measured using the flow follower technique of Porcelli and Marr.

Mersmann and Laufhutte (1985) suggested another time measure θ_{macro} to characterize macromixing. The shortest macromixing time which can be expected for standard geometries was correlated as:

$$\theta_{macro} = 7.3\left(\frac{T^2}{\bar{\epsilon}}\right)^{0.333} \tag{4.169}$$

which is another relationship between power consumption and mixing time. $\bar{\epsilon}$ is the mean energy dissipation rate or power consumption.

Other studies. McManamey (1980) developed a simple circulation model for batch mixing in a baffled tank based upon a mixing rate constant k in a decaying exponent e^{-kt} which modeled mixing. Mixing was complete at large kt values. The constant k was inversely proportional to both the mean circulation and the mixing time. The mean circulation time was proportional to the length of a representative recirculation loop divided by an average local velocity. McManamey considered the length of the loop to reach to the furthest parts of the vessel. The model required estimates for velocity data which were obtained from the literature. For radial flow impellers, the average velocity of the impeller discharge close to the wall was selected. For axial flow impellers, the average discharge velocity at the impeller was selected as being a representative velocity. Generally, the model performed well for clearances between $0.3T$ and $0.6T$.

Draft tubes and baffling. Tatterson (1982) performed a study of the effects of draft tubes and a cone bottom on circulation times for a pitched blade turbine and a disk style turbine in a 0.9-m-diameter tank in an unbaffled geometry and in the standard geometry using the technique of Porcelli and Marr (1962). The impeller Reynolds number was varied from 5×10^4 to 1.5×10^5. Two different sized draft tubes were used, 0.61 and 0.457 m in diameter equipped with 10 percent baffles, height 0.457 m, with the bottom of the draft tube placed approximately at the impeller height. The larger-diameter draft tube provided approximately equal flow areas in the tube and outside annulus. The cone bottom was 0.33 m in diameter and was placed below the impeller shaft.

In studies performed without the draft tube, the circulation times for the unbaffled tank were found to be actually lower than for the baffled tank. However, the standard deviations in the circulation times for the unbaffled case were larger than those of the baffled case, and lower mixing times were obtained in the baffled tank for the same conditions.

For the unbaffled tank, there existed two regions in the flow. Under the conditions studied, a well-mixed region surrounded the impeller. Poor circulation occurred in the bulk of the tank. For such situations, circulation times cannot be used to judge mixing performance or momentum transfer in the tank. Instead, Tatterson suggested that the standard deviation of the circulation times actually was a measure of the momentum transfer occurring in the tank. Low standard deviations indicated more uniform momentum transfer and better mixing performance.

The circulation times obtained using draft tubes in baffled tanks were as much as 20 percent higher than without the use of draft tubes. However, the standard deviations in circulation times obtained using draft tubes were lower. The use of a cone bottom reduced the circulation times and their standard deviations as well. In comparing data between the different draft tubes, the smaller-diameter draft tube performed slightly better than the larger-diameter draft tube. The baffled tank with draft tubes performed better than the unbaffled geometry. Shiue and Wong (1984) have also studied the effects of draft tubes.

Eccentricity.   Baffles are used to prevent solid body rotation and to enhance mixing. Placing an impeller off-center produces the same effect. Such arrangements also change flow patterns, circulation, and mixing times, entrained or induced flow and total circulation.

Medek and Fort (1985) studied pumping capacity and total circulation for the pitched blade turbine for various baffle-eccentricity arrangements for $2.5 < D/T < 5.0$ and $0.5 < C/D < 1.5$. No central vortex was observed in the studies, and the influence of the Froude number was not observed. Flow patterns were observed to change significantly at an eccentricity of $T/4$. Generally it was found that increasing eccentricity reduced the impeller pumping and overall circulation flow except in one geometry having three baffles. In this geometry at a small eccentricity of $T/12$, the overall circulation flow increased by about 10 percent over that of the typical standard geometry. The three-baffle arrangement at low eccentricity was also found to be the most energy efficient in terms of pumping.

Nontraditional impellers.   Nagase (1987) discussed the circulation profiles and discharge velocity for a reciprocating impeller. The discharge flow rate was found to be independent of Reynolds number in the turbulent regime and a function of reciprocation rate. The circulation constant C_R was 9.4 for this impeller which was much higher than the C_R values for paddles and disk style turbines given above. The dissipation of power was higher in the impeller region for the same circulation flow for this impeller than for paddles and turbines.

Nagase et al. (1988) studied the circulation and discharge flow of reciprocating impellers made from beams having a trianglar cross section. Such impellers can generate high shear and have been used as dispersers of high-polymer powders. The flow patterns of single- and dual-impeller systems were shown in the work. These fluctuated extensively in the tank because of the reciprocating action, but an overall circulation pattern was produced which increased with increasing reciprocating frequency. The pumping number N_Q, or Q/ND^3, was independent of Reynolds number above about 10^3 and varied between 0.15 and 0.30 as a function of the D/T ratio.

Computational models. Computational methods have been developed to simulate circulation of material in agitated tanks.

Mann et al. (1981) developed a two-dimensional network of interconnecting cell loops to model the circulation flow in a tank. The main circulation flow through the cells was assigned on the basis of circulation times for a flow follower and from velocity data. The cells were assumed to be under plugflow conditions. A switching parameter was used between cells of adjacent loops to account for the turbulent diffusion in the mixing.

The model was used to simulate the various pathways taken by a fluid particle in the tank. A number of predictions were performed, including the number of passes through the various circulation loops in the upper and lower compartments of the tank. The results of the circulation pathways were found insensitive to the switching parameter used to model turbulent mixing between the circulation loops.

Theoretical models of pumping capacities. van de Vusse (1955a; 1955b), Cooper and Wolf (1967), Todd (1967), and Revill (1982) presented theoretical discussions of impeller pumping capacities. However, until the nature of the impeller flows has been fully obtained and understood experimentally, theoretically derived models can only be viewed pedagogically.

Mixing under turbulent flow conditions

The following contains a review of mixing of tracers, mixing of two liquid volumes, and mixing in continuous flow systems. The first is by far the most studied. Mixing of liquid volumes and mixing in continuous flow systems require more study. Some discussion on transitional mixing is also included.

Mixing is the process by which a nonuniform system is made uniform. An ideal mixture is generally conceived as having the condition where the concentration or other scalars are uniformly distributed throughout the tank and at the exit. Such a situation is convenient mathematically since a lumped systems approach can be used to model the process. In some cases, an ideal mixture can occur in a process fairly easily with proper mixing equipment design and with appropriate bulk motion and turbulence levels. In other cases, an ideal mixture is only approached in practice.

Mixing becomes important in determining the residence time or the ratio of tank size to feed rate necessary to approach ideal mixing. Such information is basic to reactor design since mixing times and the desire to obtain ideal mixing will determine the physical size of the reactor and mixing equipment. The objective is to obtain ideal mixing by having the mixing time substantially less than the residence time of material in the tank.

If ideal mixing is not reached, then the process becomes a spatially and temporally distributed turbulent flow process. In many processes, ideal mixing is not obtained. The instantaneous dispersion and mixing of feed throughout a tank does not occur.

Typically, the flow field generated in an agitated tank, baffled or unbaffled, has fairly complicated circulation patterns. Generally too, the impeller or jet region is well mixed while the surrounding tank bulk is relatively unmixed. To be able to reach the surrounding fluid in the outer tank bulk and have it pass through the impeller region to accomplish mixing, several tank volumes have to pass through the impeller. It is the dilution of these relatively large quiescent regions in the tank bulk that requires mixing times be longer than the mean circulation times.

The mixing of one liquid into another liquid under turbulent flow conditions involves bulk motion or convective transport, turbulent diffusion, and molecular diffusion. Beek and Miller (1959) divided the mixing process up into three simultaneous stages: (1) the distribution of one material into another, (2) the breakup of this distribution into smaller portions, resulting in increased surface area, and (3) the molecular diffusion of the materials into each other. Convective transport can be considered macromixing, and turbulent diffusion is micromixing.

Such terms as micromixing and macromixing are not very well defined in application, and the distinctions between convective transport, bulk motion, macromixing and turbulent diffusion, and micromixing are not entirely clear. Ultimately, mixing is only achieved by molecular diffusion. However, the hydrodynamic conditions under turbulent flow greatly increase the speed at which mixing is achieved and are the time-controlling factors in achieving a uniform mixture.

In a very real sense, the turbulent hydrodynamic conditions help to maintain high concentration gradients as well as create large interfacial areas between the materials being mixed permitting diffusion to mix more effectively. In laminar mixing, mixing is achieved by the creation of interfacial area and molecular diffusion; both are very slow processes compared to turbulent mixing.

To maintain turbulence in the tank, an energy source is needed which is typically either an impeller or a jet. The product of power input and mixing time is the input of energy which accomplishes the mixing. However, there is no one-to-one correspondence between power input and mixing time since power input may go to mix material that has already been mixed.

Under turbulent flow conditions, density is a major consideration in mixing, and density difference will influence mixing times. Under the transition regime, both density and viscosity are major considerations and both will influence mixing times. Hence, the characterization of viscosity for non-Newtonian fluids is necessary in transitional regime mixing.

For axial impellers, pumping direction is not considered important in mixing of scalars and most work have the axial flow impellers pumping downward.

Another consideration in the discussion of tank flows is the existence of secondary motions and compartmentalization of the flow. Toroidal vortices are examples of secondary flows. The secondary flows are highly significant in laminar mixing but are less so in turbulent mixing due to turbulent diffusion and the ability of turbulence to transport momentum. However, secondary flows do occur in turbulent mixing. Compartmentalization of flow causes larger mixing times. Essentially the flow divides into circulation loops or compartments which come into contact only in a limited volume. Mixing between the two volumes is limited as a result.

Various mixing scales. Various scales of length, velocity, and time for the eddies were developed in turbulence theory using dimensional analysis. Various length scales for concentration can also be devised in the same manner. A length scale by Batchelor (1959) is:

$$\eta_b = \left(\frac{\nu D_L^2}{\epsilon}\right)^{1/4} \tag{4.170}$$

or by Corrsin (1964):

$$\eta_c = \left(\frac{D_L^3}{\epsilon}\right)^{1/4} \tag{4.171}$$

Both involve liquid diffusivities. In terms of order of magnitude for liquids, $\eta > \eta_b > \eta_c$ where η is the Kolmogorov length scale.

The relationship between the Batchelor microscale and the Kolmogorov microscale is:

$$\eta_b = \left(\frac{\nu D_L^2}{\epsilon}\right)^{1/4} = \left(\frac{\nu^3}{\epsilon}\right)^{1/4}\left(\frac{D_L^2}{\nu^2}\right)^{1/4} = \frac{(\nu^3/\epsilon)^{1/4}}{Sc^{1/2}} = \frac{\eta}{Sc^{1/2}} \qquad (4.172)$$

The Schmidt number for liquids is usually on the order of 500 to 2000, and the Batchelor microscale is much smaller than the Kolmogorov microscale or:

$$\eta_b \ll \eta \qquad (4.173)$$

For gases, the Schmidt number is on the order of 1 and the Batchelor microscale is approximately equal to the Kolmogorov microscale.

The time scale of the diffusion at the Kolmogorov microscale can be estimated as:

$$\theta_D = \frac{\eta^2}{2D_L} \qquad (4.174)$$

where η is either the Kolmogorov length scale or one of the concentration length scales.

Such equations can be extended to turbulent diffusion. Turbulent diffusivity has units of velocity and length and can be considered to consist of a velocity scale and a length scale. A logical velocity is the turbulent fluctuation velocity; the length being one of the concentration length scales:

$$D_{LT} \propto u'l \qquad (4.175)$$

Scaling on impeller diameter D for l and impeller speed ND for u':

$$D_{LT} \propto ND^2 \qquad (4.176)$$

In terms of a turbulent diffusion coefficient D_t and using velocity fluctuations as the characteristic velocity, a turbulent mixing time θ_t can be written as:

$$\theta_t \propto \frac{D_t}{(u')^2} \qquad (4.177)$$

If the turbulent diffusion coefficient and the velocity fluctuations vary with impeller rotational speed N then:

$$\theta_t \propto \frac{1}{N} \qquad (4.178)$$

Mixing revolutions.   The number of revolutions to accomplish a particular mixing task in an agitated tank is essentially a constant for a particular geometry. The chapter on laminar mixing discusses this further. In terms of impeller rotational speed and mixing time, this relationship becomes:

$$N\theta_m = K \tag{4.179}$$

where K has units of revolutions. The number of revolutions to accomplish mixing is a function of geometry or size of the tank and is subject to geometric scaling. The mixing time θ_m is dependent upon geometry of the process, and the determination of the mixing time as a function of geometry is central to the study and calculation of mixing behavior. Geometric effects appear mainly from the impeller and tank size. Naturally the bigger the tank, the longer it will take to mix, and the bigger the impeller, the shorter the mixing time.

Approach to uniformity.   The approach to uniformity is an important criterion in the determination of mixing times. Typically, 5 or 1 percent deviation from the mean is the usual uniformity criteria; however, other problems do exist which represent more difficult mixing problems. pH control to 0.01 percent uniformity is such an example. Such studies have not been performed. pH studies may help determine dead zones and their effects on such mixing criterion.

Mixing of tracers

Baffled tanks.   Kramers et al. (1953) performed a comparative study of the mixing times by measuring concentration differences at two points for a propeller and two types of disk style turbines under a number of different baffling arrangements, i.e., no baffles, standard wall baffles, wall baffles displaced off the wall, and a cross baffle on the tank bottom. Mixing times were found inversely proportional to the impeller rotational speed. For the propeller, the mixing times obtained with standard baffles were lower but similar in magnitude to the mixing times obtained without baffles. The results indicated wall baffles were more favorable than the other arrangements. Crossed baffles on the tank bottom had no effect on mixing times except when the propeller was placed directly above the cross. Small eccentricity and low clearance appeared to increase mixing times in unbaffled tanks. Small eccentricities increased the tendency for the formation of central surface vortices. Increased eccentricity decreased the solid body rotation and lowered mixing times. Mixing times were also independent of the direction of rotation but increased as the amount of tracer added increased.

Mixing times for flat blade turbines at 10 times the power draw were comparable to the mixing times of the propeller. The impeller geometry used by Kramer et al. had a W/D = ⅛ which is a small blade width. Mixing times were also found to be affected by scale (i.e., a larger tank required a larger number of revolutions for the same mixing) although no general correlations were given for the effect.

Norwood and Metzner (1960) studied mixing times in tanks having the standard configuration using the disk style turbine, W/D = ⅙, from transition to fully turbulent flow using a correlation developed by Fox and Gex (1956) cited below. For the transition region, $2 <$ Re $< 10^3$, they found that mixing times were correlated as:

$$\theta_m = C_1 \frac{H^{1/2}T^{3/2}}{(Re)^{1/2}} \frac{1}{(ND^2)^{2/3}} \frac{1}{g^{1/6}} \frac{1}{D^{1/2}} \tag{4.180}$$

where C_1 was approximately 200. For the fully turbulent regime, mixing time was correlated as:

$$\theta_m = C_2 \frac{H^{1/2}T^{3/2}}{(ND^2)^{2/3}} \frac{1}{g^{1/6}} \frac{1}{D^{1/2}} \tag{4.181}$$

where C_2 was approximately 6. The correlation and data are shown in Fig. 4.31. Norwood and Metzner also noted that the mixing times for turbines were about 6 percent that of propellers. Kramers et al. (1953) noted just the opposite, but the disk style turbines of their study had W/D = ⅛. Apparently, there is a substantial blade width effect on mixing times.

Marr and Johnson (1961) considered imperfect mixing problems and developed a second-order model which described the behavior. In their experimental study, they found that mixing times followed Eq. (4.179), where the constant K was independent of Reynolds number but varied

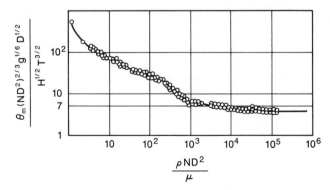

Figure 4.31   Mixing time correlation. *(From K. W. Norwood and A. B. Metzner, AIChEJ., 6, 432, 1960. Reproduced by permission of the American Institute of Chemical Engineers.)*

linearly with tank volume and decreased with increasing impeller diameter. Typical valves for $N\theta_m$ ranged from 20 to 100 for the propellers studied as shown in Fig. 4.32. Abrupt discontinuities in the mixing time data were noted and were attributed to changes in flow patterns. One pattern was noted in which two circulation loops were present in the tank. One loop directly recirculated into the impeller, and the other secondary loop was not associated with the impeller. This phenomenon, referred to earlier as compartmentalization by Reed et al. (1977), caused difficulties in establishing mixing times in a systematic manner and are not operationally desirable. The potential for the secondary loop formation was measured in terms of the average velocity of the impeller discharge to the average recirculation velocity and the size of the impeller. Secondary loops were more likely to form with small-diameter impellers having low discharge velocities and clearances. High clearances, up to $C/D = 2$ to 3, were found to minimize the amount of liquid pumped to achieve same mixing. Data were reported for turbulent, laminar, and transition regimes.

Landau and Prochazka (1961) investigated three experimental methods used to determine mixing times: a conductivity method, a thermometric method using temperature as a tracer, and colormetric method. The conductivity and colormetric results agreed; the general form of the correlation for propellers was given as:

$$N\theta_m = K\left(\frac{T}{D}\right)^{2.05} \tag{4.182}$$

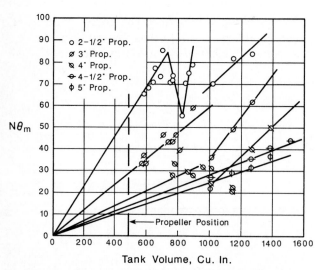

Figure 4.32   Batch mixing data for propellers operated in the turbulent regime. *(From G. R. Marr and E. F. Johnson, Chem. Eng. Progr. Symp. Ser. No. 36, 109, 1961. Reproduced by permission of the American Institute of Chemical Engineers.)*

for $10^4 < \mathrm{Re} < 10^5$ where the constant K was about 5.5. For the colormetric method, Landau and Prochazka noted that the color change of the indicator had to be suitably fast to be useful in measuring mixing times. Landau and Prochazka found that the results for the thermometric method was a strong function of Reynolds number, which indicated: (1) substantial differences in the mechanisms of temperature equalization and mixing of a material scalar and (2) different probe dynamics used in measurement. Thermal diffusivities are much higher than mass diffusivities by several orders of magnitude as indicated by the Lewis number, $k/\rho C_p D_L$.

Prochazka and Landau (1961) developed mixing time correlations for three-blade square pitch propellers, six-blade disk turbines, and four-blade pitched blade turbines in baffled turbulent stirred tanks in the standard configuration above an impeller Reynolds number of 10^4. Their correlations were given as:

For propellers:

$$N\theta_m = 3.48\left(\frac{T}{D}\right)^{2.05} \log\left(\frac{X_0}{X_c}\right) \qquad (4.183)$$

For pitched blade turbines:

$$N\theta_m = 2.02\left(\frac{T}{D}\right)^{2.20} \log\left(\frac{X_0}{X_c}\right) \qquad (4.184)$$

And for disk style turbines:

$$N\theta_m = 0.905\left(\frac{T}{D}\right)^{2.57} \log\left(\frac{X_0}{X_c}\right) \qquad (4.185)$$

where X_0 was an initial value which fluctuated between 1.0 and 3.0 in relative magnitude of the deviation. To use the equations, a X_0 value of 2.0 was recommended. The quantity X_c was the integral mean value of the local degree of inhomogeneity defined as:

$$X_c = \left[\frac{C(t) - C_x}{C_x - C_i}\right] \qquad (4.186)$$

where C_i and C_x were the initial and final concentrations, respectively. The value 2.0 appeared to be a mean value of X_c at time equals zero. The parameter X_c was used to increase the mixing time for different specified endpoints at which mixing was considered to be complete. Typically, X_c was considered to be 0.05 for most work. The amount and concentration of tracer and the feed location of the tracer had little effect on mixing times.

Biggs (1963) measured the terminal mixing times required for a tracer to reach 5 percent of its final value in a continuous flow system using a propeller and disk style, flat blade, and pitched blade turbines in the standard tank configuration. The observed mixing times were equivalent to batch mixing times since the residence time of the tracer in the tank was over 100 times the observed mixing times. The terminal mixing time was correlated as:

$$\theta = aN^b D^c \qquad (4.187)$$

where $a = 1.554$, $b = -0.843$, and $c = -1.930$ for turbines. There was little difference in the mixing times for the three different turbines studied. The mixing times for the propeller were about twice those for the turbines but the turbines drew anywhere from 4 to 20 times the power draw of the propeller. There was no effect of feed rate observed in the mixing data, and feed location was not important. Feed inlet location near the impeller gave faster initial dispersion but had little effect on the terminal mixing time. Biggs cited MacDonald and Piret (1951) as indicating that mixing times should be less than 5 percent of the average residence time of the feed for deviations in average exit concentration to be held below 5 percent.

Sykes and Gomezplata (1965) noted considerable agreement between the work by Biggs (1963) and the work by Prochazka and Landau (1961). They further suggested that the probe used by Biggs was sufficiently small so that it measured the distribution of one material into another and the breakup of this distribution into smaller portions. The probe could not measure the molecular diffusion time. Sykes and Gomezplata also suggested that the time for diffusion θ_D could be calculated from:

$$\theta_D = \frac{\eta^2}{2D_L} \qquad (4.188)$$

where η is the Kolmogorov microscale.

Moo-Young et al. (1972) investigated mixing times of miscible liquids having the same viscosity and density. Most of the experiments were conducted in tanks having the standard configuration with various types of impellers: six-blade disk style impeller, helical ribbon impellers, and an assortment of tubular agitators. The disk style turbine had the smallest diameter. The tubular agitators permitted centrifugal pumping from the blade tips which, it was hoped, would improve axial mixing. Mixing times were measured using a decolorization method.

TABLE 4.2   Values for the Constants C and a

Agitator type	Range of impeller Reynolds number	C	a
Turbine	5 to 10^3	1.75×10^4	-0.75
	10^3 to 10^5	60	0
Turbine (baffled)	10^3 to 10^5	36	0
Helical ribbon	1 to 10^4	4.3×10^2	-0.25
Tubular agitator	10 to 500	2.2×10^4	-0.85
	1000 to 10^5	44	0

Moo-Young et al. correlated their results for mixing times using:

$$N\theta_m = C\mathrm{Re}^a \qquad (4.189)$$

where the values for the constants C and a are given in Table 4.2.

General conclusions reached by Moo-Young et al. were that the different impellers performed about the same in the turbulent regime whereas the helical ribbon and tubular agitators, having much larger D/T ratios than the other impellers, performed much better in the laminar regime. Centrifugal pumping of the tubular agitators had no measurable effect on mixing times. The results also indicated that equal power per unit volume gave equal mixing times on scaleup. Mixing in unbaffled tanks was more efficient than in baffled tanks under laminar and transitional flow conditions. Under turbulent flow conditions, the reverse was true. Mixing times for shear thinning fluids increased much more rapidly with decreasing impeller Reynolds number in comparison with Newtonian fluids.

Khang and Levenspiel (1976a; 1976b) placed bounds on the response of a tracer input with a decaying exponential function, $e^{-K_a t}$ where the amplitude decay constant K_a was related to the properties of the response curves. A typical response is shown in Fig. 4.33 showing the amplitude decreasing as $2e^{-K_a t}$. For turbine mixers K_a was correlated as:

$$\frac{N}{K_a}\left(\frac{D}{T}\right)^{2.3} = 0.5 \qquad (4.190)$$

for $\mathrm{Re} > 2 \times 10^3$ and for propellers:

$$\frac{N}{K_a}\left(\frac{D}{T}\right)^{2.0} = 0.9 \qquad (4.191)$$

for $\mathrm{Re} > 10^4$. The tanks used in the experimental study were 0.56 and 1.22 m in diameter having the standard configuration. Typical K_a values were: 0.178 s^{-1} for a 0.366-m-diameter turbine at $N = 1.33$ s^{-1} in the 1.29-m-diameter tank and 0.0678 s^{-1} for a 0.254-m-diameter pro-

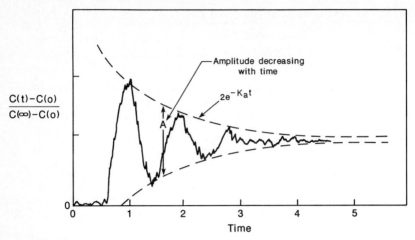

Figure 4.33 Amplitude of the concentration fluctuations as a function of time.

peller. The calculation of the mixing time requires the specification of $[C(t) - C_i]/(C_x - C_i)$ for the system, i.e., the definition of the endpoint for mixing, and the determination of the decay constant from the above equations. The equations hold for fully baffled tanks with vertically mounted impellers centrally placed in the tank.

Brennan and Lehrer (1976) investigated mixing times for low-viscosity Newtonian liquids between $10^4 < Re < 3 \times 10^5$ for flat blade open and disk turbines having D/T of 0.15 to 0.24 and W/D of 0.125 to 0.20 in baffled tanks. $N\theta_m$ was found to be a weak function of impeller Reynolds and Froude numbers. Dish-bottom tanks had higher mixing times than flat-bottom tanks. Mixing time was a minimum with the impeller clearance of $0.5H$ for flat-bottom tanks; for dished-bottom tanks, minimum mixing times occurred at $0.35H$. Wider impeller blades and the use of two impellers also reduced mixing times by 25 to 30 percent. Brennan and Lehrer noted that probes did not detect dead regions in the tank as easily as decolorization methods and that gas entrainment affected mixing times. Integral mixing time measurements do not provide information on the mixing rates in the early stages of mixing which is important in fast chemical reactions or in the selectivity of reactions. Delays in mixing times were noted for indicators which changed colors in the high pH range. The reason for the delays was unclear. They suggested the use of methyl orange (decoloring from pH 3.1 to 4.4) or methyl red (decoloring from 4.2 to 6.2) as indicators. Mixing times also increased with increasing volumes of reagents up to a certain level but remained constant after that. Feed location had little effect on mixing times for baffled tanks.

Hiraoka and Ito (1977) based their work on the dimensional analysis of the component mass balance to arrive at the dimensionless equation:

$$\frac{DC_a^*}{D(tQ_c/V)} = \frac{D_{LT}V}{Q_cD^2}\nabla^2C_a^*$$ (4.192)

where the left side of the equation is the Substantial Derivative. Q_c is the volumetric circulation rate, V is the tank volume, and D_{LT} is the turbulent mass diffusivity. From the equation, the following relationship was suggested:

$$\frac{\theta_m Q_c}{V} = f\left(\frac{D_{LT}V}{Q_cD^2}\right)$$ (4.193)

The turbulent mass diffusivity was assumed to be proportional to the tank diameter and the apparent friction velocity at the impeller tip. The friction velocity was related to a friction factor and to power input and geometry. The final expression for mixing time for paddles in unbaffled tanks, $0.3 < D/T < 0.8$, was:

$$N\theta_m = 4.8\left(\frac{T}{D}\right)^{0.82}\text{Re}^{1/6}$$ (4.194)

which contains no geometric variables describing the impeller. For baffled tanks:

$$N\theta_m = 7.2\left(\frac{n_bW}{D}\right)^{-0.35}\left(\frac{T}{D}\right)^{1.40}\left(\frac{H}{D}\right)^{0.5}$$ (4.195)

where the importance of the impeller geometry is shown. For baffled tanks, Norwood and Metzner (1960) obtained approximately the same dependencies of $N\theta_m$ on tank and impeller diameters and liquid height.

Kappel (1979) noted that although mixing times were used to determine the progress of mixing in an agitated tank, such data could not be used to judge mixing quality. He suggested the use of a degree of deviation to measure mixing quality.

Ogawa et al. (1980) obtained mixing responses in terms of a mixing index M (Ogawa and Ito, 1975) for a disk style turbine, an open flat blade impeller and a 45° pitched blade turbine, all with six blades, with the tank in the standard geometry. The mixing index response was modeled as:

$$M = 1 - \exp(KNt)$$ (4.196)

where K was found to equal -0.498 for the disk style turbine,

−0.418 for the flat open blade impeller, and −0.295 for the pitched blade turbine. The K values were indicative of the circulation path length and the nature of the impeller flow. The circulation path for the pitched blade turbines were longer than those of the flat blade impellers.

Bryant and Sadeghzaheh (1982), using a radio pill technique for flat blade impellers at low clearance in a baffled tank, suggested that an equation for the ratio of terminal mixing time to circulation θ_m/θ_c was a function of the mean circulation time θ_c and standard deviation s of the circulation time distribution:

$$\frac{\theta_m}{\theta_c} = A + B\left(\frac{s}{\theta_c}\right)^2 \tag{4.197}$$

where θ_c and s were correlated as:

$$\theta_c = k_1\left(\frac{H}{ND^3}\right) \tag{4.198}$$

$$s = k_2\left(\frac{H^{7/3}}{ND^3}\right) \tag{4.199}$$

where H is the liquid height. This relationship is not in agreement with the generally accepted concept of a constant θ_m/θ_c ratio. The relationship may, however, account for the effects of low impeller clearance and liquid height used in their study.

Shiue and Wong (1984) performed a mixing time study using a hemispherical bottomed fully baffled tank, 0.4 m in diameter. Six different agitators were studied: a disk style turbine, a curved blade disk style turbine, an open curved blade turbine, a propeller, and a two-blade and a four-blade turbine. A low clearance draft tube was also used. Mixing times were measured using a thermal tracer method with two thermistors placed diagonally in the tank. The thermal pulse was injected at a point midway between the agitator shaft and tank wall. Power data were also obtained using a torque meter. The range of Reynolds numbers was from 2×10^2 to 2×10^5.

In the turbulent regime, mixing times for the six-blade disk style turbine were correlated as:

$$N\theta_m = 5.01\left(\frac{T}{D}\right)^{2.4} \tag{4.200}$$

where the constant was for a clearance of $0.5H$. $N\theta_m$ was found to be a function of clearance: 34 at $C = 0.325H$ and 70 at $C = 0.5H$. For the two- and four-blade pitched blade turbines, the $N\theta_m$ values were 36 and 54, respectively. For the six curved blade disk style and four

curved blade open style turbines, $N\theta_m$ values were 56 and 60, respec
tively. The $N\theta_m$ values for the propeller were correlated as:

$$N\theta_m = 4.43\left(\frac{T}{D}\right)^{2.24} \tag{4.201}$$

for $0.254 < D/T < 0.382$ and a clearance of $0.5H$. Overall, the $N\theta_m$ values were dependent upon blade number, impeller clearance, and impeller geometry as well as D/T ratio. The use of a low bottom clearance draft tube, $C_t/T = 0.15$, $D_t = 0.5T$ and $L_t = 0.75T$, with the
pitched blade turbines placed at $C = 0.5H$ reduced $N\theta_m$ values by 50
percent. The draft tube also reduced the power numbers for the
pitched blade turbines by 50 percent.

Sano and Usui (1985) reported mixing time data for paddles and
turbines in baffled tanks for the turbulent region, $\text{Re} > 5 \times 10^3$. The
paddles and turbines had various dimensions, D/T from 0.3 to 0.7,
W/D from 0.05 to 0.4, and blade numbers from 2 to 8. The tanks were
0.2 m and 0.4 m in diameter with four baffles, B/T of $\frac{1}{10}$, and liquid
height H equaling the tank diameter. Mixing times of tracer injections were measured using a conductivity probe, placed on the upstream side of the baffles.

For paddles:

$$N\theta_m = 2.1\left(\frac{D}{T}\right)^{-1.67}\left(\frac{W}{T}\right)^{-0.74}n_p^{-0.47} \tag{4.202}$$

For turbines:

$$N\theta_m = 3.8\left(\frac{D}{T}\right)^{-1.80}\left(\frac{W}{T}\right)^{-0.51}n_p^{-0.47} \tag{4.203}$$

Raghav Rao and Joshi (1988) studied various geometric effects on
the homogenization number. The vessel diameter varied from 0.3 to
1.5 m with four 10 percent baffles and $H/T = 1$. Three different impellers were used: standard disk turbine, down-pumping pitched blade
turbine, and up-pumping pitched blade turbine at three different
clearances, $T/3$, $T/4$, and $T/6$. Conductivity measurements were used
to determine mixing times. The amount and location of tracer input
had no effect on mixing.

The up-pumping pitched blade turbine was found less efficient than
the down-pumping mode. For the down-pumping pitched blade turbine, mixing times increased with decreasing clearances. Lowering
the clearance on a down-pumping pitched blade turbine decreased
pumping rate and increased circulation loop length. Changes of 30 to
40 percent in mixing times were noted. $N\theta_m$ values for the different
impellers were correlated as:

For the disk style turbine:

$$N\theta_m = 9.43\left(\frac{aH + T}{T}\right)\left(\frac{T}{D}\right)^{13/6}\left(\frac{W}{D}\right) \qquad (4.204)$$

where a was determined from geometry. For the down-pumping pitched blade turbine:

$$N\theta_m = 5.0\left(\frac{2H + T}{D}\right)\left(\frac{C}{D}\right) \qquad (4.205)$$

For the up-pumping pitched blade turbine:

$$N\theta_m = 5.0\left(\frac{2H + T}{D}\right)\left(\frac{H + C}{D}\right) \qquad (4.206)$$

Pitched blade turbines at large diameters generated substantial radial flow which was related to the D/T ratio. For the effect of D/T, Raghav Rao and Joshi found:

$$N\theta_m \propto \left(\frac{T}{D}\right)^{1.83} \qquad (4.207)$$

for the pitched blade turbine pumping down. Mixing times also increased as blade width decreased for the down-pumping pitched blade turbine. For larger tanks, $N\theta_m$ values were higher.

Merzouk et al. (1988) discussed the mixing capabilities of a submersible agitator. The impeller used in the study was a 0.2-m-diameter three-bladed propeller in a ring propector. The mixing experiments were carried out in a dished-bottom tank 1.6 m diameter. Values of $N\theta_m$ for different positions were found to vary around 335. The high $N\theta_m$ values were because of the large T/D ratio.

The relationship between power and mixing times. A number of studies have related power, pumping capacity, and mixing times in the form of a general correlation. The assumption is that all power input is dissipated in performing mixing. When a fluid receives mechanical energy from a mixer, the fluid undergoes motion to dissipate the energy into internal energy. The dissipation occurs at the smallest length scales of the turbulence. If there exist concentration or scalar gradients, these are changed by the dissipation of the energy. The scale of the energy dissipation is the Kolmogorov scale. After the concentration or scalar gradients have been disrupted to this level and the energy dissipation is complete, diffusion is the only mechanism left to complete the mixing. However, diffusion is usually so rapid that it never becomes a separate step except in laminar creeping flow and transition regimes. The amount of mixing that takes place in the energy dissipation process is of obvious interest in mixing.

Corrsin (1957) developed a theoretical approach to turbulent mixing using equations for concentration fluctuations and obtained an equation for the mean-square concentration fluctuation, $\overline{c^2}$ which he reduced to:

$$\frac{d\overline{c^2}}{dt} = -12D_L\left(\frac{\overline{c^2}}{l^2}\right) \qquad (4.208)$$

where D_L is the diffusivity of the liquid and l is a length scale associated with the concentration. With additional assumptions, Corrsin integrated an equation for the mean-square fluctuation as:

$$\overline{c^2} = \overline{c_0^2} \exp\left[-3.3\left(\frac{ML^2}{P}\right)^{0.333} t\right] \qquad (4.209)$$

where L is some integral length scale of the flow, M is the mass of the material undergoing mixing, P is the power input from the agitator, and t is time. The mixing time can be determined from the specification of $\overline{c^2}$, $\overline{c_0^2}$, M, L, and P. The time constant for the decay of the mean-square fluctuation was $-3.3(ML^2/P)^{0.333}$ which has reciprocal time units. Corrsin (1958) extended the results to include chemical reactions. Separate cases were given for extremely low fluctuation levels, very slow reactions, and very fast reactions.

Landau and Prochazka (1963) studied mixing and power draw in a 0.4-m-diameter vessel in the standard geometry for a propeller, paddle, and a disk style impeller. Their final correlation, which related mixing time to power number, was given as:

$$N\theta_m = 2.18N_P^{-0.5}\left(\frac{T}{D}\right)^{\left(2+0.14N_P^{0.91}\right)} \log\left(\frac{2.0}{X_c}\right) \qquad (4.210)$$

where X_c is the final degree of inhomogeneity. The constant 2.0 is the initial degree of inhomogeneity.

Kamiwano et al. (1967), as cited by Hiraoka and Ito (1977), derived a relationship between mixing time, impeller flow, and power input. Mixing was related to eddy diffusion which was dependent upon power input N_P, flow rate N_Q, and the D/T ratio. Mixing time was correlated in terms of N_Q and N_P as:

$$\frac{1}{N\theta_m} \propto \left[\left(\frac{D}{T}\right)^3 N_Q + 0.21\left(\frac{D^2N_P}{T^2N_Q}\right)\right](1 + e^{-13(D/T)}) \qquad (4.211)$$

Mersmann et al. (1976) suggested a general correlation for mixing times given as:

$$N\theta_m = 6.7\left(\frac{T}{D}\right)^{5/3} N_P^{-1/3} \qquad (4.212)$$

The expression held for a number of different impellers including propellers, paddles, gate impellers, and pitched blade turbines.

Khang and Levenspiel (1976a; 1976b) took advantage of the magnitude of the constants in their equations, given above, to obtain expressions between power or power number and mixing times as:

For disk style turbine:

$$\frac{N}{K_a}\left(\frac{D}{T}\right)^{2.3} = 0.1 N_P \qquad (4.213)$$

for Re > 1×10^4 (i.e., $N_P = 5$); and for propellers (i.e., $N_P = 0.6$):

$$\frac{N}{K_a}\left(\frac{D}{T}\right)^{2.0} = 1.5 N_P \qquad (4.214)$$

for Re > 10^4.

Brennan and Lehrer defined the total mixing time as a sum of a mixing time due to turbulence θ_t and a mixing time due to diffusion θ_{DL} and developed correlations to calculate θ_t and θ_{DL}. For θ_t:

$$\theta_t = \frac{-Z^2}{\Psi ND[N_P H(T - D)D]^{0.33}}\left(\frac{T}{D}\right)^2 \log(1 - C^*) \qquad (4.215)$$

where Z was $(H - C - W/2)$ or $(C + W/2)$ whichever was greater. For θ_{DL}:

$$\theta_{DL} = \frac{-v^{3/2}(VD^2 W\pi/4)^{0.25}}{24(N_P N^3 D^5)^{0.5} D_L} \log(1 - C^*) \qquad (4.216)$$

where C^* is $[C(t) - C_x]/[C(0) - C_x]$ and varied around 0.90 or 0.95, and V is the tank volume. The function Ψ used in the equation for θ_t was 1.0 for noncavitating impellers. Brennan and Lehrer also suggested multiple impellers for geometries where $(H - C)/T$ was greater than 0.7 or C/T is greater than 0.6. For more efficient use of power, wider impeller blades was recommended rather than larger diameter impellers.

Sasakura et al. (1980) developed a relationship between mixing and circulation times incorporating the impeller power number for flat blade impellers of various geometries using their circulation time information and mixing time data. Their final correlation for mixing time was given as:

$$(N\theta_m^{-2} - 0.518 N\theta_c^{-2})^{0.5} = 5\left(\frac{T}{D}\right)^{1.5}(N\theta_c N_P)^{-0.5} \qquad (4.217)$$

The relationship correlated circulation and mixing times with power number and T/D.

Sano and Usui (1985) developed a relationship between power number and mixing times as:

$$N\theta_m = C\left(\frac{D}{T}\right)^{-2.5} N_P^{-0.75}\left(\frac{n_P W}{T}\right)^{0.1} \qquad (4.218)$$

where C is 9.1 and 12.2 for paddles and turbines, respectively. Specific energy consumption $E_m = (P/V)\theta_m/\rho$ during a mixing time θ_m was correlated as:

$$\frac{(E_m)^{0.5}\theta_m}{T} = 9.1\left(\frac{D}{T}\right)^{-0.56}\left(\frac{W}{T}\right)^{0.55} n_p^{-0.33} \qquad (4.219)$$

Eccentricity. Nishikawa et al. (1979) investigated power draw and mixing time for off-center mixing and the effects of eccentricity E. As the distance between the shaft and the center of the tank increased for unbaffled tanks, solid body rotation diminished and mixing was improved. The effects of baffle number and off-center distance E on mixing times for three different flat blade impellers, $\frac{1}{3} < D/T < \frac{1}{2}$, $0.125 < W/D < 0.2$, $B/T = \frac{1}{10}$, and $H/T = 1$, in flat-bottom cylindrical tanks were obtained using an acid-base decolorization. Irrespective of the eccentricity of the impeller shaft, the mixing times for two baffled systems were approximately equal to the mixing times for a fully baffled tank. The geometry for the studies using two baffles had the baffles placed symmetrically in the tank counter to the eccentricity of the impeller shaft. Regions of incomplete mixing were not observed in this geometry. Mixing times for unbaffled geometries decreased as the eccentricity increased to $E/D = 0.5$ where E was measured from the center of the tank. Above an $E/D = 0.5$, mixing times were twice the mixing times obtained for the fully baffled geometry. Mixing energy, which is the product of mixing time and power draw, was also found to be a minimum for the two baffle geometries for $\frac{1}{3} < E/D < \frac{1}{2}$. The mixing energy for three baffles was approximately the same as a fully baffled tank. For larger-diameter impellers, $D/T = 0.5$, the minimum in mixing energy occurred for $\frac{1}{4} < E/D < \frac{1}{3}$ using two baffles.

Multiple impellers. Chang (1986) performed mixing time studies, using an acid-base method, with feed entry at the top in an agitated tank $H = 2T$ with multiple impellers, three or four in number. The results were dependent upon the type of impeller, rotational speed, and the top clearance, i.e., the distance between the liquid surface and the top impeller. Radial flow impellers were found to disperse the feed more rapidly than axial flow impellers. This was attributed to less turbulence at the liquid surface generated by axial flow impellers. Longer mixing times occurred with increasing top clearance, which increased the circulation path length. In most cases, but not all, mixing times were shown to decrease with impeller rotational speed. Different combinations of impellers produced quite different flow patterns and mixing times which were not easily predicted without experimentation. In one case, a three-impeller system had a lower mixing time than a four-impeller system with all other conditions being equal. In the four-impeller system, axial circulation was reduced, which caused an increase in mixing times. Mixing time for a single impeller was inversely proportional to impeller rotational speed. For multiple-impeller systems, the flow patterns changed substantially with impeller rotational speed, and the relationship between mixing times and impeller rotational speed was not so simple. Compartmentalization of the flow and spacing between impellers was found to effect mixing times.

Komori and Murakami (1987) studied mixing times for two-impeller systems using four-blade flat blade impellers. Generally, the mixing performance of two-impeller systems was less than two single impeller systems for the same total volume. Komori and Murakami found that higher mixing efficiencies were obtained by large-scale recirculating flows and a reduction in the number of such flows (i.e., decreasing the number of compartments) in agitated tanks increased mixing performance.

Draft tubes. Tatterson (1982) performed a study of the effects of draft tubes and a cone bottom on mixing times for a pitched blade turbine and a disk style turbine for the various geometries cited above in the discussion on circulation times. There was little difference in the mixing times obtained in the various geometries. The standard deviations in the mixing times were substantially lower than those for circulation times. The standard deviations in mixing times for the standard configuration with the small-diameter draft tube and cone bottom were lower than those of the standard configuration without a draft tube. It was also noted that draft tube supports created regions of poor mixing. The magnitude of the circulation and mixing times were found to be misleading as measurements of mixing performance if their standard deviations were large. The draft tubes used by Tatterson had a very high clearance, $0.33T$, from the tank bottom. The top clearance of the draft tube was $0.17T$.

Shiue and Wong (1984), using pitched blade turbines inside a draft tube with a low bottom clearance of $0.15T$ and a top clearance of $0.1T$, obtained lower $N\theta_m$ values of 20 and 30, respectively, in comparison with $N\theta_m$ values of 40 and 60 without the use of a draft tube. However, the advantages of the draft tube became more pronounced in the transition regime. Shiue and Wong found that draft tubes reduced power draw for their geometries whereas Landau and Prochazka (1963) for their geometries found that draft tubes did not affect power draw significantly. The draft tubes used by Landau and Prochazka were not at a low bottom clearance.

Draft tubes must cause significant changes in flow patterns to have a significant effect on power and mixing times. This is done primarily by having low bottom clearances for axial impellers pumping down. The tank geometry with a low-clearance draft tube approaches that of an internal pump-around-loop. Draft tubes also minimize the effects of compartmentalization which often occurs in mixing.

Scaleup. Kipke (1983), in a general review of scaleup anomalies, noted several items of interest. He considered that the macroscale scaled with the impeller size. In large-scale vessels, large variations can be expected to occur because of the presence of larger eddies, and, as a result, the final state of mixing is achieved later in larger tanks. However, the way in which the macroscale and microscale behavior changes with scale remained unclear.

Rielly and Britter (1985) cited Kipke (1983), Middleton (1979), and Mersmann et al. (1976) as finding that traditional mixing relationships like Eq. (4.179) do not hold upon scaleup. Rielly and Britter noted that it is common for mixing times not to scaleup according to impeller rotational speed and tank and impeller diameters typically reported in the literature. Rielly and Britter attributed this to possible effects due to two characteristic lengths: the scales of the tracer volume and the scale of the probe. To investigate these effects, they performed mixing time studies using a disk style turbine in the standard configuration $C = T/3$ and a conductivity probe. $N\theta_m$ was found to increase slightly with impeller Reynolds number, from 25 at Re = 10^4 to 30 at Re = 7×10^4. The effect of the length scale of the tracer volume was negligible. The size of the probe was increased in their study by an order of magnitude which resulted in only a 12 percent reduction in mixing time θ_m. Neither the scale of the tracer volume nor the scale of the probe accounted for the anomalies occurring in the scaleup.

Transitional mixing. The transition mixing regime occurs between an impeller Reynolds number of 1 to 10,000. Mixing behavior in the transition regime can be quite different from that occurring in the turbulent regime. A major difference is in the use of baffles. In transitional mixing, baffles are not recommended and can actually increase mixing time. In addition, in the transition regime, $N\theta$ is a function of the impeller Reynolds number. Both the density and viscosity of the fluid affect mixing times. This is not the case in turbulent mixing. Another difficulty in transitional mixing is in the use of appropriate impellers. In laminar mixing, full-tank impellers are used to bring motion to the entire tank for mixing. In turbulent mixing, such impellers are not needed. However, when $N\theta_m$ values for turbulent agitators reach 200, laminar impellers should be considered.

Fox and Gex (1956), Norwood and Metzner (1960), and Godleski and Smith (1962) have presented mixing time data as reviewed elsewhere in this chapter for transitional mixing in baffled and unbaffled tanks.

Rzyski (1985) measured mixing times in the transition regime for several impellers in tanks, 0.6 and 0.075 m in diameter, with and without baffles, for Newtonian and shear thinning liquids using a decolorization method. The final correlation took the form of:

$$N\theta_m = C\mathrm{Fr}^{0.14}\mathrm{Re}^a \qquad (4.220)$$

where the exponent a varied from -0.90 to -1.11 and where C depended upon the type of impeller. The results indicated that mixing time varied with impeller rotational speed to the -1.76 power. Mixing times for dissolving salt were also found to be on the order of mixing times obtained by decolorization.

Shiue and Wong (1984) provided data for mixing in the transition region using turbulent agitators. In the transition regime, $N\theta_m$ values varied with the inverse of the impeller Reynolds number, reaching values from between 400 and 1000 at a Reynolds number of 10^2. In the transition regime, turbulent agitators are not as advantageous for mixing as laminar mixing impellers.

Voit and Mersmann (1988) divided a typical tank geometry into cells and used measured velocity profiles to determine transfer rates between the cells to calculate macromixing times in the transition regime. Experimental mean fluctuating velocities were volume averaged over the tank volume and were given as a function of impeller Reynolds number for the transition regime. From these data and simulation of the mixing, macromixing times were estimated as:

$$\theta_m = \frac{4(T^2 D)^{0.5}}{u' N_{\mathrm{Re}}} \qquad (4.221)$$

Voit and Mersmann also reported data by Laufhutte of the mean fluctuating velocities as a function of impeller Reynolds number between 150 and 5000. The data were fitted by a saturation growth rate curve as:

$$\ln \left(\frac{u'}{\pi ND} \right) = \frac{-1.28 \ln Re}{-3.85 + \ln Re} \qquad (4.222)$$

Dispersion during circulation. In the circulation of material around the tank, dispersion occurs by molecular and turbulent diffusion. Voncken et al. (1964) studied this process in the circulation loop using a model with longitudinal dispersion. The characteristic length and velocity corresponded to the average length L_c and average velocity V_c of the circulation loop. The circulation time was modeled as:

$$\theta_c = \frac{L_c}{V_c} \qquad (4.223)$$

The concentration as a function of time t at a detection point was given as:

$$\frac{C}{C_x} = \left(\frac{Bo\theta_c}{4\pi t} \right)^{0.5} \sum_{j=1}^{x} \exp \left[- \left(\frac{Bo\theta_c}{4t} \right) \left(j - \frac{t}{\theta_c} \right)^2 \right] \qquad (4.224)$$

where C_x is the final concentration and Bo is the Bodenstein number $L_c V_c / D_{LD}$, and D_{LD} is the dispersion coefficient. Supporting justification for this model can be found in Levenspiel (1962). Levenspiel noted that the model was for low longitudinal dispersion.

Data were taken of tracer injections at impeller Reynolds numbers greater than 2×10^4 in three geometrically similar tanks in the standard configuration for several D/T ratios. The Bodenstein number was found independent of impeller rotational speed N, which indicated that D_{LD} was proportional to N. The product $Bo(D/T)$ was found to be a constant value of about 12, which indicated that the relative dispersion, as measured by D_{LD}, increased as the impeller diameter increased with respect to the tank diameter. They found surprisingly little dispersion in the bulk of the tank. Voncken et al. (1964) essentially found that mixing in the bulk of the tank was relatively small in comparison to mixing occurring in the impeller region.

Efficiency studies. Landau and Prochazka (1963) studied mixing and power draw in a 0.4-m-diameter vessel in the standard geometry for a propeller, paddle, and a disk style impeller. The propeller was also studied with draft tubes. The mixing time and power data permitted an evaluation of efficiencies. In all arrangements, efficiencies decreased rapidly with increasing values of T/D, indicating that large-diameter impellers are more efficient than small-diameter impellers. The highest efficiencies were obtained with the propeller; the disk style impeller had the lowest efficiency. The efficiency of the propeller was also significantly higher when operated with a conical draft tube, particularly in the Reynolds number range of 10^3 to 10^4.

Shiue and Wong (1984), in their study cited above, measured power and mixing times, which permitted evaluations of mixing efficiency. They noted that very few, if any, efficiency studies have been performed on agitators in the turbulent flow and transition regimes. The six-blade disk style turbine and the six-curved blade disk style turbine were the least efficient at mixing. Decreases in T/D ratio and clearance to $0.325H$ increased mixing efficiency. The propeller and the open flat blade impellers had the next highest efficiency level. The four-blade pitched blade turbine had the highest efficiency of all impellers studied. With the use of a draft tube, efficiency was improved further. The two-blade pitched blade turbine in a draft tube had the highest efficiency of all configurations. Shiue and Wong observed greatly improved mixing in the transition region using a draft tube with the pitched blade impellers. Shiue and Wong used a thermal tracer technique to obtain their data.

The thermal technique used to measure mixing times has been criticized by Landau and Prochazka (1961) primarily because the effect of thermal buoyancy interferred with mixing time measurements at low impeller Reynolds numbers. Thermal diffusivities are much greater than molecular diffusivities, which caused lower mixing times in thermal studies than in concentration based studies, particularly in transitional mixing regime. A metal draft tube was used in the study by Shiue and Wong which could thermally distribute temperature differences as well.

Fort and Medek (1988) provided efficiency data based upon pumping capacity and power for a pitched blade turbine and two different types of fluidfoil impellers. The fluidfoil impellers were shown to be more efficient at pumping than the normal pitched blade turbine. The performance index used was the ratio $(N_Q)^3/N_P$, which was 0.7 for the fluidfoil impellers and 0.5 for the pitched blade turbine. Wong and Huang (1988) have also made efficiency studies about impeller types.

Lehtola et al. (1988) presented data on the effects of impeller blade angle and baffle type on mixing energy $P\theta_m$. With standard type baffles, mixing energy was correlated with blade angle α as:

$$P\theta_m = 42N^{2.20} \exp\left[1.56 \sin(2\alpha - 0.5\pi)\right] \qquad (4.225)$$

and, with twisted element baffles, as:

$$P\theta_m = 53N^{1.98} \exp\left[1.41 \sin(2\alpha - 0.5\pi)\right] \qquad (4.226)$$

where α is the blade angle to the vertical. Low-angle impeller blades were shown to be the most efficient. The twisted baffles were better at minimizing the mixing energy.

Komori and Murakami (1988) investigated mixing using single and double impellers in unbaffled tanks. Mixing efficiency and energy consumption $P\theta_m$ were determined for five different multiple impeller geometries. However, the highest efficiency was obtained with two single impeller tanks with $H = D$. The double impeller geometry with $H = 2D$ was found to be less efficient. Circulation times were approximately one-half of mixing times and were proportional to energy consumption.

Unbaffled tanks. Unbaffled tanks are often used in mixing, in the transition region, for mixing sticky materials, in processes where perfect cleaning is required, and in processes where tanks are so large that the effects of baffles are not important. Unbaffled tanks are also used in processes where it is not clear or not recognized that the baffles have an effect on mixing performance.

Kramers et al. (1953), Moo-Young et al. (1972), Hiraoka and Ito (1977), and Nishikawa et al. (1979) have commented on mixing in unbaffled tanks as cited above.

Miller and Mann (1944) studied mixing in liquid-liquid systems in unbaffled agitated tanks for an assortment of impellers including flat and pitched blade impellers. Grab samples were taken to determine the degree of mixing at a number of locations. Their particular mixing index, which was based upon a ratio of volume fractions, was found to pass through a maximum at a power per volume of 2 to 4 kW/m^3 (250 to 500 ft · lb/min ft^3). The behavior of the mixing index was fairly independent of the impeller type although a propeller showed the poorest performance. It was recommended that the impeller should be placed in the heavier phase, regardless of design. The effect of viscosity was negligible. Phase separation occurred more easily in unbaffled tanks because of centrifugal forces. Baffles or an eccentric shaft prevented separation. The use of average mixing indices as opposed to "worst-case" indices was discussed. The appropriate type of mixing index depended upon the particular application.

van de Vusse (1955b) discussed the various effects of liquid volume, impeller and tank dimensions, and liquid height and blade width and angle on mixing times for unbaffled tanks. For minimum power input at a constant mixing time, small tanks and multiple impellers pumping in the same direction were recommended. A D/T ratio of about 0.4 for turbulent mixing was recommended.

Fox and Gex (1956) developed correlations for jets and square pitch propellers for mixing in nongeometrically similar, unbaffled tanks, 0.3 and 1.5 m in diameter, for both transitional and turbulent mixing of miscible Newtonian liquids having similar densities and viscosities. The results on jet mixing will be given later.

Fox and Gex considered mixing to be a function of momentum flux generated by the mixing device. They noted that power input did not uniquely determine mixing behavior and that the same mixing performance could be obtained from different power inputs, e.g., a large, low-speed impeller or jet at lower power produced the same mixing results as a smaller, high-speed propeller or jet. They divided their mixing time data for the propellers into two regimes: (1) incipient laminar mixing $100 < \mathrm{Re} < 10^4$ and (2) turbulent mixing $\mathrm{Re} > 10^4$. In both regimes there was considerable scatter in the data, by a factor of 3 in some cases, which was attributed to use of nongeometrically similar equipment. The data are shown in Fig. 4.34. The mixing time for turbulent mixing for the propellers was correlated as:

$$\theta_m = C_1 \frac{H^{1/2}T}{(\mathrm{Re})^{1/6}} \frac{1}{(ND^2)^{2/3}} \frac{1}{g^{1/6}} \tag{4.227}$$

where C_1 is 153.4 from the data given by Fox and Gex. For incipient laminar mixing, a similar correlation was obtained:

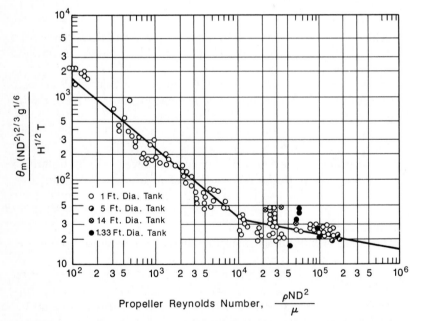

Figure 4.34 Mixing time data for a propeller as a function of impeller Reynolds number. *(From E. A. Fox and V. E. Gex, AIChEJ., 2, 539, 1956. Reproduced by permission of the American Institute of Chemical Engineers.)*

$$\theta_m = C_2 \frac{H^{1/2}T}{Re} \frac{1}{(ND^2)^{2/3}} \frac{1}{g^{1/6}} \qquad (4.228)$$

where C_2 is equal to 2.1×10^5. Considerable variation can be expected in the constants C_1 and C_2 since there was no attempt to maintain geometric similarity in the study. The time required for complete mixing was found to vary directly with square root of tank volume and inversely proportional to the momentum input raised to 5/12 power for turbulent mixing, or:

$$\theta_m \propto \frac{V^{0.5}}{(N^2 D^4)^{5/12}} \qquad (4.229)$$

For incipient laminar mixing:

$$\theta_m \propto \frac{V^{0.5}}{(N^2 D^4)^{5/4}} \qquad (4.230)$$

Data for plant size tanks, 4.3 m diameter, were about 25 percent higher than data for the smaller tanks. This was attributed to different initial conditions. Varying the amount of tracer affected mixing times up to a critical value, after which there was no effect. It was recommended that the location of the propellers be such that the central surface vortex did not reach the impeller. The agreement in the data for the different geometries was remarkable since geometric similarity was not maintained in the study.

Godleski and Smith (1962) studied mixing times of shear thinning fluids using an acid-base neutralization and an indicator color change. The work was done in tanks having the standard configuration, initially with baffles, using six-blade, flat blade turbines in the impeller Reynolds number range of 0.6 to 900.

Godleski and Smith noted that not all mixing operations need to be carried to completion and considered two mixing times: a gross mixing time and a terminal mixing time. The gross mixing time was about 75 to 90 percent that of the terminal mixing time. Terminal mixing times were very long and difficult to reproduce; terminal mixing times for identical tests differed by as much as 20 percent. Mixing times were also dependent on the acid-base volume ratio.

In unbaffled tanks, two overall flow patterns controlled mixing times: (1) the recirculation loops and (2) two toroidal vortex systems, one above the impeller and one below the impeller, which formed inside the recirculation loops. The flow inside the toroidal vortices did not pass through the impeller, and, hence, mixing into these vortices was by diffusion. Mixing times in the toroidal vortices were five times higher than mixing times in the recirculation loops. The insertion of baffles changed the flow pattern by disrupting the toroidal vortices; however, mixing times actually increased because of the formation of vortices behind the baffles. Mixing in these baffle bound vortices was by diffusion. Over the operational range where a central vortex was formed above the impeller, mixing was rapid and diffusion no longer controlled. The shape of the central vortex was used to correlate mixing times as:

$$\theta = \frac{282s}{m^{0.5}}\left(\frac{4H_v}{D^2}\right)^{-0.5}$$ (4.231)

where H_v is the vortex depth. The data and correlation are shown in Fig. 4.35. The correlation applies to the surface addition of material and not to the addition into the impeller directly. Godleski and Smith suggested that, for rapid mixing, a central vortex should be formed. Mixing times were found to be the same for fluids having different viscosities if the vortex had the same shape. Measured mixing times were also found to be 10 to 50 times the calculated values of the Norwood and Metzner correlation given above. Apparently, the mixing of the shear thinning fluids took much longer than mixing of Newtonian fluids because of a decrease in the motion of shear thinning fluids, relative to Newtonian fluids, with increasing distance from the impeller.

Brennan and Lehrer (1976) investigated mixing times in unbaffled

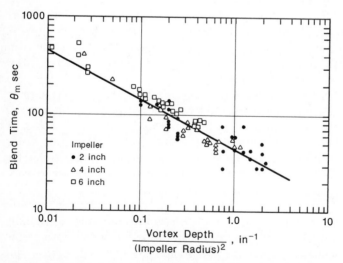

Figure 4.35   Mixing data in an unbaffled tank with a central surface vortex; surface addition of material. *(From E. S. Godleski and J. C. Smith, AIChEJ., 8, 617, 1962. Reproduced by permission of the American institute of Chemical Engineers.)*

tanks under the conditions cited earlier. Mixing times were found to decrease with increasing impeller rotational speed N. The number of revolutions to mix $N\theta_m$ varied with $Fr^{0.1}$ to $Fr^{0.3}$ depending upon the D/T ratio, until surface entrainment occurred. Impeller position had little effect on mixing times. After surface entrainment, mixing times increased with increasing impeller rotational speed. $N\theta_m$ varied with $Fr^{0.7}$. The use of probes, such as in conductivity methods, in unbaffled vessels acted as baffles to some degree and decreased mixing times. Internal coils had the same effect. Addition of material into the central vortex led to large mixing times. Apparently, in unbaffled tanks, a central solid body rotation core existed around the impeller which was the last region to be mixed. Blade width had a minor effect on mixing times, and the use of wider impeller blades did not significantly improve mixing performance.

Nishikawa et al. (1979) noted mixing times for unbaffled tanks in a study of eccentricity of the impeller shaft cited above. As the eccentricity approached an E/D of 0.5, mixing times in unbaffled tanks approached approximately twice that of fully baffled tanks.

Other geometries. Bauer and Moser (1985) obtained mixing times using a conductivity probe for a horizontal stirred tank with agitation provided by a rotating perforated cylinder equipped with baffles on both sides. Mixing times were on the order of 8 to 250 times those found in traditional geometries due to the very poor axial circulation provided by the rotating cylinder. The mixing times were found to be a strong function of power per volume at low power input and constant at high power input levels. In such a configuration it is difficult to measure axial circulation or axial circulation times since well-defined axial circulation is not present. Ramsey et al. (1988) obtained mixing times in a slab tank using up-down agitators and found considerable similarity in mixing behavior between the slab tank configuration and the traditional cylindrical tank.

Mixing of liquid volumes

Most studies of mixing times have been performed by adding a tracer and recording the time necessary to disperse the tracer. van de Vusse (1955a) performed a mixing study in which the two materials were initially segregated in the tank. The impeller-tank geometries studied included a flat blade paddle in a baffled tank, propellers and pitched blade turbines in unbaffled tanks and an impeller called the turbomixer. The turbomixer had a large number of curved radial blades, e.g., 8, 12, 18, and 20 blades, and was placed inside a stator ring, i.e., a stationary ring of vanes which directed the discharge flow of the impeller. Under the stator ring configuration, the turbomixer was considered fully baffled. van de Vusse started with two superimposed layers of miscible liquids and observed the refractive striae during mixing. Mixing was complete when the striae disappeared. van de Vusse's experiments were transient studies, e.g., the materials were added and then mixed.

van de Vusse developed the following correlation for mixing time θ_m:

$$\frac{\theta_m Q}{V} = f_1\left(\frac{ND^2}{\nu}\right)f_2\left(\frac{\rho N^2 D^2}{\Delta \rho g H}\right)f_3\left(\frac{D^3}{T^{2.5}H^{0.5}}\right) \qquad (4.232)$$

where Q is the impeller pumping capacity ND^3; V is the tank volume; the group $\theta_m Q/V$ is a dimensionless mixing time; the functions f_1, f_2, and f_3 contain the impeller Reynolds number, a density Froude number and a geometric parameter. van de Vusse studied mixing in the transition and turbulent regions, which was accounted for through the use of the Reynolds number. The Froude number accounted for the differences in density which were on the order of 10 percent of the continuous phase density. For decreasing Froude number, the dimensionless mixing time increased. The geometric group accounted for effects of unsuitable dimensions, which permitted the formation of zones of poor mixing. For the Froude number effect only, van de Vusse found that:

$$\frac{\theta_m Q}{V} \propto \left(\frac{\rho N^2 D^2}{\Delta \rho g H}\right)^{-0.3} \qquad (4.233)$$

The effect of Reynolds number on dimensionless mixing time was more complex since there was a gradual increase in dimensionless mixing time in the transition region at lower Reynolds numbers. In laminar mixing, viscosity differences rather than density differences were significant. The mixing times, found by van de Vusse, were on the same order of magnitude as those obtained by typical tracer pulse injection.

van de Vusse recommended that the geometric parameter $D^3/T^{2.5}H^{0.5}$ be between 0.001 and 0.04 to prevent the formation of dead zones. Generally, the mixing behavior was about the same for the different impellers. Eccentricity increased mixing times for radial flow impellers and decreased mixing times for axial flow impellers. Impeller shafts, set at an angle, decreased mixing times for propellers. Baffles, used with the paddle impellers, decreased mixing times by increasing the effective pumping capacity of the impeller. Clearance of $0.5H/T$ showed the lowest mixing times. For the turbomixer, increasing blade curvature increased mixing times and for inclined blades, mixing times varied with $(\sin \alpha)^{-2}$ for $40° < \alpha < 90°$ and $(\sin \alpha)^{-1}$ for $0° < \alpha < 40°$ where α is the blade angle. For a small number of blades, pumping capacity was proportional to blade number, but, as blade number increased, the effect was less. Pumping capacity was found to be proportional to blade width for $W/D < 0.5$. In scaleup of the mixing in his experiments, van de Vusse recommended that, for constant $\theta_m Q/V$ on scaling, $ND^{0.2}$ should be held constant.

Work by Wesselingh (1975), concerning initially stratified layers, is discussed below in mixing in large storage tanks by side entering propellers.

Hemrajani (1985) studied the height rise of an interface between two immiscible liquids as a function of impeller type, power per volume, baffle width, and impeller number. Liquid height and D/T ratio had no effect on the height rise. Although the study was of immiscible liquids, flow behavior of immiscible liquids should be similar to flow behavior of miscible liquids in some cases.

Rielly and Prandit (1988) studied mixing of Newtonian liquids with large density and viscosity differences in an initially stratified condition. Densities varied from 1000 to 1220 kg/m^3 and viscosity differences from 0.5 to 200 mPa · s. The experiments were performed in 0.29-m-diameter fully baffled tank with a disk style impeller in the standard configuration $D = 0.096$ m. The tank volume was 1.92×10^{-2} m^3. Power draw remained constant during the experiment at N_P of 4.4 for the disk style impeller. Standard deviations of mixing times were 5 percent of the mean. The volume fraction of the heavier liquid was always 10 percent. The clearance C of the impeller equalled the impeller diameter D.

Mixing of fluids having density differences occurred by two mechanisms. At low Richardson numbers $\Delta\rho gH/\rho_L N^2 D^2$, mixing was by engulfment of the heavier liquid by the lighter liquid. The interface was highly distorted. Liquid above and below the impeller became homogeneous and moved in the circulation loops until mixing was achieved. Mixing times were correlated as:

$$\frac{\theta_m ND^3}{V} = 1.21 \qquad \text{for } \frac{\Delta\rho gH}{\rho_L N^2 D^2} < 2.0 \qquad (4.234)$$

where H and V are the total liquid height and liquid volume, respectively. At higher Richardson numbers, the lighter liquid did not penetrate into the heavier liquid. However, the heavier liquid was picked up by filament and billows entrainment and carried into the bulk. Mixing times were correlated as:

$$\frac{\theta_m N D^3}{V} = 0.55 \left(\frac{\Delta \rho g H}{\rho_L N^2 D^2} \right)^{0.77} \quad \text{for } \frac{\Delta \rho g H}{\rho_L N^2 D^2} > 3.0 \qquad (4.235)$$

$N\theta_m$ values were typically between 20 and 100 revolutions for the data given. Viscosity differences increased mixing times. Although the data were incomplete, for a viscosity ratio of 160, mixing times increased by a factor of 2 to 4 over that of a viscosity ratio of 1.0. Another study of the mixing of two fluids with results similar to work by Rielly and Pandit is by Ahmad et al. (1985).

Smith and Schoenmaker (1988) studied mixing of a small amount of a high-viscosity material into a turbulent stirred low-viscosity bulk phase. Shear rates were limited in the high-viscosity liquid which retained its identity for a relatively long time. Laminar deformation, fine-scale turbulent diffusion, and mass transfer finally led to mixing however. Stages in the process were identified as: (1) the elongation of the clump of added fluid, (2) further distortion by large-scale turbulence, and (3) assimilation by small-scale turbulence and diffusion. The mechanism was similar to Beek and Miller's mechanism of mixing. Shear rates in the tank were much less than the average Metzner-Otto shear rate of $10N$. The data indicated that $N\theta_m$ values were not a function of the rheology of the additive and were constant for a given geometry. One and two percent CMC and syrup solutions were studied having viscosities up to 1400 mPa.

One problem noted by Smith and Schoenmaker was the adhesion of material to the walls and tank bottom. Mixing times could be extended by a factor of 60 because of this effect. Adhesion to the tank bottom was a function of feed position. High impeller speeds and low impeller clearances reduced the likelihood of adhesion. Material adhesion behind wall baffles at the tank bottom caused the longest mixing times.

Instead of mixing two layers of different densities, Pelletier and Cloutier (1973) discussed the formation of a combined mixed-plug flow configuration in a single tank where tubular portion of the reactor was unagitated and stabilized by a density gradient caused by a highly exothermic reaction. Pelletier and Cloutier showed first that it was possible to create such a flow situation in a stirred tank and then presented data on the agitated fraction of the liquid as a function of impeller rotational speed, impeller diameter, power consumption, feed rate, and specific gravity differences.

Mixing in continuous-flow stirred tanks

Typically, mixing effects will not be significant if the residence time of the material in the tank is two orders of magnitude larger than the mixing time. However, mixing effects will be encountered in any process where the residence time and the mixing time are of the same order of magnitude or where the process contains time scales on the order of the mixing time.

MacDonald and Piret (1951) studied mixing in a continuous-flow reactor using dye tracer studies and the continuous hydrolysis of acetic anhydride having first-order kinetics. The tank had a rounded bottom, 0.1 m in diameter and $D/T = 0.5$, equipped with paddle impellers and baffles. Under continuous-flow situations, the agitator and feed stream contributed to the mixing occurring in the tank. Mixing time was measured from the time of the dye injection to uniform dispersion throughout the tank. Injection time was less than 15 percent of the mixing time and density differences between the dye and the tank contents did not affect mixing times. Generally, increasing the agitator rotational speed or increasing the feed rate resulted in decreasing mixing times.

MacDonald and Piret noticed a balance between power input due to the feed stream and power input due to mechanical agitation. At high feed rates, the importance of the agitation was reduced since the main power input was from the feed stream. The power input of the feed stream was calculated in two ways: (1) from the kinetic energy of the feed stream and (2) from pressure drop calculations. MacDonald and Piret determined the power input necessary to accomplish a certain mixing time θ_{cf} from the following correlation:

$$\theta_{cf} = 12.26 \times 10^{-2}\left(\frac{P_{\text{eff}}}{V}\right)^{-0.726} \tag{4.236}$$

where the units on the constant is $s(kW/m^3)^{0.726}$. The amount of power input from the feed stream was substracted from P_{eff} to provide the amount of power needed from mechanical agitation. Using data, the mechanical agitation efficiency was correlated in terms of impeller rotational speed as:

$$\%\text{eff} = 5.4N^{-2.1} \qquad \text{for } 0.3 \text{ s}^{-1} < N < 2.33 \text{ s}^{-1} \tag{4.237}$$

Low rotational speeds were found to be more efficient at mixing than higher impeller rotational speeds as shown by this correlation. Whether the procedure outlined by MacDonald and Piret can be used in general is open to question. The constants in the correlations are specific to their particular system.

In their reactor studies, MacDonald and Piret observed that, at low agitation levels, the exit concentration of the reactor fluctuated unpredictably by as much as 20 percent above and below the value predicted by the ideal mixing reaction model over time periods of hours or minutes. The actual fluctuation level was dependent on the flow field in the reactor. However, at low values of θ_{cf}/θ_R less than 0.05, the fluctuations approached zero. Systems with large density or viscosity differences will have large fluctuations. MacDonald and Piret noted channeling in the reactor due to slight density differences from the reactor inlet to the reactor outlet. In dye studies with reaction, flow channels fluctuated in the reactor. The degree of conversion for the reaction at the outlet varied to zero in some cases as unreacted feed passed through the reactor entirely unmixed.

van de Vusse (1955b) discussed continuous mixing in an unbaffled stirred tank which was fed through a perforated inlet compartment. The mixing time θ_{cf} for the arrangement was correlated in terms of the mixing time for batch mixing θ_b and the unagitated mixing time θ_{ua} of mixing due to the feed arrangement and convection occurring in the tank not caused by agitation. The latter mixing time θ_{ua} was lower than the residence time of the material in the geometry studied and was obtained experimentally as a function of flow rate through the tank. The unagitated mixing time θ_{ua} contained effects of premixing due to the perforated compartment, circulation, and recirculation zones generated in the tank due to feed and was found to vary with -0.85 power of the flow rate. The power input to such mixing was obtained from the pressure drop data, which occurred as the feed passed through the unit. The final equation for mixing time θ_{cf} under continuous-flow operations and agitation was given as:

$$\frac{1}{\theta_{cf}} = \frac{1}{\theta_b} + \frac{1}{\theta_{ua}} + 0.5(\theta_b\theta_{ua})^{-0.5} \qquad (4.238)$$

Mixing data and the equation are shown in Fig. 4.36. The equation matches the two limits in the process; mixing time θ_{ua}, due to flow, and the batch mixing time θ_b, due to agitation only. The mixing time θ_{cf} was considered to be due to both agitation and flow. If θ_b is much smaller than the residence time, θ_R perfect mixing was approach since θ_{ua} was also considered less than the residence time θ_R.

Gutoff (1960) and Sinclair (1961) investigated the mixing of cyclic fluctuations in the feed in agitated tanks. Gutoff found that the turnover time (i.e., the tank volume divided by the impeller pumping capacity) determined the extent to which input fluctuations were reduced. His results indicated that the residence time divided by circulation or turnover time should be on the order of 100 or greater to dampen feed fluctuations. Work by Gutoff also strongly suggested that the circulation or turnover time and residence times should not occur at multiples of the cycle time in the feed fluctuations.

Using a second-order model of imperfect mixing, Marr and Johnson

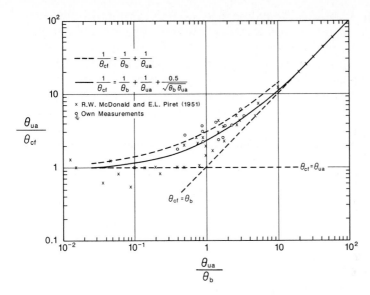

Figure 4.36 Continuous mixing: influence of the batch mixing time θ_b and the mixing time due to flow input and distribution θ_{ua} on the continuous mixing time θ_{cf}. *(From J. G. van de Vusse, Chem. Eng., Sci., 4, 209, 1955. By permission.)*

(1961) were able to predict the observed imperfect mixing recorded experimentally at low holdups or low residence times V/Q_F for their propeller agitated system. At higher holdups, perfect mixing was approached although a cyclic behavior was observed in the model. With the feed stream distributed across the top of the tank, perfect mixing was observed. Marr and Johnson considered that the circulation due to feed in the continuous flow system reduced the effective pumping of the propeller by as much as 30 percent. Although the feed stream of the magnitude studied did not affect the impeller power draw, the feed stream changed the flow patterns significantly and caused imperfect mixing to occur.

Voncken et al. (1965) studied circulation for continuous-flow systems for the disk style turbine. They considered that the disk style turbine established two recirculation zones, one above the impeller and one below the impeller. Transfer between the zones occurred only in the impeller discharge flow in a region close to the tank wall. Using a flow follower, the exchange between the two zones was not perfect in the sense that the flow follower tended to be recirculated in the zone that it was in. However, the exchange between zones was still found to be reasonable for mixing and not a function of impeller rotational speed. Feed streams influenced circulation zones by (1) either increasing or decreasing the momentum of the circulation and (2) increasing the dispersion in the circulation zone and decreasing the Bodenstein number. The manner of distribution and the momentum of the incoming stream determined the effects. In one pulse study, an appreciable amount of tracer flowed out of the tank unmixed. In another study under the same conditions and geometry, such behavior was not noted.

The introduction of feed will always cause an increase in dispersion, but to what extent cannot be easily determined. If the feed enters in the same direction as the circulation, then the circulation time is reduced. The momentum of the entry stream, however, has to be sizable in comparison with the momentum of the circulation to have any significant effect.

Hubbard and Patel (1971) provided exit concentration data for imperfect mixing in continuous-flow systems as functions of operating variables for a marine propeller and the disk style turbine.

Burghardt and Lipowska (1971) noted that: (1) there were no correlations which established the range of parameters for ideal mixing in an agitated reactor and (2) ideal mixing can occur in a reactor without stirring over a wide range of operating variables. The objective of their work was to determine the range of ideal mixing in tanks without mechanical stirring for single-phase flow. The energy for the mixing was obtained from the pressure drop occurring in the reactor. The experiments were performed in the standard configuration with and without a disk style turbine. The diameter of the tank was 0.17 m, and inlet and exit streams were 0.0066 m in diameter with the inlet at the tank bottom and the exit at top. Ideal mixing occurred when the tank Reynolds number, Re_T or $\rho T V_c/\mu$, was greater than 13.5 where V_c was the superficial liquid velocity or circulation velocity in the tank. With mechanical mixing, ideal mixing occurred when the impeller Reynolds number, Re or $\rho D^2 N/\mu$, was equal to or greater than 89.3[ln $(13.5/Re_T)$].

Lipowska (1974) determined the range of ideal mixing in a continuous-flow tank as a function of inlet flow energy, energy due to mechanical agitation, inlet tube diameter, tank diameter, liquid viscosity, and residence time. Two Reynolds numbers were defined: (1) mean residence time Reynolds number, Re_τ: $4Q_F\rho/\pi\mu T$ or $T^2\rho/\tau\mu$ where Q_F is the volumetric flow rate of the feed and τ is the residence time, and (2) the inlet tube Reynolds number, Re_i or $4Q_F\rho/\pi d_i\mu$, where d_i is inlet diameter of the feed pipe. Nine d_i/T ratios were studied for tanks in the standard configuration with and without agitation by recording the residence time distributions as a function of inlet flow rates. When the $I(t/\tau)$ approached the ideal mixing curve for $I(t/\tau) = e^{-t/\tau}$, the system was considered perfectly mixed. The results showed that the mean residence time Reynolds number had to be greater than $569.1(d_i/T)^{1.09}$ or:

$$\text{Re}_\tau = \frac{4Q_F\rho}{\pi T\mu} > 569.1\left(\frac{d_i}{T}\right)$$ (4.239)

to have perfect mixing without agitation.

With agitation, a number of effects were noted. The minimum number of revolutions necessary for ideal mixing increased with increasing liquid viscosity and decreased with decreasing tank diameter. Mean residence time and inlet tube diamter d_i had little effect on mixing times. The mixing was correlated as:

$$N\theta_{cf} = 2428.2\left(\frac{4Q_F\rho}{\pi T\mu}\right)^{-1.215}\left(\frac{d_i}{T}\right)^{0.088}$$ (4.240)

Plasari et al. (1977) found the approach and results obtained by Burghardt and Lipowska (1971; 1974) were not necessarily true in general. They demonstrated that mechanical stirring actually led to reduced mixing performance in some cases. The residence times for the experiments by Lipowska (1974) were on the order of 0.8 to 208 min; the residence times of Plasari were on the order of 2 to 30 s. The kinetic energy input of the jets used in the study by Plasari was of the same order of magnitude as the kinetic energy input by mechanical agitation. Power input by jets in the studies by Burghardt and Lipowska was small in comparison to the power input by mechanical agitation. Both effects, the differences in residence times and the power input levels of the feed streams, led to differences between these studies. In the Plasari et al. study, mixing performance was found by measuring the $E(t/\tau)$ distribution (RTD curve) which was found to be a function of the direction of impeller rotation.

Pohorecki and Baldyga (1983) considered that, for ideal mixing, the Reynolds number of the length scales of turbulence in the inertial convective subrange was greater than 1, or:

$$\text{Re}_\lambda \gg 1$$ (4.241)

From this condition, they derived the following expression as a condition for perfect mixing. They further assumed that if this condition of the inertial convective subrange in the tank was met, then the tank was also perfectly mixed macroscopically:

$$1.4N_p^{3/8}\text{Re}^{3/8}\left(\frac{D}{H}\right)^{1/8}\left(\frac{D}{T}\right)^{1/4} \gg 1$$ (4.242)

Pohorecki and Baldyga (1983) also provided a rather complex model of micromixing for an agitated tank.

Unno and Akehata (1984) studied the effect of three inlet and outlet configurations on the flow in continuous systems for the disk style turbine centrally located in the tank. The feed rates were less than 10 percent of the pumping capacity of the impeller. The flow in the tank was studied using a flow follower. An asymmetric distribution of the feed in the tank was noted for the configuration having a horizontal feed at the bottom of the tank. The configuration having the feed stream directly below the impeller had an even feed distribution and a flow pattern similar to a batch mixing tank. Unno and Akehata suggested that the local flow patterns would have to be considered in modeling processes having a time constant ⅛ of the mean residence times for feed configurations showing asymmetric behavior.

The concept of sufficient mixing to obtain ideal mixing is very dependent upon the process objectives. If very fast reactions are taking place, the criteria for ideal mixing are much more stringent than for simple mixing of a tracer. For fast chemical reactions, ideal mixing must be specified in terms of the speed of the reaction or time constant of the reaction and the mixing time. If the time constant of the reaction is smaller than that of the actual mixing, then imperfect mixing will occur.

Concentration fluctuations

The study of the concentration fluctuations is helpful in the understanding of the mixing process. In the discussion, the rms of the concentration fluctuations will be denoted by c'; the time average concentration as C_{avg}. Developmental studies of experimental techniques used to obtain concentration fluctuations included Prausnitz and Wilhelm (1956; 1957) Lamb et al. (1960), Gibson and Schwarz (1963), Lee and Brodkey (1963; 1964), Goldfish et al. (1965), Nye and Brodkey (1967), Christiansen (1969), Torrest and Ranz (1969), Tsujikawa et al. (1974), and Vavanellos and Olson (1975). Goldfish et al. (1965) discussed an experimental technique in which colored tracer was formed in situ in flowing fluids without an injection system.

Manning and Wilhelm (1963) measured local concentration fluctuations in a standard baffled agitated tank with a disk style turbine. The feed was from below the impeller. The spatial resolution of the conductivity probe was 0.3 mm^3 which was considered microscopic relative to the tank size. The time-averaged concentration was found to be uniform in the tank whereas the concentration fluctuations c'/C_{avg} decreased with radial distance from the impeller from about 30 down to 1 percent near the wall. A power law fit indicated:

$$c'/C_{avg} \propto r^{-0.7} \tag{4.243}$$

However, c'/C_{avg} data were independent of angular position except near the baffles. The highest values of the concentration fluctuations

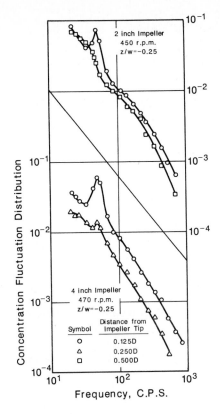

Figure 4.37 Concentration fluc-
tuation distribution functions
for 2- and 4-in-diameter im-
pellers at 450 and 470 rpm.
*(From F. S. Manning and R. H.
Wilhelm, AIChEJ., 9, 12, 1963.
Reproduced by permission of the
American Institute of Chemical
Engineers.)*

were associated with the smallest impeller and the lowest impeller
speeds. Peaks at the blade passage frequency were observed in the
spectra of the concentration fluctuations, shown in Fig. 4.37, and indicated that the discharge flow of the impeller contained relatively
unmixed concentration differences. Such fluctuations arose because of
the nature of the discharge flow discussed earlier, e.g., Fig. 4.18.

Corrsin (1964) modeled the decay of velocity fluctuations in conjunction with concentration fluctuations as:

$$\frac{dc'/c'}{du'/u'} = \frac{3}{5}\left(\frac{1}{\text{Sc}}\right)\left(\frac{l_f}{l_c}\right)^2 \tag{4.244}$$

where l_f and l_c are the microscales of the turbulence and the concentration fields, respectively. The equation originated from separate
equations for $\overline{c^2}$ and $\overline{u^2}$ which incorporated characteristic lengths and
microscales. Sc is the Schmidt number $\mu/\rho D_L$. Typically, the Schmidt
number for gases is on the order of 0.6 to 1.7; for liquids, 500 to 2000.
At high Schmidt number, the decay of the concentration fluctuations

was related to the decay of the velocity fluctuations as:

$$\frac{dc'/c'}{du'/u'} = \frac{(2/3)^{5/3}(L_f/L_c)^{2/3}}{1 + 1.1(L_f/L_c)^{2/3}\text{Re}_\lambda^{-1}\log(\text{Sc})}$$ (4.245)

where L_f and L_c are integral scales for the turbulence and concentration fluctuations, obtained from the integration of the energy and concentration spectra. Re_λ is the turbulence Reynolds number based upon the turbulence microscale. The time constant for mixing was also estimated as:

$$\tau = \frac{1}{2}\left[4\frac{L_c^{2/3}}{\epsilon^{1/3}} + \left(\frac{\nu}{\epsilon}\right)^{1/2}\log(\text{Sc})\right]^{2/3}$$ (4.246)

where ϵ is the kinetic energy dissipation rate per unit mass. This time constant was used to model the decay of concentration fluctuations as:

$$\overline{c^2}(t) = \overline{c_0^2}e^{-t/\tau}$$ (4.247)

Corrsin's work also indicated that the power requirements on scaleup for equal mixing times of equal-density materials were related as:

$$P_2 = K^5 P_1$$ (4.248)

where K is a length-scale factor.

Rosensweig (1964) presented an idealized theory for mixing in tanks in which a scalar fluctuation or unmixedness was considered to cascade down the spectrum of eddy sizes. The mean square fluctuation component was considered to pass through the various eddy sizes to be dissipated in the smallest eddies by molecular diffusion. It was assumed that the rate of transfer down to the smallest eddies was equal to the dissipation rate by molecular diffusion. The production of unmixedness was caused by the introduction of feed streams. This production was balanced by the flow out of the unmixedness by the exit streams and the dissipation of unmixedness by molecular diffusion. The balance between these quantities was given as:

$$(1 - \Gamma)\Gamma - \epsilon_\gamma\tau - \gamma^2 = 0$$ (4.249)

in which the first term is production of the unmixedness, e.g., due to flow of material into the tank; the second is the dissipation of the unmixedness; and the third is the outflow of the unmixedness. In the equation, Γ is the volume fraction of the mean concentration, γ is the local fluctuation about Γ, ϵ_γ is the concentration dissipation rate, and τ is the mean residence time. An explicit expression for unmixedness was given as:

$$\frac{\gamma^2}{(1 - \Gamma)\Gamma} = \left[1 + \left(\frac{0.477}{K}\right)(k_0^2 g_c \epsilon)^{1/3}\tau\right]^{-1} \tag{4.250}$$

where K was considered to be a universal constant varying between 0.2 and 1.8. In the equation, k_0 is a characteristic wave number, shown in Fig. 4.5, of the large eddies which varied as $1/L$ or the dimensions of the inlet port, ϵ is the mechanical energy dissipation rate and τ is the mean residence time of the input. Applications of these equations were given for a mechanical stirrer. The theory applies to a tank having effective large-scale mixing, high ratio of residence times to circulation times, and where turbulent diffusion dominates molecular diffusion.

Rosensweig (1966) extended his previous work to include the effects of multiple feed streams, transient processes, and the coupling of the scalar field to the momentum field in predicting the mean square fluctuating quantities. Both the scalar and momentum fluctuations were assumed to be spatially isotropic. The balance on the mean square scalar fluctuations was obtained as:

$$\frac{d\overline{\gamma^2}}{dt} - \sum_i \frac{(\Gamma_i - \overline{\Gamma})^2}{\tau_i} + \frac{\overline{\gamma^2}}{\tau} = -\epsilon_\gamma \tag{4.251}$$

The residence times were:

$$\tau = \frac{V}{\sum_i q_i} \tag{4.252}$$

and:

$$\tau_i = \frac{V}{q_i} \tag{4.253}$$

and the scalar dissipation rate was given as:

$$\epsilon_\gamma = 2D_L\overline{(\nabla\gamma)(\nabla\gamma)} \tag{4.254}$$

The first term on the left side of Eq. (4.251) is the accumulation of the mean square variance, i.e., $\overline{\gamma^2}$; the second term is the rate of production of variance due to input streams; the third term is the rate of removal of the variance by the output streams; and ϵ_γ is the dissipation of the variance throughout the mixer volume. It was assumed that the variance of any leaving stream was the same as the variance obtained by volume averaging.

The dissipation rate of the scalar variance was because of diffusion on the molecular level. However, this dissipation was preceded by the cascade of the variance from the large-scale eddies to the small-scale eddies. The variance was dissipated by molecular diffusion at the smallest eddy sizes. As a result, there was a relationship between the mechanical energy dissipation and variance dissipation. Rosensweig estimated the scalar dissipation rate in terms of the energy dissipation rate as:

$$\epsilon_\gamma = \frac{\overline{\gamma^2}\epsilon^{1/3}}{\Lambda_\gamma^{2/3}} \tag{4.255}$$

where Λ_γ is the integral scale of the concentration fluctuations and the mechanical energy dissipation rate is:

$$\epsilon = \frac{k^3}{\Lambda} \tag{4.256}$$

where Λ is the integral scale for the velocity fluctuations. The equation implies rapid dissipation of the scalar variance in a system having a small integral scale and high energy dissipation rates.

Rosensweig provided different case studies with solutions for the following:

1. General transient solutions for the turbulent kinetic energy and specific solutions under conditions of low and high dissipation rates

2. Asymptotic solutions for the turbulent kinetic energy for both high and low turbulent production rates, i.e., rapid and slow mixing

3. Analogous solutions for the scalar variance having equal integral scales for momentum and the scalar

4. Asymptotic solutions for the scalar variance in the presence of a constant turbulent kinetic energy field for low and high turbulent production rates

5. Solutions of transient turbulent kinetic energy and scalar fields

The application of the turbulent kinetic energy and scalar equations would require additional terms if applied to agitated tanks since the equations as they were given do not include terms which account for the presence of an impeller.

Rao and Brodkey (1972) compared the two theories presented by Corrsin (1957; 1964) and Rosensweig (1964; 1966) and provided the conditions where the two theories agreed. Rao and Brodkey noted that Rosensweig's work was suitable for mixing of gases.

Reith (1965) performed a study of the generation and decay of concentration fluctuations in a stirred baffled tank using a standard disk style turbine. The tracer was fed directly to the impeller. A conductivity probe with high spatial resolution was used to measure the fluctuations. Reith developed an expression for the rate of dissipation of the concentration fluctuations as:

$$\frac{\chi}{Nc_0^2} = \chi^*\left(\frac{r}{D}, \theta, \frac{z}{D}\right) \tag{4.257}$$

where χ is rate of dissipation of the concentration fluctuation and $\overline{c_0^2}$ is the initial concentration fluctuation. Reith considered that the decay of concentration fluctuations was dependent upon the impeller rotational speed, the initial level of the fluctuations, and the spatial location in the tank. The analysis was applied only to turbulent convective transport processes, i.e., high Reynolds numbers and high Peclet numbers. The initial concentration fluctuation was obtained from a specie balance which reduced to:

$$\frac{\overline{c_0^2}}{\overline{C^2}} = \frac{Q_F^2}{q_t Q} \tag{4.258}$$

where Q_F is the flow rate through the tank, q_t is the flow rate of the tracer, and Q is the circulation or pumping rate of the impeller. The tracer input generated the concentration fluctuation c_0; the mean concentration in the tank was $\overline{C}$. The decay of the concentration fluctuations in the impeller stream was measured and modeled as:

$$\frac{\overline{[c(r)]^2}}{\overline{c_0^2}} = \exp\left[-\frac{12T}{l_c^2}\frac{(2r/D)^2 - 1}{4\pi N}\right] \tag{4.259}$$

where $l_c^2/12T$ is the decay constant suggested by Corrsin in which l_c is the concentration microscale. The term $[(2r/D)^2 - 1]/4\pi N$ was the time elapsed since the concentration fluctuation left the impeller blade. $\overline{c_0^2}$ was found to vary inversely with impeller rotational speed N and the impeller pumping capacity Q in the above equation. The decay constant $l_c^2/12T$ was the highest near the impeller, roughly $0.08N^{-1}$ for $2r/D$ between 1 and 2, but changed abruptly to between 0.03 to $0.04N^{-1}$ for $2r/D$ between 2 and 8. As reported earlier in this chapter, the nature of the blade vortex systems and jet discharge from the disk turbine changes abruptly between $2r/D$ of 1 to 2 as shown in Fig. 4.11.

Mixing was most intense near the impeller blades. The decay constant $l_c^2/12T$ in this region could not be estimated using isotropic turbulence theory since the concentration distributions were highly anisotropic. The size of the concentration fluctuations and the space over which they were distributed determined the decay rate. Higher dissipation rates were obtained by injecting the tracer through little holes in the stirrer disk near the blades which agreed with the notion that finer initial dispersion scales yield higher decay rates. Unfortunately, the phenomenom could not be investigated systematically. The concentration spectra were found to be very similar and independent of position in the tank. No peaks in the spectra at the blade passing frequency were observed.

To simulate the transport and distribution of a scalar requires the solution of the equations of motion and the scalar transport equation. To this end, numerical studies of concentration fluctuation distributions were performed by Patterson (1970) and Waggoner and Patterson (1975). The data by Reith (1965) were calculated from the hydrodynamic model as described in the latter section on models.

Patterson et al. (1982) developed a fluorescence method for measuring concentration fluctuations with and without chemical reactions. The fluorescence was stimulated by a xenon light source and was measured using a photomultiplier. Measurements of $\overline{c^2}/c_0^2$ were made in the impeller stream of a standard disk style turbine in a stirred tank having the standard configuration. Measurements of $\overline{c^2}/c_0^2$ were also made for a reacting system at a Damkohler number k_1CV/F of 1000.

Ruszkowski and Muskett (1985) measured the concentration fluctuations using conductivity probes and modeled the decay of the rms concentration fluctuations c' in an agitated tank in the standard configuration as:

$$\frac{d\overline{c^2}}{dt} = - \beta\overline{c^2} \tag{4.260}$$

where β is the decay constant. A square pitch marine propeller and a 45° pitched blade turbine were used. Mixing times were found to be related to specific power input as:

$$\theta_m = (\overline{\epsilon})^{-0.33} \tag{4.261}$$

Since $N\theta_m$ equals a constant for a particular geometry, ϵ varies with N^3. Values for β were compared with those predicted using a modification of an expression provided by Corrsin (1964):

$$\beta = 2\left[3\left(\frac{D^2}{2\overline{\epsilon}}\right)^{0.333} + \left(\frac{2v}{\overline{\epsilon}}\right)^{0.5}\ln Sc\right]^{-1} \tag{4.262}$$

where $\overline{\epsilon}$ is the mean turbulent energy dissipation per unit mass in watts per kilogram, D is the impeller diameter in meters, and the kinematic viscosity is in square meters per second. In the calculations, the second term in the brackets is negligible. The predicted values from this expression were found to be twice the values determined experimentally.

Gaskey et al. (1988) described a method for measuring local values of concentration fluctuations, following work by Patterson et al. (1982). The method used a laser beam to excite a fluorescent dye at a given point. The dye emitted fluorescent light which was monitored at right angles to the incident beam. The concentration fluctuations caused changes in the fluorescent intensity. The spatial resolution was 10 to 50 μm at frequencies of 200 to 5000 Hz. In the experiments, large packets, 1 to 3 cm, of unmixed material were observed in the time signal.

Reaction studies

The need in reaction studies is to quantify the mixing which gives rise to the contacting of the reacting species. If the time scales of the mixing and the time scales of the chemical reactions are similar in magnitude, then the perfectly backmixed assumption cannot be made and mixing effects will be present in the system for all non-first-order reactions.

The basic problem with chemical reactions occurring in an agitated tank can be observed in the time-averaged rate equations using Reynolds rules of averaging such as:

$$\frac{dC}{dt} = kC^n \qquad (4.263)$$

In applying Reynolds time averaging to this equation, additional quantities appear. For first-order reactions where $n = 1$, mixing has no effect. For reactions other than first order, mixing effects can occur. As an example:

$$\frac{dC_A}{dt} = kC_A C_B \qquad (4.264)$$

Upon time averaging, the relationship becomes:

$$\frac{d\overline{C_A}}{dt} = k(\overline{C_A C_B} + \overline{c_A c_B}) \qquad (4.265)$$

where $\overline{c_A c_B}$ is an unknown quantity. In flow situations, $\overline{u_i c_A}$ are unknown quantities as well. Similar situations occur if such reaction equations were space averaged.

Extensive reviews of mixing and chemical reactions can be found in

Brodkey (1975), Patterson (1975; 1981), and Toor (1975). These will not be reviewed because of their breadth.

To eliminate or minimize the effects of mixing, Beek and Miller (1959) suggested that mixing should be completed in a time shorter than the halftime of the fastest significant reaction. Under such conditions, unknown quantities such as $\overline{c_A c_B}$ and $\overline{u_i c_A}$ approach zero.

Rice et al. (1964), using a disk style impeller and a tank in the standard configuration, visually observed the reaction zone in a very fast acid-base neutralization study and reported the variation in zone radius with impeller rotational speed. The feed of the reactants was from below the impeller. Using the pumping capacity of the impeller, these measurements were converted to reaction times which were found to be on the order of 20 ms. In some cases, the reaction zone shrunk into the impeller volume. They suggested that a cylindrical reaction zone was an appropriate model following work by Toor (1962).

Paul and Treybal (1971) studied the yield of R for the following competitive, consecutive reaction:

$$A + B \xrightarrow{k_1} R \qquad R + B \xrightarrow{k_2} S$$

Initially, A was charged in the stirred tank with B being added at a fixed feed rate. R was the desired product. The overall objectives of the work were to correlate the mixing effects on the yield in R and to determine scaleup methods for the reaction. The effects of mixing depended upon the absolute values of the rate constants and the magnitude of the local concentrations. The behavior of the mixing determined the local concentrations.

Paul and Treybal found that the yield of R varied from 50 to 72 percent and was a function of impeller rotational speed, impeller diameter, and feed position. Feed into the impeller discharge stream produced the highest yields for a fixed impeller rotational speed. Feed into the impeller provided lower yields and feed into the upper part of the tank had the lowest yields of all the feed positions studied. Removal of baffles increased or decreased yield depending upon feed location. Effects of feed rate and feed distribution did not occur in the operational ranges studied. Yield was shown to be well correlated with fluctuating velocity, Fig. 4.38, and with a microscale time scale parameter, Fig. 4.39, defined as:

$$\tau = \frac{\eta}{u'} = 0.882 \frac{\nu^{3/4} L_f^{3/4}}{u'^{7/4}} \tag{4.266}$$

where η is the Kolmogorov microscale and L_f was selected as either the tank diameter or the impeller diameter depending upon the actual

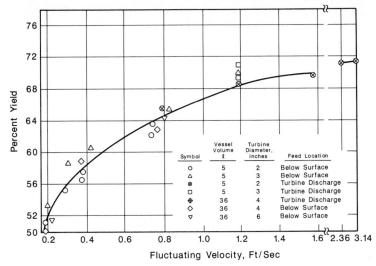

Figure 4.38   The yield of R correlated with fluctuating velocity.   *(From E. L. Paul and R. E. Treybal, AIChEJ., 17, 718, 1971. Reproduced by permission of the American Institute of Chemical Engineers.)*

feed location. The energy dissipation, used in the calculation of the Kolmogorov microscale, was the local energy dissipation rate calculated from the relationship:

$$\epsilon = 1.65 \frac{u'^3}{L_f} \tag{4.267}$$

where the u' data were obtained from the literature. The effect of the initial concentration of A was also studied. Increasing A_0 caused a reduction in yield. An effect of viscosity was noted in the experimental data although the ¾ exponent in the above equation was not confirmed.

For scaleup, Paul and Treybal suggested that the local fluctuating velocity, energy dissipation rate, and turbulent length scale were critical scaleup parameters. Local fluid and flow properties surrounding the feed position were not accounted for in scaleup based upon power per volume or equal impeller tip speed. Power per volume is an average quantity for the entire tank and, as such, is not a local parameter. Equal impeller tip speed is not the sole determining parameter of the fluctuating velocity for the bulk of the tank since u' varies with D/T in the bulk of the tank. Equal impeller tip speed on scaleup would be satisfactory only for a feed position in the impeller discharge flow. Paul and Treybal also considered that special design considerations were needed to improve local micromixing around the feed position when

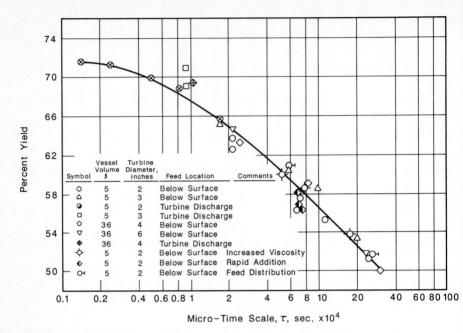

The table within the figure:

Symbol	Vessel Volume ℓ	Turbine Diameter, inches	Feed Location	Comments
○	5	2	Below Surface	
△	5	3	Below Surface	
◐	5	2	Turbine Discharge	
□	5	3	Turbine Discharge	
◇	36	4	Below Surface	
▽	36	6	Below Surface	
✦	36	4	Turbine Discharge	
◈	5	2	Below Surface	Increased Viscosity
◗	5	2	Below Surface	Rapid Addition
⊶	5	2	Below Surface	Feed Distribution

Micro−Time Scale, τ, sec. ×10⁴

Figure 4.39   The yield of R correlated with microscale time scale.   *(From E. L. Paul and R. E. Treybal, AIChEJ., 17, 718, 1971. Reproduced by permission of the American Institute of Chemical Engineers.)*

$k_1 C_A \tau$ was greater than 10^{-5}. Canon et al. (1977) have simulated some effects of feed position and impeller type for the data of Paul and Treybal (1971).

The study by Paul and Treybal points directly to the use of chemical reactions in the study of mixing, particularly micromixing. Paul and Treybal noted that the following competitive parallel reactions, as shown, have not been studied:

$$A + B \xrightarrow{k_1} R \qquad A + B \xrightarrow{k_2} S$$

Bourne and Garcia-Rosas (1985) noted that many single-phase chemical reactions are sufficiently fast that the distribution of products is affected by the rate of mixing. These mixing effects disappear or are minimized as the rate of energy dissipation increases. Conventional stirred tanks cannot necessarily provide such high-energy dissipation rates. A high-speed (400 s^{-1}) rotor stator mixer can provide much higher levels of energy dissipation in such applications. In their study, the clearance between the rotor and stator was on the order of 0.0002 m. The arrangement with the feed going directly to the rotor-stator amounts to an in-line mixer set inside an agitated tank. The results showed that slower micromixing occurred at decreasing rotor speeds and that diffusion and reaction ran to completion in a time much shorter than the mean residence time in the rotor-stator volume. The effect of mixing on the product distribution was a function of the volumetric feed rates and a bulk mixing modulus M which was proportional to the square of a length scale δ and inversely proportional to the liquid diffusivity D_L as:

$$M \propto \frac{\delta^2}{D_L} \qquad (4.268)$$

The length scale was set equal to one-half of the Kolmogorov microscale, $0.5(\nu^3/\epsilon)^{1/4}$. The local rates of energy dissipation exceeded values occurring in agitated tanks by more than one order of magnitude. Effects of mixing became unimportant as the mixing modulus approached 1. The time scale for diffusion was on the order of 0.3 to 0.9 ms. At such a diffusion time scale, reactions having a time scale of 1 ms or greater would not be affected adversely by mixing.

Takao and Murakami (1987) performed mixing studies with and without reaction in an impeller Reynolds number range from 5400 to 8100 in a system using a disk style turbine in the standard configuration. The tracer and reactants were added in the same manner, and the local concentrations were measured at 72 locations in the tank for each of the studies. The stoichiometric ratio of the two reacting species was 1 when mixing was complete. The mean concentrations, four measures of segregation, and the mean unreacted fraction for the instantaneous reaction were calculated and compared. It was found that there was an equality between one of the segregation measures D_d and the mean unreacted fraction for the instantaneous reaction. The measure of segregation was referred to as the degree of deviation and was defined as:

$$D_d = \frac{|C_a/\overline{C}_a - 1|}{2\overline{B}} \qquad (4.269)$$

where $\overline{B}$ is the mean volume fraction of liquid B. Takao and Murakami (1988) further documented the agreement between the degree of deviation and the intensity of segregation.

Micromixing

Mixing phenomena in an agitated tank are very difficult to describe. However, to help characterize mixing, two mixing states have been discussed: maximum mixedness and complete segregation. However, neither can actually take place. Maximum mixedness is where the feed mixes immediately and instantaneously with the rest of the contents of the tank. A situation of maximum mixedness is approached with the addition of material to the impeller intake stream or by having a very good initial distribution of material. However, even in these physical situations, there is a time scale in which the feed is mostly segregated.

Complete segregation is where the feed does not mix upon entry into the tank and remains unmixed until exiting the tank. Actual physical behavior of material can resemble complete segregation. Cream added slowly to coffee can demonstrate substantial segregation. If the coffee remains unmixed until exiting the cup, then complete segregation was approached. With the complete segregation concept, there is a problem. There is always some mixing which takes place, and there are time and length scales associated with it.

The key to understanding the actual mixing process lies in the fluid mechanics of the process. More experimental work is needed to determine the flow fields in mixing.

There are two extremes in mixing which occur in the material added to an agitated tank: (1) mixing controlled by the feed stream and (2) mixing controlled by the local agitated tank flow. The rate of addition and the fluid mechanics around the feed point determine the extent of segregation. High levels of segregation can result, however, which require a certain amount of time to dissipate. Micromixing effects due to jet mixing have been relatively neglected in the technical literature. Typically, without any injection energy from the feed, micromixing might easily occur in the boundary layers if the local energy dissipation in the tank is also low.

The addition of material down the sides of a tank is an example where there is very little energy added by the feed and very little locally occurring in the tank. When there is a jet entry, then jet energy dissipation rates should be used in micromixing models. Here the traditional engulfment process may take place. If the local energy dissipation in the tank is high, then jet mixing may not be significant. However, the energy of the jet is still important because the feed is also contained in the jet.

Micromixing is turbulent diffusion and proceeds by molecular diffusion and convection within the laminae of the turbulence. The laminae are embedded in the energy dissipating vortices of the turbulence. The fine-scale turbulent properties of interest are the local energy dissipation, the Kolmogorov microscales, the Batchelor concentration microscales, vortex life times, and rates of deformation in the vortices.

Micromixing is the process by which there is a further size reduction of local unmixed regions. The micromixing mechanism can be interpreted in terms of statistical turbulence theory. The concentration spectrum is divided into an inertial-convective subrange, a viscous-convective subrange, and a viscous-diffusive subrange. At the beginning stages, distributive mixing occurs where specie-rich solutions break up into large- and intermediate-sized eddies to form a macroscopically uniform mixture. This occurs at the smallest wave numbers in the concentration spectrum and in the lowest portions of the inertial-convective subrange. The inertial-convective subrange is characterized by the largest-scale eddies down to the Kolmogorov microscale. In this subrange, convection is primarily responsible for the mass transfer, and diffusion is not considered important. The scale of the unmixed material is reduced by eddy motions. This stage is also called dispersive mixing where the specie-rich eddies become smaller but high segregation still exists between species. The viscous-convective subrange extends from the Kolmogorov microscale to the concentration scale of Batchelor η_b [i.e., $(\nu D_L^2/\epsilon)^{1/4}$]. The eddies are subject to laminar strain which causes further scale reduction into the viscous-diffusive subrange which begins at the Batchelor concentration length scale. In this subrange, diffusion and laminar strain are equally important and diffusion dissipates the concentration differences. The eddies have reached the smallest scales, and diffusion occurs on the molecular level. Such a mechanism as described is a restatement of the mechanism provided by Beek and Miller (1959) but is given in terms of energy and concentration spectra.

Micromixing can be considered to proceed by small convective motions in the viscous-convective subrange near the Kolmogorov microscale and by molecular diffusion in the viscous-diffusive subrange near the Batchelor concentration length scale. Other discussions of these concepts can be found in Brodkey (1967) and Baldyga and Bourne (1984a; 1984b; 1984c).

In terms of the transport of a scalar in a turbulent flow field and using the time-averaged scalar transport equations, a measure of micromixing appears as the time average of the product of fluctuating velocity u and the fluctuating scalar component γ. There appear to be no measurements of the correlation $\overline{\gamma u}$ in the mixing literature. Such measurements would involve the construction of a combined velocity and concentration probe, e.g., hot wire with a conductivity probe.

The turbulent eddy diffusivity D_{LT} for concentration can be defined from the correlation $\overline{c_a u_j}$ using:

$$\overline{c_a u_j} = D_{LT}\left(\frac{\partial \overline{C_a}}{\partial x_j}\right) \qquad (4.270)$$

This eddy diffusivity D_{LT} is also a measure of micromixing. Different levels of micromixing occur in an agitated tank because both u_j and c_a are spatially distributed throughout the tank.

A number of processes occur quite rapidly, e.g., neutralization, precipitation, combustion, in which conversion occurs between reacting surfaces where high concentration gradients occur. As noted by Paul and Treybal (1971) and Bourne (1982), various detrimental effects occur because of the poor mixing under these conditions. Deviations from ideal mixing conditions result in the requirement of increased reactor volume and in the inability of predicting conversion. This, in turn, implies the inability to scale up the process. The production of unwanted products occurs which represents lower utilization of raw materials and increased separation duty downstream. If the reaction or process is selective where different product distributions are possible, operating at the desired selectivity is desirable. Unfortunately, the conditions of this operation are not usually known without a detailed experimental investigation.

The micromixing literature in this review is divided up into two categories. The first is based upon the use of chemical reactions and a shrinking laminae model of micromixing. The second category includes all other approaches. In the first category, the studies have dealt with the following reaction system as a means of studying micromixing. Although initially studied by Paul and Treybal as cited above, the reaction system has been dubbed the Bourne reactions and are single-phase, competitive, consecutive reactions of the type:

$$A + B \overset{k_1}{\to} R \qquad R + B \overset{k_2}{\to} S \qquad \text{where } k_1 \gg k_2$$

that show effects of operating conditions (e.g., impeller rotational speed, viscosity) and geometry (e.g., feed location) on product distributions. These reactions are sufficiently fast that the yield of R can be varied depending upon conditions (Bourne, 1982). The first reaction is essentially instantaneous relative to the micromixing with reaction times on the order of 0.2 to 2 ms and $k_1 \simeq 10^4 \, \text{m}^3 \, \text{mole}^{-1} \text{s}^{-1}$.

In a well-mixed A system, where B is in limited supply (i.e., B is the limiting reagent), the first reaction is favored since B will react with A if A is available. If R is formed, then in a well-mixed A system, R will only be with A. In a poorly mixed system, after R has been formed, B will still be available to react. A well-mixed system will form R, and a poorly mixed system will form S. With $X = 2C_S/(2C_S + C_R)$, a well-mixed system will have X approach 0, and a poorly mixed system will have X approach 1. X is the amount of B used to make S divided by the amount of B used overall. A mixing modulus is also defined as:

$$M = \frac{k_2 C_{B0} \delta_0^2}{D_L} \qquad (4.271)$$

which is low for effective mixing and high when mixing occurs by diffusion. The mixing modulus is a form of the Damkohler number. The Damkohler number is the ratio of the transport or circulation time to reaction time. The size of the laminae δ_0 can be estimated as one-half of the Kolmogorov micro length scale, $0.5(\nu^3/\epsilon)^{0.25}$. The conversion X is a function of the feed ratio N_{A0}/N_{B0}; the ratio of reaction rates k_1/k_2; the type of reactor; the Schmidt number Sc; the volume ratio V_A/V_B, the mixing modulus M, or:

$$X = f\left(\frac{N_{A0}}{N_{B0}}, \frac{k_1}{k_2}, \text{Reactor type, Sc}, \frac{V_A}{V_B}, M\right) \qquad (4.272)$$

The actual functional relationship for X is determined experimentally or calculated from models. Extensive work of importance along similar lines includes Patterson (1975; 1981) and Toor (1975).

Angst et al. (1979) used the conversion of the very fast, series-parallel chemical reaction, described above, to study micromixing in three identical agitated tanks, 0.63 m³, using a paddle-palette impeller, an anchor impeller, and a Pfaudler impeller. The Reynolds numbers ranged from 8×10^4 to 9×10^5. A study was made of the effects of feed location on the conversion X where the conversion X was considered a measure of micromixing. Feed through the tank base provided the poorest micromixing and was not significantly affected by the impeller rotational speed. For the other feed locations studied, the conversion and micromixing increased with impeller rotational speed. The Pfaudler impeller provided the highest conversion with a feed point located somewhat above the impeller tip.

Angst et al. provided a size estimate for the reaction zone from a model. These estimates decreased with increasing impeller rotational speed and were on the order of 30 to 80 μm, somewhat less than one-half the Kolmogorov microscale. Favorable feed locations were considered to be where the energy dissipation rates were higher than the average dissipation rate for the tank.

Bourne (1982) considered that in such reactions, described above, with relatively small amounts of B, a laminar B-rich sheet was formed which was thinned to the size of the Kolmogorov microscale $\eta = (\nu^3/\epsilon)^{1/4}$. The diffusion and reaction in the sheet or slab were modeled as:

$$\frac{\partial C_i}{\partial T} = (1 + \beta T^2)\frac{\partial^2 C_i}{\partial Z^2} + \frac{\delta_0^2 \, r_i}{D_L c_0} \qquad (4.273)$$

where C_i, T, and Z are dimensionless concentration c_i/c_{i0}, time tD_L/δ_0, and distance z/δ, respectively. Time began when the thin laminar sheet had a thickness δ_0. However, there was no general method given to determine this time, which probably makes up a considerable portion of the mixing time. The dimensionless constant β was modeled as

$\gamma \delta_0^2/D_L$ and was proportional to the ratio of half-lives for diffusion, i.e., δ_0^2/D_L, and for thinning γ^{-1}. The shear rate experienced by the sheet was γ.

The solution of the above equation was stated in a general form as:

$$C_i = f\left(T, Z, \frac{k_1}{k_2}, \frac{N_{A0}}{N_{B0}}, \frac{V_A}{V_B}, M, \beta\right) \qquad (4.274)$$

where k_1/k_2 is the ratio of reaction rates for R and S; N_{A0}/N_{B0} is the initial stoichiometric ratio of A to B; V_A/V_B is the volume ratio; and M is the mixing modulus $k_2 c_{B0} \delta_0^2/D_L$, proportional to ratio of half-lives for diffusion and the second reaction.

The model explained many of the effects found experimentally. As impeller rotational speed increased or the viscosity decreased, the thickness of the reacting zone decreased, giving rise to different yields. Equal product distributions were obtained for different impellers having the same power input. In all cases, the importance of the micromixing, rather than macromixing, was emphasized. Modeling difficulties were also mentioned. The Kolmogorov microscale was found to overestimate the size of the reaction zone. Of major concern also was the determination of the time needed to approach the length scales upon which the slab model was based. Such models cannot be applied in general without such a basis.

Baldyga and Bourne (1984a; 1984b; 1984c) developed several other concepts in support of their work. An important strain rate in the viscous-convective subrange was:

$$\gamma = a\left(\frac{\epsilon}{\nu}\right)^{1/2} \qquad (4.275)$$

where the constant a varied from 1 to 5. The thickness of the eddies decreased rapidly and diffusion became essentially one dimensional. The velocity of the thinning layers was considered to be:

$$u_t = \frac{(\epsilon/\nu)^{1/2}x}{(4 + \epsilon t^2/\nu)^{1/2}} \qquad (4.276)$$

The shapes of the reaction region was time dependent as well. Baldyga and Bourne used vorticity to explain micromixing. Generally, turbulence contains a high level of fluctuating vorticity which is distributed over various size eddies. Fluid deformation causes vortices to stretch, causing vorticity and kinetic energy to be transported to smaller eddies. In the model provided by Baldyga and Bourne, the vorticity of a vortex was considered to decay exponentially with time and to return to an isotropic state. Vortical eddies which returned most rapidly to an isotropic state were estimated to be on the size of

$12(\nu^3/\epsilon)^{1/4}$ or 12η, having a mean lifetime of $12.7(\nu/\epsilon)^{1/2}$. These eddies were considered to form an important part of the fine dissipating structure of the turbulence, being 12 times the Kolmogorov microscale η and were pictured as vortices which engulfed fluid to form partially segregated laminae. The vorticity and the vortices decayed to the Kolmogorov microscale while diffusion and reaction continued. After a time, however, the process repeated itself; a vortex structure was rebuilt of size 12η, engulfing more material and causing diffusion and reaction. The sequence was repetitive until the system was entirely homogeneous. Baldyga and Bourne provided a pictorial representation of the process, showing engulfment of material, formation of the vortex, vortex stretching, and decay. The vortices were considered to be about 12 times the Kolmogorov microscale and engulfed material having layers of thickness of the Kolmogorov microscale. Layers of A and B were formed in this process which permitted modeling in terms of a shrinking slab diffusion and reaction model.

Baldyga and Bourne improved the diffusion-reaction slab model to:

$$\frac{\partial C_i}{\partial T} = \left(\frac{\delta_0}{\delta}\right)^2 \frac{\partial^2 C_i}{\partial Z^2} + \frac{\delta_0^2 \, r_i}{D_L c_0} \qquad (4.277)$$

where C_i, T, and Z are dimensionless concentration c_i/c_{i0}, time tD_L/δ_0, and distance z/δ, respectively. The initial thickness δ_0 was one-half Kolmogorov microscale $(\nu^3/\epsilon)^{1/4}$, and the thickness of the slab as a function of dimensionless time and Schmidt number was given as:

$$\frac{\delta_0}{\delta} = \left\{ 1 + \frac{T^2 \text{Sc}^2}{32} - \left[\left(1 + \frac{T^2 \text{Sc}^2}{32} \right) - 1 \right]^{0.5} \right\}^{-0.5} \qquad (4.278)$$

Calculations were performed for semibatch and continuous-stirred tank reactors which showed that the measure X increased with increasing mixing modulus M; decreased with increasing V_A/V_B and was lower in the semibatch reactor than for the continuous-stirred tank.

In studies where k_1/k_2, N_{A0}/N_{B0}, V_A/V_B, and Sc were fixed, the effects of mixing were studied through the mixing modulus M, which contained the initial thickness δ_0. The initial thickness was equal to $0.5(\nu^3/\epsilon)^{1/4}$ which was used to obtain local values of the energy dissipation rate ϵ. The local energy dissipation rate ϵ was modeled as a constant ϕ times the average dissipation rate, $\bar{\epsilon}$, for the tank:

$$\epsilon = \phi \bar{\epsilon} \qquad (4.279)$$

From the model, ϕ was determined to be 8 for feed placed on the intake of the impeller, $1/8$ to $1/4$ for feed placed below the liquid surface, and 1 for feed placed in the impeller discharge flow between the impeller and tank wall. The model also showed that the time for diffusion and reaction was much smaller than the circulation time of the tank. The time for diffusion and reaction θ_{DR} was estimated to be 70 to 120 $(\nu/\bar{\epsilon})^{1/2}$.

Bourne and Dell'ava (1987) in a semibatch reactor studied various effects of the feed stream, feed pipe size, and inadequate tank turbulence for the reaction system above. When the variables N_{A0}/N_{B0}, k_1/k_2, Reactor type, Sc, and V_A/V_B were held constant, X became a function of M entirely or:

$$X = f(M) = f\left(\frac{k_2 C_{B0} \delta_0^2}{D_L}\right) \tag{4.280}$$

where
$$\delta_0 = 0.5 \left(\frac{\nu^3}{\epsilon}\right)^{1/4} \tag{4.281}$$

In their study, N_{A0}/N_{B0} was 1.05, k_1/k_2 = 3860, Sc = 1280, V_A/V_B = 50, and M was varied as C_{B0} and the impeller rotational speed varied. The time of the micromixing occurring in the tanks was on the order of 0.25 to 1.0 s as calculated from $84(\nu/\epsilon)^{1/2}$ which assumed six to eight generations of vortices and eddies for diffusion and reaction to occur completely. The time of transport was calculated from circulation time data.

Among the results observed, X increased as a function of feed conditions for large feed pipes, e.g., 3- and 4-mm-diameter feed pipes. The addition times of B were 70 to 600 s. This entrance effect on product distribution was attributed to A entering the B feed pipe and reacting with high excess B forming S. Small feed pipes did not have this effect. One can also conclude that the momentum input of the feed stream affected the product distribution.

Bourne and Dell'ava also felt that if reactions are faster than one circulation time, then fresh B solution would react faster than A could arrive. This lowers locally the A concentration in that feed region. Typically, in mixing time studies, feed location does not matter. However, in reacting systems, this is no longer the case. Mixing determines the amount of material reacted which determines the concentration. Some circulation loops are better mixed than other circulation loops. When some material enters a circulation loop which is well mixed, the product distribution is different from when material enters a poorly mixed circulation loop.

Experimentally, bulk mixing problems were noted to occur at the large scale in the work. In the large-scale tank, liquid in the top 10 percent of the tank was observed to participate irregularly in the general circulation; more so at low stirrer speeds. Turbulence was not well developed in the upper portion of the liquid which explained the differences between laboratory studies and plant tests. Laboratory studies are typically well mixed.

For their system, Bourne and Dell'ava concluded that: (1) product distributions from fast reactions should be unaffected by scaleup of a reactor if the energy dissipation rate is held constant in the reaction zone, and (2) other geometries should be selected to replace the stirred tank when rapid reactions are considered.

As any reaction system is convected, the inhomogeneity of the local energy dissipation rate in the bulk circulation affects the micromixing and has to be eventually accounted for in micromixing models. Baldyga and Bourne (1988) noted this interrelationship between micromixing and bulk circulation in their reaction studies. The reaction was actually spread out over a volume rather than a point. Their reactions needed six to eight generations of vortices and decay for completion. In their model, the energy dissipation rate was discretized to form the energy dissipation distribution function $\epsilon(r, z)$, and the mixing modulus M was given as a function of position as:

$$M(r, z) \propto Sc\ Re^{-1/2}\epsilon^{-1/2}(r, z)\ Da \qquad (4.282)$$

Using this relationship for the mixing modulus, satisfactory agreement was obtained between measured and predicted product distributions. The Damkohler number Da or $k_2 C_{A0}/N$ accounted for the inhomogeneity of the mixing field and the limited reaction times.

Bourne et al. (1988) noted that, for fast chemical reactions, selectivity depended not only upon the concentration of the reactants but also upon the concentration of the products. In the reaction system studied, Bourne et al. noted that the pH affected the product distribution. The reaction system in this study contained 13 species, each of which satisfied the convection diffusion equations to model the micromixing. The model used was the shrinking laminated slab model mentioned above. The product distribution depended upon the initial pH values of the reagent solutions, the initial buffer concentration, and the various ionic equilibrium constants. Local transient pH gradients were observed in the calculated results.

Bourne and Ravindranath (1988) described the methods used for the calculation of micromixing in the shrinking laminae model.

Studies in the second category include the following. Klein et al. (1980) studied micromixing in a continuous-stirred tank reactor using a chemical reaction as an indicator of micromixing to establish the mechanisms by which micromixing occurs. Depending upon the impeller rotational speed and space time in the reactor, they considered micromixing to occur either as an erosion process or a diffusional type process which approached molecular diffusion. The space time in their reactor was set, and the impeller rotational speed was varied in a range where perfect macromixing occurred. Under such conditions, the extent of reaction indicated the state of micromixing.

Klein et al. based their model upon the following mechanisms. The fresh feed was viewed as being made up of big aggregates of size much smaller than the macroscale of the reactor but much larger than the dimensions of the smallest eddies. These large aggregates underwent erosion or shrinking to form smaller aggregates the size of the smallest eddies as suggested by Plasari et al. (1978). The small aggregates had the capability of exchanging material by molecular diffusion and were considered to behave as either: (1) premixed feed with little diffusion or (2) well-mixed aggregates where diffusion was very rapid. The reactor was viewed as having large aggregates and small aggregates of variable composition and states of segregation. Klein et al. modeled the erosion time scale by a turbulent mass transfer correlation for mass transfer from small particles in turbulent flow. The length and time scales for the small aggregate to undergo molecular diffusion were estimated using a correlation for concentration fluctuations by Corrsin (1964).

Geisler et al. (1988) reviewed macro and micromixing in agitated tanks noting the parallel between micromixing and laminar mixing. They derived an expression from the work by Corrsin (1964) for the characteristic micromixing time as:

$$\theta_D = \left(\frac{1}{2}\right)\left(\frac{\nu}{\epsilon}\right)^{1/2}(0.88 + \ln Sc) \tag{4.283}$$

The constant 0.88 arose from the inertial convective portion of the concentration spectrum and the ln Sc term arose from the diffusive portion of the concentration spectrum. Typically, for liquids, ln Sc is much greater than 0.88. For a specific fluid then:

$$\theta_D = C\left(\frac{\nu}{\epsilon}\right)^{1/2} \tag{4.284}$$

where ϵ has units of power per mass and C varies between 7 and 14. Typical mixing times can range between 10^{-3} to 1 s for turbulent micromixing. If C was on the order of between 50 and 100, $N\theta_m$ values of 50 to 150 can be obtained from the expression which is typical of laminar mixing with an effective impeller.

Pohorecki and Baldyga (1988) considered the decay of a large jet plume as:

$$\frac{dV_{SB}}{dt} = -\frac{V_{SB}}{t_{ms}} \tag{4.285}$$

where V_{SB} is the volume of the segregated blob and where $t_{ms} = (3/2)k_{0c}\epsilon^{-1/3}$. The time t_{ms} is the characteristic time of the inertial convective destruction of large eddies. The wave number k_{0c} is associated with the integral scale of turbulence as determined by the jet dimensions and system geometry and can be estimated as $2\pi/d_p$ in which d_p is the feed pipe diameter.

Andrigo et al. (1988) and Krusch et al. (1989) have discussed micromixing concepts in laminar flow situations. Other articles concerning micromixing include Mersmann and Kind (1988) and David and Villermaux (1989).

Various systems can yield completely different results for fast-reaction systems. Mixing times for the decay of the concentration fluctuations by molecular mixing were given by Pohorecki and Baldyga as:

$$Nt_{ms} = 2.73 \left(\frac{T}{D}\right)\left(\frac{H}{D}\right)^{1/2} N_p^{-1/2} N_{Re}^{-1}[\ln (Sc) - 1.27] \qquad (4.286)$$

Results of a study of mixing effects on $BaSO_4$ precipitation were given. Induction periods for $BaSO_4$ precipitation were estimated to be 0.00015 to 0.01 s. Nucleation and growth rates were expressed as a function of concentration differences to powers ranging from 3 to 15. Large concentration fluctuations were observed to increase rapidly nucleation and growth. The majority of the nucleation occurred in the completely segregated zone. The size of the zone decreased with increasing mixing intensity, which caused a decrease in nucleation and an increase in particle size. As can be noted, micromixing effects can appear in other processes besides chemical reactions.

Pohorecki and Baldyga (1979) have discussed the influence of mixing intensity on the rate of precipitation, and Tavare (1988) reviewed the effects of micromixing in a mixed-suspension, mixed-product removal crystallizer. Detrez et al. (1988) described an extension of the $BaSO_4$ precipitation method to study local micromixing. Tosun (1988) determined the effects of the addition mode, feed tube position, and the impeller rotational speed on the particle size distribution for $BaSO_4$ precipitation. Stavek et al. (1988) reported the effect of impeller rotational speed on particle size.

In a real crystallizer, the nonlinear kinetics of nucleation and growth make micromixing effects very important. Nucleation, in particular, can vary to very high powers of concentration difference, e.g., 5, 10, and 15. Any fluctuation in the concentration can lead to substantial nucleation. The number of crystals which can grow is substantially affected.

Small eddies have much smaller time scales than large eddies. As a result, the question of how micromixing scales in the impeller region and in the bulk of the tank is unclear at the present time. It is possible that micromixing scales differently for different regions in the tank which is evident in work by Tatterson et al. (1980). Trailing vortex systems are present in the impeller flow. It is also well known that such rotational flows stabilize turbulence, however. Vortex systems contain a high level of shear and elongation, which accomplish mixing as well. In the bulk of the tank, the micromixing scales with the turbulent fluctuations. Between these two spatial regions, there is transitional volume where the vortex systems are not nearly as coherent as in the immediate region of the impeller. High-speed discharge flows also are present which add further complications to accurate scaling of micromixing.

Models

There are several approaches used to model a stirred tank. Type 1 is a lumped-systems approach in which mass and material balances are written separately for each lump. Type 2 is a distributed lumped approach and involves the application of k-ϵ and other differential equations models which are solved for the flow field and the spatial distribution of the flow and flow properties. The lumps are much smaller, and numerical simulation is typically employed using finite difference or finite element techniques. Type 3 uses experimental data for the generation of a model of the flow field.

To be useful, a model must be related to physical reality and have parameters which are physically realistic and meaningful in a mechanistic sense. Although models are outside the scope of this chapter, brief reviews of some models are listed here since the different bases of the models are of interest. A more general modeling text on nonideal mixing is Levenspiel (1962).

No model provides a totally accurate description of the flow phenomena, and no general method can be used to determine the path molecules take through a tank. Further, there is no way of controlling generally the motion of molecules through a reactor.

Type 1. Danckwerts (1957) suggested a macromixed model where groups of molecules remain together in small lumps which are much smaller than the tank size. This condition has been called complete segregation. Each lump was considered as a small perfectly mixed batch reactor. As a result, each molecule remained with the molecules with which it entered the reactor and had the same age as those surrounding molecules. The lumps were mixed together only at the exit of the reactor. If the residence time distribution of the fluid in the tank was known, the overall yield for reactions can be calculated for this condition.

Zweitering (1959) put forth a maximum mixedness or minimum

segregation model. The concept of maximum mixedness was where an entering molecule became completely mixed with surrounding molecules upon entering the reactor and remained associated with these molecules until it exited the reactor. Zwietering (1959) noted that the conversion in a continuous flow system was dependent on the mixing behavior, particularly in the case of side reactions and chain reactions. Ng and Rippin (1965) studied the two extremes of complete segregation and maximum mixedness in terms of a transfer coefficient and the mean residence time to model the conversion of a second-order reaction.

Marr and Johnson (1961) considered a model of imperfect mixing for continuous-flow systems where most of the mixing occurred in the impeller region with some additional mixing occurring during the recirculation of the material through the bulk of the tank. Circulation was characterized by some average circulation time. Parameters for their second-order model were established from batch-mixing studies, and the predictions of the model were verified with experiments.

van de Vusse (1962) considered the deviations from ideal mixing in an agitated tank as consisting of: (1) material bypass where the material left the reactor immediately, (2) dead zones which remained unmixed, and (3) streamline flow where little mixing occurred across streamlines. Dead zones effectively reduced the reactor volume and led to intermittant reductions in mixing quality. Streamline flows possessed velocities and residence times in the reactor but essentially passed through the reactor unmixed.

van de Vusse based his model on a circulating flow generated by the impeller incorporating the nonideal mixing effects into the model. At high circulation-rate-to-feed-rate ratios, the model approached an ideal mixer. Conversion of a first-order reaction, residence time distribution functions, batch mixing, and estimations of the circulation and diffusivities were discussed. Gibilaro et al. (1967) further extended the model by van de Vusse.

Rice et al. (1964) presented a model for mixing by considering that the impeller reduced the entering feed stream into very thin packets by shearing and elongation processes in the impeller region. In recirculation, no shear or elongational mixing occurred, and mixing was accomplished by molecular diffusion.

Manning et al. (1965) in their model divided the tank into two regions: (1) a perfectly micromixed region of the impeller and (2) a macromixed region in the bulk of the tank in which no micromixing occurred. Two extremes were investigated: (1) The feed, entering far from the impeller, was completely macromixed before entering the impeller, and (2) the feed, injected into the impeller, was micromixed before entering the bulk of the tank. These assumptions affected the way

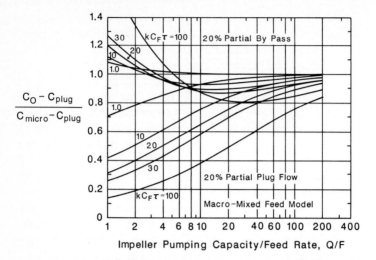

Impeller Pumping Capacity/Feed Rate, Q/F

Figure 4.40 The effect of $kC_F\tau$ on the macromixed feed model where k is the reaction rate constant for a second-order reaction, C_F is the feed concentration, and τ is the mean holdup time. C_o is the computed outlet concentration, C_{plug} is the outlet concentration of a plug flow reactor with the same holdup time, and C_{micro} is the outlet concentration of a perfectly micromixed reactor with the same holdup time. Micromixed feed is fed into the impeller; macromixed feed is fed into the bulk of the tank. *(From F. S. Manning, D. Wolf, and D. L. Keairns, AIChEJ., 11, 723, 1965. Reproduced by permission of the American Institute of Chemical Engineers.)*

the various concentration and residence time distributions were used in the calculation of mixing and yield for a second-order irreversible reaction. The results generally indicated that the ratio of circulation rate to feed rate should be on the order of 100 and in some cases 200 to 400. Figure 4.40 shows some of the results obtained.

Patterson (1981) also discussed the coalescence dispersion mixing model where packets of fluid enter and move in the tank in such a way to approximate the actual convective patterns in the tank. The packets were allowed to mix with one another, but each packet maintained its identity through the tank.

David et al. (1985) studied micromixing using a basic EDTA barium complex reacting with an acid to form barium sulfate involving two instantaneous reactions and one fast reaction in a consecutive-competitive arrangement. In their model, macromixing was accounted for by the use of average velocity profiles and two circulation loops; one having a low mixing intensity away from the impeller, zone 1; the other having intense mixing close to the impeller, called zone 2.

Upon injection of the acid, a reacting cloud of volume V was formed which was convected along the average velocity profiles. The volume of the cloud increased with time and was modeled as:

$$\frac{dV}{dt} = D_T V^{0.333} \qquad (4.287)$$

where D_T is a turbulent diffusion coefficient. Turbulent diffusivity in the impeller flow, zone 2, was assumed to be three times larger than the rest of the tank. The time during which the material remained in a zone was determined from hydrodynamics. For zone 2, the circulation time was a small fraction, 0.1, of the overall circulation time for the tank. Micromixing in the model was treated assuming an exchange between eddies of volume V_i and the reacting cloud V. The specie balance was stated as:

$$\frac{dC_i}{dt} = r_i + \frac{(\overline{C_j} - C_i)}{t_m} \qquad (4.288)$$

where i is a counter on the eddies in the reacting cloud, r_i is net appearance or loss of the particular specie due to reaction, $\overline{C_j}$ is the average concentration of the specie in the reacting cloud, and t_m is the micromixing time. The micromixing times in the two circulation zones varied with $\epsilon^{-1/3}$, and the ratio of micromixing times between zone 1 and 2 was assumed to be 2.2. The free parameters in the model were the micromixing time in zone 1 and the turbulent diffusivity. Data were fitted to the model, and the results indicated that the micromixing time in zone 1 was 0.175 s with the turbulent diffusivity being 0.1 m^2/s for zone 1. $D_T \theta_c$ was found to be 0.175 m^2 and assumed to be constant independent of impeller rotational speed. The model excluded the spatial variation of concentration in the tank but did take into account the macroscopic growth of the cloud and the different mixing states in the tank.

Gutoff (1960) and Sinclair (1961) presented models of mixing in agitated tanks as cited above. A model by Voncken et al. (1964) was also discussed above which was used to investigate dispersion in the recirculation loops.

Type 1 models have also been classified as environmental models. The different environments characterize the different mixing states which occur in the tank. Kim and Kang (1988) have performed micromixing, macromixing, and reaction studies and have developed environmental models to explain their experimental results.

Type 2. Patterson (1974) developed a hydrodynamic model in order to treat mixing behavior in a much more detailed manner. Using an instantaneous specie balance:

$$\frac{\partial C_i}{\partial t} + U_k \frac{\partial C_i}{\partial x_k} = D_L \nabla^2 C_i \qquad (4.289)$$

Patterson time-averaged the equation to:

$$\underbrace{\frac{\partial \overline{c_i^2}}{\partial t}}_{} + \underbrace{\overline{U_k} \frac{\partial \overline{c_i^2}}{\partial x_k}}_{\text{Convection of } \overline{c_i^2}} + \underbrace{2\overline{U_k c_i} \frac{\partial \overline{C_i}}{\partial x_k}}_{\text{Transport of } C_i} + \underbrace{\overline{U_k} \frac{\partial \overline{c_i^2}}{\partial x_k}}_{\text{Diffusion of } \overline{c_i^2}} = \underbrace{D_L \nabla^2 \overline{c_i^2}}_{\text{Decay of } \overline{c_i^2}} \qquad (4.290)$$

and assumed: (1) steady-state conditions and (2) transport and diffusion terms were small in comparison to the convection term. For simulation purposes, the tank was divided in segments of volume V_i and the convection term was modeled using flows into and out of the segments, Q_{ji} and Q_{ij}. The specie balance for an individual segment was then:

$$\sum_j Q_{ji} \overline{c_j^2} - \sum_j Q_{ij} \overline{c_i^2} = - V_i D_L \nabla^2 \overline{c_i^2} \qquad (4.291)$$

To model the decay of c_i^2, Patterson used an expression derived by Corrsin (1957; 1964) given as:

$$D_L \nabla^2 \overline{c_i^2} = \frac{- 2(\overline{c_i^2})}{4.1(L_i^2/\epsilon_i)^{1/3} + (\mu/\rho\epsilon)^{1/2} \ln (\text{Sc})} \qquad (4.292)$$

to provide the necessary closure for the calculation of $\overline{c_i^2}$. Estimates of the length scale L_i the energy dissipation rate ϵ_i and flow rates Q_{ij} were obtained from Cutter (1966) for simulation of the mixing performance of the disk style turbine and Porcelli and Marr (1962) for the simulation of the mixing performance of a propeller. In the simulations, feed streams were located at the top of the reactor and at the impeller. Distributions of $\overline{c_i^2}/c_0^2$ were obtained as a function of location in the tank, showing the differences between impellers. The results showed reasonable correspondence with data obtained by Reith (1965). Feed momentum or jetting effects at intermediate feed rates were observed in the calculations of the disk style turbine.

Mann and Mavros (1982) and Knysh and Mann (1984) developed a network of zones model to describe mixing. Asbjornsen (1965) presented a similar model for incomplete mixing simulated by fluid flow networks and applied it to reactions. The models by Mann et al. included the effects of convective and turbulent mixing and accepted any form of internal flow profile. Each cell in the model had a direct-through-flow convective component and an exchanged flow αQ between adjacent cells to model turbulent diffusion. Unsteady state, lumped specie balances were then performed for each cell and integrated forward in time to determine the concentration of material in each cell. The model was studied as the number of cells and the exchange coefficient α were varied. Near the impeller, α was 2 indicating high turbulent diffusion and away from the impeller in the bulk of the tank, α was 0.2. The output of the model mimicked tracer studies and provided the spatial distribution of the tracer throughout the tank as a function of time.

Since the model incorporated both convective and turbulent mixing components, effects of each of these were investigated. It was found that the assumed velocity profiles had little effect on mixing performance. The exchange coefficient α had a substantial effect on mixing performance. Impellers which enhanced the effective turbulent diffusivity throughout the tank were recommended. Mann et al. (1988) further extended their network of zones model. Experimental response curves of mixing were simulated using the model with excellent agreement between the model results and the actual mixing in the tank. The model can be used for situations where reaction times are shorter than mixing times and can provide a temporal and spatial distribution of the mixing process.

Jaworski (1985) and Jaworski et al. (1985) have also numerically simulated liquid mixing using a disk style turbine and have calculated three mixing indices as a function of time. Jaworski (1988) made comments regarding modeling of axial flow turbines.

Voit and Mersmann (1988) also divided the mixing tank up into individual volumes or cells to simulate mixing in the transition regime. Transfer between cells was determined from velocity profiles.

Type 3.   This particular category of models is limited due to the specificity which comes of basing models on a particular geometry and data. The work by DeSouza and Pike (1972) divided the flow field up into different regions: (1) the impeller discharge modeled as a jet, (2) an impinging wall jet, (3) two potential flow corners to model the circulation back to the impeller, (4) feed regions to the impeller modeled as a circular jet, and (5) "dead water" regions which were the toroidal vortices in the recirculation loops. The model was two dimensional and only qualitatively described the flow field. The model was not used to study mixing.

Residence time distribution

Residence time distributions have been developed to aid in the modeling of reactors by providing age and time information concerning material in the reactor tank or exiting the reactor. Mixing is a mechanical process and residence time information does not contain information on mixing. Age and time information is not information about mixing behavior and residence time distributions cannot be used to determine mixing. However, the longer the length of time in which a molecule remains in the mixing tank does imply that it will be mixed. However, if the time scale of mixing is much smaller than the residence time, residence time distribution contains very little information about mixing. Furthermore, this also holds true for any process of interest which occurs much faster than the residence time of the tank. Such processes include fast chemical reactions and multiphase contacting.

Uniqueness is a general problem of residence time information. Material can have the same internal and exit age distributions but can experience different states of mixing while in the tank. This can be demonstrated easily when considering two phases which can be well mixed in the tank but separate upon exiting the tank, giving the appearance of poor mixing. The internal and exit age information only implies a mixed condition but cannot specify one (Zwietering, 1959).

Rippin (1967) noted that the degree of segregation and the residence time distribution does not provide sufficient information to determine conversion in a reactor.

Tank mixing by jets and side-entering agitators

Nature of jets and general mixing comments. Folsom and Ferguson (1949) studied jet mixing of two liquids having the same density. Generally they observed that mixing outside of the turbulent jet in the bulk of the recirculating liquid was negligible. The amount of liquid entrained increased with distance, and the ratio of entrained fluid divided by power input Q_e/P_{in} was used as a measure of performance. The radius of the expanding jet shown in Fig. 4.41 was found to be:

$$R_j = 0.232X \tag{4.293}$$

where X is the centerline distance of the jet from the origin. The centerline velocity for $X/D_j > 8$ was:

$$U_{cj} = 5.13\left(\frac{D_j}{X}\right)U_0 \tag{4.294}$$

where U_0 is the axial velocity at the jet orifice. The amount of liquid entrained was:

$$Q_e = \left[0.234\left(\frac{X}{D_j}\right) - 1\right]Q_0 \tag{4.295}$$

where Q_0 is the volumetric flow rate at the orifice. The power input in the jet was:

$$P_{in} = Q_0 \rho g \left(\frac{U_0^2}{g} \right) \tag{4.296}$$

where U_0^2/g is the head of the jet.

The distance over which the jet decayed to zero was not determined. However, the jet was assumed to entrain fluid over 100 jet diameters downstream or 0.75 times the tank diameter. At this point, the induced recirculation made the entrainment process negligible. Although the jet orientation for mixing was not studied, the proper orientation for the jet was assumed to be perpendicular to the tank wall along the tank diameter.

Forstall and Gaylord (1955) noted that data on the turbulent Schmidt number ν_t/D_{LT} were not available for jets. The Schmidt number values were found to be from 0.75 to 0.85. With information about eddy viscosity, the eddy mass diffusivity can be calculated. A high eddy mass diffusivity is desirable to accomplish mixing.

Donald and Singer (1959) studied the entrainment of turbulent jets. They noted that, in previous work by Moses (1947), dye was entrained into a jet at right angles or normal to the jet, shown in Fig. 4.41, which is counter to the typically envisioned frictional drag mechanism. Since the flow entered at right angles, no momentum was added to the jet or lost to entrainment. The total momentum of the jet remained constant since very little viscous dissipation or viscous heating occurred along the length of the jet.

Donald and Singer noted the various discrepancies in the literature, particularly equations for the entrainment rate. Although jets were

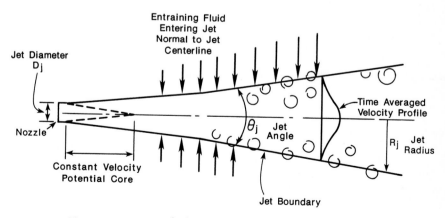

Figure 4.41   The gross structure of a jet.

known to expand at a constant cone angle, the reported cone angles for jets varied from 14 to 26° in the literature. Using different viscosity liquids in their study, the cone angle θ_j was found to vary with kinematic viscosity as:

$$\tan\left(\frac{\theta_j}{2}\right) = 0.238\nu^{0.133} \qquad (4.297)$$

where ν is in units of Stokes. Since the jet angle θ_j varied with kinematic viscosity ν, it varied slightly with jet Reynolds number as well. For water jets in water, the jet had little, if any, cone angle for a jet Reynolds number of 300 with little or no entrainment occurring along the jet. The jet became turbulent at 2000 and fully turbulent at 5000. Above 5000, entrainment occurred uniformly over the length of the jet, and the cone angle of the jet was constant at 14°. The inner geometry of the nozzle appeared to have no effect on flow behavior for the turbulent jets and the length of the potential core varied from 10 to 18 nozzle diameters.

The total volumetric flow in the jets was correlated as:

$$Q_T = Q_0\frac{(0.576\nu^{0.133})X}{D_j} \qquad (4.298)$$

where Q_0 is the discharge flow rate of the jet, Q_T is the total flow in the jet at the distance X from the nozzle, ν is the kinematic viscosity of the jet fluid, and D_j is the jet diameter. The equation reduced to:

$$Q_T = Q_0\left(\frac{0.246X}{D_j}\right) \qquad (4.299)$$

for water which is similar to the equation given above. There were no density differences between the material in the jets and the entrained fluid in the study by Donald and Singer. The kinetic energy of the jets varied inversely with distance from the nozzles.

Rushton (1954) and Rushton and Oldshue (1959) provided the following correlation:

$$Q_e = Q\left[0.23\left(\frac{X}{D}\right) - 1\right] \qquad (4.300)$$

to estimate the entrainment flow Q_e by a circular jet produced by a propeller where Q is the pumping capacity of the propeller; D, the impeller diameter (i.e., initial jet diameter); and X, the distance from the impeller. They found that the equation was sufficiently accurate for large-scale tanks with side-entering propellers for distances up to 100 jet or propeller diameters. They concluded that large-diameter streams were more advantageous for mixing than small jet streams at the same power. They also provided mixing time data on the effect of mixer entry angles suggesting that a 10° angle to the left of the tank diameter with no tilt was advantageous for mixing. The geometry is shown in Fig. 4.42.

Rushton (1980) reviewed axial velocities for submerged axially symmetric jets, generated by recirculation loops or propellers in very large tanks of 40 m³ to 4 × 10⁴ m³ (10⁴ to 10⁷ gal). Rushton noted that the total momentum remained constant for isothermal jets. The jet centerline velocity U_{cj} was modeled as:

$$\frac{U_{cj}}{U_0} = 1.41 \ \mathrm{Re}^{0.135}\left(\frac{D_j}{X}\right) \tag{4.301}$$

for distances greater than six jet diameters where U_0 is the initial jet exit velocity, D_j is the initial jet diameter and Re is based upon the initial jet conditions $D_j U_0/\nu$. Jets were also observed to expand at a constant angle which was dependent upon the liquid kinematic viscosity and velocity of the fluid. Rushton noted that mixing in such large tanks was usually best done during filling.

Other comments.   Folsum and Ferguson (1949) discussed power losses in pump-around-loops and showed in an example that a propeller design was more efficient than a recirculating pumping loop.

Rushton (1954) suggested that propellers created spiral flows which entrained surrounding fluid more effectively than simple jets. Rushton found that the best mixing orientation of the side-entering propeller caused a clockwise motion in the tank and that the side-entry angle was between 7 and 10° from the diameter with the jet being on the tank bottom, Fig. 4.42. Data which justified these angles were not

TOP VIEW

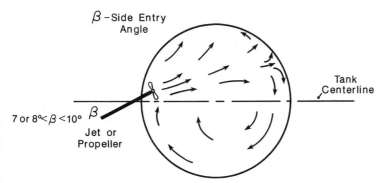

Figure 4.42   Angle for side-entering jet or propeller.

given, however. Wesselingh (1975) has provided such data. Rushton further suggested that the propeller or nozzle should be tilted upward to permit the fullest expansion of the flow and that material should be added at the propeller, noting the importance of density differences and cavitation which could occur near the impeller. He also noted that propellers were more efficient than recirculation loops for mixing and that large-diameter propellers at low rotational speeds were more effective than small-diameter propellers at high rotational speeds.

Mixing times of radial jets. Fox and Gex (1956) studied mixing using jets in unbaffled tanks and correlated mixing times based upon the momentum flux generated by the jet as:

$$\theta_m = C_1 \frac{H^{1/2}T}{\mathrm{Re}^{1/6}} \frac{1}{(U_0 D_j)^{4/6}} \frac{1}{g^{1/6}} \tag{4.302}$$

or:

$$\theta_m \propto \frac{V^{0.5}}{(U_0 D_j)^{5/6}} \tag{4.303}$$

From the turbulent mixing data, C_1 appeared to have two values: 75.4 and 163.5. For incipient jet mixing in the laminar region, $\mathrm{Re} < 2 \times 10^3$, Fox and Gex found:

$$\theta_m = C_2 \frac{H^{1/2}T}{\mathrm{Re}^{8/6}} \frac{1}{(U_0 D_j)^{4/6}} \frac{1}{g^{1/6}} \tag{4.304}$$

or:

$$\theta_m \propto \frac{V^{0.5}}{(U_0 D_j)^2} \tag{4.305}$$

where C_2 was 6.7×10^5. The correlations and data are shown in Fig. 4.43.

Any combination of jet diameter and velocity produced the same mixing times if they produced the same momentum flux. The location of the jet had little effect on mixing times provided that (1) the jet was not cycled back through the circulation system directly and (2) the jet did not induce swirl or "Taylor Walls" in the tank which dampened mixing.

van de Vusse (1955*a*; 1955*b*; 1959) assumed that the circulation time in a tank was equal to the tank volume divided by the volumetric flow rate of the jet midway between the jet origin and liquid surface. Mixing times were correlated using:

$$\theta_m = \frac{8.7 T^2 \sin \alpha}{D_j U_0} \tag{4.306}$$

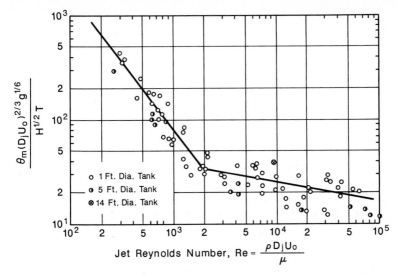

Figure 4.43 Mixing time data of jets as a function of the jet Reynolds number. *(From E. A. Fox and V. E. Gex, AIChEJ., 2, 539, 1956. Reproduced by permission of the American Institute of Chemical Engineers.)*

where α is the angle of jet inclination from the horizontal shown in Fig. 4.44. D_j and U_0 are the nozzle or jet diameter and the average velocity at the nozzle, respectively.

Okita and Oyama (1963), as cited by Maruyama et al. (1982), obtained mixing times for two different tanks sizes, $T = 0.4$ m and $T = 1.0$ m, using conductivity measurements at two different locations. Using mean circulation time and dimensional analysis, they correlated their results using:

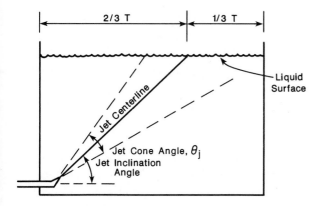

Figure 4.44 Suggested jet inclination angle for jet mixing in large tanks.

$$\theta_m = 5.5\left(\frac{D_j}{U_0}\right)\left(\frac{T}{D_j}\right)^{1.5}\left(\frac{H}{D_j}\right)^{0.5} \tag{4.307}$$

for jet Reynolds number $D_j U_0/\nu$ between 5000 and 10^5.

Lane and Rice (1981) performed mixing time studies using vertical jets in 0.31- and 0.91-m-diameter tanks with hemispherical bottoms with water and CMC solutions. Tracer concentrations were detected using a conductivity probe. Their mixing time correlations were given as:

$$\theta_m = C_1\frac{H^{1/2}T^{3/4}}{\text{Re}^{0.15}}\frac{1}{(U_0D_j)^{0.5}}\frac{1}{g^{1/4}} \tag{4.308}$$

for jet Reynolds numbers above 2×10^3. For laminar jet mixing, i.e., for laminar jet Reynolds number below 2×10^3, mixing time was correlated as:

$$\theta_m = C_2\frac{H^{1/2}T^{3/4}}{\text{Re}^{1.30}}\frac{1}{(U_0D_j)^{0.5}}\frac{1}{g^{1/4}} \tag{4.309}$$

where C_1 and C_2 were approximately 72.4 and 3.15×10^5 but were also weak functions of jet Reynolds number as well. In the correlations, U_0 and D_j are the jet exit velocity and diameter, respectively, and g is acceleration of gravity. The jet mixing performance was found to be a function of jet momentum and not jet energy. Differences in scale were not observed for the two tanks. Submerged vertical jets were found to provide more rapid mixing than inclined side-entry jets. Shorter mixing times were also obtained due to the hemispherical bottoms; flat-bottom tanks had longer mixing times. The results of Lane and Rice are in agreement with the study by Fox and Gex cited above.

Maruyama et al. (1982) studied jet mixing in tanks to determine the optimum geometrical configuration for rapid mixing. Maruyama et al. noted the commonality of work by Fossett and Prosser (1949; 1951), van de Vusse (1959), and Okita and Oyama (1963) which was simply a volume divided by a circulation flow rate. Maruyama et al., using a similar reasoning, assumed that the mean circulation time of a jet was approximately equal to:

$$\theta_c = \frac{V}{q_j} \tag{4.310}$$

where q_j is the volumetric flow rate of the jet at its termination. The termination was assumed to occur at the tank wall or liquid surface, a distance L away from the jet origin. Entrainment was also assumed to occur at a constant rate with distance. Scaling based upon these assumptions, Maruyama et al. obtained:

$$\frac{q_j}{Q_0} = \frac{kL}{D_j} \qquad (4.311)$$

where k is a constant and Q_0 is the initial flow rate through the nozzle having a diameter D_j. Using circulation time θ_c and tank residence time θ_R (i.e., V/Q_0), instead of flow rates, and including time in the expression, Maruyama et al. obtained:

$$\frac{t}{\theta_c} = \frac{k(t/\theta_R)}{D_j/L} \qquad (4.312)$$

and modeled the amplitude response to a tracer input, following Khang and Levenspiel (1976a; 1976b), as:

$$A = 2 \exp\left[- 2\pi^2\sigma_c^2\left(\frac{k(t/\theta_R)}{(D_j/L)}\right)\right] \qquad (4.313)$$

where A is the amplitude of the decaying fluctuations and σ_c is the variance of the circulation time.

Maruyama et al. measured mean circulation and mixing times in two tank sizes, 0.5 and 1.0 m in diameter, using three different nozzles diameters at different elevations in the jet Reynolds number range of 10^3 to 10^5. Their results showed that the amplitude response accurately represented their tracer response in most cases. The constant k was found to vary from 0.48 to 1.0. Above a jet Reynolds number of 3×10^4, mixing times were independent of Reynolds number. Below Re $< 10^4$, mixing time was a complicated function of geometry and $(\theta_m/\theta_R)/(D_j/L)$ was found to increase at smaller Reynolds numbers. For Re $> 3 \times 10^4$, there was a wide range of injection heights which provided optimum mixing times. As H/T (i.e., liquid height-tank diameter) decreased to 0.3, mixing time increased. Mixing time was at a minimum when the injection height was one-half the liquid height as shown in Fig. 4.45. At this height, the liquid jet developed the largest three-dimensional circulation possible with the largest variance σ_c in circulation time as shown in Fig. 4.46. Large variances σ_c provided more rapid decay of the fluctuations as the equation above indicates. At low injection heights, the jet evolved into a wall jet with two-dimensional circulation loops and low variance. Maruyama et al. suggested that the jet cone angle of $\pi/6$ be extended as far as possible into the tank without contacting a wall or the tank bottom. Maruyama et al. suggested that, since most nozzles are placed at the bottom of the tank, an elevation angle of the jet should be such that the jet did not contact the tank bottom as indicated in Fig. 4.44.

The model as presented by Maruyama does not account for the effects of jet-tank orientation on the variance σ_c of the circulation time

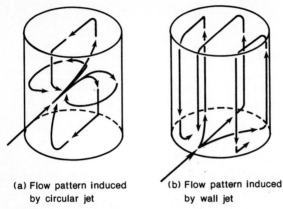

(a) Flow pattern induced
by circular jet

(b) Flow pattern induced
by wall jet

Figure 4.45 Qualitative sketch of flow pattern induced by jets at two locations. *(From T. Maruyama, Y. Ban, and T. Mizushina, J. Chem. Eng. Japan, 15, 342, 1982. By permission.)*

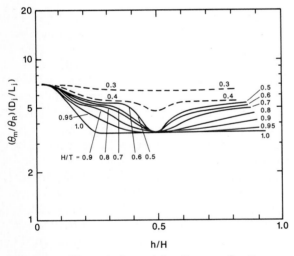

h/H

Figure 4.46 Dimensionless mixing time as a function nozzle height and liquid depth. *(From T. Maruyama, Y. Ban, and T. Mizushina, J. Chem. Eng., Japan, 15, 342, 1982. By permission.)*

which Maruyama et al. showed had a substantial effect. The variance of the circulation time is also a difficult quantity to obtain in practice.

Flow behavior and mixing times of tangential jets. Tangential jets behave differently from freely expanding radial jets just discussed and differences in flow behavior can be expected.

Fackler (1985) investigated jet mixing in a cylindrical tank with a tangential jet inlet and an axial outlet. The jet Reynolds number was above a critical value in the work, which was probably a value near 2100. Mixing times were determined as a function of inlet jet height to liquid height ratio h/H; the height to diameter ratio, H/T, of the tank and the nature of the velocity field. Fackler identified two major circulation zones at $h/H = 0.5$ in the tank; one in the top and the other in the bottom which mixed only at the jet height. This compartmentalization extended the mixing times.

Kittner (1983) found mixing times to be a function of H/T and h/H as well. For a jet inlet at h/H of 0.5, the mixing times were more than twice those found having the jet inlet at the top or bottom of the tank. The height-to-diameter ratio H/T of 2.0 provided mixing times three to five times faster than a height-to-diameter ratio of 1.0. In the study by Fackler, large toroidal circulation zones were formed in the top and bottom portions of the tank inside the circulation loops. Although not mentioned by Fackler, such zones increased the mixing times. For H/T of 2, such zones were thinner than for H/T of 1.0.

Hiraoka et al. (1987) studied circulation and mixing times for jet mixing in tanks and correlated mixing time with jet flow rate, number of jet nozzles, and liquid depth. The jet originated through one or two nozzles, oriented 30° from the radial direction, from an inner cylinder centrally located in the tank. The flow exited the tank through the same cylinder. The nozzles for the two jet system were placed opposite to each other on the inner cylinder. Tracer input was from the top of the tank and was detected with a conductivity probe. Location of the tracer input had no effect on mixing times, and the decay of the concentration response followed the model provided by Khang and Levenspiel (1976a; 1976b). The observed mixing times θ_m were correlated with the injection velocity at the jet nozzle using the same correlation method as Maruyama et al. (1982). The data showed that the correlation by Maruyama et al. predicted the mixing times for this configuration. The dimensionless mixing time was defined as:

$$\theta_m^* = \left(\frac{\theta_m}{\theta_R}\right)\left(\frac{D_j}{L_j}\right) \tag{4.314}$$

where θ_m and θ_R are the mixing and residence times, and D_j and L_j are the jet diameter at the nozzle and the jet length, respectively. For jet Reynolds numbers less than 10^4, the dimensionless mixing times varied between 5 and 13 and increased rapidly with decreasing Reynolds numbers. Above a jet Reynolds number of 10^4, the dimensionless mixing times varied between 2 and 9 and were considered not to vary with Reynolds number. The dimensionless mixing times for the two-jet system were about twice that for a one-jet system. The circulation times θ_c were found to be about $0.05\theta_R$ for the one-nozzle system and $0.077\theta_R$

for the two-nozzle system. For large jet Reynolds numbers above 10^4, the ratio of θ_m/θ_c was about 5 to 6 for both systems.

Coker and Jeffreys (1988) have studied mixing of a tracer by tangential jets in a large cone-bottom tank.

Although not a tank study, Ali and Whittington (1979) reported on the study of circulation flows in city reservoirs. Ali and Pateman (1982) found that radial jets produced better circulation and mixing than tangential jets in such reservoirs.

Flow behavior and mixing of stratified liquids. In investigations of mixing of different fluids in very large tanks (36 m in diameter and 6 m high) using jets, Fossett and Prosser (1949; 1951) assumed that the momentum in a jet was conserved and that the jet terminated at the wall of the tank. Their study used a single jet that crossed through the center of the tank forming a two-lobe flow pattern. A small high-speed jet was found to be just as effective at mixing as a large low-speed jet if the momentums of the two jets were equal. The orientation of the jet was such that tank swirl was avoided. Different geometric arrangements of jets were also possible but were not investigated.

With a second material in the tank, density differences were observed to cause stratification. A critical Froude number, used to account for this effect, was expressed as:

$$S = \frac{H_j \rho_2 \sin^2 \theta_j}{K(\rho_2 - \rho_1)} \qquad (4.315)$$

where S is the height actually obtained by the jet, H_j (i.e., U_0^2/g) is the discharge head of the jet, θ_j is the angle of the centerline of the jet plus 5°, and ρ_1 and ρ_2 are the densities of the lighter and heavier fluids, respectively. K varied with the density difference (K was 23 for a 1 percent density difference, 18 for a 2 percent difference, 15 for a 6 percent difference, and 10 for 63 percent difference). With the two densities, H_j, and K, then the height S obtained by the jet can be calculated from the equation. With the two densities, S, and K, then the head or discharge velocity of the pump necessary to obtain a height S can be calculated.

The recommended jet angle was where the jet centerline cut the top surface about ⅔ the way across the tank as shown in Fig. 4.44. The measured mixing times were correlated as:

$$\theta_m = 8T^2(Q_0 U_0)^{-0.5} \qquad (4.316)$$

for a jet Reynolds number greater than 4500 where Q_0 and U_0 are the volumetric flow rate and average velocity through the nozzle, respectively. In some cases the constant 8 in the equation reached as high as 14. Tests were also performed in large tanks, $T = 4.6$ m and $H = 4.6$

m to T = 43.9 m and H = 10.0 m, by taking grab samples at various heights.

The mixing times, obtained by Fossett and Prosser, were typically smaller than the time required for the addition of the second material to the tank. The formation of a double stratified layer during filling with two different fluids increased mixing time substantially. The jet velocity, necessary to penetrate a double-layer interface, was substantial.

Wilson (1954) studied mixing in very large crude oil storage tanks in which sedimentation and density variations caused problems for refinery units. He found that mixing while filling the tank was the best procedure for mixing. Mixing after the tank was filled was not satisfactory. Mixing for the elimination of large density differences existing initially in a large tank was impractical due to the long mixing times. Data for conservative blend times in such an application, however, can be crudely approximated as:

$$\theta_m = 6.11(\text{h}/°\text{API})\Delta\rho \qquad (4.317)$$

where $\Delta\rho$ is in °API and mixing time is in hours for 12,000 and 18,000-m^3 tanks.

Oldshue et al. (1956) studied the mixing of low-viscosity liquids using side-entering agitators mostly in the standard off-centered position, i.e., 7° < β < 10° in Fig. 4.42. Oldshue et al. found that the most difficult mixing case was the blending of a stratified condition in large tanks where the gradual erosion of the stratified condition by the flow over the interface was the only mechanism for mixing. Mixing times were correlated as:

$$\theta_m = K\left(\frac{\Delta\rho}{\rho}\right)^{0.9}(P)^{-1.0}\left(\frac{D}{T}\right)^{-2.3} \qquad (4.318)$$

for geometries having H/T = 1 where the constant K was not explicitly given. The D/T ratio ranged from 0.03 to 0.11. The data were given for conditions where the impeller was close to the tank bottom.

Wesselingh (1975) discussed mixing of liquids in cylindrical storage tanks using propellers. The studies were performed in tanks having volumes of 0.06, 4, and 33 m^3. Different water and salt solutions with density differences between 2 and 160 kg/m^3 were used in equal proportions and initially stratified at the start of the mixing. Total liquid height was 0.42 of the tank diameter. The impeller was a side-entering propeller, placed in the heavier phase close to the tank bottom, at an angle of 8° from the tank diameter, Fig. 4.42. The orientation caused flow in the clockwise direction looking down from the top of the tank. Conductivity cells were distributed throughout the tank at various locations to record the mixing. Because of the large sizes of the tanks, mixing times were on the order of hours or days in some cases.

Flow behavior and mixing times were dependent upon a critical density Froude number, Fr_c; the density Froude number Fr_ρ, being defined as $(\rho N^2 D / g \Delta \rho)$. Above Fr_c, motion occurred throughout the tank soon after the start of agitation. Density differences had little effect upon the mixing, and the responses from the conductivity cells reached the final composition at approximately the same time. Below Fr_c, the liquids remained stratified with an interface. During mixing, the interface had waves of lengths and heights which varied inversely with Froude number. Mixing between layers occurred by the breaking of these interfacial waves and trapping portions of one phase in the other. Under conditions of the experiments, the upper phase was trapped in the lower phase and the interface moved gradually upward. The final correlation for mixing time was given as:

$$N\theta = 1500 \left(\frac{H}{T}\right)^{2.4} \left(\frac{T}{D}\right)^{1.3} \left(\frac{H_s}{H}\right)\left(\frac{\mathrm{Fr}_c}{\mathrm{Fr}_\rho}\right)^{1.5} + 180\left(\frac{T}{D}\right)^{1.3} \left(\frac{H}{T}\right)^{0.8} \qquad (4.319)$$

where H is the total liquid height, H_s is the initial height of the thinnest liquid layer, and where the critical Froude number Fr_c was correlated as:

$$\log_{10} \mathrm{Fr}_c = 1.40 + 0.04\left(\frac{T}{D}\right) \qquad (4.320)$$

The off-center angle of the propeller shaft, in this case 8°, was found to be very important. At positive angles, the propeller jet traveled across the tank and, upon striking the tank wall, rose up into the second stratified layer and caused mixing. Negative angles induced counterclockwise circulation in the tank bottom with little axial circulation and little mixing. The impeller Reynolds numbers of the experiments were 3×10^4, 10^5, and 3×10^5. Mixing times were independent of the impeller Reynolds number above 10^4; below 10^4, mixing times increased sharply. The use of two impellers reduced mixing times by two-thirds. For some continuous-flow studies having no stratification, mixing times were two to three times lower than mixing times for batch systems, initially stratified. Agreement with similar data by Oldshue et al. (1956) and Wilson (1954) was noted although observed mixing times by Wilson were 30 to 50 percent lower than calculated by the above correlation.

Most of the jet mixing studies have been done in the turbulent regime. Whether jet mixing is practical in the laminar regime is questionable. Models of imperfect mixing for jet mixing systems have not been published. For mixing some materials, incomplete mixing by jets may be useful for the initial dispersion.

Sample Problem: Calculation of Jet Properties and Mixing Times in a Storage Tank
In this example, the jets can originate from inlet streams, pump-around-loops, or other prime movers. The jet source is relatively unimportant. Two jets will be considered. One discharges at 600 gpm through a 1.5-in nozzle to form jet 1; the other discharges at 2000 gpm through a 3-in nozzle to form jet 2. Table P1 and Table P2 list the calculated properties of the two jets and the power required by the jets. Table P3 contains the calculated mixing times from some of the correlations which apply. The calculations in these tables are based upon water as the process fluid and assume that the jets are freely expanding. Bounded jets will lead to longer mixing times. The tank is assumed to be 40 ft in diameter and 35 ft high.

jet 1   Data used in the calculations for the first jet are listed in the top of Table P1. The calculations indicate that the power required to operate at the stated jet conditions is 55.9 hp. Column 1 in Table P1 provides the distance the jet has traveled. Column 2 contains the expanding jet radius. Column 3 contains the jet centerline velocity. Column 4 contains the ratio of the jet centerline velocity to the initial average jet nozzle velocity. Column 5 contains the average velocity of the jet at the point. This average velocity is calculated from the total volumetric flow in the jet (Column 7) divided by the jet area calculated using the jet radius (Column 2). Entrained flow rate is listed in Column 6, and two estimates of total flow rate are listed in Columns 7 and 8.

Inspection of the calculations show several interesting properties of the jet. First, the jet expands rapidly with distance, e.g., from 4 in initially to 4.62 ft in diameter at a distance of 10 ft and to 9.2 ft at 20 ft. One hundred jet diameters is 12.5 ft. Second, the centerline velocity drops rapidly with distance, e.g., from 108.9 ft/s initially to 6.98 ft/s at 10 ft to 3.49 ft/s at 20 ft. The average velocity of the jet drops to 1.54 ft/s at 10 ft to 0.75 ft/s at 20 ft. The entrained flow increases rapidly with distance, e.g., from 0.0 initially to 24.7 ft^3/s at 10 ft and to 49.7 ft^3/s at 20 ft. The total volumetric flow behaves in the same way. The initial volumetric flow rate from the pump is 1.34 ft^3/s. Essentially, the jet is mostly entrained flow. Estimated average circulation rates and circulation times are listed at the bottom of the Table P1.

jet 2   Table P2 contains similar calculations. In comparison between the two jets, jet 2 has a lower initial discharge velocity but a larger circulation flow than jet 1. This appears as differences in the circulation times: 11 to 14 min for jet 1 in comparison to 7 to 8 min for jet 2.

The correlations as cited in Table P3 were used to calculate mixing times for the configurations listed in Tables P1 and P2. Taking an average of overall values obtained from the correlations, the mixing time for jet 1 is 12.6 min; the mixing time for jet 2 is 7.7 min, as listed in Table P3. These values are comparable to those listed in Table P1 and P2 for circulation times.

Summary.   Jets perform mixing by entrainment, so it is expected that mixing times should be the same order of magnitude as the circulation times. This is not the same as in agitated tanks where, typically, mix-

TABLE P1   Jet 1: Calculations for One Nozzle

Jet volumetric flow rate: Q_0, 600 gpm/nozzle	Given
Initial jet diameter: D_j, 1.5 in	Given
Jet volumetric flow rate: 1.34 ft^3/s	Given
Initial jet diameter: D_j, 0.125 ft	Given
Kinematic viscosity: ν, 0.00001 ft^2/s	Fluid property
Jet Reynolds number: $U_0 D_j/\nu$, 1.36 × 10^6	Calculated
Density: ρ, 62.4 lb$_m$/ft^3	Fluid property
Average jet nozzle velocity: U_0, 108.9 ft/s	Calculated
Tank diameter: D_T, 40 ft	Assumed
Liquid height: H_L, 35 ft	Assumed
Power input: 55.9 hp, 30,770 lb$_m$ft/s	Calculated
Jet angle, 14.7°	Calculated

Dist. X, ft (1)	Jet R. R_j, ft (2)	Jet C. Vel. U_j, ft/s (3)	U_{cj}/U_0 (4)	Avg. vel., ft/s (5)	Ent'ned flow Q_e, ft^3/s (6)	Total flow Q_T, ft^3/s (7)	Total flow Q_T, ft^3/s (8)
2.0	0.46	34.92	0.593	8.91	4.7	6.0	6.6
4.0	0.92	17.46	0.297	4.08	9.7	11.0	13.2
6.0	1.39	11.64	0.198	2.63	14.7	16.0	19.8
8.0	1.85	8.73	0.148	1.94	19.7	21.0	26.4
10.0	2.32	6.98	0.119	1.54	24.7	26.0	33.0
12.0	2.78	5.82	0.099	1.27	29.7	31.0	39.6
14.0	3.24	4.99	0.085	1.08	34.7	36.0	46.2
16.0	3.71	4.36	0.074	0.94	39.7	41.0	52.9
18.0	4.17	3.88	0.066	0.84	44.7	46.0	59.5
20.0	4.64	3.49	0.059	0.75	49.7	51.0	66.1
22.0	5.10	3.17	0.054	0.68	54.7	56.0	72.7
24.0	5.56	2.91	0.049	0.62	59.7	61.0	79.3
26.0	6.03	2.68	0.046	0.57	64.7	66.0	85.9
28.0	6.49	2.49	0.042	0.53	69.7	71.0	92.5
30.0	6.96	2.32	0.040	0.50	74.7	76.1	99.1
32.0	7.42	2.18	0.037	0.46	79.7	81.1	105.8
34.0	7.88	2.05	0.035	0.44	84.7	86.1	112.4
36.0	8.35	1.94	0.033	0.41	89.7	91.1	119.0
40.0	9.28	1.74	0.030	0.37	99.7	101.1	132.2

	From column:	
	7	8
Average circulation rate, ft^3/s	51.3	66.5
Tank volume, ft^3	44,000 ft^3	44,000 ft^3
Circulation time, min	14	11

ing time is three to four times the circulation time for efficient turbulent mixing equipment.

Mixing times in impeller agitated tanks are independent of kinematic viscosity as long as the impeller Reynolds number is in the turbulent regime. The same is practically true for jet mixing in tanks. The exponents on kinematic viscosity of 0.133 and ⅙ in the correlations given above indicate a very weak dependency. Mixing times in the laminar and transition mixing regimes will be affected by viscosity.

TABLE P2   Jet 2

Jet volumetric flow rate: Q_0, 2000 gpm/nozzle	Given
Initial jet diameter: D_j, 3.0 in	Given
Jet volumetric flow rate: 4.46 ft^3/s	Given
Initial jet diameter: D_j, 0.25 ft	Given
Kinematic viscosity: v, 0.00001 ft^2/s	Fluid property
Jet Reynolds number: $U_0 D_j/v$, 2.27 × 10^6	Calculated
Density: ρ, 62.4 lb$_m$/ft^3	Fluid property
Average jet nozzle velocity: U_0, 90.8 ft/s	Calculated
Tank diameter: D_T, 40 ft	Assumed
Liquid height: H_L, 35 ft	Assumed
Power input: 129 hp, 71,200 lb$_m$ft/s	Calculated
Jet angle: 14.7°	Calculated

Dist. X, ft (1)	Jet R. R_j, ft (2)	Jet C. Vel. U_j, ft/s (3)	U_{cj}/U_0 (4)	Avg. vel., ft/s (5)	Ent'ned flow Q_c, ft^3/s (6)	Total flow Q_T, ft^3/s (7)	Total flow Q_T, ft^3/s (8)
2.0	0.46	58.21	1.27	17.38	7.2	11.7	11.0
4.0	0.92	29.10	0.63	7.42	15.6	20.0	22.0
6.0	1.39	19.40	0.42	4.67	23.9	28.4	33.0
8.0	1.85	14.55	0.31	3.39	32.3	36.7	44.0
10.0	2.32	11.64	0.25	2.66	40.6	45.1	55.1
12.0	2.78	9.70	0.21	2.19	49.0	53.4	66.1
14.0	3.24	8.31	0.18	1.86	57.3	61.8	77.1
16.0	3.71	7.27	0.15	1.62	65.6	70.1	88.1
18.0	4.17	6.46	0.14	1.43	74.0	78.4	99.1
20.0	4.64	5.82	0.12	1.28	82.3	86.8	110.2
22.0	5.10	5.29	0.11	1.16	90.7	95.1	121.2
24.0	5.56	4.85	0.10	1.06	99.0	103.5	132.2
26.0	6.03	4.47	0.09	0.97	107.4	111.8	143.2
28.0	6.49	4.15	0.09	0.90	115.7	120.2	154.2
30.0	6.96	3.88	0.08	0.84	124.0	128.5	165.3
32.0	7.42	3.63	0.07	0.79	132.4	136.8	176.3
34.0	7.88	3.42	0.07	0.74	140.7	145.2	187.3
36.0	8.35	3.23	0.07	0.70	149.1	153.5	198.3
40.0	9.28	2.91	0.06	0.62	165.8	170.2	220.4

	From column	
	7	8
Average circulation rate, ft^3/s	87.3	110.8
Tank volume, ft^3	44,000	44,000
Circulation time, min	8	7

The correlations used to calculate mixing times were all for free expanding jets. However, jets can become bound to the tank bottom or wall and become a boundary layer flow rather than free jets. This attachment hinders entrainment which performs the mixing. To what degree mixing times are increased because of jet attachment requires further study.

Compartmentalization occurs often in systems which have multiple impellers. Essentially, the multiple impellers establish individual cir-

TABLE P3   Mixing Times Calculated from Correlations

Source	Mixing time (min)	
	Jet 1	
Fox and Gex	3.4	$(C_1 = 75.4)$
Fox and Gex	7.3	$(C_1 = 163.5)$
van de Vusse	17.0	$\sin \alpha = 1.0$
Okita	10.1	
Lane and Rice	2.1	
Fossett	17.7	$(C = 8)$
Fossett	30.9	$(C = 14)$
Average	12.6 min	
	Jet 2	
Fox and Gex	2.2	$(C_1 = 75.4)$
Fox and Gex	4.8	$(C_1 = 163.5)$
van de Vusse	10.2	$\sin \alpha = 1.0$
Okita	6.0	
Lane and Rice	1.5	
Fossett	10.6	$(C = 8)$
Fossett	18.6	$(C = 14)$
Average	7.7 min	

culation compartments of their own. There is very poor exchange or cross-mixing between the compartments and mixing times are substantially increased. For jet mixing, the same phenomena may occur with multiple jets.

Mixing by gas sparging

Gas sparging has been used to mix liquids in tanks although not many literature articles have been published in the chemical processing area. Kauffman (1930) and Lamont (1958) are some of the earliest articles. Liquid mixing in aerated agitated tanks has also been discussed in the chapter on gas dispersion.

Work done by injected gas is from: (1) gas expansion work and kinetic energy transfer during gas injection and (2) the gas bubbles rising through the liquid. The first is assumed to be fairly small, e.g., 0.06 of the total, and is typically neglected. The latter power input P is modeled as pressure/volume work input per time or:

$$P = \dot{n} R T_L \ln \left(\frac{P_1}{P_2}\right) \tag{4.321}$$

where $\dot{n}$ is the molar gas flow rate or $Q(P_m/RT_L)$. The gas flow rate Q is the gas flow at the liquid temperature T_L and the log mean pressure P_m. The log mean pressure is $(P_1 - P_2)/\ln (P_1/P_2)$. The pressure difference $(P_1 - P_2)$ is $\rho g H$. Making the proper substitutions, the power input P becomes $Q\rho g H$. Under steady-state conditions, power input is equal to the total energy dissipation.

The proper evaluation of Q is important in any power calculation. In conditions away from boiling, the vapor pressure of the liquid is low and Q is typically assumed to be the amount of gas sparged. In near-boiling conditions, such an assumption cannot be made, and the evaporation of liquid into the rising bubbles must be accounted for usually through a general heat flux-vapor evolution calculation. Boiling systems are well mixed without any gas sparging. Under such conditions, the distribution of power is higher nearer the liquid surface. Gas sparging is also possible over the full height of liquid.

For mixing by gas sparging, mixing time and energy dissipation for turbulent mixing are related as:

$$\theta_m \propto \epsilon^{-0.33} \qquad (4.322)$$

and for laminar mixing as:

$$\theta_m \propto \epsilon^{-0.5} \qquad (4.323)$$

This result is typical of mixing in mechanically agitated tanks where $N\theta_m$ is a function of system geometry including tank and impeller size. For a fixed geometry, $N\theta_m$ is a constant. The impeller rotational speed N varies with $\epsilon^{0.33}$ from the definition of the power number for turbulent flow and with $\epsilon^{0.5}$ for laminar flow. Such a result is not as obvious in gas sparged or jet mixing systems since there is no parameter in gas sparged mixing that is analogous to the impeller rotational speed. Other direct parallels also exist between mixing by gas sparging and mixing as described in other sections. This will become apparent in the articles reviewed below. Most mixing by gas sparging is performed in unbaffled tanks from the center of the tank. The main thrust of gas sparge mixing is in metallurgical processing.

Nakanishi et al. (1975) studied the relationship between energy dissipation and power input for four different processes involving gas sparging. Mixing times were related to energy dissipation as:

$$\theta_m \propto \epsilon^{-0.4} \qquad (4.324)$$

Asai et al. (1983), using a conductivity probe, measured mixing times in unbaffled tanks of diameters 0.4, 0.2, and 0.1 m at various liquid heights from 0.5 to 1.0 m. From theoretical considerations, Asai et al. showed that mixing times varied as:

$$\theta_m \propto \epsilon^{-0.5} \qquad (4.325)$$

for laminar flow and:

$$\theta_m \propto \epsilon^{-0.33} \tag{4.326}$$

for turbulent flow where ϵ is power input due to sparging. Experimentally, they obtained:

$$\theta_m \propto \epsilon^{-0.68} \tag{4.327}$$

for laminar flow and:

$$\theta_m \propto \epsilon^{-0.32} \tag{4.328}$$

for turbulent flow. There was no effect of vessel size on mixing times for laminar flow condition, but mixing time varied as:

$$\theta_m \propto \frac{T^{1/3}}{H} \tag{4.329}$$

for turbulent flow conditions where T is the tank diameter and H is the liquid height. The transition between the flow regimes occurred between $\epsilon = 0.8$ and 10 kg/m $\cdot$ s^3 using water as the process fluid. For the same power input, mixing times decreased with increasing gas injection depth and increasing eccentricity of the injection location from the tank center. Buoyancy work W calculated from adiabatic expansion of the sparged gas, was:

$$W = \xi P_1 V_1 \frac{1 - (P_2/P_1)^{(\xi-1)}}{\xi - 1} \tag{4.330}$$

where ξ is the ratio of specific heats, and subscripts 1 and 2 indicate conditions at the sparger location and tank surface, respectively. The power input for mixing was calculated from the buoyancy work and from the kinetic energy of the gas assuming 0.06 as the minor loss coefficient.

Sano and Mori (1983) studied the mixing behavior in gas-stirred molten melt baths. Gas holdup, liquid velocity in the plume, the energy dissipation in the bubble plume, and the associated downward liquid flow occurring in the melt were discussed. A correlation for mixing time was given as:

$$\theta_m = 100 \left[\frac{(T^2/H)^2}{\epsilon} \right]^{0.337} \tag{4.331}$$

where θ_m is in seconds and $[(D^2/H)^2/\epsilon]$ is in ton m^2/W. The correlation is dependent upon the diameter of the plume being $0.37H$. The correlation shows that mixing time is a function of the tank geometry and power input. The liquid velocity and gas holdup in the plume and liquid circulation flow were also correlated as simple functions of gas flow rate, liquid depth, and the cross-sectional areas of plume and tank.

Krishna et al. (1988), using a conductivity probe, measured mixing times in water agitated by injecting air through a circular nozzle at the bottom center of the tank. Mixing times were not dependent upon location or size of the measurement probes, the location of the tracer injection, or the amount of tracer injected. Mixing times decreased with increasing gas flow rate and bath height, but decreased with decreasing nozzle diameter. The two-phase plume was observed to swirl above a certain gassing rate which enhanced mixing. The energy, imparted to the flow which caused the mixing, was because of the buoyancy of the gas bubbles and gas kinetic energy. The gas kinetic energy was negligible at low liquid heights but increased substantially at larger liquid heights. The rate of buoyancy energy per unit volume of liquid was given as:

$$\epsilon_B = \frac{4QP_{atm}T_L}{298.2\ \pi T^2 H}\ln\left(1 + \frac{\rho_L g H}{P_{atm}}\right) \qquad (4.332)$$

where T_L is the liquid temperature and Q is the gas flow rate. The kinetic energy dissipation of the gas was given as:

$$\epsilon_K = \frac{32\rho_G\,Q^3}{\pi^3 d_n^4 T^2 H} \qquad (4.333)$$

where Q is the gas flow rate. The total kinetic energy input to the flow was ϵ which was modeled as:

$$\epsilon = \epsilon_B + x\epsilon_K \qquad (4.434)$$

where x varied from 0 to 1.0 and is the fraction of ϵ_K contributing to the mixing. At high gassing rates and small nozzle diameters, ϵ_K was significant.

The contribution of ϵ_K to mixing was found to be much higher at higher liquid height since the gas had more time to transfer its kinetic energy to the liquid. Hence, there were two important considerations in how ϵ_K affected mixing times: first, the relative magnitude of ϵ_K to ϵ_B and, second, the ability of ϵ_K to be transferred to the liquid. The fraction x was essentially zero below $H = 0.25$ m and increased linearly to 1.0 at $H = 0.43$ m.

Mixing times were expressed in terms of a power law of ϵ as:

$$\theta_m = C\epsilon^{-n} \qquad (4.335)$$

for the data where x was known. C and n were empirical parameters dependent upon liquid height and swirl in the two-phase plume. The transition to swirl occurred at about ϵ = 16 to 20 kg/m · s^3 and increased with increasing liquid height and nozzle diameter. A smaller-diameter column was also investigated which essentially became a bubble column at high gas rates with multiple circulation cells.

Beishon and Robertson (1988) discussed mixing in large batches of liquid metals having specific gravities of around 7 at temperatures from 1300 to 1700°C. Thermal and chemical homogenization were desired. Alloying materials used in treatment had specific gravities ranging from 2 to 11 which gave rise to buoyancy or negative buoyancy forces which opposed mixing. Methods of mixing included pouring from one vessel into another, subsurface gas injection or electromagnetic stirring. Mechanical agitation was difficult to perform due to the high temperature and density. Gas sparging with a lance was enhanced by the liquid density and the high temperature which increased the gas volume. The efficiency of mixing was not dependent upon the method of injection although the height of gas injection was very important. Off-centered sparging reduced homogenization time. A heavy slag cover increased mixing times. Power per unit mass was not a useful criterion upon which to base mixing times since simulations using liquid mercury showed that circulation times were fairly constant over a broad range of power input. Beishon and Robertson also found the use of water as a simulant for liquid metals inadequate in mixing time studies.

Salcudean and Lai (1988) presented the computation of three-dimensional flow with heat and mass transfer in gas agitated tanks. The k-ϵ model was used to calculate the flow patterns, isotherms, and concentration profiles.

Chung and Lange (1989) studied the nature of bubble plumes in mixing molten metal baths. Mixing times were correlated as:

$$\theta_m = 0.3745 \frac{H^2}{D_L} \tag{4.336}$$

where H is the height of the gas plume and D_L is an axial diffusion coefficient. This parameter was correlated in a manner similar to bubble columns using a Peclet number:

$$\text{Pe} = \frac{\bar{u}D}{D_L} \tag{4.337}$$

where $\bar{u}$ is the average circulation velocity and D is the bath diameter. The axial dispersion coefficient D_L for steel melts ranged between 140 cm^2/s and 400 cm^2/s. Peclet numbers were found to be on the order of 6 to 15. Mixing time was approximately equal to the circulation time.

The energy dissipated in the bubble plume was from 50 to 70 percent of the total energy input. 20 to 25 percent of the energy was dissipated in the surface flow generated by the bubble plume. Energy dissipation per unit volume increased with increasing bath depth above the sparger and increasing gas flow rate. The mean size of the energy containing eddies increased with bath size and entrainment, reaching 17 percent of the bath diameter at high gas flow rates. Chung and Lange (1989) also provided an extensive reference list for further reading on gas plume liquid mixing.

Oeters (1989) discussed the importance of mixing in metallurgical systems. Koh and Batterham (1989) discussed liquid splashing from gas sparging. The work was primarily for molten melt baths. The average splash height (millimeters) was correlated with the bath height (millimeters) and gas flow rate (liters per minute) as:

$$h = 53.73H^{-0.44}Q^{0.79} \qquad (4.338)$$

Interfacial area and splash volume were also correlated. Other articles which discuss flow and turbulence characteristics for gas stirred tanks include Guthrie (1982) and Chung and Lange (1989a; 1989b).

In an article quite apart from metallurgical applications, Smith et al. (1982) discussed the mixing of large ponds using bubble columns. The bubble columns were unconstrained by the walls of the ponds. The relationship between gas flow rate and liquid flow rate at the surface was:

$$\frac{Q_L^3}{QgH^5} = 4 \times 10^{-4} \qquad (4.339)$$

where Q_L is the volumetric flow rate of the water at the surface, Q is the gassing rate, and H is the liquid depth plus the virtual origin of the bubble plume. Mixing times were estimated as five times the bulk circulation time. This was assumed from work in mixing in agitated tanks. In jet mixing, mixing time is approximately the bulk circulation time. In the work, the bulk circulation time was given as:

$$\theta_c = 50 \left(\frac{H^4}{gQ}\right)^{1/3} \text{s} \qquad (4.340)$$

This relationship holds only for unconfined bubble columns. Mixing and circulation times were estimated by dividing the volume of the pond by the circulation flow.

Chisti and Moo-Young (1988) discussed gas void fraction in pneumatically agitated reactors for rheologically complex fluids. Gas holdup was expressed as a function of superficial gas velocity as:

$$\phi = \alpha V_s^\beta \qquad (4.341)$$

where β was a function of flow index n.

Other Articles of Interest

Norin (1958) discussed the mixing of different temperature fluids which occurs during a change in operations for a nuclear reactor. Temperature changes in an emergency shutdown can be as much as 3000°F in 1 s where the mixing determines the thermal stresses experienced by the containing structure. Beerbower et al. (1957) discussed the use of a radioactive iodine tracer in the evaluation of mixing efficiency of process equipment for waxes, greases, and a lime soap base in nonstandard mixing geometries. Response curves showed that mixing time was not obtained in three or four circulation times but instead 10 to 20 times the circulation times. White (1974) discussed the use of radioactive tracers to study mixing and residence times in lakes and large-scale digesters (5000 m^3).

Kipke (1979) and King (1985) discussed the effects of hydraulic forces on different impellers and fluid structure interactions in mixing processes. King and Muskett (1985) presented results of fluid loading and effects of eccentricity on the pitched blade turbine.

Strek and Karcz (1988) have studied heat transfer in different liquids in unaerated and aerated tanks.

Nomenclature

A	generalized component, volumetric flow parameter, universal constant, amplitude
B	generalized component, baffle width
Bo	Bodenstein number, $L_c V_c / D_{LD}$
a	rms fluctuations in generalized component, source parameter, radius of cylindrical tangential jet, constant
b	rms fluctuations in generalized component, mean volume fraction of B, constant
C	off-bottom clearance, concentration or mean concentration, constant
C_{avg}	time-averaged concentration
C_F	feed concentration
C_i, C_∞	initial and final concentrations
C_i	dimensionless concentration in the slab model
C_{micro}	outlet concentration of a perfectly micromixed reactor with the same holdup τ
C_o	computed outlet concentration

C_P	pressure coefficient, heat capacity
C_{plug}	outlet concentration of a plug flow reactor with the same holdup time τ
C_R	circulation constant
C_t	bottom clearance of the draft tube
C_1, C_2	constants
c	concentration fluctuation
c'	rms concentration fluctuation, constant
c_0	initial rms concentration fluctuation
c_i	concentration in the slab
c_{i0}	initial concentration in the slab
c_μ	constant in the k-ϵ model
$c_{1\epsilon}$	constant in the k-ϵ model
$c_{2\epsilon}$	constant in the k-ϵ model
D	impeller diameter
Da	Damkohler number, $kC_{A0}\delta^2/D_L$, $kC_{A0}\tau$
D_d	degree of deviation
D_j	initial jet diameter
D_L	molecular diffusivity, mass diffusivity
D_{LD}	dispersion coefficient
D_{LT}	turbulent diffusivity
D_T	turbulent diffusion coefficient
D_t	draft tube diameter
d_i	inlet tube diameter
d_n	nozzle diameter
E	energy spectrum, eccentricity, exit age distribution
$E_1(k)$	energy spectrum of u_1
E_m	specific energy consumption during mixing
$E_0(T)$	energy spectrum value below k_0
$E_c(k)$	concentration spectrum
F	feed rate
F_s	other forces
f	friction factor
Fr	Froude number, ND^2/g
Fr_c	critical Froude number
Fr_ρ	density Froude number, $\rho N^2 D/g\,\Delta\rho$
f_w	wall friction factor
f_{max}	maximum wall friction factor
$f(x), f(r)$	longitudinal velocity correlation

$f_c(r)$	longitudinal concentration correlation
g	acceleration of gravity
g_c	gravitational constant
$g(x), g(r)$	lateral velocity correlation
H	liquid height
H_v	vortex depth
H_j	discharge head of jet
H_s	height of thinnest liquid layer
h	inlet jet height, splash height
I	internal age distribution
I_D	limiting diffusion current
j	counter
K	constant, power law fluid consistency coefficient, length scale factor, discharge coefficient, mixing revolutions
K_a	amplitude decay constant
K_1	constant
k	turbulent kinetic energy, wave number, reaction rate constant, mixing rate constant, constant, thermal conductivity
k_1	friction factor
k_1, k_2	kinetic rate constants
$k_0(T)$	spectrum cutoff wave number
k_0	wave number of the largest eddies
k_K	Kolmogorov wave number, $(\epsilon/\nu)^{1/4}$
k_B	Batchelor wave number, $(\epsilon/\nu D_L^2)^{1/4}$
L	characteristic length $(T/2) \ln (T/D)$, length scale in the eddy viscosity
L_c	integral length scale for concentration fluctuations, circulation length
L_f	integral length scale for turbulence
L_i	integral length scale of i
L_j	jet length
L_s	integral length scale for turbulence
L_t	draft tube length
l	distance along the vortex axis, length scale
l_f	microscale of turbulence
l_c	length scale of concentration fluctuations
M	mixing modulus, $k^2 c_{Bo} \delta_0^2 / D_L$
m	number of circulation times
N	impeller rotational speed, revolutions per second

N_{A0}	initial moles of A
N_{B0}	initial moles of B
N_Q	pumping number, Q/ND^3
N_{QC}	total circulation number, Q_T/ND^3
N_P	power number, $Pg_c/\rho N^3 D^5$
N_{PL}	laminar power number
N_{PT}	turbulent power number
N_{QF}	feed number, Q_F/ND^3
n	power law fluid index, order of reaction, exponent
$\dot{n}$	molar gas flow rate
n_p	blade number
P	static pressure, power, pressure
Pe	Peclet number, $\bar{u}D/D_L$
P_{in}	power input
$P_{ij}(r)$	pressure velocity correlation
P_m	log mean pressure
P_w	wall pressure
ΔP	pressure drop between the front and back of blade
Q	flow rate between cells, volumetric flow rate through the impeller or impeller pumping capacity, gas flow rate
Q_c	volumetric circulation rate
Q_L	liquid circulation rate
Q_e	flow rate entrained by the impeller discharge or by jet
Q_F	feed volumetric flow rate
Q_0	initial volumetric flow rate at jet orifice
Q_T	total discharge flow of the impeller
Q_{ij}	flow rates between volume segments
$Q_{ij}(r)$	correlation of velocity fluctuations
q_j	volumetric flow rate of jet at its point of termination
q_t	flow rate of tracer
R	impeller radius, ideal gas constant
Re	impeller Reynolds number $\rho ND^2/\mu$, jet Reynolds number $\rho U_0 D_j/\mu$
Re_λ	Reynolds number based upon Taylor microscale and rms velocity fluctuation, $\rho \lambda u/\mu$
Re_T	tank Reynolds number, $\rho T V_c/\mu$
Re_τ	residence time Reynolds number, $4\rho Q_F/\pi\mu T$ or $\rho T^2/\tau\mu$
Re_i	inlet tube Reynolds number, $4\rho Q_F/\pi d_i\mu$
R_j	jet radius
$R_{ij}(r)$	normalized Eulerian correlation function

r	separation distance in velocity correlations, radius from the center of tank, vortex radius
r_i	reaction rate of i, net rate of appearance or loss due to reaction
r_0	impeller radius
S	source term, height obtained by jet
Sc	Schmidt number, $\mu/\rho D_L$
$S_{ijk}(r)$	triple velocity correlation
s	standard deviation of circulation time distribution
T	tank diameter, dimemsionless time tD_L/δ_0^2
T_L	liquid temperature
$T_{ij}(r)$	triple velocity correlation
t	time
t_K	Kolmogorov time scale
t_m	micromixing time
U	Eulerian velocity; subscripts: i, j, k are for Cartesian coordinates $x, y, z; r, \theta, z$ are for cylindrical coordinates, average velocity in the bulk
$\bar{u}$	average circulation velocity
U_{cj}	centerline velocity of a jet
U_0	initial axial velocity of jet at orifice
U_T	impeller tip velocity
U_z	mean axial velocity for the entire tank
u	velocity fluctuations
u'	rms velocity fluctuation
u_t	slab thinning velocity
u_w^*	friction wall velocity
$\overline{u_i u_j}$	Reynolds stress terms
V	volume, tank volume, volume of reacting cloud
V_c	circumferential velocity, circulation velocity
V_L	longitudinal velocity along the axis of the vortex
V_t	impeller tip velocity
V_r	radial velocity
V_0	tangential velocity, characteristic velocity $(\pi/2)ND\beta$
V_i	volume
V_A, V_B	volumes of A and B
$\hat{V}$	velocity scale in eddy viscosity
v	local tangential velocity, Kolmogorov velocity scale
v^+	dimensionless velocity
v^{2+}	dimensionless velocity

W	blade width, work
$W(k)$	Fourier transform of $T_{ij}(r) - P_{ij}(r)$
$W_c(k)$	counterpart to $W(k)$ in the concentration spectrum equation
X	centerline distance along a jet, dimensionless distance in slab, x/δ_0
X_0, X_c	initial and final values of the integral mean value of the local degree of inhomogeneity
x	distance
y	distance
Z	dimensionless axial distance, z/W
z, z_0	distance between impeller centerline and jet centerline

Greek symbols

α	exchange coefficient, ratio of torques, blade angle, angle of jet inclination, volume ratio, diffusional current parameter, proportional sign
δ	length scale in mixing modulus, slab thickness
δ_0	initial slab thickness
δ_{ij}	Kronecker delta
θ_b	batch mixing time
θ_c	circulation time
θ_{cf}	mixing time for continuous-flow systems
θ_D	diffusion time scale
θ_{DR}	time for diffusion and reaction
θ_j	jet cone angle
θ_m	mixing time
θ_{macro}	macromixing time
θ_R	residence time
θ_t	turbulent mixing time
θ_{DL}	mixing time due to diffusion
θ_{ua}	unagitated mixing time due to feed stream
θ^*	dimensionless mixing time
β	geometric parameter, $2 \ln (T/D)/[(T/D) - (D/T)]$; decay constant, jet orientation angle, dimensionless constant $\gamma\delta_0^2/D_L$
Γ	scalar quantity, volume fraction of the mean concentration, vortex circulation
Γ_∞	vortex circulation at infinity
γ	rms fluctuation in scalar component, shear rate, geometric parameter: $(D/T)^5 \ln (T/D)/\beta^5$, shear rate experienced by slab
γ_w	wall shear rate

ϵ	local or mean energy dissipation rate, energy dissipation distribution function, power input
$\bar{\epsilon}$	mean energy dissipation rate
ϵ_B	energy dissipation by buoyancy forces
ϵ_K	energy dissipation by kinetic energy of sparged gas
ϵ_i	energy dissipation in the impeller region
ϵ_b	energy dissipation in the bulk of the tank
ϵ_γ	concentration dissipation rate
ϕ	ratio of $\epsilon/\bar{\epsilon}$
ξ	distance parameter $(T/4)[(T/2r) - (2r/T)]$, ratio of specific heats
ξ^+	dimensionless distance parameter
ξ^2	dimensionless distance parameter
σ	jet half-width parameter, universal constant
σ_0	dimensionless variance of mixing time
σ_c	variance of the circulation time
σ_k	constant in k-ϵ model
σ_ϵ	constant in k-ϵ model
σ_t	Prandtl or Schmidt number for modeling heat or mass transport using the k-ϵ model
η	Kolmogorov microscale
η_b	Batchelor concentration microscale
η_c	Corrsin concentration microscale
Λ	integral macroscale, macroscale
λ_f	microscale determined from $f(r)$, Taylor microscale
λ_g	microscale determined from $g(r)$
λ_c	concentration microscale determined from $f_c(r)$
ρ	density
ν	kinematic viscosity
ν_L	laminar kinematic viscosity
ν_t	eddy kinematic viscosity, turbulent kinematic viscosity
ν_{eff}	effective viscosity
ν_z	zero shear kinematic viscosity
ϑ_t	eddy diffusivity
μ	viscosity
μ_t	turbulent viscosity, eddy viscosity
τ	stress tensor or stress tensor component, time constant for the decay of rms concentration fluctuations, mean residence or holdup time

τ_w wall shear stress

ω angular velocity in vortex, vorticity

χ dissipation rate of rms concentration fluctuations

Subscripts

i, j, k, l component counters for direction

r, θ, z components in the cylindrical coordinate system

n direction normal to the flow

Mathematical operators

∇ del operator $\partial/\partial x_i \mathbf{i} + \partial/\partial x_j \mathbf{j} + \partial/\partial x_k \mathbf{k}$

∇_r $-\nabla_A$ or ∇_B

D/Dt substantial derivative

∇^2 Laplacian operator $\partial^2/\partial x_i^2$

Note:. Different symbols are used to identify different definitions which the various authors have developed for the same concept.

References

Adrian, R. J., *Proc. International Sym. on Fluid Control and Measurement, Tokyo*, September, 2–5, 1985, Pergamon, Oxford, England, 1985.
Adrian, R. J., *Applied Optics*, **23**, 1690, 1984.
Adrian, R. J., and C. -S. Yao, *Proc. 8th Biennial Sym. on Turbulence*, University of Missouri, Rolla, edited by X. B. Reed, G. K. Patterson, and J. L. Zakin, 1984, p. 170.
Adrian, R. J., and C. -S. Yao, *Applied Optics*, **24**, 44, 1985.
Ahmad, S. W., B. Latto, and M. H. I. Baird, *Chem. Eng. Res. Des.*, **63**, 157, 1985.
Aiba, S., *AIChEJ.*, **4**, 485, 1958.
Ali, K. H. M., and R. B. Whittington, *Proc. 3rd Eur. Conf. on Mixing, University of York, York, England*, BHRA Fluid Eng., Cranfield, England, A3, 1979.
Ali, K. H. M., and D. R. Pateman, *Proc. 4th Eur. Conf. on Mixing, Noordwijkerhout, Netherlands*, BHRA Fluid Eng., Cranfield, England, 81, 1982.
Andrigo, P., R. Bagatin, P. Cavalieri d'Oro, C. Perego, and L. Raimondi, *Chem. Eng. Sci.*, **43**, 8, 1923, 1988.
Angst, W., J. R. Bourne, and F. Kozicki, *Proc. 3rd Eur. Conf. on Mixing, University of York, York, England*, BHRA Fluid Eng., Cranfield, England, A4, 1979.
Armstrong, S. G., and S. Ruszkowski, *Proc. 6th Eur. Conf. on Mixing, Pavia, Italy*, BHRA Fluid Eng., Cranfield, England, 29, 1988.
Asai, S., T. Okamoto, J. -C. He, and I. Muchi, *Trans. Iron Steel Institute Japan*, **23**, 1, 43, 1983.
Asbjornsen, O. A., *Proc. AIChE—I. Chem. E. Joint Sym.*, No. 10, *Mixing—Theory Related to Practice*, AIChE, New York, p. 40, 1965.
Baldyga, J., and J. R. Bourne, *Chem. Eng. Commun.*, **28**, 231, 1984a.
Baldyga, J., and J. R. Bourne, *Chem. Eng. Commun.*, **28**, 243, 1984b.
Baldyga, J., and J. R. Bourne, *Chem. Eng. Commun.*, **28**, 259, 1984c.
Baldyga, J., and J. R. Bourne, *Chem. Eng. Res. Des.*, **66**, 33, 1988.
Batchelor, G. K., *J. Fluid Mech.*, **5**, 113, 1959.
Bauer, A., and A. Moser, *Proc. 5th Eur. Conf. on Mixing, Wurzburg, Germany*, BHRA Fluid Eng., Cranfield, England, 171, 1985.

Beek, J. J., and R. S. Miller, *Chem. Eng. Progr. Sym. Ser. No. 25*, **55**, 23, 1959.
Beerbower, A., E. O. Forster, J. J. Kolfenbach, and H. G. Vesterdal, *Ind. Eng. Chem.*, **49**, 1075, 1957.
Beishon, D. S., and T. Robertson, *Fluid Mixing III, I. Chem. Eng. Sym. Ser. No. 108*, 1, 1988.
Bertrand, J., J. P. Couderc, and H. Angelino, *Chem. Eng. Sci.*, **35**, 2157, 1980.
Biggs, R. D., *AIChEJ.*, **9**, 5, 636, 1963.
Bourne, J. R., *Chem. Eng. Commun.*, **16**, 79, 1982.
Bourne, J. R., and J. Garcia-Rosas, *Proc. 5th Eur. Conf. on Mixing, Wurzburg, Germany*, BHRA Fluid Eng., Cranfield, England, 81, 1985.
Bourne, J. R., and P. Dell'ava, *Chem. Eng. Res. Des.*, **65**, 181, 1987.
Bourne, J. R., H. Gablinger, and K. Ravindranath, *Chem. Eng. Sci.*, **43**, 8, 1941, 1988.
Bourne, J. R., and K. Ravindranath, *Proc. 6th Eur. Conf. on Mixing, Pavia, Italy*, BHRA Fluid Eng., Cranfield, England, 145, 1988.
Bowers, R. H., *Proc. AIChE—I. Chem. E. Joint Sym., No. 10, Mixing—Theory Related to Practice*, AIChE, New York, p. 8, 1965.
Braginskii, L. N., and V. I. Begachev, *Theo. Fund. Chem. Eng.*, **6**, 231, 1972.
Brasseur, J. G., and I-D. Chang, *AIAA 13th Fluid & Plasma Dynamics Conference*, July 14–16, 1980, Snowmass, Color., AIAA-80-1330.
Brennan, D. J., and I. H. Lehrer, *Trans. Instn. Chem. Engrs.*, **54**, 139, 1976.
Brodkey, R. S., *The Phenomena of Fluid Motions*, The Ohio State Bookstore, The Ohio State University, Columbus, Ohio, 1967.
Brodkey, R. S., ed., *Turbulence in Mixing Operations*, Academic, New York, 1975.
Bryant, J., and S. Sadeghzadeh, *Proc. 3rd Eur. Conf. on Mixing, University of York, York, England*, BHRA Fluid Eng., Cranfield, England, 325, 1979.
Bryant, J., and S. Sadeghzadeh, *Proc. 4th Eur. Conf. on Mixing, Noordwijkerhout, Netherlands*, BHRA Fluid Eng., Cranfield, England, 49, 1982.
Burghardt, A., and L. Lipowska, *Chem. Eng. Sci.*, **27**, 1783, 1972.
Canon, R. M., K. W. Wall, A. W. Smith, and G. K. Patterson, *Chem. Eng. Sci.*, **32**, 1349, 1977.
Castellana, F. S., M. I. Friedman, and J. L. Spencer, *AIChEJ*, **30**, 207, 1984.
Chandrasekhar, S., *Proc. Roy. Soc. (London)*, **A200**, 20, 1949.
Chang, T. P. K., Y. H. E. Sheu, G. B. Tatterson, and D. S. Dickey, *Chem. Eng. Commun.*, **10**, 215, 1981.
Chang, T. P. K., T. A. Watson, and G. B. Tatterson, *Chem. Eng. Sci.*, **40**, 269, 1985a.
Chang, T. P. K., T. A. Watson, and G. B. Tatterson, *Chem. Eng. Sci.*, **40**, 277, 1985b.
Chang, T. P., E. I. du Pont de Nemours & Co., Savannah River Laboratory, Report No.: DP 1721, April, 1986.
Chen, C. -P., and P. E. Wood, *Numerical Heat Transfer*, **9**, 115, 1986.
Chisti, Y., and M. Moo-Young, *Chem. Eng. J.*, **39**, B31, 1988.
Christiansen, D. E., *I&EC Funds.*, **8**, 263, 1969.
Chung, S. -H., and K. W. Lange, *Steel & Metals Mag.*, **27**, 78, 1989a.
Chung, S. -H., and K. W. Lange, *Steel & Metals Mag.*, **27**, 190, 1989b.
Chung, S. -H., and K. W. Lange, *Steel Research*, **60**, 2, 49, 1989.
Clark, R. A., J. H. Ferziger, and W. C. Reynolds, *J. Fluid Mech.*, **91**, Part 1, 1, 1979.
Coker, A. K., and G. V. Jeffreys, *Fluid Mixing III, I. Chem. Eng. Sym. Ser. No. 108*, 105, 1988.
Cooper, R. G., and D. Wolf, *Canad. J. Chem. Eng.*, **45**, 197, 1967.
Corrsin, S., *AIChEJ.*, **3**, 329, 1957.
Corrsin, S., *Phy. Fluids*, **1**, 42, 1958.
Corrsin, S., *AIChEJ.*, **10**, 870, 1964.
Costes, J., and J. P. Couderc, *Proc. 4th Eur. Conf. on Mixing, Noordwijkerhout, Netherlands*, BHRA Fluid Eng., Cranfield, England, 25, 1982.
Cutter, L. A., *AIChEJ.*, **12**, 35, 1966.
Danckwerts, P. V., *Chem. Eng. Sci.*, **2**, 1, 1953.
Danckwerts, P. V., *Chem. Reaction Engr. (Chem. React. Eng. 12th Meeting Europ. Fed. Chem. Engng., Amsterdam)*, **1**, 93, 1957.
David, R., J. P. Barthole, and J. Villermaux, *Proc. 5th Eur. Conf. on Mixing, Wurzburg, Germany*, BHRA Fluid Eng., Cranfield, England, 433, 1985.

David, R., and J. Villermaux, *Chem. Eng. Commun.*, **78**, 233, 1989.
Deardorff, J. W., *J. Fluid Mech.*, **41**, Part 2, 453, 1970.
DeSouza, A., and R. Pike, *Canad. J. Chem. Eng.*, **50**, 15, 1972.
Detrez, C., R. David, and J. Villermaux, *Proc. 6th Eur. Conf. on Mixing, Pavia, Italy*, BHRA Fluid Eng., Cranfield, England, 153, 1988.
Donald, M. B., and H. Singer, *Trans. Instn. Chem. Engrs*, **37**, 255, 1959.
Dosanjh, D. S., E. P. Gasparek, and S. Eskinazi, *The Aero. Quart.*, **13**, 167, 1962.
Drbohlav, J., I. Fort, and J. Kratky, *Coll. Czech. Chem. Commun.*, **43**, 696, 1978a.
Drbohlav, J., I. Fort, K. Maca, and J. Placek, *Coll. Czech. Chem. Commun.*, **43**, 3148, 1978b.
Fackler, R., *Proc. 5th Eur. Conf. on Mixing, Wurzburg, Germany*, BHRA Fluid Eng., Cranfield, England, 541, 1985.
Faraday, A. G., *Fluid Mixing III, I. Chem. Eng. Sym. Ser. No. 108*, **108**, 63, 1988.
Ferziger, J. H., "Higher-level Simulations of Turbulent Flows," *Computational Methods for Turbulent, Transonic and Viscous Flows*, edited by J. A. Essens, Springer Verlag, Berlin, 1983, pp. 93–182.
Folsom, R. G., and C. K. Ferguson, *Trans. Amer. Soc. Mech. Eng.*, **71**, 73, 1949.
Forstall, W., and E. W. Gaylord, *J. Appl. Mechanics*, **22**, 161, 1955.
Fort, I., "Flow and Turbulence in Vessels with Axial Impellers," *Mixing: Theory and Practice*, Vol. III, edited by V. W. Uhl and J. B. Gray, Academic, Orlando, Fla., 1986, chap. 3.
Fort, I., J. Drbohlav, J. Kratky, M. Grospicova, and Z. Krouzilova, *Coll. Czech. Chem. Commun.*, **37**, 222, 1972.
Fort, I., J. Placek, J. Kratky, J. Durdil, and J. Drbohlav, *Coll. Czech. Chem. Commun.*, **39**, 1810, 1974.
Fort, I., H-O. Mockel, J. Drbohlav, and M. Hrach, *Coll. Czech. Chem. Commun.*, **44**, 700, 1979.
Fort, I., and J. Medek, *Proc. 6th Eur. Conf. on Mixing, Pavia, Italy*, BHRA Fluid Eng., Cranfield, England, 51, 1988.
Fossett, H., and L. E. Prosser, *Proc. I. Mech. Eng.*, **160**, 224, 1949.
Fossett, H., *Trans. Instn. Chem. Engrs.*, **29**, 322, 1951.
Fox, E. A., and V. E. Gex, *AIChEJ.*, **2**, 4, 539, 1956.
Funahashi, H., H. Harada, and H. Taguchi, *J. Eng. Chem. Japan*, **20**, 3, 1987.
Gaskey, S., P. Vacus, R. David, J. C. Andre, and J. Villermaux, *Proc. 6th Eur. Conf. on Mixing, Pavia, Italy*, BHRA Fluid Eng., Cranfield, England, 129, 1988.
Geisler, R., A. Mersmann, and H. Voit, *Chem. Ing. Tech.*, **60**, 12, 947, 1988.
Gibson, C. H., and W. H. Schwarz, *J. Fluid Mech.*, **16**, 357, 365, 1963.
Gilbilaro, L. G., H. W. Kropholler, and D. J. Spikins, *Chem. Eng. Sci.*, **22**, 517, 1967.
Godleski, E. S., and J. C. Smith, *AIChEJ.*, **8**, 5, 617, 1962.
Goldfish, L. H., J. A. Koutsky, and R. J. Adler, *Chem. Eng. Sci.*, **20**, 1011, 1965.
Graichen, K., *Chem. Eng. J.*, **20**, 1, 1980.
Gunkel, A. A., R. P. Patel, and M. E. Weber, *I&EC Funds*, **10**, 627, 1971.
Gunkel, A. A., and M. E. Weber, *AIChEJ.*, **21**, 931, 1975.
Guthrie, R. I. L., *Iron & Steel Maker*, **9**, 41, January, 1982.
Gutoff, E. B., *AIChEJ.*, **6**, 347, 1960.
Harvey, P. S., and M. Greaves, *Trans. I. Chem. Eng.*, **60**, 195, 1982a.
Harvey, P. S., and M. Greaves, *Trans. I. Chem. Eng.*, **60**, 201, 1982b.
Heisenburg, W., *Z. fur Phys.*, **124**, 628, 1948.
Heisenburg, W., *Proc. Roy. Soc. (London)*, **A195**, 402, 1948.
Hemrajani, R. R., *Proc. 5th Eur. Conf. on Mixing, Wurzburg, Germany*, BHRA Fluid Eng., Cranfield, England, 63, 1985.
Hinze, O. J., *Turbulence*, 2d ed., McGraw-Hill, New York, 1975.
Hiraoka, S., R. Ito, I. Yamada, K. Sawada, M. Ishiguro, and S. Kawamura, *J. Chem. Eng. Japan*, **8**, 156, 1975.
Hiraoka, S., and R. Ito, *J. Chem. Eng. Japan*, **8**, 323, 1975.
Hiraoka, S., and R. Ito, *J. Chem. Eng. Japan*, **10**, 75, 1977.
Hiraoka, S., Y. Tada, I. Yamada, T. Takahashi, and S. T. Koh, *Proc. SCEJ-CIChE Joint Seminar on Mixing Technology*, edited by W-L Lu, Taipei, Taiwan, July 14–15, 1987, p. 38.

Hjelmfelt, A. T., and L. F. Mockros, *Appl. Sci. Res.*, **16**, 2, 149, 1966.
Hoffman, E. R., and P. N. Joubert, *J. Fluid Mech.*, **16**, 395, 1963.
Holmes, D. B., R. M. Voncken, and J. A. Dekker, *Chem. Eng. Sci.*, **19**, 201, 1964.
Hubbard, D. W., and H. Patel, *AIChEJ.*, **17**, 1387, 1971.
Hughes, R. R., *Ind. Eng. Chem.*, **49**, 947, 1957.
Ito, R., Y. Hirata, K. Sakata, and I. Nakahara, *International Chem. Eng.*, **19**, 605, 1979.
Ito, S., S. Urushiyama, and K. Ogawa, *J. Chem. Eng. Japan*, **7**, 462, 1974.
Ito, S., K. Ogawa, and N. Yoshida, *J. Chem. Eng. Japan*, **8**, 206, 1975.
Jaworski, Z., *Proc. 5th Eur. Conf. on Mixing, Wurzburg, Germany*, BHRA Fluid Eng., Cranfield, England, 469, 1985.
Jaworski, Z., *Proc. 5th Eur. Conf. on Mixing, Wurzburg, Germany*, BHRA Fluid Eng., Cranfield, England, 377, 1985.
Jaworski, Z., *Proc. 6th Eur. Conf. on Mixing, Pavia, Italy*, BHRA Fluid Eng., Cranfield, England, 57, 1988.
Ju, S. Y., T. M. Mulvahill, and R. W. Pike, Paper presented at Mixing XI, Engineering Foundation Conference on Mixing, New England College, Henniker, N. H., August 2–7, 1987.
Kamiwano, M., K. Yamamoto, and S. Nagata, *Kagaku Kogaku*, **31**, 365, 1967.
Kappel, M., *International Chem. Eng.*, **19**, 196, 1979.
Kauffman, H. L., *Chem. Met. Eng.*, **37**, 178, 1930.
Kawamura, T., and K. Kuwahara, AIAA 23rd Aerospace Sciences Meeting, January 14–17, 1985, Reno, Nevada, AIAA-85-0376.
Keey, R. B., *British Chem. Eng.*, **12**, 7, 1081–1085, July 1967.
Keller, D. B. A., *Proc. 5th Eur. Conf. on Mixing, Wurzburg, Germany*, BHRA Fluid Eng., Cranfield, England, 475, 1985.
Kent, J. C., and A. R. Eaton, *Applied Optics*, **21**, 904, 1982.
Khang, S. J., and O. Levenspiel, *Chem. Eng.*, 141, October 11, 1976a.
Khang, S. J., and O. Levenspiel, *Chem. Eng. Sci.*, **31**, 569, 1976b.
Kim, T. O., and W. K. Kang, *Intern. Chem. Eng.*, **28**, 690, 1988.
Kim, W. J., and F. S. Manning, *AIChEJ.*, **10**, 5, 747, 1964.
King, R., *Process Engng.*, February 1985, p. 50.
King, R., and M. J. Muskett, *Proc. 5th Eur. Conf. on Mixing, Wurzburg, Germany*, BHRA Fluid Eng., Cranfield, England, 285, 1985.
Kipke, K. D., *Proc. 3rd. Eur. Conf. on Mixing, University of York, York, England*, BHRA Fluid Eng., Cranfield, England, E1, 1979.
Kipke, K., *Ger. Chem. Eng.*, **6**, 119, 1983.
Kittner, R., R. Fackler, and V. Denk, *Brauwissenschaft*, 316, August 1983.
Klaboch, L., M. Ptacnik, J. Lamka, and I. Fort, *Proc. 6th Eur. Conf. on Mixing, Pavia, Italy*, BHRA Fluid Eng., Cranfield, England, 35, 1988.
Klein, J. -P., R. David, and J. Villermaux, *I&EC: Funds.*, **19**, 373, 1980.
Knysh, P., and R. Mann, *Fluid Mixing II, I. Chem. Eng. Sym. Ser. No. 89*, 127, 1984.
Koh, P. T. L., and R. J. Batterham, *Chem. Eng. Res. Des.*, **67**, 211, 1989.
Kolar, V., P. Filip, and A. G. Curev, *Chem. Eng. Commun.*, **27**, 313, 1984.
Kolar, V., P. Filip, and A. G. Curev, *Proc. 5th Eur. Conf. on Mixing, Wurzburg, Germany*, BHRA Fluid Eng., Cranfield, England, 483, 1985.
Kolmogorov, A. N., *Comptes rendus (Doklady) de l'Academie des sciences de l'U.R.S.S.*, **30**, 301, 1941; **31**, 538, 1941; **32**, 16, 1941.
Komasawa, I., R. Kuboi, and T. Otake, *Chem. Eng. Sci.*, **29**, 641, 1974.
Komori, S., and Y. Murakami, *Proc. SCEJ-CIChE Joint Seminar on Mixing Technology*, edited by W-L Lu, Taipei, Taiwan, July 14–15, 1987, p. 94.
Komori, S., and Y. Murakami, *Proc. 6th Eur. Conf. on Mixing, Pavia, Italy*, BHRA Fluid Eng., Cranfield, England, 63, 1988.
Kovasznay, L. S. G., *J. Aeronaut. Sci.*, **15**, 745, 1948.
Kramers, H., G. M. Baars, and W. H. Knoll, *Chem. Eng. Sci.*, **3**, 35, 1953.
Kratky, J., I. Fort, and J. Drbohlav, *Coll. Czech. Chem. Commun.*, **39**, 3238, 1974.
Krishna Murthy, G. G., S. P. Mehrotra, and A. Ghosh, *Metallurgical. Trans. B*, **19B**, 6, 839, 1988.
Kuboi, R., I. Komasawa, and T. Otake, *Chem. Eng. Sci.*, **29**, 641, 1974a.
Kuboi, R., I. Komasawa, and T. Otake, *Chem. Eng. Sci.*, **29**, 651, 1974b.

Kudrna, V., V. Koza, and J. Hajek, *Collect. Czech. Chem. Commun.*, **54**, 633, 1989.
Kusch, H. A., J. M. Ottino, and D. M. Shannon, *Ind. Eng. Chem. Res.*, **28**, 302, 1989.
Kwak, D., W. C. Reynolds, and J. H. Ferziger, Report No. TF-5, Thermoscience Division, Dept. Mech. Eng., Stanford University, Stanford, Calif., May 1975.
Laity, D. S., and R. E. Treybal, *AIChEJ.*, **3**, 176, 1957.
Lamb, D. E., F. S. Manning, and R. H. Wilhelm, *AIChEJ.*, **6**, 682, 1960.
Lamont, A. G. W., *Canad. J. Chem. Eng.*, 153, August 1958.
Landau, J., and J. Prochazka, *Coll. Czech. Chem. Commun.*, **26**, 1976, 1961.
Landau, J., and J. Prochazka, *Coll. Czech. Chem. Commun.*, **28**, 1866, 1963.
Lane, A. G. C., and P. Rice, *Fluid Mixing, I. Chem. Eng. Sym. Ser. No. 64*, K1, 1981.
Laufhutte, H. D., and A. Mersmann, *Proc. 5th Eur. Conf. on Mixing, Wurzburg, Germany*, BHRA Fluid Eng., Cranfield, England, 331, 1985.
Laufhutte, H. D., and A. Mersmann, *Chem. Eng. Technol.*, **10**, 56, 1987.
Lee, J., and R. S. Brodkey, *Rev. Sci. Instr.*, **34**, 1086, 1963.
Lee, J., and R. S. Brodkey, *AIChEJ.*, **10**, 187, 1964.
Lehtola, T., J. Soderman, and J. Laine, *Proc. 6th Eur. Conf. on Mixing, Pavia, Italy*, BHRA Fluid Eng., Cranfield, England, 85, 1988.
Leonard, A., *Advances in Geophysics*, **18A**, 237, 1974.
Leslie, D. C., and G. L. Quarini, *J. Fluid Mech.*, **91**, Part 1, 65, 1979.
Levenspiel, O., *Chemical Reaction Engineering*, Wiley, New York, 1962, p. 266.
Levins, D. M., and J. R. Glastonbury, *Trans. Instn. Chem. Engr.*, **50**, 32, 1972.
Lin, C. C., *First Symp. Appl. Math.*, Am. Math. Soc., 1947.
Lin, C. C., *Proc. Seventh Int. Congr. Appl. Math.*, London, Vol. 2, Part 1, 1948.
Lipowska, L., *Chem. Eng. Sci.*, **29**, 1901, 1974.
MacDonald, R. W., and E. L. Piret, *Chem. Eng. Progr.*, **47**, 363, 1951.
Magni, F., J. Costes, J. Bertrand, and J. P. Couderc, *Proc. 6th Eur. Conf. on Mixing, Pavia, Italy*, BHRA Fluid Eng., Cranfield, England, 7, 1988.
Mahouast, M., R. David, and G. Cognet, *Proc. 6th Eur. Conf. on Mixing, Pavia, Italy*, BHRA Fluid Eng., Cranfield, England, 23, 1988.
Mann, R., P. P. Mavros, and J. C. Middleton, *Trans. I. Chem. Eng.*, **59**, 271, 1981.
Mann, R., and P. Mavros, *Proc. 4th Eur. Conf. on Mixing, Noordwijkerhout, Netherlands*, BHRA Fluid Eng., Cranfield, England, 35, 1982.
Mann, R., P. Knysh, M. Didari, and E. A. Rasekoala, *Fluid Mixing III, I. Chem. Eng. Sym. Ser. No 108*, 49, 1988.
Manning, F. S., and R. H. Wilhelm, *AIChEJ.*, **9**, 1, 12, 1963.
Manning, F. S., D. Wolf, and D. L. Keairns, *AIChEJ.*, **11**, 4, 723, 1965.
Marr, G. R., and E. F. Johnson, *AIChEJ.*, **9**, 3, 383, 1963.
Marr, G. R., and E. F. Johnson, *Chem. Eng. Progr. Symp. Ser., No. 36*, 109, 1961.
Martin, J. J., *Trans. Amer. Inst. Chem. Eng.*, **42**, 777, 1946.
Maruyama, T., Y. Ban, and T. Mizushina, *J. Chem. Eng. Japan*, **15**, 342, 1982.
McManamey, W. J., *Trans. I. Chem. E.*, **58**, 271, 1980.
Medek, J., and I. Fort, *Proc. 3rd. Eur. Conf. on Mixing, Vol. 2, University of York, York, England*, BHRA Fluid Eng., Cranfield, England, G1, 1979.
Medek, J., and I. Fort, *Proc. 5th Eur. Conf. on Mixing, Wurzburg, Germany*, BHRA Fluid Eng., Cranfield, England, 263, 1985.
Mersmann, A., W.-D. Einenkel, and M. Kappel, *Intern. Chem. Eng.*, **16**, 590, 1976.
Mersmann, A., and H. D. Laufhutte, *Proc. 5th Eur. Conf. on Mixing, Wurzburg, Germany*, BHRA Fluid Eng., Cranfield, England, 273, 1985.
Mersmann, A., and M. Kind, *Chem. Eng. Technol.*, **11**, 264, 1988.
Merzouk, K., C. Ramel, and J. Bertrand, *Proc. 6th Eur. Conf. on Mixing, Pavia, Italy*, BHRA Fluid Eng., Cranfield, England, 91, 1988.
Metzner, A. B., and J. S. Taylor, *AIChEJ.*, **6**, 1, 109, 1960.
Middleton, J. C., *Proc. 3rd. Eur. Conf. on Mixing, University of York, York, England*, BHRA Fluid Eng., Cranfield, England, A2, 1979.
Middleton, J. C., F. Pierce, and P. M. Lynch, *Chem. Eng. Res. Des.*, **64**, 18, 1986.
Midler, M., and R. K. Finn, *Biotech. Bioeng.*, **8**, 71, 1966.
Miller, S. A., and C. A. Mann, *Trans. Amer. Inst. Chem. Eng.*, **40**, 709, 1944.
Mitchell, J. E., and T. J. Hanratty, *J. Fluid Mechanics*, **26**, 199, 1966.
Mitschka, P., and J. J. Ulbrecht, *Coll. Czech. Chem. Commun.*, **30**, 2511, 1965.

Mizushina, T., R. Ito, S. Hiraoka, A. Ibusuki, and I. Sakaguchi, *J. Chem. Eng. Japan*, **2**, 98, 1969.

Mochizuki, M., and I. Takashima, *Kagaku Kogaku*, **38**, 3, 249, 1974.

Mochizuki, M., and I. Takashima, *Kagaku Kogaku Ron.*, **8**, 487, 1982.

Mochizuki, M., and I. Takashima, *Kagaku Kogaku Ron.*, **10**, 3, 399, 1984a.

Mochizuki, M., and I. Takashima, *Kagaku Kogaku Ron.*, **10**, 4, 539, 1984b.

Moin, P., and J. Kim, *J. Fluid Mech.*, **118**, 341, 1982.

Moo-Young, M., K. Tichar, and F. A. L. Dullien, *AIChEJ.*, **18**, 1, 178, 1972.

Mujumdar, A. S., B. Huang, D. Wolf, M. E. Weber, and W. J. M. Douglas, *Canad. J. Chem. Eng.*, **48**, 475, 1970.

Mukataka, S., H. Kataoka, and J. Takahashi, *J. Ferment. Technol.*, **58**, 155, 1980.

Mukataka, S., H. Kataoka, and J. Takahashi, *J. Ferment. Technol.*, **59**, 303, 1981.

Nagase, Y., T. Iwamoto, S. Fujita, and T. Yoshida, *Kagaku Kogaku*, **38**, 519, 1974.

Nagase, Y., and M. Kikuchi, *J. Chem. Eng. Japan*, **10**, 2, 164, 1977a.

Nagase, Y., and S. Sawada, *J. Chem. Eng. Japan*, **10**, 3, 229, 1977b.

Nagase, Y., and M. Kikuchi, *Chem. Eng. J.*, **26**, 13, 1983.

Nagase, Y., *Proc. SCEJ-CIChE Joint Seminar on Mixing Technology*, edited by W-L Lu, Taipei, Taiwan, July 14–15, 1987, p. 49.

Nagase, Y., H. Hirofuji, and M. Imiya, *Proc. 6th Eur. Conf. on Mixing, Pavia, Italy*, BHRA Fluid Eng., Cranfield, England, 97, 1988.

Nagata, S., N. Yoshioka, and T. Yokoyama. *Mem. Fac. Eng., KyotoUniv.*, **17**, 175, 1955.

Nagata, S., K. Yamamoto, K. Hashimoto, and Y. Naruse, *Mem. Fac. Engng., Kyoto Univ.*, **22**, 68, 1960.

Nagata, S., M. Nishikawa, A. Inoue, and Y. Okamoto, *J. Chem. Eng. Japan*, **8**, 243, 1975.

Nagata, S., *Mixing: Principles and Applications*, Kodansha Ltd., Tokyo, Wiley, New York, 1975.

Nakanishi, K., T. Fujii, and J. Szekely, *Ironmaking and Steelmaking* (Quarterly), **3**, 193, 1975.

Newitt, D. M., G. C. Shipp, and C. R. Black, *Trans. Instn. Chem. Engrs.*, **29**, 279, 1951.

Newman, B. G., *The Aer. Quart.*, **10**, 149, 1959.

Ng, D. Y. D., and D. W. T. Rippin, *Proc. 3rd Eur. Sym. Chem. React. Eng.*, Pergamon, Amsterdam, 1965, p. 161.

Nienow, A. W., and R. Kuboi, *Fluid Mixing II, I. Chem. Eng. Sym. Ser. No. 89*, 97, 1984.

Nishikawa, M., Y. Okamoto, K. Hashimoto, and S. Nagata, *J. Chem. Eng. Japan*, **9**, 489, 1976.

Nishikawa, M., K. Ashiwake, N. Hashimoto, and S. Nagata, *Intern. Chem. Eng.*, **19**, 1, 153, 1979.

Norin, M., *J. Franklin Inst.*, **226**, 229, September, 1958.

Norwood, K. W., and A. B. Metzner, *AIChEJ.*, **6**, 3, 432, 1960.

Novak, V., P. Ditl, and F. Rieger, *Proc. 4th Eur. Conf. on Mixing, Noordwijkerhout, The Netherlands*, BHRA Fluid Engineering, Cranfield, Bedford, England, C1, 57, 1982.

Nye, J. O., and R. S. Brodkey, *Rev. Sci. Instr.*, **38**, 26, 1967.

Obeid, A., I. Fort, and J. Bertrand, *Coll. Czech. Chem. Commun.*, **48**, 568, 1983.

Obukhoff, A. M., *Comptes rendus (Doklady) de l'Academie des sciences de l'U.R.S.S.*, **30**, 19, 1941.

Oeters, F., *Steel Research*, **60**, 185, 1989.

Ogawa, K., S. Ito, and Y. Matsumura, *J. Chem. Eng. Japan*, **13**, 324, 1980.

Ogawa, K., and S. Ito, *J. Chem. Eng. Japan*, **8**, 148, 1975.

Okamoto, Y., M. Nishikawa, and K. Hashimoto, *International Chem. Eng.*, **21**, 88, 1981.

Okita, N., and Y. Oyama, *Kagaku Kogaku*, **27**, 252, 1963.

Oldshue, J. Y., H. E. Hirschland, and A. T. Gretton, *Chem. Engng. Progr.*, **52**, 481, 1956.

Onsager, L., *Phys. Rev.*, **68**, 286, 1945.

Patterson, G. K., *Proc. of Chemeca '70, Melbourne/Sydney, Australia*, 1970, Butterworths, 1971, p. 129.

Patterson, G. K., *Proc. 1st Eur. Conf. on Mixing and Centrifugal Sep., Churchill College, Cambridge, England*, BHRA Fluid Eng., Cranfield, England, A4-33, 1974.

Patterson, G. K., "Simulating Turbulent-Field Mixers and Reactors or Taking the Art

Out of the Design," *Turbulence in Mixing Operations*, edited by R. S. Brodkey, Academic, New York, 1975.

Patterson, G. K., *Chem. Eng. Commun.*, **8**, 25, 1981.

Patterson, G. K., W. Bockelman, and J. Quigley, *Proc. 4th Eur. Conf. on Mixing, Noordwijkerhout, The Netherlands*, BHRA Fluid Engineering, Cranfield, Bedford, England, J1, 303, 1982.

Patterson, G. K., and H. Wu, *Proc. 5th Eur. Conf. on Mixing, Wurzburg, Germany*, BHRA Fluid Eng., Cranfield, England, 355, 1985.

Paul, E. L., and R. E. Treybal, *AIChEJ.*, **17**, 718, 1971.

Pelletier, M. P., and L. Cloutier, *AIChEJ.*, **19**, 3, 474, 1973.

Placek, J., and L. L. Tavlarides, *AIChEJ.*, **31**, 1113, 1985.

Placek, J., L. L. Tavlarides, G. W. Smith, and I. Fort, *AIChEJ.*, **32**, 1771, 1986.

Plasari, E., R. David, and J. Villermaux, *Chem. Eng. Sci.*, **32**, 1121, 1977.

Plasari, E., R. David, and J. Villermaux, *ACS Symp. Ser.*, **65**, 11, 1978.

Platzer, B., *Chem. Technik.*, **33**, 16, 1981.

Plion, P., J. Costes, and J. P. Couderc, *Proc. 5th Eur. Conf. on Mixing, Wurzburg, Germany*, BHRA Fluid Eng., Cranfield, England, 341, 1985.

Pohorecki, R., and J. Baldyga, *Proc. Sym. on Industrial Crystallization, Warsaw, Poland*, September 27–28, 1978.

Pohorecki, R., and J. Baldyga, *I&EC Fund.*, **22**, 4, 398, 1983.

Pohorecki, R., and J. Baldyga, *Chem. Eng. Sci.*, **43**, 8, 1949, 1988.

Popiolek, Z., J. H. Whitelaw, and M. Yianneskis, *Proc. 2nd International Conf. on Applications of Laser Doppler Anemometry to Fluid Mechanics, Lisbon, Portugal*, 1–6, July 2–5, 1984, paper 17.1.

Porcelli, J. V., and G. R. Marr, *I&EC Funds.*, **1**, 3, 172, 1962.

Praturi, A. K., and R. S. Brodkey, *J. Fluid Mech.*, **89**, Part 2, 251, 1978.

Prausnitz, J. M., and R. H. Wilhelm, *Rev. Sci. Instr.*, **27**, 941, 1956.

Prausnitz, J. M., and R. H. Wilhelm, *Ind. Eng. Chem.*, **49**, 978, 1957.

Prochazka, J., and J. Landau, *Coll. Czech. Chem. Commun.*, **26**, 2961, 1961.

Raghav Rao, K. S. M. S., and J. B. Joshi, *Chem. Eng. Commun.*, **74**, 1, 1988.

Ramsey, C. G., E. A. Kyser, and G. B. Tatterson, *AIChEJ.*, **35**, 1219, 1989.

Rao, M. A., and R. S. Brodkey, *Chem. Eng. Sci.*, **27**, 137, 1972.

Rao, M. A., and R. S. Brodkey, *Chem. Eng. Sci.*, **27**, 2199, 1972.

Rayleigh, Lord, *Proc. Roy. Soc. A.*, **93**, 148, 1916.

Reed, X. B., M. Princz, and S. Hartland, *Proc. 2nd Eur. Conf. on Mixing, St. John's College, Cambridge, England*, BHRA Fluid Eng., Cranfield, England, B1–1, 1977.

Reith, T., *Proc. AIChE—I. Chem. E. Joint Sym., No. 10, Mixing—Theory Related to Practice*, AIChE, New York, 1965, p. 14.

Revill, B. K., *Proc. 4th Eur. Conf. on Mixing, Noordwijkerhout, Netherlands*, BHRA Fluid Eng., Cranfield, England, 11, 1982.

Rice, A. W., H. L. Toor, and F. S. Manning, *AIChEJ.*, **10**, 1, 125, 1964.

Rieger, F., P. Ditl, and V. Novak, *Chem. Eng. Sci.*, **34**, 397, 1979.

Rielly, C. D., and R. E. Britter, *Proc. 5th Eur. Conf. on Mixing, Wurzburg, Germany*, BHRA Fluid Eng., Cranfield, England, 365, 1985.

Rielly, C. D., and A. B. Prandtl, *Proc. 6th Eur. Conf. on Mixing, Pavia, Italy*, BHRA Fluid Eng., Cranfield, England, 69, 1988.

Rippin, D. W. T., *Chem. Eng. Sci.*, **22**, 247, 1967.

Robertson, B., and J. J. Ulbrecht, AIChE Annual Meeting, Miami Beach, Fla., paper 8c, November 1986.

Rodi, W., *Turbulence Models and Their Application in Hydraulics*, 2d ed., Intern. Ass. Hydraulic Res., Rotterdamseweg 185, 2660 MH Delft, the Netherlands, 1984.

Rosensweig, R. E., *AIChEJ.*, **10**, 91, 1964.

Rosensweig, R. E., *Canad. J. Chem. Eng.*, 255, October 1966.

Rushton, J. H., E. W. Costich, and H. J. Everett, *Chem. Eng. Progr.*, **46**, 395, 1950*a*.

Rushton, J. H., E. W. Costich, and H. J. Everett, *Chem. Eng. Progr.*, **46**, 467, 1950*b*.

Rushton, J. H., *Petroleum Refiner*, **33**, 8, 101, August 1954.

Rushton, J. H., and J. Y. Oldshue, *Chem. Eng. Progr.*, **49**, 4, 161, 1953.

Rushton, J. H., and J. Y. Oldshue, *Chem. Eng. Progr. Sym. Ser., No. 25*, **55**, 181, 1959.

Rushton, J. H., *AIChEJ.*, **26**, 6, 1038, 1980.

Ruszkowski, S. W., and M. J. Muskett, *Proc. 5th Eur. Conf. on Mixing, Wurzburg, Germany*, BHRA Fluid Eng., Cranfield, England, 89, 1985.
Rzyski, E., *Chem. Eng. J.*, **31**, 75, 1985.
Sachs, J. P., and J. H. Rushton, *Chem. Eng. Prog.*, **50**, 597, 1954.
Saffman, P. G., *Physics of Fluids*, **16**, 1181, 1973.
Salcudean, M., and K. Y. M. Lai, *Num. Heat Tran.*, **14**, 97, 1988.
Sano, M., and K. Mori, *Trans. Iron Steel Inst. Japan*, **23**, 169, 1983.
Sano, Y., and H. Usui, *J. Chem. Eng. Japan*, **18**, 47, 1985.
Sasakura, T., Y. Kato, S. Yamamuro, and N. Ohi, *International Chem. Eng.*, **20**, 251, 1980.
Sato, K., *Chem. Eng. Commun.*, **7**, 45, 1980.
Schwartzberg, H. G., and R. E. Treybal, *I&EC Funds.*, **7**, 1, 1968.
Sheu, Y. H. E., T. P. K. Chang, G. B. Tatterson, and D. S. Dickey, *Chem. Eng. Commun.*, **17**, 67, 1982.
Shiue, S. J., and C. W. Wong, *Canad. J. Chem. Eng.*, **62**, 602, 1984.
Sinclair, C. G., *AIChEJ.*, **7**, 709, 1961.
Smagorinsky, J. S., *Monthly Weather Review*, **91**, 99, 1963.
Smith, J. M., L. H. J. Goossens, and M. van Doorn, *Proc. 4th Eur. Conf. on Mixing, Noordwijkerhout, Netherlands*, BHRA Fluid Eng., Cranfield, England, C2, 71–79, 1982.
Smith, J. M., and A. W. Schoenmaker, *Chem. Eng. Res. Des.*, **66**, 16, 1988.
Squire, H. B., *The Aer. Quart.*, **16**, 302, 1965.
Stavek, J., I. Fort, J. Nyvlt, and M. Sipek, *Proc. 6th Eur. Conf. Proc. on Mixing, Pavia, Italy*, BHRA Fluid Eng., Cranfield, England, 171, 1988.
Stoots, C. M., Ph.D. dissertation, Chemical Engineering Dept., University of Maryland at College Park, College Park, 1989.
Strek, F., and J. Karcz, *Proc. 6th Eur. Conf. on Mixing, Pavia, Italy*, BHRA Fluid Eng., Cranfield, England, 375, 1988.
Sykes, P., and A. Gomezplata, *AIChEJ.*, **11**, 1, 174, 1965.
Sykes, P., *Chem. Eng. Sci.*, **20**, 1145, 1965.
Takao, M., and Y. Murakami, Proc. SCEJ-CIChE Joint Seminar on Mixing Technology, edited by W-L Lu, Taipei, Taiwan, July 14–15, 1987, p. 105.
Takao, M., and Y. Murakami, *Proc. 6th Eur. Conf. on Mixing, Pavia, Italy*, BHRA Fluid Eng., Cranfield, England, 171, 1988.
Takashima, I., and M. Mochizuki, *J. Chem. Eng. Japan*, **4**, 66, 1971.
Takeda, K., and T. Hoshino, *Kagaku Kogaku*, **4**, 394, 1966.
Tatterson, G. B., H.-H. Yuan, and R. S. Brodkey, *Chem. Eng. Sci.*, **35**, 1369, 1980.
Tatterson, G. B., J. T. Heibel, and R. S. Brodkey, *I&EC Funds.*, **19**, 175, 1980.
Tatterson, G. B., *Chem. Eng. Commun.*, **19** (1–3), 141, 1982.
Tavare, N. S., *Chem. Eng. Technol.*, **12**, 1, 1989.
Tay, M., and G. B. Tatterson, *AIChEJ.*, **31**, 1915, 1985.
Tchen, C. M., *J. Res.*, National Bureau of Standards, **50**, 1, 51, 1953.
Tennekes, H., and J. L. Lumley, *A First Course in Turbulence*, MIT Press, Cambridge, Mass., 1972.
Tily, P. J., S. Porada, C. B. Scruby, and S. Lidington, *Fluid Mixing III, I. Chem. Eng. Sym. Ser. No. 108*, 75, 1988.
Todd, F. H., "Resistance and Propulsion," *Principles of Naval Architecture*, edited by J. P. Comstock, Society of Naval Architects and Marine Engineers (74 Trinity Place, New York, 10006), 1967, Chap. 7, pp. 373–385.
Toor, H. L., *AIChEJ.*, **8**, 70, 1962.
Toor, H. L., "The Non-premixed Reactions: $A + B \rightarrow$ Products," *Turbulence in Mixing Operations*, edited by R. S. Brodkey, Academic, New York, 1975.
Torrest, R. S., and W. E. Ranz, *I&EC Funds.*, **8**, 810, 1969.
Tosun, G., *Proc. 6th Eur. Conf. on Mixing, Pavia, Italy*, BHRA Fluid Eng., Cranfield, England, 161, 1988.
Tsujikawa, H., S. Mishima, H. Nagamoto, O. Miyawakin, and Y. Uraguchi, *J. Chem. Eng. Japan*, **7**, 299, 1974.
Unno, H., and T. Akehata, *J. Chem. Eng. Japan*, **17**, 356, 1984.
van de Vusse, J. G., *Chem. Eng. Sci.*, **4**, 178, 1955a.

van de Vusse, J. G., *Chem. Eng. Sci.*, **4**, 209, 1955*b*.
van de Vusse, J. G., *Chemie. Ing. Tech.*, **31**, 583, 1959.
van de Vusse, J. G., *Chem. Eng. Sci.*, **17**, 507, 1962.
van der Molen, K., and H. R. E. van Maanen, *Chem. Eng. Sci.*, **33**, 1161, 1978.
van't Riet, K., and J. M. Smith, *Proc. 1st Eur. Conf. on Mixing and Centrifugal Sep.*, Churchill College, Cambridge, England, BHRA Fluid Eng., Cranfield, England, B2–17, 1974.
van't Riet, K., and J. M. Smith, *Chem. Eng. Sci.*, **30**, 1093, 1975.
van't Riet, K., W. Bruijn, and J. M. Smith, *Chem. Eng. Sci.*, **31**, 407, 1976.
Vavanellos, T. D., and J. H. Olson, *IE&C Funds.*, **14**, 355, 1975.
Viot, H., and A. B. Mersmann, *Proc. 6th Eur. Conf. on Mixing, Pavia, Italy*, BHRA Fluid Eng., Cranfield, England, 15, 1988.
von Essen, J. A., Paper presentation, 9th Eng. Found. Conf. on Mixing, New England College, Henniker, N. H., 1983.
von Karman, T., *Proc. Natl. Acad. Sci. (USA).*, **34**, 530, 1948.
von Karman, T., and L. Howarth, *Proc. Roy. Soc. (London)*, **A164**, 192, 1938.
von Karman, T., and C. C. Lin, *Rev. Modern Phy.*, **21**, 516, 1949.
von Weizsacker, C. F., *Z. fur Phys.*, **124**, 614, 1948.
Voncken, R. M., D. B. Holmes, and H. W. den Hartog, *Chem. Eng. Sci.*, **19**, 209, 1964.
Voncken, R. M., J. W. Rotte, and A. TH. Ten Houten, *Proc. AIChE—I. Chem. E. Joint Sym., No. 10*, Mixing—Theory Related to Practice, AIChE, New York, 1965, p. 24.
Waggoner, R. C., and G. K. Patterson, *ISA Trans.*, **14**, 4, 331, 1975.
Weetman, R. J., and J. Y. Oldshue, *Proc. 6th Eur. Conf. on Mixing, Pavia, Italy*, BHRA Fluid Eng., Cranfield, England, 43, 1988.
Wessenlingh, J. A., *Chem. Sci. Eng.*, **30**, 973, 1975.
White, K. E., *Proc. 1st Eur. Conf. on Mixing and Centrifugal Sep., Churchill College, Cambridge, England*, BHRA Fluid Eng., Cranfield, England, A6–57, 1974.
Wichterle, K., M. Kadlec, L. Zak, and P. Mitschka, *Chem. Eng. Commun.*, **26**, 25, 1984.
Wichterle, K., L. Zak, and P. Mitschka, *Chem. Eng. Commun.*, **32**, 289, 1985.
Wichterle, K., P. Mitschka, J. Hajek, and L. Zak, *Chem. Eng. Res. Des.*, **66**, 102, 1988.
Wilson, N. G., *Oil Gas J.*, 165, November 8, 1954.
Wolf, D., and F. S. Manning, *Canad. J. Chem. Eng.*, 137, June 1966.
Wong, C. W., and C. T. Huang, *Proc. 6th Eur. Conf. on Mixing, Pavia, Italy*, BHRA Fluid Eng., Cranfield, England, 29, 1988.
Wu, H., and G. K. Patterson, *Chem. Eng. Sci.*, **44**, 2207, 1989.
Yianneskis, M., Z. Popiolek, and J. H. Whitelaw, *J. Fluid Mech.*, **175**, 537, 1987.
Yuän, H. H. S., and G. B. Tatterson, *Chem. Eng. Commun.*, **4**, 531, 1980.
Yuu, S., and T. Oda, *Chem. Eng. J.*, **20**, 35, 1980.
Zweitering, T. N., *Chem. Eng. Sci.*, **8**, 244, 1958.
Zweitering, T. N., *Chem. Eng. Sci.*, **11**, 1, 1959.

5

Laminar Mixing

Introduction

In creeping flow laminar mixing, the types of problems encountered in turbulent mixing do not arise. The inertial terms in the equations of motion for incompressible fluids are either not present or not important because of their magnitude. In such fluids, the viscosity is very large, and the viscous terms in the equations of motion dominate the flow behavior. Mixing of very viscous fluids is difficult since there are no turbulent eddies which enhance mixing. Accompanying the large viscosities are relatively low diffusion coefficients. As a result, creeping flow laminar mixing essentially becomes an area-controlled process. The increase in interfacial area during mixing provides additional area over which diffusion can act so that mixing can be accomplished.

The impellers used in creeping flow laminar mixing are usually full-tank impellers as shown in Fig. 5.1. The diameter of such impellers approaches the diameter of the tank. Their primary objective is to bring motion to the entire tank. The impellers used in turbulent mixing rely on turbulence to cause motion throughout the tank. The nomenclature and the major geometries for discussion in this chapter are shown in Fig. 5.1. As can be noted, the impeller-tank geometries are quite complex.

Other major references which treat the general area of laminar mixing include Patton (1979) who discussed mixing in media mills and powder dispersion, Skelland (1983) who discussed impeller power draw and circulation and mixing times, Ranz (1985) who discussed lamellar mechanical mixing, Ulbrecht and Carreau (1985) who emphasized rheological aspects of mixing, flow patterns, power consumption, and mixing efficiency, and Godfrey (1985) who emphasized present industrial practice.

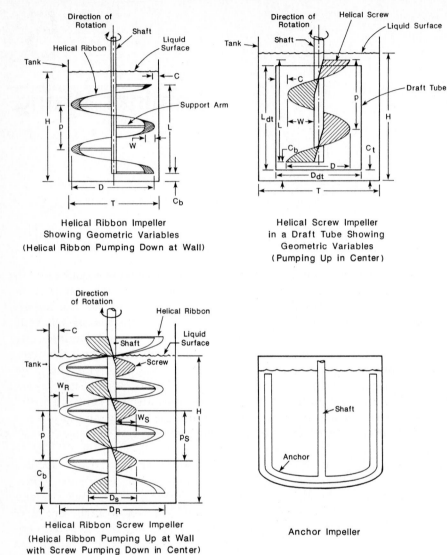

Figure 5.1   Creeping flow laminar mixing agitators.

Extensive Mixing

There are basically two processes being accomplished in mixing of viscous materials under creeping flow laminar mixing: extensive mixing and diffusion. Mixing cannot be completed unless diffusion occurs. Ironically, however, if diffusion becomes the rate controlling step, the mixer is poorly designed. A mixer operating in the creeping flow laminar mixing regime is primarily a device which causes extensive mixing. The various aspects of diffusion have been discussed in many texts and will not be discussed in this chapter.

Extensive mixing is the creation of new surface area. However, extensive mixing is not mixing in the sense of the creation of a homogeneous mixture since there is always complete segregation between the phases. Extensive mixing is an appropriate term since surfaces or material lines are being extended. Convective and advective mixing are terms synonymous with extensive mixing.

Two measures of extensive mixing are surface area and striation thickness. In this discussion, surface area or interfacial area, will represent any type of surface of interest under the extensive mixing. Striation thickness is the separation distance between areas or layers. Striation thickness is related to interfacial area since the product of the interfacial area and mean striation thickness is related to the fluid volume.

The objective of extensive mixing is to create and increase the interfacial surface area and reduce the thickness between areas. Much depends upon the kinematics of the mixing motion, the orientation of initial material or interfacial areas, and any reorientation of the material or areas in the mixing process. As will be shown, these measures can be calculated and, as a result, mixing efficiencies as well.

Spencer and Wiley (1951) noted that the phenomena of turbulence cannot be relied upon for mixing in laminar flow. The problems of modeling the mixing of very viscous materials become much simpler than in turbulent mixing phenomena. The deformation and redistribution sequences required to accomplish the desired mixing can actually be designed in principle. In actual mixers, however, this task is very difficult.

Several tasks which are necessary to accomplish effective extensive mixing are: (1) the creation and extension of surface area in the material as shown in Fig. 5.2 and Fig. 5.3 and (2) the distribution and the redistribution of this extended material, as shown in Fig. 5.4, for further extensive mixing. The problem in extensive mixing is one of determining the deformed interfacial area from the deformation process and the original surface and the orientation and the redistribution of new surfaces in the new volume. The distribution of the geometry of the volume throughout a deformed new volume is the basic task which is repeated over and over again.

Interfacial area

There are two ways in which the deformation and the creation of new surface area can occur in a material: shear deformations and normal

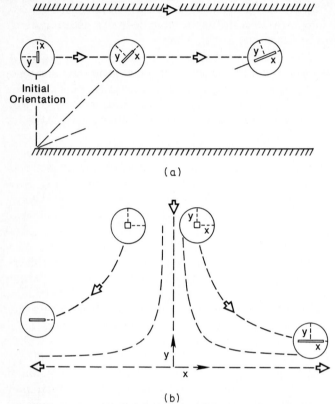

Figure 5.2   Stretch of fluid volumes in shear and stagnation flows.
(a) Stretch of fluid volume in shear flow. Mixing frame rotates at
decreasing rate. (b) Stretch of fluid volume in stagnation flow. Mixing frame does not rotate.   *(From W. E. Ranz, "Fluid Mechanical
Mixing—Lamellar Description," Mixing of Liquids by Mechanical
Agitation, edited by J. J. Ulbrecht and G. K. Patterson, Gordon
Breach Science Publishers, New York, 1985. By permission.)*

deformations (tensile, elongational, or extensional deformations). For
simple shear, the deformed surface is directly proportional to the original surface; the proportionality factor depends upon the orientation of
the new surface with respect to the original surface. The orientation
requiring minimum shear is $\tan \alpha = A/A'$ where A is the original surface and A' is the new deformed surface. The minimum shearing
strain corresponding to this orientation is $\partial u_1/\partial x_2 = A'/A - A/A'$
(Spencer and Wiley, 1951). Efficient production of new surface occurs
due to normal deformations or elongational flows, i.e., when the surface to be extensively mixed is parallel to the normal components of
the displacement vectors. Shear deformations are less efficient.

Spencer and Wiley (1951) provided a method by which the new in-

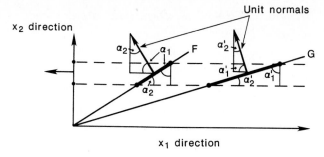

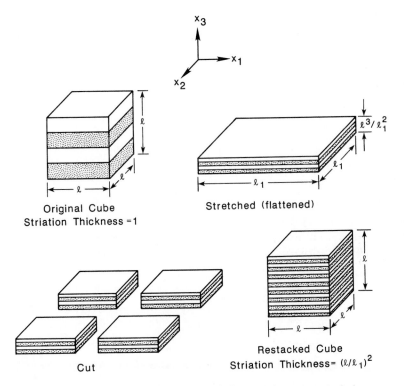

Figure 5.3 Stretch of a surface in a shear flow.

Figure 5.4 Three-dimensional extensional flow mixing; steps include stretching, cutting, and restacking as shown. *(From M. G. Reichman and R. J. Adler, AIChE Annual Mtg., November 1980. By permission.)*

terfacial area after a shear deformation can be calculated in terms of the old interfacial area and directional cosines which take into account surface orientations. Such a relationship solves the practical problem of computing the interfacial area from information concerning the original surface and the deformation process.

The new and the old interfacial surfaces will be denoted as $G(x_1', x_2', x_3')$ and $F(x_1, x_2, x_3)$, respectively, or:

$$G(x_1', x_2', x_3') = 0 \tag{5.1}$$

$$F(x_1, x_2, x_3) = 0 \tag{5.2}$$

The displacement vector l between the same material point in the two surfaces will be denoted by $l(l_1, l_2, l_3)$ where the components of l are functions of (x_1, x_2, x_3) or:

$$l_i = f(x_1, x_2, x_3) \tag{5.3}$$

The relationship between x_i and x_i' is:

$$x_i = x_i' - l_i \tag{5.4}$$

The definitions of the directional cosines of a unit normal for the surfaces G and F for the same material point are:

$$\cos \alpha_k' = \frac{\partial G / \partial x_k'}{[\Sigma(\partial G / \partial x_i')^2]^{1/2}} \tag{5.5}$$

$$\cos \alpha_k = \frac{\partial F / \partial x_k}{[\Sigma(\partial F / \partial x_i)^2]^{1/2}} \tag{5.6}$$

where $[\Sigma(\partial G / \partial x_i')^2]^{1/2} = [\Sigma(\partial F / \partial x_i)^2]^{1/2} = 1$ is required of the directional cosines of a unit normal vector. As a result of the displacement vector l between the same material point located in the surfaces G and F, the surface gradients between G and F, which are used in the calculation of the directional cosines, are related by:

$$\frac{\partial G}{\partial x_k'} = \frac{\partial F}{\partial x_k} + \sum_i \left(\frac{\partial F}{\partial l_i} \right) \left(\frac{\partial l_i}{\partial x_k} \right) \tag{5.7}$$

or:

$$\frac{\partial G / \partial x_k'}{\partial F / \partial x_k} = 1 + \sum_i \left(\frac{\partial l_i}{\partial x_k} \right) \left(\frac{\partial F / \partial l_i}{\partial F / \partial x_k} \right) \tag{5.8}$$

The ratios $(\partial G / \partial x_k')/(\partial F / \partial x_k)$ and $(\partial F / \partial l_i)/(\partial F / \partial x_k)$ are the ratios of the directional cosines of a unit normal vector or $\cos \alpha_k'/\cos \alpha_k$ and $\cos \alpha_i/\cos \alpha_k$, respectively. Thus, the directional cosines in the surface G are related to the directional cosines in surface F by:

$$\cos \alpha_k' = \cos \alpha_k + \sum_i \left(\frac{\partial l_i}{\partial x_k} \right) \cos \alpha_i \tag{5.9}$$

Since $\Sigma [(\cos \alpha_i')^2] = 1$, the directional cosine $\cos \alpha_k'$ becomes:

$$\cos \alpha'_k = \frac{\left[\cos \alpha_k + \sum_i (\partial l_i/\partial x_k) \cos \alpha_i \right]}{\left[1 - 2\sum_i \sum_j (\partial l_i/\partial x_j) \cos \alpha_i \cos \alpha_j + \sum_i \sum_j (\partial l_i/\partial x_j)^2 \cos \alpha_i^2 \right]^{0.5}}$$

(5.10)

The deformed surface is then given by:

$$A' = \int\int \left(\frac{dx' \, dy'}{\cos \alpha'_k} \right)$$

(5.11)

where the integration is over dA'_{ij}.

Example Problem Apply the above equation to unidirectional simple shear as shown in Fig. 5.3 such that the displacements $l_2 = l_3 = 0$, $dx_3 = 1$ (i.e., unit width), $\partial l_1/\partial x_2$ is constant, $\partial l_1/\partial x_1 = \partial l_1/\partial x_3 = 0$. The displacement l_1 and two directions x_1 and x_2 need to be considered.

solution For the figure shown, the displacement occurs in the x_1 direction:

$$x'_1 = x_1 + l_1$$

However, the magnitude of l_1 is only a function of x_2 and time. For a fixed time of deformation:

$$l_1 = f(x_2)$$

x_3 is not considered in the problem and l_1 is not uniquely determined by x_1 since for any x_1 position, l_1 cannot be specified. There is an infinite number of displacements for any given x_1 position only one of which is shown in the figure. Therefore:

$$\frac{\partial l_1}{\partial x_1} = 0 \quad \text{and} \quad \frac{\partial l_1}{\partial x_2} = K \neq 0$$

where K is not a constant. To clarify this further, under simple shear as shown in the figure above, the velocity V_{x1} is a function of x_2 only:

$$V_{x1} = f(x_2)$$

and l_1 is then:

$$l_1 = \int V_{x1} \, dt = \int f(x_2) \, dt$$

which upon integration provides:

$$l_1 = f'(x_2, t)$$

At any fixed deformation time, l_1 is then only a function of x_2 or:

$$l_1 = f''(x_2)$$

F-plane surface area is:

$$A = \frac{\int\int dx_2(1)}{\cos \alpha_1}$$

G-plane surface area is:

$$A' = \frac{\iint dx_2'(1)}{\cos \alpha_1'}$$

Applying the relationship between the directional cosines, the following relationship between the surfaces can be obtained:

$$A' = \iint \left[1 - 2\left(\frac{\partial l_1}{\partial x_2}\right) \cos \alpha_1 \cos \alpha_2 + \left(\frac{\partial l_1}{\partial x_2}\right)^2 \cos \alpha_1^2 \right]^{0.5} \left[\frac{dx_2(1)}{\cos \alpha_1}\right]$$

The directional cosines and partial derivatives of the displacement vector are independent of the coordinate system and can be taken out of the double integral. This provides the final relationship between A and A' which is:

$$A' = A \left[1 - 2\left(\frac{\partial l_1}{\partial x_2}\right) \cos \alpha_1 \cos \alpha_2 + \left(\frac{\partial l_1}{\partial x_2}\right)^2 \cos \alpha_1^2 \right]^{0.5}$$

The following relationships can also be obtained directly from the figure:

$$\cos \alpha_1 = \sin \alpha_2$$

$$(\sin \alpha_2)dF = dx_2 \Rightarrow dF = A = \frac{dx_2}{\cos \alpha_1}$$

$$(\sin \alpha_2)dG = dx_2 \Rightarrow dG = A' = \frac{dx_2}{\cos \alpha_2}$$

It should also be noted that the projections of A and A' onto the x_1 plane are equal to the same constant given as:

$$\frac{\iint dx_2(1)}{\cos \alpha_1} = \iint dx_2(1) = \text{constant}$$

where $\cos \alpha_1$ of the x_1 plane is 1.

Under unidirectional shear, the amount of interfacial area increases; however, the effect of shear in the generation of new area decreases with time, and the change in interfacial area decreases as the interfacial area is oriented parallel to the shearing plane as shown in Fig. 5.3. There is no growth of interfacial area when the interface is parallel to the shear plane, since all material in a shear plane moves at a constant velocity. In elongational flows, this does not occur.

Mohr et al. (1957) and Erwin (1978; 1979; 1981) have provided further studies of the growth of interfacial areas. Erwin (1978), in particular, provided another derivation for the relationship of interfacial areas before and after deformation which is somewhat easier to understand. For simplicity, the strain tensor is expressed as principal values and principal directions. In this derivation, the element of the surface in an initial state (x, y, z) undergoes a strain having elongation ratios λ_x, λ_y, λ_z. Initially, the surface element is formed by vectors $\mathbf{a}(a_x, a_y, a_z)$ and $\mathbf{b}(b_x, b_y, b_z)$, and the area of the surface element is another vector $\mathbf{c}$ formed by the cross product of $\mathbf{a}$ and $\mathbf{b}$. The vector $\mathbf{c}$ is perpendicular to the element and has a magnitude twice the area, formed by vectors $\mathbf{a}$ and $\mathbf{b}$. After the deformation, vectors $\mathbf{a}'$ and $\mathbf{b}'$ are formed as well as $\mathbf{c}'$ and a new area magnitude. The objective is to note the change in areas and direction as a result of the deformation. The vectors $\mathbf{a}'$ and $\mathbf{b}'$ are related as:

$$\mathbf{a}' = (a_x\lambda_x, a_y\lambda_y, a_z\lambda_z) \tag{5.12}$$

$$\mathbf{b}' = (b_x\lambda_x, b_y\lambda_y, b_z\lambda_z) \tag{5.13}$$

and the new area magnitude is formed from the cross product as:

$$A' = (\tfrac{1}{2})\{[\lambda_y\lambda_z(a_yb_z - a_zb_y)]^2 + [\lambda_z\lambda_x(a_zb_x - a_xb_z)]^2$$
$$+ [\lambda_x\lambda_y(a_xb_y - a_yb_x)]^2\}^{0.5} \tag{5.14}$$

The ratio of new area to old area can be written as:

$$\frac{A'}{A} = \left\{\frac{[(\lambda_y\lambda_z c_x)^2 + (\lambda_z\lambda_x c_y)^2 + (\lambda_x\lambda_y c_z)^2]}{|c^2|}\right\}^{0.5} \tag{5.15}$$

and in terms of directional cosines this becomes:

$$\frac{A'}{A} = \left\{\frac{[(\lambda_y\lambda_z \cos \alpha_x)^2 + (\lambda_z\lambda_x \cos \alpha_y)^2 + (\lambda_x\lambda_y \cos \alpha_z)^2]}{|c^2|}\right\}^{0.5} \tag{5.16}$$

For a constant volume deformation, the elongation ratios must obey continuity:

$$(\lambda_x)(\lambda_y)(\lambda_z) = 1 \tag{5.17}$$

When there is no elongation in a direction, then the elongation ratio for that direction is equal to 1. The sum of the directional cosines equals 1 also.

Example Problem Extensional steady flow mixer has $\lambda_x = \lambda_0$ and $\lambda_y = \lambda_z$. Find A'/A for various values of λ_0 (Erwin, 1978).

solution

$$\lambda_y = \lambda_z = \left(\frac{1}{\lambda_0}\right)^{0.5}$$

$$\frac{A'}{A} = \left\{\lambda_0 + \cos \alpha^2\left[\left(\frac{1}{\lambda_0}\right)^2 - \lambda_0\right]\right\}$$

For λ_0 above 1: If $\cos \alpha = 0$, the interface is aligned with the deformation initially and $A'/A = \lambda_0^{0.5}$. If $\cos \alpha \neq 0$, then the interface takes time to align with the deformation during which $\cos \alpha$ has to be calculated. After this, A'/A proceeds as $\lambda_0^{0.5}$. For the worst-case orientation, $A'/A \cong 0.8\,\lambda_0^{0.5}$. The flow of this nature resembles pulling a fluid cylinder from the top and bottom.

For λ_0 less than 1: Biaxial elongational flow results. This can be thought of as two jets colliding perpendicularly and the resulting flow expanding outward radially. If $\cos \alpha = 1$, then $A'/A = 1/\lambda_0$. For the worst-case orientation, $A'/A \cong 0.8/\lambda_0$.

One of the most complete discussions of extensive mixing is the study by Cheng (1979). Cheng reviewed three simple flows: simple shear, unaxial extension, and planar extension and provided the basic kinematics, velocity fields, the vorticity, stresses, viscosity types, and energy dissipation for these flows. Comparisons were provided between the three different flows as to their performance in dispersion (i.e., the breakup and separation of particle clusters) and emulsification (i.e., the breakup of drops and threads). Whereas extensional flows have been emphasized in the general literature as being more effective than simple shear flow, Cheng noted exceptions where exactly the opposite is true, particularly for non-Newtonian fluids. The effectiveness of a particular flow field depends greatly upon the criteria used in judging effectiveness as well as the qualities of the flow field. Simple extensional flows have no vorticity, which may be detrimental to mixing. Cheng cited various performance criteria by which flow fields can be compared. Quantitative statements about which flows are more effective than other flows require statements of the performance criteria used in the analysis.

Ottino, Ranz, and Macosko (1981) provided a mathematically elegant unified continuum view of the mixing process, having noted the lack of a theoretical basis for mixing due to the complex geometry and time dependencies of mixing fluids. The kinematics of mixing were presented in which material surfaces and lines are characterized according to their original orientation and to their motion. Equations for the material line length and surface area were given as functions of time. Applications given by Ottino et al. (1981) include local and global efficiency indices for mixing as applied to a shear flow and an elongational flow. Elongational flows with either the material line or area in the proper orientation were found more efficient than shear flows based upon the local efficiency index. Other references of interest in this area of extensional mixing include Ottino (1989a; 1989b).

Redistribution and reorientation

Redistribution and reorientation are important in the extensive mixing process and make up the concept of repetitive mixing which permits blocks of mixed material to be redistributed or reorientated for further mixing. Distribution matrices (Spencer and Wiley, 1951), cutting-stacking procedures (Reichman and Adler, 1980), or transformations (Ottino et al., 1981) can be used to describe the redistribution process. However, distribution is usually vaguely defined and quite general, specific only in applications. An example of a redistribution process is shown in Fig. 5.4.

Good initial distribution should have the interfacial surface of interest cut across as many flow streamlines (Shearer, 1973) or material lines (Ottino et al., 1981) as possible. The feed rate of the material should be weighted according to the local flow velocity. Streamlines or material lines can be systematically subdivided and redistributed. The redistribution can be modeled and the degree of subdivision can be determined although experimentation is necessary for model verification. Shearer (1973) provided three examples: helical flights, scraper blades, and planetary rollers. For effective redistribution, impellers should generate an asymmetric effect upon streamlines.

Spencer and Wiley (1951) noted an interesting property of distribution matrices. Distribution matrices simply model "cutting and stacking" procedures, and distribution matrices, such as $D[i, j]$, describe in digital form the redistribution which takes place in mixing. In work by Spencer and Wiley, the elements (i, j) of D were considered as the fraction of the ith cell before mixing found in the jth cell after mixing. For n distributions:

$$(a_j^n) = (a_i^0)[D(i,j)]^n \tag{5.18}$$

Spencer and Wiley noted that the variance of the elements of the matrix $D(i, j)^n$ was much smaller than the variance of the elements in the matrix, $D(i, j)$. Any matrix of distribution values becomes more homogeneous after multiplication. It should also be noted that redistribution alone can produce extensive mixing. The stacking process causes the creation of new surface area.

With regard to redistribution, Erwin (1981) also noted that for very large unidirectional shears s, the expression by Spencer and Wiley reduces to:

$$\frac{A}{A_0} = s \cos \alpha \tag{5.19}$$

As is indicated by Eq. (5.19), extensive mixing is directly proportional to the shear s. This relationship will appear later in experimental measurements of mixing or blending time for agitated tanks where the mixing or blending time varies inversely with area, and shear rate γ is proportional to the impeller rotational speed N. If the interface is arranged such that $\alpha = 0$, then:

$$\frac{A'}{A} = s \tag{5.20}$$

If this mixture, after some shear, is reorientated and sheared again, then:

$$\frac{A'}{A} = s^2 \tag{5.21}$$

For N such reorientations:

$$\frac{A'}{A} = s^N \qquad (5.22)$$

As indicated by this equation, extensive mixing is greatly increased due to the reorientation of the material and interface. Mixing, as measured by A'/A, is no longer linear with shear, and orientation of the interface is important in directing how the reorientation is done. Ignoring the fact that redistribution has been done with n reorientations, extensive mixing can be accomplished at higher powers of shear than 1. In an idealized mixer, extensive mixing may be exponential with shear and dramatic improvements in extensive mixing may be quite possible with proper design which includes frequent redistribution and reorientation.

There are very few studies which have made comparisons over the spectrum of real and idealized mixers. Le Goff (1977) considered three elementary mixing processes: diffusion, shearing, and viscous elongation for which the minimum amount of energy required to obtain a mixture was calculated. The study of Cheng (1979) is available for simple flow fields but not for mixers. Reichmann and Adler (1980) performed a study which established a set of parameters to judge all types of batch and continuous-flow mixers from the point of energy efficiency. The dimensionless group used for comparison of batch and continuous systems was given as:

$$M = \frac{PV}{\mu V'^2} \qquad (5.23)$$

where P is the power consumed, V is the mixer volume, V' is the volume of fluid processed per unit time, and μ is the fluid viscosity. For batch systems, V/V' is a time measure such as mixing time or circulation time. Reichmann and Adler provided a ranking of mixers and performed design calculations for several general types of mixers showing the different efficiencies of the mixing equipment.

Real mixers are complex devices whose geometries are difficult to characterize, and the fluid fields generated in such devices are even more difficult to treat. To extend the difficulties in analysis, mixtures are added whose entrance orientation and geometry lead to further complications in mixing characterization. In the mixer, the deformation and redistribution further develop the mixture, and redistribution causes further characterization problems. Overall the entire process appears to be overwhelming. However, in principle, the process is deterministic given the entrance orientation and nature of the mixture and the knowledge of the entire fluid field and any redistribution mechanisms in the mixer. Unfortunately these are not typically available during design.

It should also be recognized that the geometry of the mixer is important in the study of mixing and is a difficult subject to treat for a number of mixers. The area of mixing involves the characterization of the geometry and the resulting fluid mechanics. It is also difficult to select the optimum mixing geometry and methods of extensive mixing, redistribution, and reorientation.

Several areas require further study. First, the extensive mixing concepts given above have been primarily concerned with interfacial area. The concept of volume deformation and the creation of new volumes on the interface surface have not been studied. Although the role of the strain or the deformation rate upon extensive mixing is understood, the role of vorticity in mixing is not clear. The effect of vorticity upon the disruption of concentration gradients requires investigation.

Flow Phenomena and Velocities under Laminar Mixing

Although the basis of extensive mixing was provided above, the design of a mixer cannot be accomplished with this information. The characterization of extensive mixing requires velocity information to calculate the displacements, deformation, and elongation rates occurring in the mixer. However, in a realistic mixer, the displacements, deformation, and elongation rates which a surface element experiences are functions of position in the mixer. The modeling involved in extensive mixing requires both the velocity flow field as well as the displacement of the surface to be calculated. Essentially, surfaces must be followed through the mixer to be able to calculate the mixing performance of the mixer.

Nichols, Hirt, and Hotchkiss (1980), using SOLA VOF computer codes, provide examples where surfaces have been tracked as the velocity flow field was calculated. McFarland and Tatterson (1985) have used these codes to study the effects of fluid properties on mixing indices for flow near rising drops. However, such calculations have not been done for flows occurring in agitated tanks to date although the capability is presently available in two dimensions. In order to discuss true mixing, the unsteady diffusion problem, as the interface is extended, must also be included.

The major item of interest has been the velocity flow field. In principle, the three-dimensional laminar velocity flow field in agitated tanks can be calculated. In practice, two-dimensional calculations are available.

Most of the work on velocity flow fields can be divided into two categories: experimental observation of the actual flow phenomena and the numerical simulation of the flow behavior. A variety of flows can occur in agitated tanks.

Fluid rheology

Different fluids and multiphase mixtures are encountered in mixing. The study of their mixing requires information concerning their deformation behavior. The common types of fluids include Newtonians, Bingham plastics, shear thinning fluids (pseudoplastics), shear thickening fluids (dilatants), and viscoelastic fluids. Many multiphase mixtures are also classified under one of these categories. Basic shear data, shown in Fig. 5.5, should be obtained before the mixing equipment is designed. Elongational data, such as normal stresses, may also be needed. Rheological data are needed before a material can be classified into the primary categories of Newtonian fluids, shear thinning and shear thickening fluids, Bingham plastics, and viscoelastic fluids. A more complete review of these different fluid types can be found in Brodkey (1967).

Shear thinning fluids have lower viscosities at higher shear rates; shear thickening fluids have higher viscosities at higher shear rates. Viscosities in both categories are shear dependent. Bingham plastics exhibit a yield shear stress below which the fluid acts as a solid. To bring motion to the entire batch, the shear stress everywhere in the tank must be maintained above this yield shear stress. Above the yield point, Bingham plastics may exhibit shear thinning or shear thickening effects. Viscoelastic fluids have a combined viscous and elastic behavior and behave differently than other non-Newtonian flu-

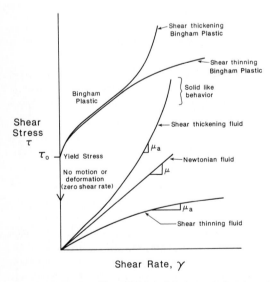

Figure 5.5 Time-independent deformation behavior of various fluids. Most show shear rate dependent viscosity.

ids. Fluids in the other categories may exhibit elastic behavior to some degree but not to the extent where the elastic behavior dominates their deformation behavior. Hence, they are not classified as viscoelastic fluids. Fluids also exhibit thixotropic and rheopetic phenomena (Brodkey, 1967; Chavan et al., 1975) which describe time-dependent behavior. Such time behavior is relatively unimportant in mixing since most phenomena do not occur in the time frame of the time-dependent behavior. However, if a process has important phenomena occurring in this time frame, such as a fast chemical reaction, then thixotropic and rheopetic phenomena require consideration.

Of these classifications, shear thinning materials are, by far, the most common category after Newtonian fluids. Shear thickening materials as well as rheopetic phenomena are rarely encountered. Viscoelastic materials are typically placed in the shear thinning category if the elastic behavior is usually of secondary importance.

A power law relationship is most often assumed in fitting rheological data for shear thinning and shear thickening fluids. Shear stress and shear rate are related as:

$$\tau = K\gamma^n \tag{5.24}$$

and apparent viscosity as:

$$\mu_a = K\gamma^{n-1} \tag{5.25}$$

For Bingham plastics:

$$\tau = \tau_0 + \eta\gamma \tag{5.26}$$

where τ_0 and η are the yield point and plastic viscosity.

Despite the distinctly different flow behavior of the fluid types given above, a common design solution exists in laminar mixing which is to have low shear rates applied more or less uniformly throughout the tank. This operation has to occur above any yield points of Bingham plastics and below solidification shear rates which the fluids may have. This solution has led to the development of full-tank impellers such as the helical ribbon impeller, the screw impeller, and the helical ribbon screw impeller. These can be considered close wall clearance full-tank impellers.

Since power input influences flow behavior and mixing times, a brief mention of rheological behavior on power number should be noted. Shear thickening fluids cause higher power numbers due to the growth of the solidified fluid effectively causing a larger diameter impeller and higher power draw. For shear thinning fluids, the laminar to turbulent transition is usually delayed to a higher impeller Reynolds number due to the tendency of the fluid to set up after leaving the impeller region. This causes lower power numbers for shear thinning fluids in the transition between laminar and turbulent flow. At the onset of turbulent flow, the power number quickly returns to its Newtonian level for turbulent flow and becomes independent of viscosity. For Bingham plastics, power must be supplied initially to break down the structure causing the yield stress before effective mixing can occur. The breakdown of the structure must be maintained during mixing. A more complete discussion of laminar power is given in Chap. 3.

Experimentally observed flow patterns in agitated tanks

Flow patterns are essentially solutions to the governing equations of motion and continuity for flow in very complex geometries and for fluids, some of which have very complex rheology. The flow fields of the agitated tank perform the mixing. Unfortunately, calculations of the two- and three-dimensional flow fields for rheologically complex fluids are difficult, and, as a result, flow information has to be experimentally obtained. The observed flow patterns can be used to check numerical solutions for the flow fields as a result.

Generally, any impeller design for mixing should be such that the fluid is forced to reorientate itself during flow and, at the same time, cause large shear and normal deformations. Flow patterns in laminar mixing are generated by helical ribbon impellers, screw impellers, and helical ribbon screw impellers or other full-tank impellers. Other impellers are also used such as turbulent type impellers and anchors. There is also recent interest in up-down rotating impellers which generate complex flow patterns. For the most part, the flow fields generated by these impellers are difficult to describe. Overall, laminar impellers are large-diameter, close wall clearance impellers which can have high shear rates at the wall even at modest rotational speeds. Their primary objective is to bring motion to the entire contents of the tank and to reorientate the fluid.

Wall baffles are not generally important in the area of laminar mixing due to the close clearance of laminar impellers to the wall. In fact, wall baffles may generate dead zones and hinder mixing. If mixing is occurring in the transition region, solid body rotation may occur and should be prevented. Wall and finger baffles disrupt solid body rotation and can be incorporated to aid mixing where impeller geometry permits.

There is a subtle division to be noted concerning the impeller types. Some impellers, for example, the helical ribbon screw impeller, reorientate the fluid because of an axial circulation flow before the fluid reenters the volume of the impeller blades again. Other impellers, for example, the anchor impeller, do not. The flow pattern or streamlines leaving the blades of the anchor are similar to the streamlines approaching the blades; very little reorientation occurs. Generally, impellers which impart an axial flow component have the tendency to cause the fluid to reorientate before circulation back into the deformation region of the impeller blade. Impellers which have an asymmetry blade configuration and position also have the tendency to reorientate the fluid more.

Flow patterns are very much dependent upon the impeller; however, the geometry of the impeller does not exclusively indicate the flow generated. In Chap. 3, the various flow analogies used in the development of power number correlations for laminar mixing were reviewed. These, to some degree, indicate the flow patterns which occur in laminar mixing in an agitated tank.

The terms primary and secondary are applied to flow patterns and require some definition. Primary flow patterns are those which have high velocity and momentum and large dimensions on the order of the tank. Secondary flow patterns are those flows not categorized as primary flow patterns and include any other observed flow phenomena. The terms primary and secondary are vague. It is important to obtain quantitative results to make true comparisons between flow phenomena.

Ford and Ulbrecht (1976) provided somewhat different definitions for primary and secondary flow patterns. Primary flow patterns were flow patterns generated by the impeller due to viscous or frictional drag of the impeller on the fluid. Secondary flow patterns were generated by inertia caused by centrifugal and/or elastic forces. The primary flow patterns are typically "cut" by the secondary flow patterns which aid mixing by causing reductions in striation thickness and redistribution and reorientation of the fluid geometry.

The following discusses the details of flow patterns of the principal impellers as well as other impellers used in laminar mixing. Their geometries, tank location, and the tank geometries have been discussed already in Chaps. 1 and 3.

Newtonian, shear thinning, and shear thickening fluids

Turbulent type agitators. Metzner and Otto (1957), Foresti and Liu (1959), Metzner et al. (1961), Godleski and Smith (1962), and Wichterle and Wein (1981) have given fairly complete accounts of the flow behavior of the different fluid types in agitated tanks using turbulent agitators. Shear thinning fluids experience high shear rates and low viscosities near an impeller blade. As the fluid leaves the impeller region, the fluid exhibits an increase in viscosity and the motion of the fluid decreases in comparison to that of a Newtonian fluid having the same viscosity level near the blade. The shear thinning fluid tends to set up more so than its Newtonian counterpart. Shear thickening fluids exhibit exactly the opposite flow behavior. If the shear rates near the impeller blade are high, solidification of a shear thickening fluid occurs as the fluid nears the impeller blade. Solidification increases as the impeller rotational speed increases, leading to poor mixing. The net effect of the shear thickening nature on power draw is as if the impeller diameter is increased. Fortunately, shear thickening fluids are not commonly encountered.

Wichterle and Wein (1981), noting the importance of turbulent type agitators for general mixing, provided an analysis of the threshold of mixing to obtain complete motion throughout a tank for shear thinning fluids. The experiments were performed in square vessels using turbine and propeller impellers. The size D_c of the well-mixed region around the impeller shown in Fig. 5.6 was determined and expressed as a function of Reynolds number as:

$$\frac{D_c}{D} = 1 \qquad \text{for } \text{Re}_{nN} < \frac{1}{a^2} \qquad (5.27)$$

and:
$$\frac{D_c}{D} = a(\text{Re}_{nN})^{0.5} \qquad \text{for } \text{Re}_{nN} > \frac{1}{a^2} \qquad (5.28)$$

where Re_{nN} is the non-Newtonian Reynolds number defined as:

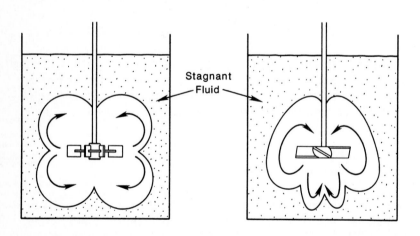

A radial mixer   An axial agitator pumping downward

Figure 5.6 Shape of the mixing cavity in a shear thinning suspension. *(From K. Wichterle and O. Wein, Intern. Chem. Eng., 21, 116, 1981. By permission.)*

$$\mathrm{Re}_{nN} = \frac{\rho N^{2-n}D}{K} \qquad (5.29)$$

As the equations show, the well-mixed region around the impeller is the same size as the impeller at low Reynolds numbers. D_c grows to the tank diameter or liquid height as the Reynolds number increases. The parameter a was found to be 0.3 for propellers, 0.6 for turbines, and approximately equal to $0.375(N_{PT})^{0.333}$ for any agitator where N_{PT} is the turbulent power number of the agitator. Solomon et al. (1981) have also provided a similar analysis which will be discussed in a later section. Elson et al. (1986) and Elson (1988) have discussed cavern formation. Etchells et al. (1988) have used the methods of Elson et al. and Solomon et al. in design.

Godfrey (1985) noted that turbulent impellers are used in mixing high-viscosity fluids due to their availability. Under such circumstances, it is apparent that optimum mixing is not of major concern. If turbulent agitators are to be used for mixing in the laminar regime, then planetary motion is recommended for effective mixing.

Anchor and gate impellers. Peters and Smith (1967) noted the singular lack of published information concerning the fluid mechanics and flow patterns generated by impellers in laminar mixing of viscous and non-Newtonian fluids. They performed a study of the anchor impeller, using neutrally bouyant tracer particles, to determine the flow field generated by the vertical blades of the anchor impeller. This work followed work by Beckner, Jolly, and Smith as cited by Peters and Smith (1967). Bechner et al. had studied the primary and toroidal secondary circulation flows of the anchor impeller as shown in Fig. 5.7.

Peters and Smith generated streamlines for the flow patterns and

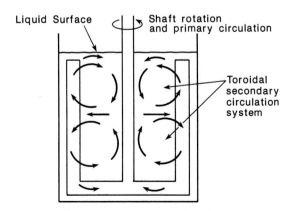

Figure 5.7 Secondary circulation in an anchor agitated tank. *(From D. C. Peter and J. M. Smith, Trans. Instn. Chem. Engrs., 45, T360, 1967. By permission.)*

showed how these changed as the rotational speed of the anchor impeller increased. Local fluid velocities were calculated. A vertical downward flow occurred behind the impeller blades for glyercol and polyacrylamide (PAA) solutions. The downward flow velocity to blade velocity for the PAA solutions was substantially higher than for the glyercol solution, 50 percent compared to 3.8 percent.

Viscoelastic fluids, such as PAA, have the tendency to flow toward the center of rotation in an attempt to relax back to their predeformed condition, causing participation in downward (or upward) flows behind impeller blades as observed by Peters and Smith. The downward flow increased the bulk circulation in the tank. The streamlines indicated the formation of a trailing vortex system behind and inside of the vertical anchor blades away from the wall. Smaller wall clearances led to more stable trailing vortex systems. Peters and Smith measured the velocity profiles in front of and behind the blade and computed the shear rate profiles, as shown in Fig. 5.8, as a function of radius and impeller Reynolds number. These profiles clearly showed

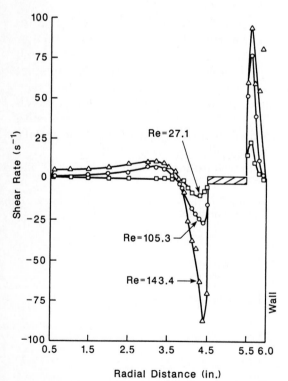

Figure 5.8 Shear rate profiles for an anchor impeller. *(From D. C. Peter and J. M. Smith, Trans. Instn. Chem. Engrs., 45, T360, 1967. By permission.)*

that much of the fluid in the tank was unaffected by the blade passage and that the vortex system was not shed by the blade.

Mass flow rates between the wall and the blade in the clearance area were substantially smaller than those occurring on the inside of the blade. The mass, which flowed inside the blade, remained with the blade considerably longer than the fluid passing through the clearance area because of the wall drag. The flow patterns indicated very little disturbance of the fluid in the wall layer (i.e., clearance area) which has important implications to heat transfer in this region. There was relatively little surface renewal occurring in the wall region, and there was no forceable removal of material away from the wall. The anchor agitation, due to the shear around the blade, mixed material. However, the volume mixed was relatively small in comparison to the tank contents.

Smith (1970) noted that the anchor agitator was viewed as having the capability of removing material from the wall and promoting heat transfer. This apparently does not occur. The laminar to turbulent transition occurred roughly around an impeller Reynolds number of 200; however, the transition was gradual. The results reported by Smith (1970) were similar to that by Peters and Smith (1967); however, at a Reynolds number of 200, a second vortex formed behind the blade, resulting in a dual vortex system which remained attached to the blade. The flow in these vortices did not participate in the bulk circulation in the tank.

Smith (1970) reported that the toroidal secondary circulation systems (Beckner et al.), shown in Fig. 5.7, were clearly observed in the dye neutralization studies. These clearly indicated that the flow occurring in the tank was three dimensional. Transport of dye between the primary and secondary circulation systems was very low which indicated compartmentalization of the flow. During neutralization studies, the wall flow and the central core region of the tank were decolorized first. The last regions to decolorize were the toroidal secondary circulation regions.

Deep tanks were thought to have more than two of these toroidal circulation zones. The size of these zones was affected by the impeller rotational speed, and improved homogenization rates were obtained by increasing the impeller rotational speed which increased the transfer of material between the circulation loops. Smith also discussed deflector plates placed on the vertical blades of the anchor which caused a reduction in mixing times. The deflector plates probably helped the fluid in the axial flow inside the vortex to leave the blade.

Murakami et al. (1972b) have also given some observations of the flow field produced by the anchor impeller in a comparative study of different mixers.

Helical ribbon impellers.   The helical ribbon impellers produce both axial and tangential flow due to the ribbon, and, generally, two ribbons
are used on an impeller. The studies by Nagata et al. (1955) and
Bourne and Butler (1965; 1969a; 1969b) are the basis for the following
discussion.

The primary flow pattern, Fig. 5.9, of the helical ribbon impeller
can be described as follows. As a point of reference, Fig. 5.1 shows the
helical ribbon geometry and the direction of rotation for pumping upward and downward at the wall. In pumping the fluid, the ribbon
moves through the fluid with an angle of attack and does not push the
fluid. The fluid flows around the ribbon as the ribbon moves forward,
crosses the ribbon almost normal to the plane of the ribbon and is
pulled behind the ribbon in the wake region of the ribbon. The ribbon
moves at a higher velocity than the surrounding fluid. The resultant
fluid velocity vector determines the vertical pumping capacity. The
fluid bulk motion moves up or down along the wall and down or up
through the interior of the ribbon helix in the bulk of the tank depending upon the geometry of the ribbon and the direction of rotation. Axial velocity measurements, obtained for the core region of the ribbon,
are discussed below. The primary top-to-bottom circulation, which
causes mixing, is due only to the axial pumping of the ribbons.

The shear for the helical ribbon impeller is localized in the inside
and outside the ribbon blade. The shear between the wall and the bulk
is periodic due to the ribbon passage. The fluid interior to the ribbon
moves tangentially in solid body rotation with very little shear.

Nagata et al. (1956) recommended that the ribbon impeller pump
upward at the wall, with pitch p/T of 1, i.e., the ribbons rise one tank
diameter in height for every 360° of rotation.

Although there was considerable scatter in the data by Bourne and
Butler (1965; 1969a; 1969b), the velocity data were found to be inde-

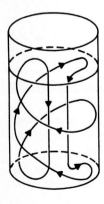

Figure 5.9 Flow pattern produced by a helical ribbon impeller. *(From S. Nagata, M.
Yanagimoto, and T. Yokoyama,
Memoirs Fac. Eng. Kyoto Univ.,
18, 444, 1956. By permission.)*

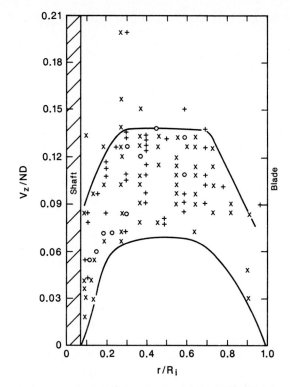

Figure 5.10   The distribution of axial fluid velocities in the core region of helical ribbon impellers pumping downward in 6- and 160-gal tanks. *(From J. R. Bourne and H. Butler, Trans. Instn. Chem. Engrs., 47, T11, 1969. By permission.)*

pendent of scale, fluid type, viscosity, and elastic effects. The vertical velocity in the interior of the ribbon helix, shown in Fig. 5.10, varied between 0.04 and 0.18 of the impeller ribbon speed. No radial flows were observed except at the top and bottom regions of the tank. By continuity, the radial flow in these regions was equal to the pumping of the helical impeller. A surface vortex formed in the center of the tank when the helix ribbon was pumping upward. The shear rates, important in mixing, were those surrounding the ribbon blade and in the volume between the solid body central core and the wall. Shear rates also increased with smaller wall clearances.

Chavan (1983) correlated the data on the tangential fluid velocity as a function of pitch of the ribbon as:

$$V_0 = \left[0.532 - 0.238\left(\frac{p}{D}\right) \right] \pi ND \qquad (5.30)$$

The correlation holds for $0.36 < p/D < 1.25$ and can also be used for the ribbon screw impeller.

The description above can be considered the primary flow pattern of the helical ribbon impeller; however, secondary flow loops can form at high impeller rotational speeds. These flow loops were similar to those described for the anchor impeller by Peters and Smith (1967).

The secondary flow loops found by Beckner et al. for anchor impellers were attributed to the vortex flow behind the blade by Peters and Smith. However, Bourne and Butler (1969) provided another mechanism for the secondary loops for the helical ribbon impeller. For the pumping upward mode, Bourne et al. (1969) noted that the tank bottom lowered the tangential velocity of fluid and decreased the radial pressure gradient. Fluid further from the bottom did not experience such a reduction in the radial pressure gradient, and, as a result, a secondary flow loop was established due to the pressure differential. For a pumping upward primary flow configuration, the secondary loop at the tank bottom had fluid moving inward at the base and upward at the center region of the ribbon. At some point above the tank bottom, the fluid flowed radially outward due to centrifugal forces through the helix to feed the flows at the wall. With the helical ribbon pumping downward at the wall, the secondary loop formed at the top of the tank. Asymmetry in secondary flow loops existed due to differences in boundary conditions between the tank bottom and liquid surface.

Bourne did not observe the vortex systems behind the ribbons of the impeller due to the difficulties of flow visualization. However, such systems are most likely present in the transition regime and may affect the pumping capacity of the impeller. Bourne and Butler (1965) have discussed other complex flow patterns of a ribbon impeller which involved flow loops as found by Peters and Smith (1969). The secondary flow patterns were quite different depending upon the direction of the fluid pumping. In pumping upward, there were six different secondary flow patterns discussed. Pumping downward had two such secondary flow patterns. In a pumping upward configuration, a center vortex was also formed.

Carreau, Patterson, and Yap (1976) have made studies of the flow patterns for the helical ribbon impeller and noted effects due to rotation. A portion of their circulation data is shown in Fig. 5.11. Overall, their results were similar to those just presented.

Helical screw impellers. There is a lack of published studies on the flow patterns generated by the screw impeller, probably because of the complex geometry of the screw. However, the screw impeller produces axial and tangential flows and is typically operated with and without

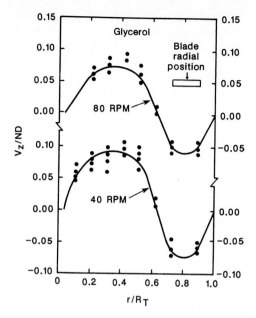

Figure 5.11   Axial velocity profiles for a helical ribbon impeller in glycerol. *(From P. J. Carreau, I. Patterson, and C. Y. Yap, Canad. J. Chem. Eng., 54, 135, 1976. By permission.)*

draft tubes. Chapman and Holland (1965) have published pictures of dye flow patterns of a off-centered, helical screw impeller pumping upward with no draft tube. A study of these showed a dispersive flow between the flights of the screw, the dispersion becoming fairly complete at the top of the screw. Trailing vortices occurring at the outer rim of the screw were not discussed. The flow into the off-centered screw impeller was from the other side of the tank and appeared as very ordered flights of dye sheets. Other regions of the tank appeared to be fairly stagnant.

Generally, there is a lack of information concerning how flow patterns change with wall clearance and other geometric variables for the screw impeller.

Other impeller types.   Jensen and Talton (1965) reported studies of helical cone impeller although no flow patterns were reported. Murakami et al. (1980; 1981) have described the flow patterns produced by up-down rotating paddles, turbines, disks, pitched blade impellers, and rotating prisms. The flow patterns were quite complex. The diameter of the up-down paddle was approximately that of the tank and the impeller went through 19 vertical strokes during 20 revolutions. The velocity flow fields for such impellers were functions of time and space making the entire system very difficult to analyze. However, this type of mixer reorientated the flow considerably and mixing times were low.

Hall and Godfrey (1968) and Kappel (1979) provided details concerning flow patterns generated by the sigma blade mixer, although further work is needed for a more complete understanding of this mixer. The sigma blade mixers have thick s- or z-shaped blades which resemble a high-pitch helical ribbon impeller. Two of these are usually positioned horizontally in separate troughs inside a mixing container. The blades rotate in opposite directions at different speeds. Earle (1957) and Michaels and Puzinauskas (1954) have studied sigma blade and pronged blade mixers as well.

Cheng and Schofield (1974) have reported on a positive displacement mixer, designed for heavy pastes and very viscous fluids, which has advantages over helical ribbon and sigma blade mixers in such applications. Nagata et al. (1956) and Murakami et al. (1972b) have made studies of the large paddles. Murakami et al. (1972b) have studied a multidisk impeller and Cable and Hakim (1972) performed a study of the flow patterns of a circular disk impeller in the mixing of glass. Hirata and Ito (1988) studied the flow and mixing in tanks using multistaged, large-diameter disk impellers. The disk diameters were 0.08, 0.10, and 0.12 m, and the tank diameter was 0.14 m. In such systems, Taylor and toroidal vortices form. Mixing into such vortices was very slow below a critical Reynolds number of $4.8 \times 10^2[D/(T - D)]^{0.4}$ where the Reynolds number was defined as $(\pi/2)ND^2/\nu$. Above the critical Reynolds number, mixing was greatly improved. Exchange coefficients were reported for mass transfer between toroidal vortex cells which were formed in multistage systems.

Flow phenomena for other fluid types

Bingham plastics. Nagata et al. (1970), in a power study of impellers in mixing Bingham plastics, described the flow patterns generated by turbulent agitators. The flow patterns and behavior clearly indicated why turbulent impellers are a poor choice for mixing a Bingham plastic. The description provided by Nagata et al. began with the fluid in a solid-like state and the stress levels below the yield stress. The power draw from the impeller was very high due the viscosity of the solid-like structure. Initially, the power input was taken up in overcoming the yield stress and became such that the internal structure was broken. At this point, motion was brought to the contents of the entire tank. The rotational speed of the impeller remained constant however. The viscosity of the Bingham plastic fluid was low, and the power input to the tank dropped to lower values as a result. With less power input to cause the motion of the fluid in the bulk of the tank, the fluid became solid-like again with the stress field falling below the yield stress. Power input increased because of the increase in viscosity

of the solid-like material, and the cycle repeated itself, cycling between a solid-like state and flow. A surface vortex was also observed
during the cycling. Essentially, when there was a flow pattern, it was
always at or near the state of collapse to solidification. In some cases,
the formation of a cavity structure around the impeller, as discussed
by Wicherle and Wein above, occurred where motion took place. The
rest of the fluid in the tank was solid-like.

Nagata et al. (1970) noted that such behavior was drastically reduced or not observed with laminar impellers such as helical ribbon
and screw impellers. Such impellers are large-diameter impellers; the
power input is more evenly distributed in the flow with less of a
chance for internal structure to have sufficient time to rebuild.

Miscible-phase flow patterns. Flow patterns of miscible phases have
not been studied to the same extent as that of single phase. However,
the flow behavior of miscible systems should be similar to single-
phase behavior but exact duplication of single-phase behavior cannot
be expected due to effects of viscosity, density, and interfacial tension
differences occurring between miscible phases. Cable and Hakim
(1972) studied glass blending using a circular disk impeller.
Takahashi et al. (1985) have studied mixing of different density and
viscosity liquids using the helical ribbon impeller. Murakami et al.
(1972) described viscosity effects on the distribution of miscible phases
during mixing.

Numerically Simulated Flow Patterns

As discussed above, the flow field and flow patterns are three dimensional, requiring three-dimensional computations for a complete description. At the present time, however, the numerical simulation of
the flow in agitated tanks has been primarily two dimensional, rely-
ing primarily upon the stream function-vorticity approach which
eliminates the need to know the pressure distributions. Simulations
using the Navier Stokes equations and Marker and Cell techniques
(Nickols et al., 1980) have not been done.

The equations for vorticity are obtained by taking the curl of the
Navier Stokes equations. For a rotating frame of reference, the dimensionless equation for vorticity (Greenspan, 1968) can be written as:

$$\frac{\partial \omega}{\partial t} + \nabla \times [(\epsilon \omega + 2\mathbf{k}) \times \mathbf{V}] = - E \nabla \times \nabla \times \omega \qquad (5.31)$$

where E, $(\nu/R^2\Omega)$, is a dimensionless number called the Ekman number. The Ekman number is the ratio of viscous forces to Coriolis forces
and can be viewed as the Reynolds number for the flow. Typically, the

Ekman number is usually small, on the order of 10^{-5}. The Rossby number, ϵ or $U/\Omega R$, is the ratio of inertial forces to Coriolis forces and is typically of unity or less in magnitude. In the equation, $\mathbf{k}$ is the unit vector of rotation. To interrelate vorticity and the stream function, the Laplacian of the stream function is set equal to the vorticity, as will be shown later.

Vorticity is a principal quantity of the flow. For a small fluid element, vorticity is its angular momentum. The changes in vorticity and the transport of vorticity occur more slowly than other more primitive quantities and, hence, are easier to model (Greenspan, 1968).

A primary reason for doing the numerical simulation is to provide accurate information concerning the properties of the flow which cannot be obtained otherwise or are difficult to obtain experimentally (e.g., stress distributions). The velocities of interest are the tangential velocity V_θ and the radial velocity V_r which are related to the stream function as:

$$V_\theta = \frac{\partial \psi}{\partial r} \tag{5.32}$$

and

$$V_r = - \left(\frac{1}{r}\right) \frac{\partial \psi}{\partial \theta} \tag{5.33}$$

Example Problem   Show that the stream function satisfies continuity.

solution   Continuity in the r and θ directions at steady state and for a constant density fluid is:

$$\frac{(1/r)\partial(rV_r)}{\partial r} + \frac{(1/r)\partial V_\theta}{\partial \theta} = 0$$

In terms of the stream function ψ, continuity becomes:

$$\left(\frac{1}{r}\right)\frac{\partial\{r[\,-\,(1/r)\partial\psi/\partial\theta]\}}{\partial r} + \left(\frac{1}{r}\right)\frac{\partial(\partial\psi/\partial r)}{\partial \theta} = 0$$

or

$$- \left(\frac{1}{r}\right)\frac{\partial\partial\psi}{\partial\theta\,\partial r} + \left(\frac{1}{r}\right)\frac{\partial\partial\psi}{\partial r\,\partial\theta} = 0$$

The stream function has physical meaning since differences in the stream function provide information about flow rates.

Example Problem   Show how the stream function is related to flow rates.

solution   The flow rate in the θ direction between radii r_1 and r_2 is:

$$Q = \int_{r_1}^{r_2} V_\theta\, dr(1) = \int_{r_1}^{r_2} \frac{\partial \psi}{\partial r}\, dr = \psi_2 - \psi_1$$

The stream functions are the streamlines for the flow, and the fluid flows parallel to them. When the stream functions approach each other, the velocity and shear rates increase.

The following describes the use of vorticity and the stream function equation to model the flow field. Much of the work assumes that the partials with respect to the z coordinate are zero and that the axial velocity V_z is zero. As a result, axial pumping capacities cannot be determined.

Anchor and paddle impellers.   Work by Hiraoka et al. (1978; 1979) gave the elements of the stress tensor in terms of an apparent viscosity μ_a and the stream function as:

$$\tau_{rr} = -\mu_a\left[\frac{2\partial(V_r)}{\partial r}\right] = \frac{-2\mu_a\partial[-(1/r)\partial\psi/\partial\theta]}{\partial r}$$

$$= 2\mu_a\left[-\left(\frac{1}{r^2}\right)\frac{\partial\psi}{\partial\theta} + \left(\frac{1}{r}\right)\frac{\partial\partial\psi}{\partial r\,\partial\theta}\right] \tag{5.34}$$

$$\tau_{\theta\theta} = -\mu_a\left\{2\left[\frac{V_r}{r} + \frac{1}{r}\frac{\partial V_\theta}{\partial\theta}\right]\right\} = -2\mu_a\left[-\left(\frac{1}{r^2}\right)\frac{\partial\psi}{\partial\theta} + \left(\frac{1}{r}\right)\frac{\partial\partial\psi}{\partial r\,\partial\theta}\right] \tag{5.35}$$

$$\tau_{\theta r} = \tau_{r\theta} = -\mu_a\left[\frac{r\partial(V_\theta/r)}{\partial r} + \frac{(1/r)\partial V_r}{\partial\theta}\right]$$

$$= -\mu_a\left[\frac{\partial^2\psi}{\partial r^2} - \left(\frac{1}{r}\right)\frac{\partial\psi}{\partial r} - \left(\frac{1}{r^2}\right)\frac{\partial^2\psi}{\partial\theta^2}\right] \tag{5.36}$$

Using these terms, the stream function equation becomes:

$$\frac{\text{Re }\partial(\psi, \nabla^2\psi)}{r\,\partial(r, \theta)} = -\left[\left(\frac{2}{r^2}\frac{\partial}{\partial\theta} + \frac{2}{r}\frac{\partial^2}{\partial r\,\partial\theta}\right)\tau_{\theta\theta} + \left(\frac{\partial^2}{\partial r^2} + \frac{3}{r}\frac{\partial}{\partial r} - \frac{1}{r^2}\frac{\partial^2}{\partial\theta^2}\right)\tau_{r\theta}\right]$$

$$\tag{5.37}$$

The vorticity equation is:

$$\nabla^2\psi = -\omega \tag{5.38}$$

where $\nabla^2 = \partial^2/\partial r^2 + (1/r)\partial/\partial r + (1/r^2)\partial^2/\partial\theta^2$. Vorticity is then:

$$\omega = \frac{(1/r)\partial V_r}{\partial\theta} - \frac{\partial(rV_\theta)}{\partial r} \tag{5.39}$$

In terms of the stream function and vorticity, the equations of motion are:

$$\frac{\text{Re }\partial(\psi, \nabla^2\psi)}{r\,\partial(r, \theta)} = \mu_a\nabla^2\omega + F(\psi, \mu_a) \tag{5.40}$$

where:

$$F(\psi, \mu_a) = - \left(\frac{2}{r^2} \frac{\partial \mu_a}{\partial \theta} + \frac{2}{r} \frac{\partial^2 \mu_a}{\partial r \, \partial \theta} \right) \left(\frac{2}{r^2} \frac{\partial \psi}{\partial \theta} - \frac{2}{r} \frac{\partial^2 \psi}{\partial r \, \partial \theta} \right) + \left(\frac{\partial^2 \mu_a}{\partial r^2} + \frac{3}{r} \frac{\partial \mu_a}{\partial r} \right.$$

$$\left. - \frac{1}{r^2} \frac{\partial^2 \mu_a}{\partial \theta^2} \right) \left(- \frac{\partial^2 \psi}{\partial r^2} + \frac{1}{r} \frac{\partial \psi}{\partial r} + \frac{1}{r^2} \frac{\partial^2 \psi}{\partial \theta^2} \right) \quad (5.41)$$

$F(\psi, \mu_a)$ accounted for the non-Newtonian behavior of the fluid. For Newtonian fluids, $F(\psi, \mu_a)$ is equal to zero. The equations by Hiraoka et al. (1979) have not assumed any constitutive relationship for the apparent viscosity.

Hiraoka et al. (1978; 1979) developed the finite difference equations of the stream and vorticity equations and solved these for the velocity flow field. The coordinate system was attached to the rotating impeller blade with the tank rotating at an angular velocity $2\pi N$. The boundary conditions, which are coordinate system specific, were:

At the impeller shaft and along the impeller blade:

$$V_\theta = \frac{\partial \psi}{\partial r} = 0 \quad (5.42)$$

and
$$V_r = - \left(\frac{1}{r} \right) \frac{\partial \psi}{\partial \theta} = 0 \quad (5.43)$$

and at the tank wall:

$$V_\theta = \frac{\partial \psi}{\partial r} = 1 \quad (5.44)$$

and
$$V_r = - \left(\frac{1}{r} \right) \frac{\partial \psi}{\partial \theta} = 0 \quad (5.45)$$

The conservation of angular momentum was required to fix the actual values of the stream function.

The equation for angular momentum, in terms of vorticity and the stream function, was written as:

$$\frac{d}{dr} \int_0^{2\pi} \omega \, d\theta + \frac{\text{Re}}{r} \int_0^{2\pi} \left(\frac{\partial \psi}{\partial \theta} \right) \omega \, d\theta = 0 \quad (5.46)$$

which holds in the region between the impeller and the wall. Successive overrelaxation was applied to solve the finite difference equations with the specified boundary conditions until the equation for angular momentum was satisfied.

Hiraoka et al. (1978; 1979) performed calculations of the flow field for a flat blade large-diameter paddle and an anchor impeller for Newtonian and power law fluids. Since the entire velocity flow field

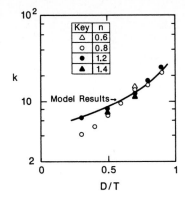

Figure 5.12 Numerical results of the relationship between k and D/T for various values of n. *(From S. Hiraoka, I. Yamada, and K. Mizoguchi, J. Chem. Eng. Japan, 12, 56, 1979. By permission.)*

was calculated in the simulation in the V_r and V_θ directions, power input, streamline and dissipation functions, normal and shear stress fields, velocity profiles, and pressure gradients were obtained. The work numerically established the basis of the Metzner Otto postulation for agitated tanks where the average shear rate is proportional to the impeller rotational speed, i.e., $\gamma_{av} = kN$. The model results concerning k are shown in Fig. 5.12.

Kuriyama et al. (1982), following equations provided by Greenspan (1968) and a computational procedure by Daiguji and Kobayashi (1979), were able to compute the flow field of an anchor impeller in a highly viscous Newtonian fluid. The calculated shear distribution is shown in Figs. 5.13 and 5.14. The shear rates in Fig. 5.13 are similar to those obtained by Peters and Smith shown in Fig. 5.8. The work

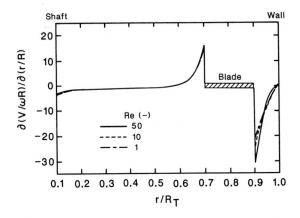

Figure 5.13 Shear rate distributions along the radial direction for an anchor impeller. *(From M. Kuriyama, H. Inomata, K. Arai, and S. Saito, AIChEJ., 28, 385, 1982. Reproduced by permission of the American Institute of Chemical Engineers.)*

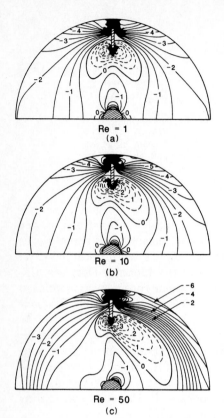

Re = 1
(a)

Re = 10
(b)

Re = 50
(c)

Figure 5.14 Distributions of shear rate $[\partial(V_\theta/\omega R)]/\partial(r/R)$ throughout the entire flow field as a function of impeller Reynolds number. *(From M. Kuriyama, H. Inomata, K. Arai, and S. Saito, AIChEJ., 28, 385, 1982. Reproduced by permission of the American Institute of Chemical Engineers.)*

could be extended to non-Newtonian fluids easily. The numerical scheme was based upon simulation of the unsteady-state vorticity equation-stream function problem which converged to a steady-state solution. The vorticity equation was:

$$\frac{D(\omega)}{Dt} = \frac{\partial \omega}{\partial t} + \left(\mathbf{V} \cdot \mathbf{\nabla}\right)\omega = \left(\frac{\nu}{R^2\Omega}\right)\nabla^2\omega \qquad (5.47)$$

The stream function was:

$$\nabla^2\psi = 2\Omega - \omega \qquad (5.48)$$

These are dimensionless equations where ω was fixed to the wall coordinate system and ψ to that fixed to the rotating impeller. The boundary conditions were similar to that by Hiraoka et al. (1979). An additional restriction placed on the solution required that the closed integral of pressure around the impeller blade and vessel wall was zero. The stream function at the shaft was fixed at zero where the values for the blade and wall were determined from calculations. The stream function near the blade obtained by Kuriyama et al. (1982) was similar to Peters and Smith (1967) at a Reynolds number of 50. The shear stress distributions on the inner and outer edges of the blade and radial and tangential velocity distributions were obtained as well.

Other noteworthy numerical studies include work by Sweeney and Patrick (1977), Bertrand and Couderc (1981; 1985), Walton et al. (1981), Arai et al. (1982), and Ohta et al. (1985). Sweeney and Patrick (1977) were perhaps the first to use the stream function approach to calculate the flow field around an impeller blade in mixing. Bertrand and Couderc (1981) provided distributions of velocity, shear stress, and viscous dissipation as well as a power correlation. Bertrand and Couderc (1985), using a Carreau viscosity model, provided stream and vorticity distributions, velocity profiles, shear and normal stress distributions, dissipation distributions, and a power correlation for the gate impeller. Walton et al. (1981) performed similar calculations and provided the same type of results using power law and Eyring models for viscosity. Ohta et al. (1985) have carried out a simulation of the flow field of an anchor impeller. The flow patterns were obtained from stream function-vorticity approach in the r and z directions. Flow patterns were found similar to that reported by Peters and Smith (1967) and Smith (1970).

Spragg et al. (1985) studied the flow fields and mixing caused by a spinning disk impeller in an unbaffled tank. The flow was numerically simulated using the stream function and vorticity equation. Mixing of a tracer was numerically simulated as well using the flow field information.

Lafon and Bertrand (1988) presented calculation results of flow and mixing fields for the anchor and paddle impellers. The Navier Stokes equations and concentration equations were discretized into a finite volume formulation having a staggered grid and mixing was stimulated by numerically injecting a tracer into the grid. Lines of equal concentration were obtained as a function of time for the anchor and paddle impellers. The mixing efficiency of the two impellers was qualitatively shown to be poor far from the impeller blades. Additional impeller blades, e.g., the gate impeller, were recommended to improve mixing. The numerical methods used in the study could be easily extended to three dimensions and non-Newtonian fluids.

Arai et al. (1982) have used a B-spline function to fit the velocity data obtained numerically and experimentally for the anchor impeller in an attempt to express quantitatively the entire flow field.

Votator and a general study.   The geometry of a votator in the study by Niida and Yoshida (1978) consisted of a large-diameter shaft having multiple vertical flat blades where the blades extended out to the wall to the point of scraping the wall in some designs. Such equipment is often used as heat exchangers and reactors in processing viscous materials and heat-sensitive materials. The flow rate through the blade clearance and the vortex circulation between the blades were calculated and comparable to the velocity data of Trommelen and Beek (1971). The center and outer boundaries of the vortex occurring between the blades were given.

Laidler and Ulbrecht (1978) performed a general study of close clearance mixers based upon numerical analysis. Streamlines and shear rate distributions were computed for Newtonian and power law fluids between an agitator arm and mixer wall. The geometry was determined to be the major factor in the extent and distribution of shear, and the width of the clearance controlled the average shear rate. Inertia and shear dependent viscosity had a minor influence upon the nature of the flow.

Summary.   The systems studied have been primarily two dimensional and have involved only two-dimensional agitators like the anchor impeller. These simulations are well worth the effort and are very impressive. However, they must be considered only approximately correct and, in a sense, misleading since the true flow fields are three dimensional. The simulations are two dimensional and cannot be used to solve for the axial flow behind the impeller blades that was observed by Peters and Smith (1967). The numerical simulations have been limited to the calculation of the vortex system which remained stationary with the blade.

Computational schemes should be developed to calculate the true three-dimensional flow fields, and flows for other agitators such as the ribbon and screw agitators should be simulated. There has been a definite lack of the study of mixing and of mixing indices which takes place in an agitated tank using numerical simulation techniques.

Circulation Times and Pumping Capacities

The integral kinematic quantities, circulation times, and pumping capacities are measures of the fluid mechanical performance of the agitator to create surface area. The volume of the tank divided by the circulation flow rate or pumping capacity is the mean circulation time in the tank. This circulation time can be considered as a time measure of extensive mixing and the ability of the impeller to mix material.

Typically, circulation time has been correlated as:

$$N\theta_c = k_c \qquad (5.49)$$

where the product of $N\theta_c$ is expressed in revolutions and represents the number of revolutions required for one circulation. Pumping capacity or circulation flow rate is the volume of liquid passing through the impeller with time or:

$$\frac{Q}{ND^3} = k_Q \qquad (5.50)$$

where Q is the volume of fluid passing through the impeller. k_Q is typically called the circulation number. In terms of tank volume and circulation flow rate, circulation time is:

$$\theta_c = \frac{V}{Q} \qquad (5.51)$$

The dimensionless circulation and pumping capacity numbers k_c and k_Q are dependent upon the geometry of the impeller and tank, and reasonable agreement between studies can be expected for similar geometries. For the laminar and turbulent regimes, k_c and k_Q are constants for a particular geometry and are independent of the impeller Reynolds number and fluid properties. This only holds for good circulation behavior, however. Under poor conditions of circulation and in the transition regime between laminar and turbulent conditions, k_c and k_Q are not independent of impeller Reynolds number and fluid properties.

Radial, tangential, and axial circulation times are present in an agitated tank since circulation can occur in all three directions. Although not commonly considered, these data form a vector integral quantity which characterizes the circulation state in an agitated tank. However, since only the final mixed state is of interest, only the longest circulation times are of interest and are the data reported in the literature. These data are typically the axial circulation times. With regard to magnitude, however, low circulation times do not necessarily indicate optimum mixing conditions, since an agitated tank can have regions of low circulation times and regions of high circulation times. The distributions of circulation times should be examined in studies but are not usually reported in the literature.

Circulation times are different from mixing times which involve actual mixing. Typically, mixing times are two to four times larger than circulation times due to the time required for the relatively quiescent regions of the tank to pass through the impeller and become mixed. For design purposes, circulation times are very important. For perfect mixing, the residence time of the material should be considerably larger than three to four times the circulation time or the mixing time. Mixing and circulation times provide a way of specifying the vessel size necessary for good mixing performance. Circulation times and pumping capacities also contain information as to flow velocities which are important in correlations, scaleup, scaledown, and heat and mass transfer.

In the following, studies of circulation times are given according to impeller type. Viscoelastic fluids will be treated separately in another section.

Anchor and gate impellers.   The vortex system behind the blade is the primary flow for axial circulation for an anchor impeller. Tangential flow should not be considered in determining pumping capacity due to the inability of the solid body rotation of the anchor to mixing. As a result, the pumping capacity for anchor or gate impellers are difficult to quantify due to the difficulty of modeling the axial flow in the vortex behind the vertical blades of the anchor.

Helical ribbon impellers.   Nagata et al. (1956) found that the ribbon impellers had circulation times inversely proportional to impeller rotational speed or:

$$N\theta_c = k_c \qquad (5.52)$$

where k_c was found to be 11 to 12.7 for their particular geometry.

Bourne et al. (1969) determined the pumping capacity of a ribbon impeller by assuming that the fluid, relative to the blade, flowed around the blade normal to the ribbon. The resultant fluid velocity was normal to the ribbon blade and followed the blade as shown in Fig. 5.15. If the velocity of the blade is the tangential velocity V_{tb} of the impeller blade rotation, then the axial velocity and pumping capacity are:

$$V_z = V_{tb} \sin \theta \cos \theta \qquad (5.53)$$

and: $$Q = 2\pi r W V_z = 2\pi r W V_{tb} \sin \theta \cos \theta \qquad (5.54)$$

where W is the ribbon width, r is the average impeller radius, and D is the ribbon outside diameter. For the geometry under study by Bourne et al. (1969):

$$\frac{V_z}{V_{tb}} = 0.125 \qquad (5.55)$$

and: $$k_Q = \frac{Q}{ND^3} = 0.046 \qquad (5.56)$$

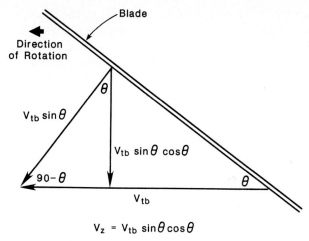

$$V_z = V_{tb} \sin\theta \cos\theta$$

Figure 5.15   The pumping velocity component V_z of a helical ribbon impeller.

The equation for V_z was checked against experimental data which gave V_z/V_{tb} to be 0.117 instead of 0.125. The equation for pumping capacity was also checked against the axial flow data obtained for the core region of the helical ribbon and was found to be too high by 10 percent. From this work, the projected ribbon angle for optimum pumping capacity is 45° which gives an approximate pitch ratio p/T of 2. Typically, a pitch ratio of 1 is recommended from mixing time studies which has an approximate projected angle of 22.5° with the horizontal. Pumping capacities of the secondary circulation loops were not reported since secondary flows are difficult to quantify.

Bourne et al. (1979) studied circulation times for Newtonian fluids under continuous blending by a helical ribbon impeller. For their particular geometry ($p/D = 0.57$, $D/T = 0.91$, $W/D = 0.12$, $H/D = 1.17$), circulation times were correlated as:

$$\theta_c = 18N^{-1.02} \qquad (5.57)$$

for pumping downward and:

$$\theta_c = 22N^{-0.99} \qquad (5.58)$$

for pumping upward. The pumping capacity was correlated as:

$$k_Q = \frac{Q}{ND^3} = 0.065 \qquad (5.59)$$

Carreau, Patterson, and Yap (1976) divided the flow field generated by the helical ribbon impeller into two regions, a high circulation region surrounding a low secondary toroidal loop similar to the ones reported by Peters and Smith (1967). The transfer between these two loops controlled mixing. Carreau et al. found circulation and mixing times for viscous (glycerol), slightly viscoelastic (CMC), and viscoelastic (PAA) fluids. For the viscous fluid, k_Q (i.e., Q/ND^3) varied between 0.05 and 0.15 depending upon geometry. For the slightly viscoelastic CMC fluid, k_Q varied between 0.047 and 0.093. Only one k_Q value, 0.014, was reported for the viscoelastic fluid PAA. Viscoelastic fluids will be discussed in a later section.

Takahashi et al. (1982), in a study similar to Carreau et al. (1976), developed a complex model for the circulation flows of the helical ribbon impeller, dividing the flows into two regions: (1) the primary circulation zone which consisted of the inner core and wall flows and (2) a poorly mixed region inside the primary circulation loop. A mixing time model was developed which assumed that mixing was dependent upon the exchange of material between these two regions. Velocity profiles for the z direction were given for the tank integrated to provide the circulation rate. The pumping number k_Q varied between 0.07 and 0.18 for the helical ribbons studied. From this view, three flow rates required calculation: the primary and secondary circulations and the exchange flow between these. The circulation and mixing times were given as complex functions of the primary circulation rate and the exchange rate between the two regions.

Ryan et al. (1988) described a polymerization reactor as having a set of marine impellers inside a draft tube with a helical ribbon impeller outside the draft tube in the annulus. The impeller systems rotated in opposite directions. The marine impellers were considered to contribute nothing to the overall pumping. Ryan et al. developed a relationship for circulation time from the Navier Stokes equations. For their double ribbon impeller, the circulation time was correlated as:

$$N\theta_c = 9.5 \pm 0.5 \tag{5.60}$$

for Newtonian fluids with Re < 500. The value from theory was 8.2. For the CMC solutions studied, $N\theta_c$ values ranged from 10 to 20 for Re < 1000. A simplified expression for $N\theta_c$ was given as:

$$N\theta_c = \frac{V}{\pi^2 D^2 CF_D \sin\theta \cos\theta} \tag{5.61}$$

where F_D is a shape correction factor from McKelvey (1962).

Takahashi et al. (1988) studied circulation times for a helical ribbon impeller using Newtonian and shear thinning fluids. Circulation times θ_c increased slightly with increasing impeller Reynolds number and were correlated as:

$$N\theta_c = \alpha \, \mathrm{Re}_a^\beta \qquad (5.62)$$

where β ranged from 0.04 to 0.12 and α was given as a function of impeller geometry and flow index. α increased as the flow index decreased, causing increased circulation times as the fluid became more shear thinning. β decreased with an increase in clearance or pitch and with a decrease in blade width. The flow index n affected circulation time by reducing the viscosity of the fluid near the impeller blade. This permitted the fluid to move around the impeller blade more easily than Newtonian fluids and made the circulation and pumping of the fluid more difficult at higher impeller rotational speeds. Overall, however, the shear thinning effect on θ_c was fairly small.

Helical screw impellers.   Nagata et al. (1956) found that the screw impeller had circulation times proportional to impeller rotational speed or:

$$N\theta_c = k_c \qquad (5.63)$$

where k_c was found to be 15 to 16.1 for a screw mixer in a draft tube, $0.7T$ in diameter, and 36.7 for a screw mixer in a draft tube, $0.5T$ in diameter.

In a more detailed analysis, Seichter (1971; 1981) studied the pumping efficiency η_p (i.e., $Q\,\Delta p/P$) of a screw impeller. Drawing an analogy to extruder screw theory, Seichter divided power usage of the screw impeller into that fraction α used for overcoming shear and that fraction $k_p\beta$ used for pumping or liquid circulation as:

$$P = (\alpha + k_p\beta)\mu N^2 D^3 \qquad (5.64)$$

or:
$$P = \alpha\mu N^2 D^3 + \beta N D^3 \, \Delta p \qquad (5.65)$$

where k_p is a number accounting for the pressure change which will be defined below. The first term is the power consumed in overcoming normal and shear stresses and the second is the power consumed in pumping which is simply $Q\Delta p/k_Q$. The pressure change, due to pumping, from the top of the screw to the bottom was related to the viscosity and rotational speed of the screw by:

$$\Delta p = k_p\mu N \qquad (5.66)$$

which defines the pressure change or pressure drop number k_p for the screw impeller. The pumped volume is given by:

$$Q = k_Q N D^3 \qquad (5.67)$$

k_Q and k_p are related to the geometries of the screw, draft tube, and tank.

For a screw in a draft tube operating in a tank of infinite extent, k_Q approached a fixed value k'_Q and became independent of tank geometry. Under such conditions, the pumped volume reached a maximum. For a screw operating in a draft tube with no pumped volume, k_p approached a fixed value k'_p and became independent of tank geometry. Under such conditions, the pressure drop Δp reached a maximum.

The pumped volume and pressure change, due to pumping, in a finite tank size were interrelated by:

$$\Delta p = - \left(\frac{\Delta p_{max}}{Q_{max}}\right)Q + \Delta p_{max} \tag{5.68}$$

This equation was rearranged to provide:

$$Q\,\Delta p = \Delta p_{max}Q\left(1 - \frac{k_Q}{k'_Q}\right) \tag{5.69}$$

The ratio k_Q/k'_Q is the volumetric efficiency of the screw, the actual volume pumped divided by the maximum volume pumped. As stated above, k_Q was found to be a function of geometry and correlated as:

$$k_Q = 0.25\left(\frac{T}{D}\right)^{0.67}\left(\frac{C_t}{D}\right)^{0.26} \tag{5.70}$$

for $k_Q < 0.5$ where C_t is the height of the draft tube from the vessel bottom. k_Q data, shown in Fig. 5.16, are in the range of 0.2 to 0.5 and different from values of 0.01 to 0.15 mentioned earlier for the helical ribbon impeller. The diameter of a helical ribbon impeller is typically larger than the screw impeller. The above equation for k_Q shows that pumping increased as the tank to screw diameter ratio increased and as the height of the draft tube from the tank bottom increased. The pumping efficiency $k\eta_p$ (i.e., $Q\Delta p/P$) was found to vary from 2 to 9 percent as a function of T/D.

Chavan and Ulbrecht (1973), following a similar analysis as Seichter, related the flow rate through a screw impeller inside a draft tube to the rotational speed and pressure as:

$$Q = A_1 N - B_1\left(\frac{\Delta p}{\mu}\right) \tag{5.71}$$

(This equation can be derived from the Navier Stokes equations. See Ryan et al., 1988. It is left as an exercise for the reader to demonstrate this.) Maximum pressure drop occurred at $Q = 0$ which was similar to the analysis provided by Seichter above. In terms of the constants, the maximum pressure drop is:

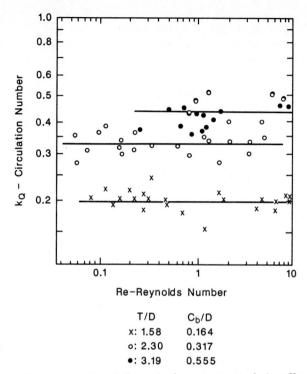

Figure 5.16 The influence of geometry and impeller Reynolds number on the circulation number for a screw centered in a draft tube. *(From P. Seichter, Trans. Instn. Chem. Engrs., 49, 117, 1971. By permission.)*

$$\frac{\Delta p_{\max}}{\mu N} = \frac{A_1}{B_1} \qquad (5.72)$$

The same flow outside the draft tube in the annulus was related to pressure drop only as:

$$Q = C_1\left(\frac{\Delta p}{\mu}\right) \qquad (5.73)$$

A_1 and B_1 are functions of the geometry inside of the draft tube. C_1 is a function of the geometry outside the draft tube. Chavan and Ulbrecht subsequently found k_Q and k_p in terms of A_1, B_1, and C_1:

$$k_Q = \frac{Q}{ND^3} = \frac{A_1 B_1}{B_1 + C_1} \qquad (5.74)$$

$$k_p = \frac{\Delta p}{\mu N} = \frac{A_1}{B_1 + C_1} \qquad (5.75)$$

For shear thinning fluids, Chavan and Ulbrecht replaced the viscosity with an apparent viscosity, assumed a power law relationship, and related shear rate to impeller rotational speed by a constant of proportionality (Metzner and Otto, 1957).

Chavan and Ulbrecht measured circulation time as the time it took for a flow follower to pass through the same plane in the annulus outside the draft tube. k_Q data varied between 0.01 and 0.2 as a function of geometry and were found to be independent of impeller Reynolds number between 0.02 and 10. The power law exponent or the flow index n varied between 0.28 and 1.0 for the fluids studied. k_Q data for the viscoelastic fluid PAA showed the effect of elasticity and were as much as 50 percent lower for some geometries than that for shear thinning fluids of similar viscosities. This effect was attributed to the sudden contractions and expansions in the circulation loop near the entrance and exit of the draft tube, which had higher-than-usual pressure drops for viscoelastic fluids. The various constants to calculate the pumping capacity number k_Q for shear thinning and Newtonian fluids are functions of geometry only:

$$A_1 = 3.18\left(\frac{W}{D}\right)f_p f_g^{\,2} \tag{5.76}$$

$$B_1 = 9.46 \times 10^{-3} \left(\frac{p}{D}\right)^{1.25} (f_g)^{-3.8} \left(\frac{D}{L}\right) \tag{5.77}$$

$$C_1 = 0.242\left(\frac{C}{D}\right)^{1.8} \left(\frac{(T - D_{dt})}{D}\right)^{0.56} \left(\frac{D}{L_{dt}}\right) \tag{5.78}$$

where f_p and f_g are defined as:

$$f_p = \frac{\pi(p/D)}{(p/D)^2 + \pi^2} \tag{5.79}$$

and
$$f_g = 1 - \frac{D_{dt} - D}{2D} \tag{5.80}$$

The ratio between the k_p and $k_{p,\,max}$ was found to be roughly independent of impeller Reynolds number and varied between 0.1 and 0.45. Circulation times were affected by the draft tube clearances from the liquid surface and tank bottom. For draft tubes, these clearances should be on the order of $0.1D$ or greater.

Rieger, Ditl, and Novak (1974) and Reiger and Novak (1985) have also developed correlations for calculations of pumping capacities of screw agitators. Rieger and Novak (1988) studied the pumping efficiency of a screw impeller in a draft tube using the parameter $PT^3/\mu Q$, where Q is the impeller pumping capacity. The most efficient geometry using this parameter was $T/D = 2.0$, $D_t/D = 1.1$, $p/D = 1.5$, and $D_{ss}/D_s = 0.2$. D_{ss} is the diameter of the screw shaft.

Summary. Although shear thinning fluids set up during flow away from high shear regions relative to Newtonian fluids, such phenomena do not appear to affect mixing and circulation times. The distribution of circulation times indicate the uniformity of the flow in the tank. Wide distributions in circulation times are not desirable.

Distributions or the standard deviations of the circulation times should be reported in circulation studies. Circulation times and pumping capacities should be obtained for other impeller types as well.

Mixing Times

Mixing in the turbulent and laminar flow regimes occur by two distinctly different mechanisms, but both rely upon the creation of interfacial area or extensive mixing concepts mentioned earlier in this chapter. Mixing times for both these regions are correlated by the simple relationship:

$$N\theta_m = k_m \tag{5.81}$$

which can be justified by extensive mixing arguments. k_m will be referred to as the mixing number or the homogenization number which is only a function of geometry for a properly designed mixer. For a specific geometry, the mixing number k_m is a constant. The mixing number has units of revolutions, and the numerical value represents the number of revolutions required to complete a specific mixing task. As observed above, such relationships also hold for circulation times as well, circulation times being about one-third to one-fourth of the time required for mixing in a properly designed mixer.

Typically, mixing time is the time required to obtain a fairly complete dispersion of dye or to obtain a color change by a reaction throughout the entire tank. Mixing time can be determined from monitoring a probe output after a tracer injection, mixing time being the time at which the probe output reaches a certain percentage of the new steady state. Time records taken at a fixed position in mixing studies typically resemble a decaying exponential or exponential sine function which can be used to determine mixing times. The distribution of mixing times should also be examined in mixing time studies. Unfortunately, these are not usually reported in the literature.

The constant k_m is different for the different impellers used in laminar and turbulent flow. However, if effective impellers are used in both mixing regimes, the magnitudes of k_m in laminar and turbulent mixing regimes are approximately the same.

In the transition region, no simple relationship holds as that given above. A considerable portion of mixing industrially occurs in the transition region. Mixing times in the transition regime are typically correlated by:

$$k_m = N\theta_m = CN_{\text{Re}}^a \qquad (5.82)$$

where a may be equal to -1.0.

Mixing times provide a rough estimate of the average residence time which is required of continuous flow systems. Typically, residence times should be anywhere from 50 to 200 times the mixing times for mixing effects not to appear in process results. Such information specifies tank volume for continuous systems. Mixing times are important in judging the capability of the impeller to mix.

Qualitative studies of mixing rates and the quality of mixing were performed by Metzner and Otto (1957). The major result of this work was the postulated relationship between average shear rate γ_{av} and impeller rotational speed N given as:

$$\gamma_{\text{av}} = kN \qquad (5.83)$$

This relationship directly indicates that mixing time varies inversely with impeller rotational speed which can be based upon the discussion of extensive mixing arguments given above. High shear rates imply low mixing times assuming no reorientation is being done in the mixing. Under conditions of considerable reorientation, a linear relationship would not be expected, and improved mixing times can be expected. A mixer which alternates between mixing and reorientation is very effective.

Several parameters are available for judging mixing performance. The first is k_m which is the number of revolutions necessary to accomplish a mixing task. Second is mixing energy which is:

$$E = P\theta \propto (\mu N^2 D^3)\left(\frac{1}{N}\right) = \mu N D^3 \qquad (5.84)$$

The equation shows that mixing energy is directly proportional to viscosity of the fluid and the pumping capacity. Another is mixing energy per volume or:

$$\frac{E}{D^3} = \mu N \qquad (5.85)$$

By lowering the viscosity, the energy per volume to perform the mixing is reduced. Other parameters are: (1) $P\theta_m^2/D^3$ which is mixing energy $P\theta_m$ divided by volumetric performance or pumping D^3/θ_m and (2) $P\theta_m^2/\mu D^3$ which is formed by combining the relationship between the power number and impeller Reynolds number $N_P N_{Re}$ and the mixing number k_m squared. The latter quantity $P\theta_m^2/\mu D^3$ contains the viscosity of the fluid which requires evaluation.

Murakami et al. (1980) noted that $(P_v\theta^2/\mu)^{0.5}$ and $(P_v/\mu N^2)^{0.5}$ can be used to evaluate mixing performance as well. The first is the characteristic amount of deformation of a fluid element for complete mixing, and the second is the characteristic amount of deformation of a fluid element during one revolution. Reichman and Adler (1980) have also discussed quantities useful in the evaluation of mixers which was mentioned earlier under extensive mixing.

Mixing time studies typically cover many impeller types. General reviews of different impeller types include Nagata et al. (1956), Hoogendoorn and den Hartog (1967), Novak and Rieger (1969; 1975), Moo Young et al. (1972), Nagata et al. (1972), Murakami et al. (1972b), Yuge and O'Shima (1975), and Kappel (1979). Various methods to obtain mixing times have been reviewed by Hoogendoorn and den Hartog (1967) and Ford et al. (1972).

The main purpose of such studies was to find geometries which provide good mixing performance while noting those impeller designs which have poor mixing performance. The studies were not concerned with the fluid mechanics of the impellers. Additional work has been accomplished in determining the effects of variations in geometry on mixing for those impellers having good performance. As such, the presentation of the work here has not been divided according to impeller type except for the helical ribbon impeller. The effects of geometric variations in the design of the helical ribbon impeller have been accomplished and are discussed in a separate section.

Data on mixing times vary from author to author due to geometric differences, different definitions, different measurement techniques and fluid properties. However, there emerges from the work a general consensus concerning the impeller designs which are effective in laminar mixing.

General laminar mixing studies of impellers of all types

Nagata et al. (1956) performed one of the first quantitative studies of mixing using highly viscous liquids. Nagata et al. noted that rapid homogeneous mixing was not easily attained for viscous liquids. Common solutions, used to solve problems in turbulent mixing such as increasing rotational speed and/or extending mixing time, were inadequate in laminar mixing. Turbulent type impellers were found inadequate in laminar mixing by Nagata et al. Preliminary studies eliminated the anchor impeller from consideration as well.

Nagata et al. focused primarily on the helical ribbon impeller, the screw impeller with and without draft tubes, and a large-area paddle agitator in flat- and dished-bottom tanks. The gate impeller was not studied. Millet jelly at various dilutions was used as the fluid and a reaction between sodium thiosulphate and iodine was used to measure mixing times.

Inadequate impellers showed dead volumes or toroidal rings which remained unreacted long after the rest of the tank was mixed. Effects of impeller rotational speed, liquid viscosity, the impeller and vessel diameters, pitch, blade width, and pumping direction on mixing times were noted. The ribbon and screw impellers had mixing times proportional to impeller rotational speed where k_m was found to be 33 for helical ribbon mixer, 45 for a screw mixer in a draft tube $0.7T$ in diameter, and 100 to 102 for a screw mixer in a draft tube $0.5T$ in diameter. The data are shown in Figs. 5.17 and 5.18, respectively. For the large-area paddle, the relationship was more closely represented by:

$$\theta = k'N^{-1.4} \tag{5.86}$$

As indicated, there were obvious differences between the paddle and the

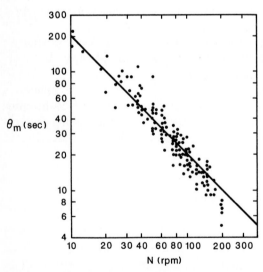

Figure 5.17 $N\theta_m$ diagram for different helical ribbon impellers. (From S. Nagata, M. Yanagimoto, and T. Yokoyama, Memoirs Fac. Eng., Kyoto Univ., 18, 444, 1956. By permission.)

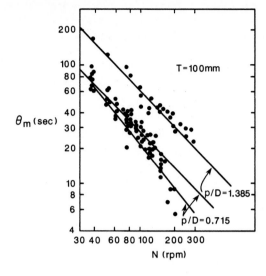

Figure 5.18 $N\theta_m$ diagram for different screw impellers with different pitches. *(From S. Nagata, M. Yanagimoto, and T. Yokoyama, Memoirs Fac. Eng., Kyoto Univ., 18, 444, 1956. By permission.)*

ribbon and screw impellers. The various paddle configurations studied performed poorer than the helical ribbon and screw impellers.

There was no effect of vessel diameter upon the mixing time for the helical ribbon impeller. For the paddle, the effect was:

$$\theta = k''T^{-0.56}N^{-1.4} \qquad (5.87)$$

Although no data were given, Nagata et al. recommended that the helical ribbon impeller pump upward along the wall and that the screw impeller pump downward for lower mixing times. In both cases, the same circulation pattern was established: upward flow along the wall and downward flow in the tank center. Ribbon pitches (p/D = 0.745, 1.12, and 1.5) had little influence on mixing; however, large pitch ribbons showed a tendency toward poor mixing. (Pitch is the height of rise for one rotation of the ribbon divided by the impeller diameter.) As the pitch of a ribbon impeller increases, an anchor impeller geometry is approached, and, hence, poor mixing is expected. At lower pitches, the ribbon impeller approaches a solid cylinder and generally has higher power consumption. Nagata et al. recommended a pitch ratio of 1 but provided no efficiency data for this selection.

Two screw impellers, having different pitches (p/D = 0.715 and 1.385), in draft tubes were also tested by Nagata et al. The higher-pitched screw showed more nonuniformity between screw flights; however, the differences in mixing times between the two were small. Mixing times on the whole were found to be approximately three times the circulation time.

Mixing energy was also discussed by Nagata et al. as $P\theta$. The screw impeller in a draft tube (D_{tb} = 0.7T, D = 0.93D_{tb}) was shown to be the most energy-efficient impeller using this parameter. Nagata's work also showed that an efficient impeller can also be an effective mixing impeller.

Metzner et al. (1961) studied mixing times of turbulent impellers in non-Newtonian fluids. Minimum values for the impeller Reynolds numbers were obtained for initial fluid motion at the liquid surface. These Reynolds numbers were in the transition regime and clearly indicated that stationary turbulent impellers were not capable of mixing non-Newtonian fluids under the conditions studied. Metzner and Taylor (1960) had previously shown that high shear and energy dissipation rates occurred only in the impeller region for turbulent impellers under laminar mixing conditions. Metzner et al. (1961) noted several advantages in the use of multiple impellers over a single impeller in terms of liquid circulation and power usage. The results also clearly indicated that the impeller diameter had to approach the tank diameter for effective mixing to occur. Pitched blade impellers were found to be more advantageous than flat blade impellers at low agitation rates. Using flat blade impellers alone, it was difficult to obtain motion throughout the entire tank.

Gray (1963) investigated the helical ribbon impeller of Nagata's experiments using dye dispersion to measure mixing time. A helical ribbon impeller was found to mix more rapidly by an order of magnitude at lower power input than a centered screw impeller using no draft tube, a curved rod turbine, and a curved blade turbine. An off-center screw impeller had approximately the same mixing capability as the helical ribbon impeller. Gray's study confirmed the results obtained by Nagata et al.

Screw impellers were shown to be more effective mixers than a disk style flat blade turbine by Chapman and Holland (1965). They also found that baffled centered screw impellers were more effective than off-centered screw impellers which is in disagreement with Nagata's studies. However, the centered unbaffled screw impellers were the least efficient in mixing. Unfortunately, mixing times were reported as efficiency data by Chapman and Holland (1965). The range of Reynolds numbers was from 0.1 to 8000, spanning both the laminar and transition regimes. Their data showed that the mixing times increased with viscosity to 0.5 power. A curved rod turbine was also studied but was shown to perform poorly.

Different intermeshing blade geometries, blade configurations, helix cone impellers, and different tank geometries were studied by Jensen and Talton (1965) in a mixing study of propellants. They found that any region not having an impeller blade pass through during the impeller rotation was a dead volume. Their study consisted of various trial studies. By the third set of trials, Jensen et al. were using a helical blade impeller to achieve the necessary pumping for mixing. The tank geometry was changed to fit the helical impeller blade in a low-clearance geometry in the fourth set of trials. The fifth set of trials had a conical-shaped tank with a conical-shaped intermeshing blade helical impeller. These last two geometries were found to give the best mixing performance. The conical-shaped tank and impeller had the advantage of adjustable wall clearance which affected power consumption. Jensen and Talton provided a correlation for mixing time dependent upon batch volume, number of blades, and clearance. However, a number of the parameters were not well defined or were subjective in interpretation. Overall, however, the work is important in that tank geometry was a variable in the study and was shown to affect mixing times. They also stated that an increase in blade volume to batch volume improved mixing which indicates the importance of redistribution, the blades "cutting and stacking" material. Several general conclusions of the work were: (1) impeller blades should sweep out a large part of the vessel during rotation, (2) intermeshing blades are more effective than single blades, and (3) vertical pumping and low blade clearances improve mixing performance.

Hoogendoorn and den Hartog (1967) performed a mixing time study for a number of impeller geometries in the laminar and transition regimes. Disk style flat blade turbines, pitched blade turbines with and without draft tubes, screw with and without draft tubes, helical ribbon impellers, propellers in draft tubes, and anchor impellers were studied—all placed on a single stationary shaft. Four different tracer methods (i.e., concentration (conductivity), temperature, color addition, and decolorization) were used to measure mixing times; temperature tracers and decolorization were used most. For the disk style flat blade turbine, the k_m data varied inversely with impeller Reynolds number from 170 to 4000 and reached a constant above a Reynolds number of 6000. Below an impeller Reynolds number of 170, the disk style turbine was inadequate for mixing. No differences were found in the data due to effects of scale for geometrically similar vessels in the standard geometry. Dead zones were apparent behind baffles, and the removal of the baffles lowered mixing times and improved mixing performance. Toroidal vortices formed below Re < 150, causing poor mixing and a steep increase in mixing times at lower Reynolds number for the disk style turbine. For a system of three pitched blade turbines in a draft tube, lower mixing times were obtained by establishing axial flow. Such a geometry approaches that of a screw impeller in a draft tube. In the unbaffled geometry with no draft tube, the three pitched blade turbines caused local mixing zones to form with little transfer of fluid from zone to zone. The k_m data for both geometries are shown in

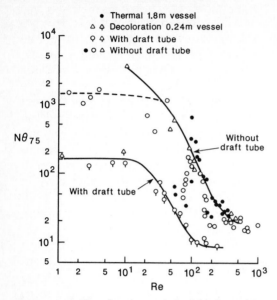

Figure 5.19 $N\theta_{75}$ for three pitched blade turbines with and without a draft tube as a function of impeller Reynolds number; vessel diameter, 0.24 and 1.8 m. *(From C. J. Hoogendoorn and A. P. den Hartog, Chem. Eng. Sci., 22, 1689, 1967. By permission.)*

Fig. 5.19. Pitched turbines in a baffled geometry gave better mixing performance than the unbaffled case. Helical ribbon impellers and screw impellers in a draft tube had k_m constants of 65 and 130, respectively, for both the laminar and transition region up to an impeller Reynolds number of 100 as shown in Fig. 5.20. Screw impellers without a draft tube performed poorly in the laminar and transition regions. For the propeller without a draft tube, large toroidal vortices were formed in the tank causing large mixing times. These were again minimized using a draft tube. Low clearance between the propeller and draft tube improved mixing times by removing dead zones in and outside the tube. An effect of impeller number for the pitched blade turbine in a draft tube was also noted; multiple turbines had lower mixing times. The anchor impeller was found inadequate for mixing. For all studies involving draft tubes, the diameter of the draft tube was $0.51T$, and all impellers were pumping downward.

A mixing time Reynolds number $(\rho T^2/\theta\mu)$ was defined by Hoogendoorn and den Hartog. Performance ratings $P\theta^2/\mu T^3$ of the different impellers and geometries were given as a function of this Reynolds number as shown in Fig. 5.21. Reichmann and Adler (1980) have recommended the use of $PV/\mu V'^2$ as a performance or merit in-

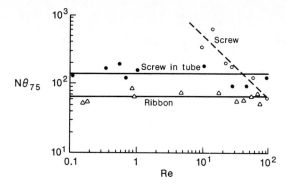

Figure 5.20 $N\theta_{75}$ for helical ribbon and helical screw impellers with and without a draft tube as a function of impeller Reynolds number. *(From C. J. Hoogendoorn and A. P. den Hartog, Chem. Eng. Sci., 22, 1689, 1967. By permission.)*

dex which is similar to that of Hoogendoorn and Hartog but can be used for both batch and continuous-flow systems.

The flow patterns of a good mixer, as noted by Hoogendoorn and Hartog (1967), should include: (1) axial flow, (2) all streamlines passing through the impeller region, (3) no closed streamlines occurring outside the impeller region, and (4) frequent disruption of fluid along the wall. The draft tube was found to eliminate or minimize dead zones.

Johnson (1967) performed a study of mixing times for a flat blade disk style turbine, propeller, and a helical ribbon impeller using dye dispersion. The flat blade turbine and propeller were found inadequate in the mixing, and toroidal vortices under laminar flow conditions were observed in the flow similar to the toroidal vortices found by Peters and Smith (1967). Diffusion of dye into these ring vortices was the rate-controlling step for the mixing process, and actual mixing times were difficult to obtain. The helical ribbon gave good mixing performance with k_m being roughly constant with impeller Reynolds number covering a range from 0.1 to 10. No unmixed toroidal zones were observed for the helical ribbon impeller. However, k_m data were found to increase as viscosity decreased, and a scale dependency was noted in the data. Nagata et al. (1956) found no effect of viscosity, and Chapman and Holland (1965) found just the opposite effect of viscosity. Johnson also noted that, as the helical ribbon diameter decreased, axial pumping decreased.

Even though the anchor impeller has been shown to be inadequate in mixing in comparison with other agitators, the flow field generated

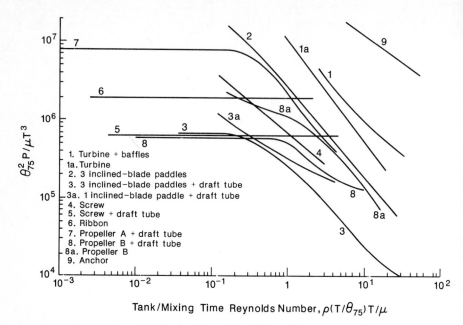

1. Turbine + baffles
1a. Turbine
2. 3 inclined–blade paddles
3. 3 inclined–blade paddles + draft tube
3a. 1 inclined–blade paddle + draft tube
4. Screw
5. Screw + draft tube
6. Ribbon
7. Propeller A + draft tube
8. Propeller B + draft tube
8a. Propeller B
9. Anchor

Tank/Mixing Time Reynolds Number, $\rho(T/\theta_{75})T/\mu$

Figure 5.21   Comparison of mixer performance.   *(From C. J. Hoogendoorn and A. P. den Hartog, Chem. Eng. Sci., 22, 1689, 1967. By permission.)*

by the anchor impeller appears to be simplistic enough to warrant an attempt to model mixing performance. Peters and Smith (1969), in a study of the anchor impeller, characterized the two-dimensional flow field in the $r\theta$ plane in terms of a blade tip Reynolds number and the wall clearance which permitted the calculation of mixing in the horizontal plane. The results did not agree with measured mixing times which indicated that the vertical secondary flows behind the blade played an important role in mixing. Peters and Smith concluded that three-dimensional flow field modeling was necessary to calculate true mixing performance. Shear mixing in the wall layer was found to dominate the mixing produced by the anchor impeller. Mixing time data versus impeller rotational speed passed through a minimum and was a function of wall clearance. Larger wall clearances caused lower mixing times. Toroidal vortices were present in the flow and were found to be the last regions to be mixed.

Novak and Rieger (1969) covered a range of impeller Reynolds numbers from laminar to turbulent conditions in a study of the helical screw impeller (with $H = D$, pitch $p/D = 1$, and $T/D = 0.63$) in various arrangements: (1) in a draft tube ($D_{dt}/D = 1.1$), (2) in a standard baffled tank, (3) in an unbaffled tank, and (4) in an off-centered position ($e/D = \frac{1}{6}$). The data for the off-centered screw impeller are shown in

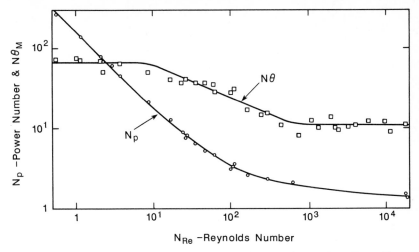

Figure 5.22   Power number and $N\theta_m$ data for an off-centered screw impeller. *(From V. Novak and F. Rieger, Trans. Instn. Chem. Engrs., 47, T335, 1969. By permission.)*

Fig. 5.22. The objective of the study was to determine the influence of geometry on mixing time and power consumption. The data for the k_m results are listed in Table 5.1.

Noteworthy items in the study by Novak and Rieger included: (1) k_m was shown to be a constant for both laminar and turbulent regimes, (2) geometry effects were present in the change in magnitude of k_m between the laminar and turbulent regimes, (3) the impeller Reynolds numbers at which the changes in k_m occurred were not constant, and (4) the differences in k_m between geometries in the turbulent regime were not as large as in the laminar regime.

The screw impeller in a draft tube or in the off-centered position provided low values for k_m over the entire range of impeller Reynolds numbers. Following Hoogendoorn and den Hartog (1967), Novak and

TABLE 5.1   The Effect of Geometry on Mixing Time for a Screw Impeller

Geometric configuration	k_m	Impeller Reynolds number range
A. In a draft tube	62.9	$N_{Re} < 30$
	22	$6,000 < N_{Re} < 12,000$
B. In a standard baffled tank	391.2	$N_{Re} < 6$
	$3063N_{Re}^{-1.09}$	$15 < N_{Re} < 90$
	10.5	$700 < N_{Re} < 18,000$
C. In an unbaffled tank	518.4	$N_{Re} < 3$
	$1246N_{Re}^{-0.756}$	$3 < N_{Re} < 300$
D. In an off-centered position	67.5	$N_{Re} < 6$
	$159.1N_{Re}^{-0.41}$	$10 < N_{Re} < 600$
	11.5	$1,000 < N_{Re} < 1,800$

SOURCE: V. Novak and F. Rieger, *Trans. Instn. Chem. Engrs.*, **47**, T335, 1969.

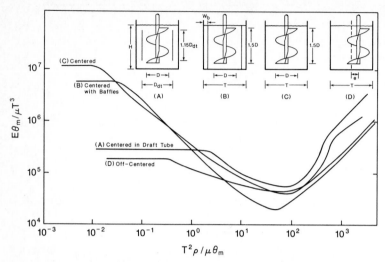

Figure 5.23   Mixing energy versus mixing time Reynolds number. *(From V. Novak and F. Rieger, Trans. Instn. Chem. Engrs., 47, T335, 1969. By permission.)*

Rieger also provided plots of the performance rating $P\theta_m^2/\mu T^3$ versus the mixing time Reynolds number, $\rho T^2/\mu\theta_m$, as shown in Fig. 5.23. Using this performance rating, the configuration having the screw in the off-centered position performed the best over the entire impeller Reynolds number range from laminar to turbulent regimes. The advantage of this configuration was already noted by Nagata et al. (1956) and Gray (1963). There is also disagreement between this study by Novak and Rieger and that of Chapman and Holland (1965) over the best screw configuration. Novak and Reiger have provided considerably more mixing time and performance data than Chapman and Holland (1965).

Novak and Rieger (1969) found that the screw in a baffled tank without a draft tube performed the best in the transition and turbulent regimes for $5 < \rho D^2/\mu\theta_m < 5000$. The performance of the screw in the draft tube was nearly as good as the performance in the off-centered position for the laminar regime for $\rho D^2/\mu\theta_m < 1$. The screw placed in the center of the tank was the least efficient mixing configuration.

Coyle et al. (1970) studied the helical ribbon screw impeller which combines a screw impeller and a helical ribbon impeller on the same shaft. Coyle et al. (1970) provided some data of the effect of viscosity upon mixing time although not explicitly citing an impeller Reynolds number range. The ribbon screw configuration lowered turnover time (i.e., circulation time) for shear thinning fluids and, hence, probably lowered mixing times as well.

Further study of the helical ribbon screw impeller was reported by Nagata et al. (1972). In this work, Nagata et al. (1972) studied a variety of impeller geometries including helical ribbon, screw, combined helical ribbon screw impellers, a multipaddle impeller, a half-ellipsoidal impeller in a draft tube as well as up-down impellers. These latter two designs were original to this work. The half-ellipsoidal impeller was made from an elliptical plate cut into two parts and mounted vertically crosswise on a shaft. The up-down agitator consisted of a perforated disk cut into six blades, resulting in a fan shape. Valve plates opened and closed according to the impeller up-down motion. The particular geometry of the agitator was difficult to describe, and no power measurements were reported for it. The k_m data for the some of the impellers are listed in Table 5.2.

Nagata et al. found that the multiple-blade paddle was inadequate for mixing because of poor vertical circulation flow. Of all the impellers studied, the up-down agitator provided the lowest mixing times and k_m values. The helical ribbon impeller with two ribbons having a pitch of 1, $p/D = 1$, was recommended. Nagata et al. also indicated that mixing time scaled inversely with pitch and with the number of ribbon or paddles. Single-ribbon impellers were found not suitable for mixing high-viscosity liquids. Various geometries of the combination ribbon screw impeller were also studied but provided no particular improvement in mixing performance over the double-ribbon geometry. The addition of the screw in some cases increased the k_m value and did not improve mixing. The addition of the screw to the helical ribbon was also found to increase the power consumption by 20 percent. The distance between the screw and ribbon had no effect on mixing performance. The ellipsoidal impeller in a draft tube had lower mixing times

TABLE 5.2   Mixing Time and Power Data

Impeller description	k_m	$N_P N_{Re}$
Double-ribbon impeller	40	330
Single-ribbon impeller	100 to 150	210 to 350
Ribbon screw impeller	30 to 57	250 to 400
Ribboned coil impeller	57	—
Screw impeller	—	210
Screw impeller with draft tube	65	330
Half-ellipsoidal impeller with draft tube	57	360
Up-down mixer	16 to 30	—

SOURCE: S. Nagata, M. Nishikawa, T. Katsube, and K. Takaish, *International Chem. Eng.*, **12**, 175, 1972.

than a screw in a draft tube and was recommended as a replacement to a screw impeller. The flow of the off-centered screw was shown to have a dead zone in the corner of the tank.

Murakami et al. (1972b) studied the anchor, large paddle, helical ribbon impeller, and a multidisk impeller. The work noted experimentally the correspondence between the different impellers and the various terms involved with energy dissipation. Similar work was performed numerically as reported above. The energy distributions were obtained from the flow patterns and given for the various impellers. Maldistribution of the energy dissipation occurred near the impeller blades. Power number and mixing time data were interrelated through strain rate and power consumption using the two coaxial cylinder analogy by noting that:

$$N \propto \left(\frac{P}{D^3 \mu}\right)^{0.5} \tag{5.88}$$

and, as a result of this relationship:

$$\left(\frac{P}{D^3 \mu}\right)^{0.5} \theta_m = k' \tag{5.89}$$

or
$$\frac{(P/D^3 \mu)^{0.5}}{N} = k'' \tag{5.90}$$

A number of different impellers were rated according to these parameters.

The work by Murakami et al. (1972b) pointed to the fact that energy dissipation can occur without mixing but mixing cannot occur without energy dissipation. Energy used in accomplishing mixing is typically less than the energy dissipated by the flow. For effective mixing, it is necessary to have material flow into and out of high-dissipation regions.

Novak and Rieger (1975) performed a homogenization efficiency study of the anchor, the pitched blade anchor, and the helical ribbon impeller. The k_m data are given in Table 5.3. As the data indicate, the anchor impeller had the lowest k_m values. However, the data for the anchor impeller only held for the transition region, and k_m values for

TABLE 5.3   The Effect of Geometry on Mixing Time for Various Impellers

Geometric configuration	k_m	Impeller Reynolds number range
A. Helical ribbon	62.1	$N_{Re} < 40$
	36.6	$1,500 < N_{Re} < 40,000$
B. Anchor	29.2	$1,000 < N_{Re} < 20,000$
C. Pitched blade anchor	23.3	$1,000 < N_{Re} < 20,000$

SOURCE: V. Novak and F. Rieger, *Chem. Eng. J.*, **9**, 63, 1975.

the anchor impeller increased as the conditions moved away from the transition region into the laminar and turbulent mixing regimes. The reason for such behavior was not given. However, the data indicated that the anchor impeller was an effective impeller in the transition regime. These results may be attributed to the vortex systems as noted by Peters and Smith (1967). Typically, the anchor impeller is not recommended for mixing. Further work is needed to resolve the differences.

The highest value for k_m for the helical ribbon impeller was approximately 200 in the transition region and varied between 20 and 80 outside this region as shown in Fig. 5.24. No reason was given for the abrupt change in k_m values for the helical ribbon impeller in the transition regime. Work by Zlokarnik (1967) should also be consulted. For the helical ribbon impeller in the laminar regime, Nagata et al. (1956) provided a value of k_m of 33 and Hoogendoorn and den Hartog (1967) provided a value of 65.

A general plot of the performance rating $P\theta_m^2/\mu T^3$ (mixing energy $P\theta_m$ divided by volumetric performance, T^3/θ_m) versus the mixing time Reynolds number $\rho T^2/\mu\theta_m$ was given by Novak and Rieger for the various impellers as shown in Fig. 5.25. Values from the literature have been used in the figure. Again the helical ribbon, screw in a draft

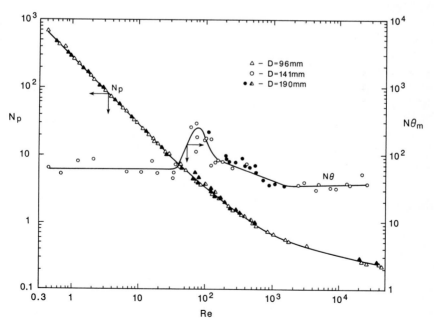

Figure 5.24 Plot of N_P and $N\theta_m$ versus Re for a helical ribbon impeller, 96 mm < D < 190 mm. *(From V. Novak and F. Rieger, Chem. Eng. J., 9, 63, 1975. By permission.)*

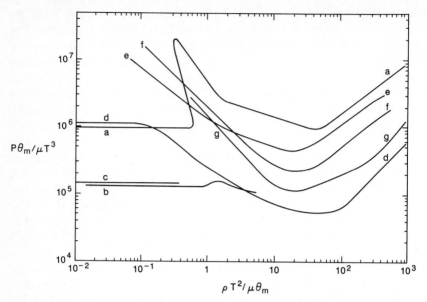

Figure 5.25 Plot of $P\theta_m/\mu T^3$ versus the tank mixing time Reynolds number $\rho T^2/\mu\theta_m$ for various systems. (*a*) Helical ribbon agitator; (*b*) screw in a draft tube; (*c*) off-centered screw; (*d*) off-centered multiple paddle; (*e*) anchor agitator; (*f*) pitched blade anchor; and (*g*) six-blade turbine. (*From V. Novak and F. Rieger, Chem. Eng. J., 9, 63, 1975. By permission.*)

tube, and an off-centered screw were shown to be appropriate in laminar mixing. The off-centered multiple-paddle impeller with pitched blades was also noted as providing good mixing performance. In the transition region, the six-blade turbine and the off-centered multiple paddle had the best performance rating although no data were given for the screw and helical impellers in this range for comparison. The anchor impeller had generally a poor performance rating.

Yuge and O'Shima (1975) performed an analysis of the concentration response curves from tracers injections and a mixing time study in the Reynolds number range of 1 to 1000 for a paddle, turbine, anchor, helical ribbon, and screw impellers. The amplitude and frequency of the response curves were given as a function of impeller type and impeller Reynolds number. Convective and diffusive time constants for the turbine were also given and used to explain the poor behavior of the turbine below an impeller Reynolds number of 10. Diffusive mixing was the rate-controlling step for turbulent agitators in this Reynolds number range. The helical ribbon and screw impellers performed well in the laminar regime, and the paddle and turbine in the transition regime.

Yuge and O'Shima were able to calculate the concentration response curves which showed reasonable agreement with experimental tests. Typically, the circulation time was the time between peaks in the response curves and, typically, three to four peaks occurred in an undampened response before a mixed condition was reached. The work by Yuge and O'Shima is important in its attempt to relate the convective and diffusive time constants and the other parameters of the response curves to the impeller Reynolds number in the transition regime.

Carreau, Patterson, and Yap (1976) studied the helical ribbon impeller, dividing the flow field into two regions: a high circulation region and a toroidal vortex region. The exchange between these controlled mixing. Carreau et al. reported their data in terms of the ratio of mixing time to circulation time, θ_m/θ_c. Depending upon geometry, θ_m/θ_c varied between 2.8 and 3.9 for a viscous fluid, glycerol, and between 4.27 and 9.4 for a slightly viscoelastic fluid, CMC. Results for the viscoelastic fluid PAA will be discussed later. To calculate mixing times, previous k_Q data cited above can be used.

Kappel (1979) performed detailed studies on the helical ribbon impeller, the MIG impeller, and the sigma impeller. Kappel used the concept of mixing work $P\theta_m^2/\text{volume}$ to judge the various designs. Kappel also noted that impellers should also be judged according to their application and not necessarily on energy efficiency alone.

The results for the helical ribbon impeller will be discussed in a later section. The MIG impeller is a multistage pitched blade impeller with the blade pitch switched between the inner and outer portions of the blades. The impeller blade width is also tapered. For the particular MIG studied, the flow was pumped downward at the wall and upward at the core. k_m data increased from 120 to 200 with increasing wall clearance between $0.01 < C/D < 0.06$ for Reynolds numbers less than 10. $P\theta_m^2/\text{volume}$ for the MIG impeller also increased with increasing wall clearance for $0.01 < C/D < 0.06$.

The sigma blade mixer is a complex mixer usually having two different size blades in troughs which pump in different directions at different rotational speeds. The outer edges of the blades pass very close to the wall which causes high shear useful in the break-up of agglomerates. Data on k_m and mixing work were obtained as a function of the speed ratio between the two impeller blades and were found to have minimums at a speed ratio of 1 when the blades were rotating in opposite directions. k_m data varied between 45 and 60 for impeller Reynolds numbers between 0.17 to 2.0. Another study of the sigma blade mixer is by Hall and Godfrey (1968) which is discussed later in this chapter in the section on sample variance.

Portner et al. (1988) measured mixing times for Newtonian fluids and shear thinning CMC solutions (0.5 to 2.0 percent) using turbulent type impellers: a paddle impeller, an Intermig 06, and a propeller in a draft tube. For the Newtonian and CMC solutions, $N\theta_m$ values, e.g., 100 to 1000, in the laminar regime decreased with increasing impeller Reynolds number to roughly a constant value of 10 to 20 in the transition region, $100 < \text{Re} < 1000$. The point at which $N\theta_m$ became constant was a function of the flow behavior index n of the power law, i.e., the lower the value of n, the higher the impeller Reynolds number at which $N\theta_m$ became constant. At impeller Reynolds numbers of 10^4 to 10^5, $N\theta_m$ values reached 10.

Pastes and very high viscosity liquids.   Cheng et al. (1974) studied a two-rotor positive displacement mixer specifically designed for the mixing of pastes and other very high viscosity fluids. The curved surfaces of the two rotors were in contact with each other in a close-fitting mixing chamber which trapped material. As the rotor revolved, the material was squeezed through narrow gaps. No dead zones existed in such motion, and the material experienced the same level of mixing. The rotors also served as good heat sinks, and the machine was self-cleaning and self-discharging. Cheng et al. related a generalized mixing index to shear strain and k_m and provided preliminary data on the effects of material properties, geometry, and operating conditions.

Specific geometric studies of the helical ribbon impeller

Roughly between 1975 and 1980, the helical ribbon impeller was recognized as a major impeller for laminar mixing. Specific studies of the variations in the helical ribbon geometry were performed to determine their effect on mixing times.

Kappel (1979) studied 12 different geometric arrangements for the helical ribbon impeller. Both single and double ribbons were used, and the effects of pitch and wall clearance on mixing times were studied. Ribbon width was not varied. These geometric variations affected the pumping capacity of the helical impeller and mixing times. The Reynolds number was from 1.2 to 28. k_m varied from 56 to 80 for a wall clearance $1.02 < T/D < 1.1$, and k_m was approximately proportional to -0.1 power of wall clearance. However, for some ribbons, $p/D = 0.5$, k_m increased with wall clearance. In any case, the relationship between wall clearance and mixing time was found to be weak. Kappel provided some reasons for these results and recommended an impeller design with $T/D = 1.1$ and $p/D = 1$ and with two helical ribbons. Further studies should be performed at larger wall clearances, noting Johnson's comment (1967) concerning decreased pumping at smaller impeller diameters. Kappel considered that large wall clearances enhanced pumping, which is opposite to Johnson's comment.

Bourne et al. (1979) studied mixing times for Newtonian fluids under batch and continuous-flow conditions using eight helical ribbon impellers having three different pitches and various wall clearances. The values for k_m under batch conditions were found to be of the order of 45 to 53 for the laminar regime. Only a slight reduction in these values was noted in the transition regime. The k_m values were independent of pumping direction, and wall clearance had little effect on the values. A pitch of 0.57 gave slightly better results than a pitch of 0.87. A particular geometry—$p/D = 0.57$, $D/T = 0.91$, $W/D = 0.12$, and $H/D = 1.17$—was recommended for optimal blending. The k_m values under continuous flow were also found to be slightly larger than batch values, around 65. This is approximately 3 to 3.5 times the circulation times cited above for continuous-flow situations. Mean residence times were also studied and found to be substantially affected by impeller rotational speed.

Takahashi et al. (1982) studied the effects of geometry of the helical ribbon impeller on mixing times and developed a model to describe the fluid circulation patterns. k_m data went through a minimum of around 30 to 40 at a clearance ratio C/T of 0.06, at a pitch ratio p/T of 0.901, and a blade-width ratio W/T between 0.15 to 0.18 for two-ribbon impellers. Some of the data are shown in Fig. 5.26. Lower pitches to 0.4 also gave satisfactory results. The effect of wall clearance upon k_m was different from that reported by Kappel. Kappel found no minimums and little effect of wall clearance on k_m. Poorly mixed zones were observed by Takahashi et al. for all single-ribbon impellers even after 500 revolutions, and secondary circulation was inadequate for mixing

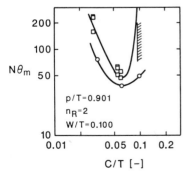

o measured by liquid crystal method
o measured by decoloration method
⚯ range where unmixed zones were observed

Figure 5.26 Mixing time as a function of impeller clearance. *(From K. Takahashi, M. Sasaki, K. Arai, and S. Saito, J. Chem. Eng. Japan, 15, 217, 1982. By permission.)*

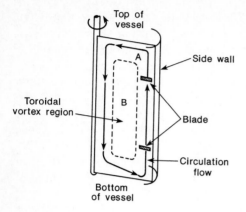

Figure 5.27 The circulation and toroidal vortex regions. *(From K. Takahashi, M. Sasaki, K. Arai, and S. Saito, J. Chem. Eng. Japan, 15, 217, 1982. By permission.)*

due to the long periods between blade passage. Takahashi et al. (1982), in a manner similar to Carreau et al. (1976), divided the flow into two regions: a high circulation region and a toroidal vortex region as shown in Fig. 5.27. The transfer between the two regions was often the controlling step in the mixing process.

Up-down and planetary impellers

Up-down and planetary impellers fulfill the requirements of full-tank impellers by their motion rather than by their geometry. Nagata et al. (1972) have studied this type of mixer as mentioned above and noted good mixing performance. Murakami et al. (1980) have performed a more extensive study by considering a number of different geometries: turbine and disk impellers having large holes to reduce resistance to vertical motion, pitched blade and cross-paddle impellers, and two- and four-blade triangular prisms. These impellers swept out the entire volume of the tank in their motion. Mixing and power requirements were measured. Generally, shorter mixing times, at equal or less power requirements than those for helical ribbon impellers, were found. k_m was found to be constant and independent of liquid viscosity. The turbine had the shortest mixing time for a single-stage impeller, whereas the cross paddle had the lowest times for a double-stage impeller. One double-stage cross paddle had a k_m value of 10 or approximately one-third that of a helical ribbon impeller. Double staging of impellers doubled the volume swept out by the impellers, improved mixing, and helped to eliminate stagnant regions which occurred at the liquid surface. Doubling the number of blades was found not to improve mixing significantly. Power was found proportional to the number of stages and blades but was not affected by stage spacing or rotational direction. Overall, the cross-paddle impeller in a double-

stage configuration with stage spacing equal to stroke length was recommended.

Transition mixing

Mixing times in the transition region between laminar and turbulent
flow, impeller Reynolds numbers between 1 and 10,000, are difficult to
characterize since there is a change in transport mechanisms in this
range, laminar to turbulent transport. These transport mechanisms
mix at different rates, which leads to difficulty and uncertainty in the
characterization of mixing.performance. Further, different impellers
are used, and the transition from laminar to turbulent mixing varies
with geometry and fluid rheology.

Fox and Gex (1956) developed a mixing time factor for jet and propellers in agitated tanks using a decolorization acid base reaction. In
the transition region, mixing time varied as:

$$\theta_m \propto \mu N^{-1.66} \tag{5.91}$$

which was similar to Nagata's results for large paddles reported
above. Chapman and Holland (1965) found a similar viscosity effect
for helical screw impellers: mixing time was found to be to the 0.5
power of viscosity for baffled and off-centered screw geometries. An
explanation of the data by Fox and Gex was given in terms of a momentum flux mechanism.

Lee, Finch, and Wooledge (1957) performed a dye dispersion mixing
time study for shrouded and unshrouded flat blade impellers. Mixing
time data were given as a function of impeller Reynolds number, but,
since the impeller diameter and fluid were fixed, mixing times were
given as a function of impeller rotational speed to various exponents, -0.37 to -64. There were also abrupt changes in these
exponents obtained for unbaffled vessels. Other effects were noted: (1)
smaller impellers had longer mixing times, (2) varying impeller spacing had little effect on mixing times, (3) unbaffled vessels required
higher impeller Reynolds numbers for the same mixing times, and (4)
the volume near the baffles was the last to be mixed.

Norwood and Metzner (1960) established a model to relate fraction
mixed to time for agitated tanks by assuming that the mixing occurred in a small volume in the vicinity of the impeller. However, the
model required information concerning the size of the volume and the
volumetric flow rate through this volume.

Moo-Young et al. (1972) performed a study from the transition to
turbulent regimes for a disk style turbine, helical ribbon, and an assortment of tubular agitators. The purpose behind the use of tubular
agitators was to investigate their possible use as heat exchangers.

However, the performance of tubular agitators was generally found to be poor and similar in performance to rod type agitators (Gray, 1963; Nagata et al., 1972). Only one tubular agitator was found to have the same capability as a helical ribbon impeller.

Mixing times in the transition regime require further characterization.

Mixing times for miscible liquids of equal volumes

The mixing of miscible liquids of similar volumes has not been extensively studied. For the helical ribbon impeller, Takahashi et al. (1985) identified three different mixing regimes responsible for mixing high-viscosity liquids having different viscosities, densities, and volume ratios from 0.3 to 10. k_m data were given as a function of impeller Reynolds number and flow regime. For liquids having the same physical properties, k_m was found to be 40.2 for Re < 10, which serves as a basis for their system.

Mixing time was found almost independent of the volume ratio for the different volume ratios studied. For the different density liquids, the dimensionless group N_M or $\Delta\rho L_v g/\mu_m N$ was used where L_v is the height of the more viscous phase, $\Delta\rho$ is the density difference, and μ_m is the viscosity of the mixture. The Reynolds number based upon mixture properties spanned from 0.01 to 100 into the transition regime. For the first mixing regime, liquid in the upper layer penetrated and spread or permeated into the lower liquid. The interface between the two liquids rose slowly. Mixing proceeded only in the lower liquid, and the permeating flow was the rate controlling step. k_m data were correlated as:

$$k_m = 0.0176 N_M^{1.56} \tag{5.92}$$

For optimum mixing conditions, N_M was correlated as:

$$N_M \mathrm{Re}^{-0.352} = 206 \tag{5.93}$$

for the volume ratio $V_B/V_A < 1$ and viscosity ratio $\mu_B/\mu_A > 60$ where the subscripts A and B are for the less viscous and more viscous liquids, respectively. For the volume ratio $V_B/V_A > 1$ and the viscosity ratio $\mu_B/\mu_A < 60$, N_M was a constant:

$$N_M = 151 \tag{5.94}$$

The second mixing regime was similar to that of mixing tracers in single-phase liquids. k_m was found to be 44.0 or almost equal to the single-phase value. In the third regime, drops were formed. The mixing in the third regime was controlled by the drop size of the more vis-

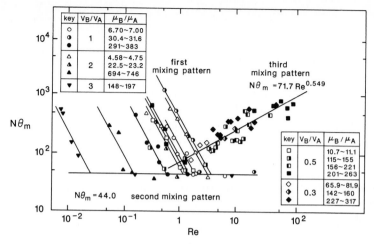

Figure 5.28 Measured mixing times for two miscible liquids with different physical properties. *(From K. Takahashi, Y. Takahata, T. Yokota, and H. Konno, J. Chem. Eng. Japan, 18, 159, 1985. By permission.)*

cous liquid which formed initially at the start of the mixing. The drop size and mixing time increased with increasing Reynolds number. k_m was correlated as:

$$k_m = N\theta_m = 71.7 \mathrm{Re}^{0.549} \qquad (5.95)$$

Mixing was complete when these drops were no longer present. The mixing time data for the three regimes are shown in Fig. 5.28. The mixing of low-viscosity miscible liquids has also been studied by Zlokarnik (1970). Cable and Hakin (1972) have also studied mixing of miscible liquids of similar volumes as discussed in the next section.

Studies using other mixing indices and effects of fluid properties

Mixing times are useful as a measure of mixing and are related to the circulation time and fluid mechanics occurring in a mixer. However, there are a whole host of other mixing indices which have been used to measure mixing as well. Summaries and discussion of these are available. Most are not related to the fluid mechanics and have not been emphasized in this chapter. Few mixing studies actually calculate the mixing indices by modeling the flow field except for the simplest cases. Although k_m has been well studied as a mixing index, k_m is hardly a universal mixing index. In fact, there is no single applicable mixing index, and the choice of which mixing index to use in a particular study is dependent upon the mixing objective to be accomplished.

Mixing indices are specific to processes and to the specific mixing task.

One of the most important requirements in the application of mixing indices is to have the appropriate scale of examination or scrutiny. The time required for the completion of mixing is very much a matter of scale of scrutiny. Many mixtures appear to be mixed macroscopically but unmixed microscopically.

Another important consideration is the desired level of mixing. Mixing times are typically expressed absolutely or as a percentage (i.e., 75, 90, or 95 percent) from the final level. However, some mixing problems require a very fine level of scrutiny. pH control and homogenization of glass melts (Cable and Hakim, 1972) are examples which require higher levels of examination. In such processes, mixing may be difficult to obtain and to analyze. In application, the scale of examination and the level of mixing need to be clearly stated and understood. Many mixing indices have been misapplied when these conditions have not been stated.

Intensity of segregation. Danckwerts (1951) developed the intensity and scale of segregation as mixing indices which have been used by many in the study of mixing. The dimensionless intensity of segregation, given by Lacey (1954), is defined by:

$$I = \frac{S^2 - S_R^2}{S_{max}^2 - S_R^2} \tag{5.96}$$

and is a measure of diffusion occurring during mixing. The scale of segregation is the size measure of the unmixed material.

Cable and Hakim (1972), in quantitative study of homogenizing glass at 1200 to 1600°C, reported intensity of segregation data, noting the need to have the silica content in glass to within $\pm$ 0.07 percent. The mixing of equal volumes of different density glass melts showed an exponential decrease in the intensity of segregation (1 to 0.001) with time and a definite density effect due to the initial placement of the heavier glass. Density driven flow improved the mixing when the higher-density glass was placed on the top. Changes in the slope of the intensity of segregation data with time were also observed and indicated changes in mixing mechanisms, possibly the disruption of the interface and the gross mixing of the two distinct glass melts. Such abrupt changes in slope caused difficulty in the prediction of mixing quality. The intensity of segregation after 1200 rev was also found to be a function of impeller rotational speed, decreasing as impeller rotational speed increased. Density and viscosity data were not available but the impeller Reynolds number was estimated to be about 0.1.

Sample variance. Hall and Godfrey (1968) performed a study of the mixing performance of a sigma blade mixer using a sample variance mixing index for Newtonian (corn syrup) and non-Newtonian fluids (methyl cellulose). The sigma mixer has close clearance, fast- and slow-moving blades in separate compartments. The variance index, from sampling and conductivity measurements, was found to decrease exponentially (i.e., a first-order rate process) with the number of revolutions. However, changes in the slope of the data curves of the mixing index were noted. Rotational speed had no effect on the mixing index, and higher concentrations of methyl cellulose in solution caused a reduction in mixing time. For corn syrup, no effect of viscosity on the mixing index was found. In comparison to the Newtonian fluids, the methyl cellulose solution mixed more slowly. Generally, initial mixing rates increased and final mixing rates decreased with increasing non-Newtonian behavior of the fluid which was attributed to the elastic nature of the fluid.

Visual observations of the mixing suggested that the transfer from blade compartment to blade compartment was very important to mixing, transfer being a form of redistribution and reorientation. This transfer was characterized using a transfer index, which was found to be a first-order rate process and dependent upon the number of revolutions (Weydanz, 1960). Differences in this transfer rate between two different mixers were noted although the mixer with the best transfer rate between compartments had poor mixing inside the compartments. Mixing rate and transfer rate decreased as the volume of the batch increased. Mixing rate was found to be dependent upon the nature of both the primary and secondary flows. Studies by Earle (1957) and Michaels and Puzinauskas (1954) on sigma mixers and pronged blades are also worth noting.

Other mixing indices. Bourne (1964) and Schofield (1974) have provided very lengthy discussions of mixing indices and measures of mixture quality for mixing of powders, paste, and non-Newtonian fluids.

Viscosity and rheology. Generally the experimental results and extensive mixing concepts indicate that k_m should be a constant for good mixing performance where the constant is a function of geometry and not of fluid properties in laminar mixing. These results were obtained from tracer studies alone where one phase was present. In the transition regime, k_m is likely to be a function of viscosity.

Mixing of two miscible phases having different fluid properties and volume ratios is not as simple as the ratio of interfacial areas would indicate, given the extensive mixing concepts discussed previously in this chapter. Under such mixing, tracking of the interface is required.

Murakami et al. (1972b) studied the effect of viscosity ratios on striation thickness in a cone rotor mixer where the materials experienced a uniform shear. Viscosity differences between two miscible liquids, where $\mu_d/\mu_c > 3$ or $\mu_d/\mu_c < 0.6$ showed uneven accumulation of material in streaks and roll vortices which prevented good mixing performance. Such behavior was noted in work by Takahashi et al. (1985) discussed above.

Fluid elasticity (Ford et al., 1972) affects mixing times through the generation of secondary flows as indicated by Smith (1970). Mixing of viscoelastic fluids will be discussed in the next section. Electrolytes (Ford et al., 1972) used in mixing time studies can have a significant effect on rheological properties which may affect mixing times. Chavan and Ulbrecht (1973) have studied the effect of rheology on circulation times where mixing times can be considered approximately three times the circulation times. As discussed above, the shear thinning nature of a fluid for n between 0.28 and 1 had no effect on circulation times. Cheng (1979) has discussed the various types of viscosities which need to be considered and their effects on various aspects of the flow fields and mixing. Ford et al. (1976) found that mixing times were strongly dependent upon the viscoelasticity of the fluid but fairly independent of the shear thinning nature of the fluid for mixing of polymer solutions using a helical screw impeller centrally positioned in a draft tube. Circulation times depended upon the viscoelasticity of the fluid.

Density and temperature.   Cable and Hakim (1972) and Takahashi et al. (1985) have studied effects of density. Obviously, denser material should be placed on the top of less dense material. Temperature affects significantly the viscosity of the fluid and the fluid mechanics. As a result, mixing energy and efficiency indices are in many cases very dependent upon temperature. Viscous dissipation is high for viscous fluids and affects mixing as noted by Jensen and Talton (1965). Ford et al. (1972) noted that thermal tracers cannot be used to obtain mixing times if temperature differences are large and viscosity is highly dependent upon temperature. These temperature effects have implications for processes which have highly endothermic and exothermic chemical reactions. In fact, any tracer method, used to obtain mixing times, should be investigated for the effect of the tracer method upon the viscosity and flow behavior of the fluid.

Viscoelastic Fluids

Viscoelastic fluids have a behavior different from other non-Newtonian fluids due to the combined properties of elasticity and high viscosity.

This fluid type has been well studied although further work is still needed in mixing of viscoelastic fluids. This summary discusses experimentally observed flow patterns, circulation times, and mixing times. Another excellent review on the same subject is by Ulbrecht and Carreau (1985).

The rheology of viscoelastic fluids has been discussed in Chap. 3, and considerable literature on the rheology of viscoelastic fluids is available. Mashelkar et al. (1972) have reported on the determination of material parameters of viscoelastic fluids; Bogue and Doughty (1966) on constitutive equations for viscoelastic fluids; Metzner et al. (1966) on constitutive equations for short deformation times, rapidly changing flows, and the significance of the Deborah number; and Hamersma et al. (1983) on a three-parameter shear stress-shear rate model for viscoelastic polymer solutions.

Some of the literature on the flow of viscoelastic fluids include: Denn and Porteous (1971) on a dimensionless quantity which characterized the influence of elasticity in a number of flows and which reduced to the Weissenberg number, Deborah number, or the elasticity number depending upon the flow; Pearson and Middleman (1977) on an elongational viscoelastic flow behavior during bubble collapse; and Ultman and Denn (1971) on slow motion flow past submerged objects. This particular article has bearing on the behavior of viscoelastic flow around impeller blades.

When faced with a mixing problem, non-Newtonian elastic properties are difficult to determine and require measurement. Typically such properties are not available during design. Viscoelastic effects can have substantial negative influence on mixing performance. It should also be clear that further research is necessary to establish a reasonably optimum impeller type for mixing of viscoelastic fluids and that the basic understanding of the interactions between impeller design and viscoelastic behavior requires further study.

The lack of computer simulations of flow fields involving viscoelastic fluids is because of the combined elastic and viscous phenomena, the complex agitator geometries and the three-dimensional secondary motions established because of the elastic nature. Modeling of the flow of viscoelastic fluids requires three-dimensional flow field modeling which has not been accomplished as yet.

Experimentally observed flow patterns and modeling of the phenomena

Most of the modeling efforts have been analytical studies of simple agitator geometries like spheres and disks, and most of these studies have had little to do with studies of mixing. However, analogies have been drawn to more complex agitator geometries with reasonable success. It is also apparent from the literature that the most optimum impeller for mixing viscoelastic fluids has not been found.

Typically, viscoelastic fluids exhibit non-Newtonian behavior, normal stresses, elastic recoil, and stress overshoot phenomena which can have important effects in mixing (Ide and White, 1974). The mechanical nature of the flow fields creates mechanically different behavior in viscoelastic fluid than other fluids. In a true sense, the viscoelastic fluid changes structurally as it enters the flow field of the agitator and has properties different from the viscoelastic fluid away from the impeller. It should also be noted that flow of viscoelastic fluids into the impeller may not be steady or smooth in appearance. Choatic jerking (Ulbrecht, 1974) often occurs which indicates the unsteady nature of the rheology and shear levels which may lead to difficulty in modeling. A viscoelastic fluid may be quite heterogeneous structurally after formulation which is an important issue by itself.

Two important properties of viscoelastic fluids are the shear dependent viscosity and the first normal stress difference coefficient. Although various models can be used to describe these properties, power laws are typically assumed. The shear dependent viscosity or apparent viscosity which was discussed above is:

$$\mu_a = K(\gamma)^{n-1} \tag{5.97}$$

The first normal stress difference coefficient is modeled as:

$$\sigma_1 = \frac{\tau_{11} - \tau_{22}}{\gamma^2} = h(\gamma)^{m-2} \tag{5.98}$$

where n, m, h, and K are parameters of the curve fit. Since viscosity and the first normal stress difference are functions of shear rate, they are distributed throughout the agitated tank as the shear rates are distributed. Typical inelastic and viscoelastic rheological data are shown in Figs. 5.29 and 5.30.

These rheological properties can be measured using a Weissenberg rheogoniometer in which a small liquid sample is placed between a plate and cone. The torque T and axial thrust F are measured as the fluid is sheared. The torque is related to the apparent viscosity as:

$$\mu_a = \frac{3T}{2\pi R^3 \gamma} \tag{5.99}$$

and the primary normal stress coefficient as:

$$\sigma_1 = \frac{2F}{\pi R^2 \gamma^2} \tag{5.100}$$

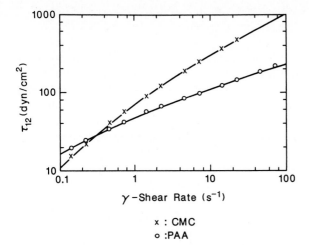

x : CMC
o :PAA

Figure 5.29   Typical shear stress—shear rate data for inelastic and viscoelastic liquids: CMC and PAA. *(From J. V. Kelkar, R. A. Mashelkar, and J. J. Ulbrecht, Trans. Instn. Chem. Engrs., 50, 343, 1972. By permission.)*

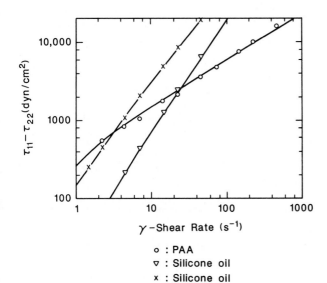

o : PAA
∇ : Silicone oil
x : Silicone oil

Figure 5.30   Typical normal stress—shear rate data for silicone fluids and aqueous PAA solutions. *(From J. V. Kelkar, R. A. Mashelkar, and J. J. Ulbrecht, Trans. Instn. Chem. Engrs., 50, 343, 1972. By permission.)*

where γ (= ω/α) is the shear rate in the conical gap; ω, the angular velocity of rotation; R, the radius of cone and plate; and α, the gap angle.

Viscoelastic fluids may be structurally different from point to point due to the initial formulation of the material (Hocker et al., 1981), the history of the material, or other reasons. As a result, the mixing of a viscoelastic fluid is similar to mixing of a miscible nonhomogeneous multiphase material having no distinct interfaces and significant variations in structural properties. Such phenomena may require kinetic modeling of the structural behavior (Denny and Brodkey, 1962; Kim and Brodkey, 1968; Jachimiak et al., 1974). This review will not consider the problem of the initial formulation of viscoelastic fluids which is actually a mixing problem.

One important phenomenon to understanding viscoelastic flow behavior is elasticity. Elasticity is a property of the material through which the shape or structure of the material is maintained or, if the material is deformed, will cause the material to return to its original shape. As a result, as a viscoelastic fluid flows, elasticity of the material attempts to maintain its shape and remain unmixed. If the deformation rates are high, the shape of the material will change. At the same time, however, the elasticity of the material generates forces which move the material in such a manner as to return it to its original shape. In flows in an agitated tank near the impeller, the fluid is typically stretched circumferentially. As a result, the fluid attempts to return to its original shape by moving to lower radii because of the elasticity and the normal forces generated by the strain rate. Material in viscoelastic flows which have high elasticity seek lower radii generally in flow situations. This has important implications for flows in agitated tanks, particularly for vertical circulation and the various vortex systems which occur. Viscoelastic fluids will also climb an impeller shaft in seeking lower radii, which is known as the Weissenberg effect.

Generally, shear stresses and the kinematics of the flow for viscoelastic fluids always create normal forces which act normal to the shear surfaces, producing secondary flow patterns. This also has implications to scale up: the need to maintain kinematic similarity as well as geometric similarity. Elastic forces which cause material to seek lower radii are countered by centrifugal forces which expel fluid outward. As a result, in certain cases the centrifugal forces dominate; in other cases elastic forces dominate. In still other cases, a balance occurs between the two.

The extent to which elasticity affects the flow field depends upon the relative magnitudes of the shear rates γ; the apparent viscosity μ_a; first normal stress difference σ_1; and the shear and normal stresses

(Chavan, Ford, and Arumugam, 1975). The first normal stress difference and apparent viscosity can give rise to a natural time for the
fluid given as:

$$\theta_n = \frac{\sigma_1}{\mu} = \frac{(\tau_{11} - \tau_{22})/\gamma^2}{\tau_{12}/\gamma} = \frac{h(\gamma)^{m-2}}{k(\gamma)^{n-1}} \tag{5.101}$$

$$= \left(\frac{h}{k}\right)\gamma^{m-n-1} \tag{5.102}$$

The degree of viscoelasticity is related to the shear thinning nature of
the solution. Elasticity becomes important at low apparent viscosity.

Spheres and turbulent agitators.   Flow patterns have been reported by
Giesekus (1965a; 1965b) and Walters and Savins (1965). Of particular
interest to mixing is work by Vermura et al. (1967), Kelkar et al.
(1972), and Ide and White (1974). While primarily studying power
consumption for viscoelastic fluids, Kelkar et al. (1972) have given an
excellent summary of flow patterns. The power consumption correlations have been discussed in Chap. 3.

Ide and White (1974) discussed flow patterns generated by a sphere,
a disk, and turbulent agitators including a flat blade turbine and propeller using various polystyrene-styrene solutions in which the concentration of polystyrene was varied. The study illustrated how
rheological properties of the fluid can be used to predict flow field behavior. A model was presented for a second-order fluid, and the solutions of the equations of motion were given which described the fluid
mechanics of flow around a sphere, following work by Bickley (1938),
Giesekus (1965a; 1965b), Thomas and Walters (1964), and Langlois
(1963). The rheological properties appeared in the solution, and the
relative magnitude of the properties determined the flow pattern. The
Reynolds number Re or $\rho NR^2/\mu$, the Weissenberg number Wi or $\sigma_1 N/\mu$,
and the viscoelastic ratio number VR or σ_2/σ_1 are of interest where
σ_2 is the second normal stress coefficient and where the mechanical
properties μ, σ_1, and σ_2 were evaluated at zero shear conditions. Typically, the second normal stress difference is not large in comparison
to the first normal stress difference. As a result, the secondary normal
stress coefficient σ_2 and the viscoelastic ratio number are relatively
unimportant.

The solution provided by Ide and White (1974) for flow around a rotating sphere contained the Aberystwyth number Ab, defined as (Wi/
Re) (1 − VR) (White et al., 1977), which subsequently controlled the

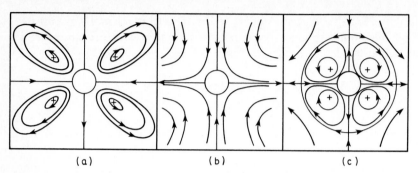

(a)   (b)   (c)

Figure 5.31   Secondary flow phenomena around a sphere rotating in a viscoelastic liquid. (a) Normal forces dominating; (b) centrifugal forces dominating; and (c) normal forces and centrifugal forces comparable. (From J. V. Kelkar, R. A. Mashelkar, and J. J. Ulbrecht, Trans. Instn. Chem. Engrs., 50, 343, 1972. By permission.)

secondary flow patterns as shown in Fig. 5.31. When Ab was less than $\frac{1}{12}$, centrifugal forces controlled the overall flow pattern. The flow was expelled at the sphere's equator and moved in at the poles. When Ab was greater than $\frac{1}{4}$, the flow was dominated by the normal stresses. Under these conditions, the viscoelastic fluid moved in at the equator and out at the poles. The material sought lower radii during this flow. Toroidal ring vortices occurred around the sphere at latitudes of 45 and 135°. When Ab was between $\frac{1}{12}$ and $\frac{1}{4}$, the centrifugal and the normal forces were balanced and both flow patterns occurred. The flow caused by the normal forces (i.e., the ring vortices and the flow in at the equator) occurred near the sphere. The centrifugal or inertial flow occurred farther away from the sphere with very little mixing between these two zones.

Spheres are not used for mixing, however. Ide and White (1974) and White et al. (1977) performed studies of disks, turbines, anchor, propellers, and multiple agitators. For the disk and turbine impellers, the flow patterns were qualitatively similar to that of the sphere for the same rheological fluid state. The analogy between a sphere and more complex impeller geometries was reasonably successful as a result. Three flow regimes were again founded for the different impeller configurations: (1) centrifugal dominated flow field at low polymer concentrations, (2) vortex patterns and segregated regions at intermediate concentrations, and (3) normal stress dominated flow at high polymer concentrations. For multiple spheres, disks, and turbine impellers, the basic flow pattern, as determined by the magnitude of the Ab number, was just repeated.

Toroidal vortices with closed streamlines existed in the bulk of the tank, Fig. 5.31a, which prolonged mixing times while reducing circulation times. The flow pattern, where both centrifugal and normal forces were balanced, created segregated zonal flow patterns with closed streamlines as shown in Fig. 5.31c. Both flow patterns lead to extremely long mixing times if mixing between the segregated zones is required.

For the propeller, which usually is considered an axial flow impeller, Ide and White (1974) found that the effect of the normal forces was to reinforce this axial flow. Segregated flow patterns were not observed. Axial flow impellers never have a dominant centrifugal force except at high D/T ratios, and, as a result, the appearance of centrifugal force component for viscoelastic fluids is not expected.

For polymerization reactors, segregated flows may occur as the polymerization proceeds if radial discharge impellers are used (Ide and White, 1974). Furthermore, flow patterns will probably be a function of the extent of reaction for such radial flow impellers, and segregation may occur as the reaction proceeds (Ide and White, 1974). Segregation is unlikely if axial flow impellers are used.

Hocker et al. (1981) in power studies of different impellers in viscoelastic fluids noted that the orientation of the flow field was important and should be reported in flow studies. Obviously flow direction is important in the flow of viscoelastic fluids. In their study, high concentrations of polymeric molecules had considerable cross linking between the molecular chains. During flow conditions, the cross linking was disrupted by the shear, and a reduction of the apparent viscosity of the solution occurred. Some energy from the shear stress was stored in the coiled molecules while aligning in the flow direction and was returned to the flow when the shear stress was relaxed. In their study, Hocker et al. noted that polymer aggromerates were formed, which made it difficult to establish a laminar flow pattern which had a describable order.

In comparing the two types of studies, i.e., Ide and White (1974) and Hocker et al. (1981), completely different types of flow patterns can result with viscoelastic fluids. The flow patterns can be very ordered or they can be very disordered depending upon the nature, past history, and formulation of the viscoelastic material.

Ranade and Ulbrecht (1978) noted flow reversals in the discharge flow of the flat blade impeller. Greene and Carpenter (1982) have studied the flow patterns generated by a pitched blade impeller which is an axial flow impeller and typically has a discharge angle of 30° from the vertical. Greene et al. found, however, that the discharge angle increased to 90° as a function of viscoelasticity for PAA solutions. Data were given for the angle of discharge as a function of the Weissenberg number, the elasticity number, and concentration of PAA in solution. The change in the flow field was postulated to be a function of the high fluid elongational rates which occurred as the fluid moved through the impeller. The elongational viscosity generated by PAA resisted axial flow accelerations. High values of the axial velocity gradient $\partial V_z/\partial z$ caused the flow field to have higher discharge angles.

Anchor and gate impellers. The study by Peters and Smith (1967) with a PAA solution has already been noted above. Although the extent of the elasticity of the PAA solution was not given, the viscoelastic nature of the PAA solution caused the considerable axial flow along the blade length of the anchor. Ironically, homogenization of viscoelastic fluids may be possible because of this phenomena. In the flow, PAA was stretched by the flow past the blade and in the trailing vortex. To gain their original size, the PAA molecules had the tendency to flow inward toward the center of the vortex which, in turn, generated the axial flow in the vortex. Smith (1970) also noted that it was easier to establish the toroidal loops for the viscoelastic fluids than for viscous fluids due to the same reason. This effect has a negative impact upon mixing times as noted above by Ide and White (1974).

The flow pattern given for the anchor impeller by White et al. (1977) did not indicate tank size. Apparently, the wall clearance was very large, in which case, the flow behavior of the anchor was similar to a radial discharge impeller. Peters and Smith (1967; 1969) and Smith (1970) have reported flow patterns for a close-clearance anchor and emphasized the blade-bound vortices as an important mixing flow pattern.

Helical screw impellers. As in the case of fluids with simpler rheological characteristics, the flow patterns generated by this impeller for viscoelastic fluids are quite complex and require study. Blade tip vortex systems and flow around the entrance and exit of the draft tubes should be investigated.

Helical ribbon impellers. Carreau, Patterson, and Yap (1976) have discussed the flow patterns for the helical ribbon impeller for viscoelastic fluids. The flow patterns for viscous shear thinning fluids have already been discussed above. For all viscoelastic fluids studied, the main fluid motion in the tank was tangential with very little axial flow. The central core region behaved as in solid body rotation, and large deformations were observed in the gap between the blade and wall for the viscoelastic PAA solution. Maximum tangential velocity occurred at a radius of about 0.7 to 0.8 of the tank diameter. The fluid, having the highest tangential velocity, was the viscoelastic fluid (PAA) followed by the slightly viscoelastic fluid (CMC or carboxymethylcellulose). Viscoelasticity tended to dampen any veloc-

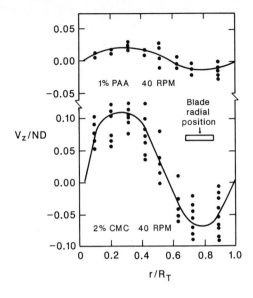

Figure 5.32 Axial velocity profiles in 2 percent CMC and 1 percent Separan. *(From P. J. Carreau, I. Patterson, and C. Y. Yap, Canad. J. Chem. Eng., 54, 135, 1976. By permission.)*

ity fluctuations, and the magnitude of the axial velocity was greatly reduced due to the elastic nature of the fluid. Axial velocities for the helical ribbon impeller are shown in Fig. 5.32 for a viscoelastic PAA solution and for slightly viscoelastic CMC solution.

Circulation times and pumping capacities

The property of elasticity affects mixing and circulation times in a different manner than has been discussed so far for other fluid types. The major division in impeller-tank geometries can be based upon whether geometric restrictions exist in the circulation of the fluid. Due to the elastic properties of viscoelastic fluids, stationary toroidal vortices are often present in the flow in the vicinity of restrictions in the flow. For sudden expansions or contractions, viscoelastic fluids also have higher pressure drops than other fluid types (Sylverster and Rosen, 1970). High stress levels cause the viscoelastic fluid to seek regions having lower radii, which leads to standing vortices and axial flow through the vortex, a reduction in area available to fluid flow, lower pumping capacities, and higher circulation times for some geometries. Mixing times may also increase because diffusion into these standing vortices become the rate-controlling step in the mixing. Such regions can have closed streamlines which implies diffusional mixing. However, if these closed regions are periodically disturbed, as might be generated in planetary or up-down mixers, then mixing times can be substantially lowered in many cases.

Turbines.   Given the possibility of segregated flow patterns as demonstrated by Ide and White (1974), turbines are not recommended for mixing viscoelastic fluids. Since the stream functions for these impellers have been given and resemble those of a sphere, flow rates and pumping capacities can be calculated.

Propellers.   Ide and White (1974) indicated that the vertical circulation pattern was enhanced by the normal forces for propellers. This was for an open propeller with no geometric restrictions in the vicinity of the impeller.

Helical screw impellers.   Chavan and Ulbrecht (1973) found that circulation numbers, k_Q or Q/ND^3, for some geometric arrangements of helical screws with draft tubes were reduced by 50 percent of that of inelastic fluids due to the effects of sudden contractions and expansions. Such effects were very apparent when the clearance between the draft tube and tank bottom was less than 10 percent of the tank diameter (Ulbrecht, 1974). Ford (Ulbrecht, 1974) correlated the effect of viscoelasticity upon circulation times as:

$$k_{Q,v} = \frac{k_{Q,i}}{(1 + \text{Wi})^{0.3}} \tag{5.103}$$

where $k_{Q,i}$ is obtained in the same mixer using an inelastic fluid. The Weissenberg number in the equation varied in the range from 0.1 to 7 and was based upon zero shear viscosity and the zero shear normal stress coefficient. To include effects of geometry, Chavan, Ford, and Ulbrecht (Ulbrecht, 1974) correlated $k_{Q,v}$ as a function of geometry and $k_{Q,i}$ for power law parameters varying $0.18 < n < 0.86$, $0.53 < m < 0.73$, $0.8 < K < 500$, and $120 < h < 1000$. K and h have units of dynes, centimeters, and seconds necessary for the curve fit of the rheological data.

Chavan, Ford, and Arumugan (1975) for helical screw impellers found that elasticity damped axial circulation for viscoelastic fluids and reduced circulation numbers. However, elasticity had little effect on power consumption in the laminar region. Chavan et al. (1975) repeated the analysis of Seichter (1971; 1981) and Chavan and Ulbrecht (1973) for the pumping capacity of a helical screw impeller in a viscoelastic fluid and arrived at the following:

$$\frac{k_{Q,i} - k_{Q,v}}{k_{Q,v}} = R_1(R_2)^{m-n-1} \tag{5.104}$$

where R_1 and R_2 were related to geometry and flow conditions.

$(k_{Q,i} - k_{Q,v})/k_{Q,v}$ data for this relationship spanned from 2 to 0.2 for $0.4 < m - n - 1 < 0.8$. The circulation numbers for elastic fluids varied between 0.33 and 0.86 of the circulation numbers for inelastic fluids.

The effects of viscoelasticity on circulation time have been correlated by Ford and Ulbrecht (1976) as:

$$N\theta_c\left(\frac{\mu_0}{\mu_a}\right)^{-0.3}(1 + 0.45\text{Wi})^{-0.3} = K' \qquad (5.105)$$

where K' is a constant. As the equation shows, circulation times were affected by both the shear thinning and elastic properties of the viscoelastic fluids. Typically, circulation times were not dependent upon shear thinning properties for inelastic fluids. The effects of elasticity for a helical screw impeller in a draft tube appeared at the inlet and outlet of the draft tube. The helix of the screw may also retain material in secondary vortices on the outer rim of the screw.

Helical ribbon impellers. Carreau, Patterson, and Yap (1976) have presented circulation time data for the three fluid types. For the viscoelastic fluid (PAA), the particles, used to obtain circulation times, tended to be trapped in the $R - \theta$ plane with very low axial velocity as indicated in Fig. 5.32. As a result, circulation numbers were not obtained for the PAA solutions except for one geometry where $k_{Q,v} = 0.014$. The circulation numbers for the slightly viscoelastic CMC (carboxymethycellulose) varied between 0.047 and 0.098 depending upon the helical impeller geometry.

Helical ribbon screw impellers. Chavan, Arumugam, and Ulbrecht (1975) found that the elastic properties of the viscoelastic fluids significantly reduced the axial circulation of the fluid and increased the angular circulation to the point that the viscoelastic fluid behaved as a solid body. In one extreme case, the angular circulation equaled the impeller rotational speed. The circulation times were increased by two to six times that for inelastic fluids. However, the data showed a constant k_c value for the viscoelastic fluids in laminar mixing. The flow regimes cited above by White et al. did not occur for the helical ribbon. The average axial circulation and mixing times were unaffected by the shear thinning properties. Generally, it took much longer to circulate viscoelastic fluids than inelastic fluids.

Mixing times

Generally, mixing times for viscoelastic fluids are larger than inelastic liquids.

Turbulent impellers. Collias and Prud'homme (1985) found that the mixing of viscoelastic fluids with a disk style turbine was very difficult. k_m values were on the order of 200 to 500. The fluid above the disk remained almost completely segregated from the fluid below the disk. Their results indicated that turbulent impellers such as the disk style turbine should not be used in mixing viscoelastic fluids.

Portner et al. (1988) measured mixing times for highly viscoelastic PAA solutions (0.01 to 0.5 percent) using turbulent type impellers: a paddle impeller, an Intermig 06, and a propeller in a draft tube. Portner et al. indicated that, when the ratio of the primary normal stress difference to the shear stress was greater than 10, the viscoelastic fluid developed a shear thickening nature under shear stress. The $N\theta_m$ data for the PAA solutions, e.g., 10 to 1000, were much higher than the other fluids studied at the same Reynolds number and the data curves for PAA solutions above 0.1 percent PAA showed local minimums and maximums as a function of the PAA concentration. This behavior was attributed to $\tau_{11} - \tau_{22}$ being 10 times greater than τ_{21}. Of the impellers studied, the paddle impeller and propellers performed the best in mixing the PAA solutions. Portner et al. (1988) recommended that $N\theta_m$ values be correlated as:

$$N\theta_m = f\left(\frac{N_{\text{Re}}}{1 + C\text{Wi}}\right) \tag{5.106}$$

for viscoelastic fluids.

Anchor impellers. Peters and Smith (1967; 1969) reported increases of 10 to 25 percent in mixing times for the anchor impeller when the clearance was reduced by four times for PAA. Mixing time data (Peters and Smith, 1969) were plotted as a function of impeller rotational speed showing a minimum of k_m of about 40 to 60.

Helical screw impellers. Ford (Ulbrecht, 1974) and Ford and Ulbrecht (1976) found that toroidal vortices occurred in the flow and caused increases in mixing times. Mixing times of viscoelastic fluids were correlated as:

$$\theta_{m,v} = \theta_{m,i}(1 + 0.45\text{Wi})^{0.8} \tag{5.107}$$

up to a Reynolds number of 10^3. The data showed that the influence of elasticity upon mixing times was not linear. This occurred when the normal stresses became significant and the Weissenberg number increased. Ford (Ulbrecht, 1974) also found that the blending of different volumes of viscoelastic fluids, having different polymer concentrations, was very difficult. k_m data were on the order of 1000 in some cases. For blending of viscoelastic fluids with polymer concentrations of not more than 2 to 1, Ford (Ulbrecht, 1974) found:

$$[N\theta_{m,v}(1 + 0.45\text{Wi})^{-0.8}] = 200 \qquad (5.108)$$

which decreased to 80 as turbulent conditions were approached. For the ratio of the mixing time to circulation time, Ford (Ulbrecht, 1974) found:

$$\frac{\theta_m}{\theta_c} = 7 \qquad N_{\text{Re}} < 10 \qquad (5.109)$$

and

$$\frac{\theta_m}{\theta_c} = 3.5 \qquad N_{\text{Re}} = 1000 \qquad (5.110)$$

At a fixed Reynolds number, θ_m/θ_c varied with $(1 + 0.45\text{Wi})^{0.5}$. Ford (Ulbrecht, 1974) also found that shear thinning properties of the viscoelastic fluids had negligible influence upon mixing times. Chavan, Ford, and Arumugan (1975) also provided a rather complicated discussion of mixing times for viscoelastic fluids based upon the pumping capacity model for helical screw impellers (Seichter, 1971 and 1981; Chavan and Ulbrecht, 1973) by relating the total shear for one circulation to the creation of new surface area in one circulation. The analysis was dependent upon functions of geometry and rheology.

Helical ribbon impellers. Carreau, Patterson, and Yap (1976) were unable to report any θ_m/θ_c for their viscoelastic PAA solutions because of the flow behavior of the PAA solutions and the lack of reproducible data. However, for the slightly viscoelastic CMC solution, θ_m/θ_c was found to be on the order of 4.27 to 9.4 depending upon the geometry of the helical ribbon impeller. Mixing times for the PAA solution studied were approximately five times those for Newtonian fluids as indicated by data in Fig. 5.33.

Helical ribbon screw impellers. Chavan, Arumugam, and Ulbrecht (1975) found that k_m was constant for viscoelastic fluids. However, it took twice as long to mix viscoelastic fluids than inelastic fluids. Although there were no stagnant regions in the tank, recirculating vortices were observed and were probably the cause of the long mixing times.

Geometric effects

Very few specific geometric studies have been done in comparing mixing times for viscous inelastic and viscoelastic fluids.

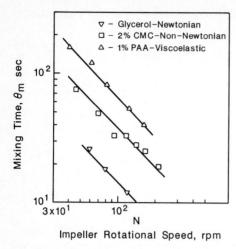

Figure 5.33 Mixing time versus impeller rotational speed, revolutions per minute, for various liquids. *(From P. J. Carreau, I. Patterson, and C. Y. Yap, Canad. J. Chem. Eng., 54, 135, 1976. By permission.)*

Helical ribbon impellers. For the helical ribbon impeller pumping downward at the wall, Carreau, Patterson, and Yap (1976) studied a number of geometric effects and their effect on mixing times for three different types of fluids: viscous (glycerol), CMC (slightly viscoelastic), and PAA (viscoelastic). Doubling the blade width-diameter ratio W/D, Carreau et al. found that the mixing time decreased by 100 percent for glycerol and CMC. For the PAA solutions, mixing times decreased by only 20 percent. An increase in pitch by 45 percent resulted in a slight increase in mixing time. However, at lower rotational speeds, an increase in pitch lowered mixing time. Doubling the clearance between the wall and ribbon caused only a slight increase in mixing times for glycerol, 50 to 100 percent increase for the CMC solution and no change for the PAA. There was a 20 percent increase in mixing times for a single-ribbon impeller over that of a double-ribbon impeller for all three fluids. Circular tubing, used in place of a flat ribbon, caused an increase in mixing times by 30 to 70 percent. The direction of rotation for the ribbon impeller appeared to have a substantial effect upon the flow. In a pumping-upward mode at the wall, axial circulation rates were lower for all three fluid types, and large unmixed zones were formed in the central core. These effects were given without any statements as to the details of the flow regime, and the nature of the secondary flows was not discussed. Nagata et al. (1956) had recommended that the helical ribbon impeller pump upward at the wall as discussed earlier.

Shear thinning, viscoelastic, Bingham plastic fluids

Solomon et al. (1981) performed a study of shear thinning, viscoelastic, Bingham plastic materials, Xanthan gum and Carbopol, and shear thinning, viscoelastic material, CMC (carboxymethyl-cellulose), in an agitated tank using turbulent type agitators. This followed the study by Wichterle and Wein (1981) of shear thinning, plastic suspensions. Given the high-yield stress and the use of turbulent agitators, local highly mixed regions, called caverns, were formed around the impellers. The rest of the tank remained fairly stagnant, resulting in the tank containing both laminar and turbulent flow. Both Wichterle and Wein (1981) and Solomon et al. (1981) generated relationships to determine cavern size. Solomon et al. measured the cavern size as a function of the impeller Reynolds number using a hot-film anemometer and visual observation. Cavern sizes were predicted at known yield stresses. Stagnant regions below the impeller and behind baffles existed in the tank even at the highest impeller rotational speeds. Comments and cavern sizes were also reported for aerated tank conditions by Solomon et al. (1981).

Other Articles of Interest

Articles of general interest discuss mixing from a practical viewpoint. Richards (1961), in a review of mixing methods for non-Newtonian fluids, provided several design examples. Bourne (1964) provided a discussion of mixing of powders, pastes, and non-Newtonian fluids. Industrial problems, a general review, and recommendations for further work were also given. Flagg et al. (1977) discussed the economics of agitator design and provided a case study. Muschick (1988) discussed the mixing of glass melts by stirrers at high temperatures.

Khang and Fitzgerald (1975) developed a probe which permitted the measurement of multiple mixing times in the same tank. Khang and Levenspiel (1976) provided correlations and example problems for the determination of mixing times in the transition and turbulent regimes. Other such articles treating transition and turbulent mixing can be found in Chap. 4. Bourne et al. (1976) discussed a photochromic tracer system for residence time measurement of highly viscous liquids. Quraishi et al. (1977) have studied the influence of drag reducing additives on mixing and solid suspension which provide results useful in the transition regime. Ottino and Macosko (1980) discussed a mixing efficiency index. Nienow and Kuboi (1984) studied mixing rates between circulation zones using a flow marker technique and established impeller pumping rates and circulation times for a dual impeller system.

Further Research Needs

The majority of studies of mixing times have primarily been studies of geometry. Specific flow mechanisms which cause the mixing are generally not known. The effects of wall clearance on circulation and mixing times have been studied by Peters and Smith (1969), Bourne et al. (1979), Kappel (1979), and Takahashi et al. (1982). The results, however, tend to be inclusive. For mixing of two nearly equal volumes, effects of the interface and density driven flow warrant further study although Takahashi et al. (1985) have provided an initial study. Effects of viscosity (Murakami et al., 1972b) and other rheological properties (Ford et al., 1972; Ford et al., 1976) need further study. The various types of viscosity (Pearson and Middleman, 1977b; Cheng, 1979) have been mentioned, but their effects are generally unknown. The effects of entrance and exit flows and their distribution in continuous processes on mixing need further study. There appears to be little intermediate analysis of mixing from the time responses of probes after tracer injections although Yuge and O'Shima (1975) have provided an initial study. Draft tube Reynolds numbers should be reported. Blade and impeller number effects are not well understood. Mixing times for planetary mixers and multiple impeller systems need study. Beckner and Smith (1966) noted that the effect of breaker bars on mixing efficiency should be studied. Different tank geometries, different internal geometries, mixing mechanisms in vortices, and effects of a secondary low-density phase such as a gas or solids on mixing times require further study. Generally, the various flow regimes and behavior occurring in mixing need to be established. The formation of the toroidal vortex systems, which were observed by Johnson (1967) to be so detrimental to rapid mixing, needs to be correlated and described using relevant dimensionless geometric groups. Combined mixing and heat transfer studies are needed especially where viscosity is a strong function of temperature. Whereas mixing efficiency studies have been performed, other types of transport studies have not been performed. Efficiency studies of which impellers provide the best heat transfer have not been performed. Studies of the effects of elasticity upon circulation and mixing times for the screw and helical ribbon impeller are needed.

Mixing behavior will be eventually computed or simulated numerically by solving the equations of motion and diffusion in three dimensions. Such solutions will be quite diverse, as can be observed in the discussion of flow patterns, and experimentation will be needed to check such solutions.

Nomenclature

A, A'	original area, new area
Ab	Aberystwyth number, (Wi/Re) $(1 - VR)$
A_1	constant

a	exponent
a_i	interfacial area in ith cell
a, **b**, **c**	vectors
B, B_1	constants
C	clearance, wall or draft tube clearance
C_b	bottom clearance
C_t	draft tube clearance
C, C_1	constants
D	impeller diameter
D_c	cavern or critical diameter
D_{dt}	draft tube diameter
$D(i, j)$	distribution matrix
D_R	helical ribbon diameter
D_S	screw diameter
D_{ss}	screw shaft diameter
E	Ekman number, $\nu/R^2\Omega$
E	energy, $P\theta$
e	eccentricity, distance from tank center
F	axial thrust of plate and cone viscometer
F, G	surfaces
F_D	shape correction factor
f_g, f_p	geometric parameters
H	liquid height
h	constant in power law for first normal stress coefficient
I	intensity of segregation
K	fluid consistency index constant in power law
K', K''	constant
k	unit vector of rotation, constant in the Metzner and Otto postulation
k_c	circulation time number, $N\theta_c$
k_Q	circulation number, Q/ND^3
$k_{Q,v}$	circulation number for viscoelastic fluids, Q/ND^3
$k_{Q,i}$	circulation number for inelastic fluids, Q/ND^3
k_m	mixing time number, $N\theta_m$
k_p	pressure change number, $\Delta p/\mu N$
l	displacement
L	length of screw
L_{dt}	draft tube length
L_v	height of more viscous phase

M	mixing performance index, $PV/\mu V'$
m	exponent in power law for first normal stress coefficient
N	impeller rotational speed
N_M	dimensionless group, $\Delta\rho L_v g/\mu_m N$
N_P	power number, $P g_c/\rho N^3 D^5$
N_{PT}	turbulent power number
N_{Re}	impeller Reynolds number, $\rho N D^2/\mu$
n	flow index exponent in power law for viscosity
n, n_R	number of ribbons or blades
P	power
P_v	power per volume
p	pitch, height rise for 360° of rotation
P_s	screw pitch
Δp	pressure drop
Q	volumetric flow rate through the impeller
Re	impeller Reynolds number, $\rho N D^2/\mu$
Re_{nN}	non-Newtonian impeller Reynolds number, $\rho N D^2/\mu_a$
Re_a	impeller Reynolds number based upon apparent viscosity
R	radius, radius of plate and cone viscometer
R_1, R_2	functions of geometry and flow conditions
R_i	inside radius of impeller
R_T	tank radius
s	shear
S^2	variance
S_R^2	residual variance of homogeneous system
$S_{\max}^2$	maximum variance of fully segregated system
T	torque, tank diameter
t	time
V	tank volume, volume of mixer
V'	processing flow rate
V_r, V_0, V_z	velocities in r, θ, and z directions
VR	viscoelastic ratio number, σ_2/σ_1
V_{tb}	tangential velocity of ribbon blade
W	blade width, screw width
W_{II}	helical blade width
W_S	screw width
Wi	Weissenberg number, $\sigma_1 N/\mu$

Greek symbols

α	directional cosines, gap angle of plate and cone viscometer, constant
β	constant
σ_1	first normal stress difference coefficient
σ_2	second normal stress difference coefficient
θ_c	circulation time
$\theta_{c,v}$	circulation time for viscoelastic fluids
$\theta_{c,i}$	circulation time for inelastic fluids
θ_m	mixing time
$\theta_{m,v}$	mixing time for viscoelastic fluids
$\theta_{m,i}$	mixing time for inelastic fluids
θ_n	natural time of fluid
θ_{75}	mixing time at 75 percent of completely mixed system
Ψ	stream function
λ	elongation ratios
ρ	density
η_p	pumping efficiency
μ	viscosity
μ_a	apparent viscosity
μ_c	viscosity, continuous phase
μ_d	viscosity, discrete phase
μ_m	viscosity, mixture
μ_0	zero shear viscosity
ν	kinematic viscosity
η	plastic viscosity
γ	shear rate
γ_{uv}	average shear rate
ϵ	Rossby number, $U/\Omega R$
Ω	angular velocity
ω	vorticity, angular velocity
τ	shear stress
τ_0	yield shear stress

References

Arai, K., K. Takahashi, and S. Saito, *J. Chem. Eng. Japan*, **15**, 383, 1982.
Beckner, J. L., and J. M. Smith, *Trans. Instn. Chem. Engrs.*, **44**, T224, 1966.
Bertrand, J., and J. P. Couderc, *Proc. 5th Eur. Conf. on Mixing, Wurzburg, Germany*, BHRA Fluid Eng. Center, Cranfield, England, 313, 1985.

Bertrand, J., and J. P. Couderc, *Fluid Mixing I. Chem. E. Sym. Series: No. 64*, Bradford, England, B1, 1981.

Bickley, W. G., *Phil. Mag.*, **25**, 746, 1938.

Bogue, D. C., and J. O. Doughty, *Ind. Eng. Chem. Funds.*, **5**, 243, 1966.

Bourne, J. R., *The Chemineer Engineer*, CE202, September 1964.

Bourne, J. R., and H. Butler, "Mixing—Theory Related to Practice," A. I. Ch. E.—I. Chem. E. Symposium No. 10, Instn. Chem. Engrs., London, 89, 1965.

Bourne, J. R., and H. Butler, *Trans. Instrn. Chem. Engrs.*, **47**, T11, 1969*a*.

Bourne, J. R., and H. Butler, *Trans. Instrn. Chem. Engrs.*, **47**, T263, 1969*b*.

Bourne, J. R., G. K. Giger, W. Richarz, and R. Riesen, *Chem. Eng. J.*, **12**, 159, 1976.

Bourne, J. R., W. Knoepfli, and R. Riesen, *Proc. 3rd. Eur. Conf. on Mixing, University of York, York, England*, BHRA Fluid Eng., Cranfield, England, A1-1, 1979.

Brodkey, R. S., *The Phenomena of Fluid Motion*, Addison-Wesley, Reading, Mass., p. 423, 1967.

Cable, M., and J. Hakim, *Chem. Eng. Sci.*, **27**, 409, 1972.

Carreau, P. J., I. Patterson, and C. Y. Yap, *Canad. J. Chem. Eng.*, **54**, 135, 1976.

Chapman, F. S., and F. A. Holland, *Trans. Instn. Chem. Engrs.*, **43**, T131, 1965.

Chavan, V. V., and J. J. Ulbrecht, *Chem. Eng. J.*, **6**, 213, 1973.

Chavan, V. V., A. K. Deysarkar, and J. Ulbrecht, *Chem. Eng. J.*, **10**, 205, 1975.

Chavan, V. V., M. Arumugam, and J. J. Ulbrecht, *AIChEJ.*, **21**, 613, 1975.

Chavan, V. V., D. E. Ford, and M. Arumugam, *Canad. J. Chem. Eng.*, **53**, 628, 1975.

Chavan, V. V., *AIChEJ.*, **29**, 177, 1983.

Cheng, D. C. -H., C. Schofield, and R. J. Janes, *Proc. 1st Eur. Conf. on Mixing and Centrifugal Sep., Churchill College, Cambridge, England*, BHRA Fluid Eng., Cranfield, England, C2-15, 1974.

Cheng, D. C. -H., *Proc. 3rd. Eur. Conf. on Mixing, University of York, York, England*, BHRA Fluid Eng., Cranfield, England, A5-73, 1979.

Collias, D. J., and R. K. Prud'homme, *Chem. Eng. Sci.*, **40**, 1495, 1985.

Coyle, C. K., H. E. Hirschland, B. J. Michel, and J. Y. Oldshue, *AIChEJ.*, **16**, 903, 1970.

Daiguji, H., and S. Kobayashi, Preprint of JSME No. 793-7,18,1979 (in Japanese).

Danckwerts, P. V., *Appl. Sci. Res.*, **A3**, 279, 1951/3.

Denn, M. M., and K. C. Porteous, *Chem. Eng. J.*, **2**, 280, 1971.

Denny, D. A., and R. S. Brodkey, *J. Appl. Phys.*, **33**, 2269, 1962.

Earle, R. L., *Trans. Instn. Chem. Engrs.*, **37**, 297, 1959.

Elson, T. P., D. J. Cheesman, and A. W. Nienow, *Chem. Eng. Sci.*, **41**, 19, 2555, 1986.

Elson, T. P., *Proc. 6th Eur. Conf. on Mixing, Pavia, Italy*, BHRA Fluid Eng., Cranfield, England, 485, 1988.

Erwin, L., *Polymer Engng. & Sci.*, **18**, 738, 1978.

Erwin, L., *Polymer Engng. & Sci.*, **18**, 1044, 1978.

Etchells, A. W., W. N. Ford, and D. G. R. Short, *Fluid Mixing III, I. Chem. Eng. Sym. Ser. No. 108*, 271, 1988.

Flagg, P. L., D. J. Hill, K. J. Molineux, and N. Sampson, *Proc. 2nd Eur. Conf. on Mixing, St. John's College, Cambridge, England*, BHRA Fluid Eng., Cranfield, England, A3-31, 1977.

Ford, D. E., R. A. Mashelkar, and J. Ulbrecht, *Process Technology International*, **17**, 803, 1972.

Ford, D. E., and J. J. Ulbrecht, *Ind. Eng. Chem. Process Des. Dev.*, **15**, 321, 1976.

Foresti, R., and T. Liu, *Ind. Eng. Chem.*, **51**, 861, 1959.

Fox, E. A., and V. E. Gex, *AIChEJ.*, **2**, 539, 1956.

Giesekus, H., *Rheol. Acta*, **4**, 85, 1965*a*.

Giesekus, H., *Proc. 4th Int. Rheol. Congr.*, **1**, 249, 1965*b*.

Godfrey, J. C., "Mixing of High-Viscosity Fluids," *Mixing in the Process Industries*, edited by N. Harnby, M. F. Edwards, and A. W. Nienow, Butterworth, Stoneham, Mass., 1985, Chap. 11.

Godleski, E. S., and J. C. Smith, *AIChEJ.*, **8**, 617, 1962.

Gray, J. B., *Chem. Eng. Progr.*, **59**, 3, 55, 1963.

Greenspan, H. P., *The Theory of Rotating Fluids*, Cambridge Press, Cambridge, England, 1968.

Greene, H. L., C. Carpenter, and L. Casto, *Proc. 4th Eur. Conf. on Mixing, Noordwijkerhout, Netherlands*, BHRA Fluid Eng., Cranfield, England, D1-109, 1982.

Hall, K. R., and J. C. Godfrey, *Trans. Instn. Chem. Engrs.*, **46**, T205, 1968.

Hamersma, P. J., J. Ellenberger, and J. M. H. Fortuin, *Chem. Eng. Sci.*, **38**, 819, 1983.

Hiraoka, S., I. Yamada, and K. Mizoguchi, *J. Chem. Eng. Japan*, **11**, 487, 1978.

Hiraoka, S., I. Yamada, and K. Mizoguchi, *J. Chem. Eng. Japan*, **12**, 56, 1979.

Hirata, Y., and R. Ito, *Proc. 6th Eur. Conf. on Mixing, Pavia, Italy*, BHRA Fluid Eng., Cranfield, England, 109, 1988.

Hocker, H., G. Langer, and U. Werner, *Ger. Chem. Eng.*, **4**, 113, 1981.

Hoogendoorn, C. J., and A. P. den Hartog, *Chem. Eng. Sci.*, **22**, 1689, 1967.

Ide, Y., and J. L. White, *J. Appl. Poly. Sci.*, **18**, 2997, 1974.

Ito, R., Y. Hirata, K. Sakata, and I. Nakahara, *International Chem. Eng.*, **19**, 4, 605, 1979.

Jachimiak, P. D., Y. S. Song, and R. S. Brodkey, *Rheol. Acta.*, **13**, 745, 1974.

Jensen, W. P., and R. T. Talton, *A. I. Ch. E.—I. Chem. E. Sym. Series No. 10*, **82**, 1965.

Johnson, R. T., *I&EC Proc. Des. Dev.*, **6**, 340, 1967.

Kappel, M., *International Chem. Eng.*, **19**, 571, 1979.

Khang, S. J., and O. Levenspiel, *Chem. Eng.*, 141, October 11, 1976.

Khang, S. J., and T. J. Fitzgerald, *Ind. Eng. Chem. Funds.*, **14**, 208, 1975.

Kim, H. T., and R. S. Brodkey, *AIChEJ.*, **14**, 61, 1968.

Kuriyama, M., H. Inomata, K. Arai, and S. Saito, *AIChEJ.*, **28**, 385, 1982.

Lafon, P., and J. Bertrand, *Proc. 6th Eur. Conf. on Mixing, Pavia, Italy*, BHRA Fluid Eng., Cranfield, England, 493, 1988.

Laidler, P., and J. J. Ulbrecht, *Chem. Sci. Eng.*, **33**, 1615, 1978.

Langlois, W. E., *Q. Appl. Math.*, **21**, 61, 1963.

Le Goff, P., *Proc. 2nd Eur. Conf. on Mixing, St. John's College, Cambridge, England*, BHRA Fluid Eng., Cranfield, England, D1-1, 1977.

Mashelkar, R. A., D. D. Kale, J. V. Kelkar, and J. J. Ulbrecht, *Chem. Eng. Sci.*, **27**, 973, 1972.

McFarland, A., and G. Tatterson, *Chem. Eng. Commun.*, **35**, 1–6, 29, 1985.

McKelvey, J. M., *Polymer Processing*, Wiley, New York, 1962.

Metzner, A. B., and R. E. Otto, *AIChEJ.*, **3**, 3, 1957.

Metzner, A. B., and J. S. Taylor, *AIChEJ.*, **6**, 109, 1960.

Metzner, A. B., R. H. Freehs, H. L. Ramos, R. E. Otto, and J. D. Tuthill, *AIChEJ.*, **7**, 3, 1961.

Metzner, A. B., J. L. White, and M. M. Denn, *AIChEJ.*, **12**, 863, 1966.

Mohr, W. D., R. L. Saxton, C. H. Jepson, *Ind. Eng. Chem.*, **49**, 11, 1957.

Michaels, A. S., and V. Puzinauskas, *Chem. Engng. Prog.*, **50**, 604, 1954.

Moo-Young, M., K. Tichar, and F. A. L. Dullien, *AIChEJ.*, **18**, 178, 1972.

Murakami, Y., K. Fujimoto, and S. Uotani, *J. Chem. Eng. Japan*, **5**, 85, 1972a.

Murakami, Y., K. Fujimoto, T. Shimada, A. Yamada, and K. Asano, *J. Chem. Eng. Japan*, **5**, 297, 1972b.

Murakami, Y., T. Hirose, T. Yamato, H. Fujiwara, and M. Ohshima, *J. Chem. Eng. Japan*, **13**, 318, 1980.

Murakami, Y., T. Hirose, M. Takao, T. Yamato, H. Fujiwara, and M. Ohshima, *J. Chem. Eng. Japan*, **14**, 488, 1981.

Muschick, W., *Proc. 6th Eur. Conf. on Mixing, Pavia, Italy*, BHRA Fluid Eng., Cranfield, England, 527, 1988.

Nagata, S., M. Yanagimoto, and T. Yokoyama, *Memoirs Fac. Eng.*, Kyoto Univ., **18**, 444, 1956.

Nagata, S., M. Nishikawa, H. Tada, H. Hirabayashi, and S. Gotoh, *J. Chem. Eng. Japan*, **3**, 237, 1970.

Nagata, S., M. Nishikawa, T. Katsube, and K. Takaish, *International Chem. Eng.*, **12**, 175, 1972.

Ng, K. Y., and L. Erwin, *Polymer Engng. & Sci.*, **21**, 213, 1981.

Nienow, A. W., and R. Kuboi, *Fluid Mixing II, I. Chem. Eng. Sym. Ser.*, **89**, 97, 1984.

Nickols, B. D., C. W. Hirt, and R. S. Hotchkis, "SOLA VOF: A Solution Algorithm for Transient Fluid Flow with Multiple Free Boundaries," LA-8355, Los Alamos Scientific Laboratory, Los Alamos, N.M., 1980.

Niida, T., and T. Yoshida, *J. Chem. Eng. Japan*, **11**, 7, 1978.
Norwood, K. W., and A. B. Metzner, *AIChEJ.*, **6**, 432, 1960.
Novak, V., and F. Rieger, *Trans. Instn. Chem. Engrs.*, **47**, T335, 1969.
Novak, V., and F. Rieger, *Chem. Eng. J.*, **9**, 63, 1975.
Ohta, M., M. Kuriyama, K. Arai, and S. Saito, *J. Chem. Eng. Japan*, **18**, 81, 1985.
Ottino, J. M., and C. W. Macosko, *Chem. Eng. Sci.*, **35**, 1454, 1980.
Ottino, J. M., W. E. Ranz, and C. W. Macosko, *AIChEJ.*, **27**, 565, 1981.
Ottino, J. M., *The Kinematics of Mixing: Stretching, Chaos, and Transport*, Cambridge University Press, Cambridge, England, 1989.
Ottino, J. M., *Scientific American*, **56**, January 1989.
Patton, T. C., *Paint Flow and Pigment Dispersion*, 2d ed., Wiley-Interscience, Wiley, New York, 1979.
Pearson, G., and S. Middleman, *AIChEJ.*, **23**, 714, 1977a.
Pearson, G., and S. Middleman, *AIChEJ.*, **23**, 722, 1977b.
Peters, D. C., and J. M. Smith, *Trans. Instn. Chem. Engrs.*, **45**, T360, 1967.
Peters, D. C., and J. M. Smith, *Canad. J. Chem. Eng.*, **47**, 268, 1969.
Portner, R., G. Langer, and U. Werner, *Proc. 6th Eur. Conf. on Mixing, Pavia, Italy*, BHRA Fluid Eng., Cranfield, England, 521, 1988.
Quraishi, A. Q., R. A. Mashelkar, and J. J. Ulbrecht, *AIChEJ.*, **23**, 487, 1977.
Ranade, V. R., and J. J. Ulbrecht, *Proc. 2nd Eur. Conf. on Mixing, St. John's College, Cambridge, England*, BHRA Fluid Eng., Cranfield, England, F6-83, 1977.
Ranz, W. E., "Fluid Mechanical Mixing—Lamellar Description," *Mixing of Liquids by Mechanical Agitation*, edited by J. J. Ulbrecht and G. K. Patterson, Gordon Breach Science, New York, 1985.
Reichmann, M. G., and R. J. Adler, "A Framework of Variables and Dimensionless Numbers for Comparing the Power Consumption and Processing Capacity of Mixers Regardless of Type," *AIChE Annual Meeting*, November 1980.
Richards, J. W., *Brit. Chem. Eng.*, **6**, 454, 1961.
Rieger, F., P. Ditl, and V. Novak, *Proc. 1st Eur. Conf. on Mixing and Centrifugal Sep., Churchill College, Cambridge, England*, BHRA Fluid Eng., Cranfield, England, D5-47, 1974.
Rieger, F., and V. Novak, *Proc. 5th Eur. Conf. on Mixing, Wurzburg, Germany*, BHRA Fluid Eng., Cranfield, England, 491, 1985.
Rieger, F., and V. Novak, *Proc. 6th Eur. Conf. on Mixing, Pavia, Italy*, BHRA Fluid Eng., Cranfield, England, 509, 1988.
Ryan, D. F., L. P. B. M. Janssen, and L. L. van Dierendonck, *Chem. Eng. Sci.*, **43**, 1961, 1988.
Schofield, C., *Proc. 1st Eur. Conf. on Mixing and Centrifugal Sep., Churchill College, Cambridge*, England, BHRA Fluid Eng., Cranfield, England, C1-1, 1974.
Seichter, P., *Trans. Instn. Chem. Engrs.*, **49**, 117, 1971.
Seichter, P., *Coll. Czech. Chem. Commun.*, **46**, 2021, 1981.
Shearer, C. J., *Chem. Engng. Sci.*, **28**, 1091, 1973.
Skelland, A. H. P., "Mixing and Agitation of non-Newtonian Fluids," *Handbook of Fluids in Motion*, edited by N. P. Cheremisinoff and R. Gupta, Ann Arbor Science, Ann Arbor, Mich., 1983, Chap. 7.
Smith, J. M., *The Chemical Engineer*, CE45, March 1970.
Solomon, J., T. P. Elson, A. W. Nienow, and G. W. Pace, *Chem. Eng. Commun.*, **11**, 143, 1981.
Spencer, R. S., and R. M. Wiley, *J. Colloid Sci.*, **6**, 133, 1951.
Spragg, A. J. P., S. J. Maskell, and M. A. Patrick, *Proc. 5th Eur. Conf. on Mixing, Wurzburg, Germany*, BHRA Fluid Eng., Cranfield, England, 507, 1985.
Sweeney, E. T., and M. A. Patrick, *Proc. 2nd Eur. Conf. on Mixing, St. John's College, Cambridge, England*, BHRA Fluid Eng., Cranfield, England, A4-43, 1977.
Sylverster, N. D., and S. L. Rosen, *AIChEJ.*, **16**, 967, 1970.
Takahashi, K., M. Sasaki, K. Arai, and S. Saito, *J. Chem. Eng. Japan*, **15**, 217, 1982.
Takahashi, K., Y. Takahata, T. Yokota, and H. Konno, *J. Chem. Eng. Japan*, **18**, 159, 1985.
Takahashi, K., M. Iwaki, T. Yokota, and H. Konno, *Proc. 6th Eur. Conf. on Mixing, Pavia, Italy*, BHRA Fluid Eng., Cranfield, England, 515, 1988.

Thomas, R. H., and K. Walters, *Q. J. Mech. Appl. Math.*, **17**, 39, 1964.
Trommelen, A. M., and W. Y. Beek, *Chem. Eng. Sci.*, **26**, 1933, 1971.
Ulbrecht, J. J., *Chem. Engr.*, 286, 347, 1974.
Ulbrecht, J. J., and P. Carreau, "Mixing of Viscous non-Newtonian Liquids," *Mixing of Liquids by Mechanical Agitation*, edited by J. J. Ulbrecht and G. K. Patterson, Gordon Breach Science, New York, 1985.
Ultman, J. S., and M. M. Denn, *Chem. Eng. J.*, **2**, 81, 1971.
Vemura, T., E. O'Shima, and M. Inoue, *Kagaku Kogaku*, **5**, 142, 1967.
Walters, K., and J. G. Savins, *Trans. Soc. Rheol.*, **9**, 407, 1965.
Walton, A. C., S. J. Maskel, and M. A. Patrick, *Fluid Mixing I. Chem. E. Sym. Series: No. 64*, Bradford, England, C1, 1981.
Weydanz, W., *Chem. Ing. Tech.*, **32**, 343, 1960.
White, J. L., S. Chankraiphon, and I. Ide, *Trans. Soc. Rheol.*, **21**, 1, 1977.
Wichterle, K., and O. Wein, *Int. Chem. Eng.*, **21**, 116, 1981.
Yuge, K., and E. O'Shima, *J. Chem. Eng. Japan*, **8**, 151, 1975.
Zlokarnik, M., *Chem. Ing. Tech.*, **39**, 218, 1967.
Zlokarnik, M., *Chem. Ing. Tech.*, **42**, 1009, 1970.

Chapter

6

Gas Dispersion in Agitated Tanks

Introduction

Approximately 25 percent of all chemical reactions occur between a gas and liquid. A good percentage of these reactions are performed in agitated tank reactor configurations. In general, a considerable number of the processes are area-controlled and depend on transport across the gas-liquid interface. Improper transfer can cause failure of the process, yield reduction, and production of unwanted by-products.

The mechanically agitated tank is a very effective device for contacting a gas with a liquid. The impeller typically used is the disk style turbine or a vane disk impeller. An aerated, agitated tank generally has high mass and heat transfer coefficients, good mixing capability which is documented, a wide range of the liquid residence times, and the capability of handling a wide range of superficial gas velocities. Gas dispersion in agitated tanks is also an active research area because of the interest in biotechnology. Efficiency is of interest, given the cost of compression in gas-liquid contacting. Hence, there is a need for studies of how gas is dispersed in an agitated tank.

The areas of interest for gas dispersion in agitated tanks are: (1) the hydrodynamic flow regimes occurring around the impeller and in the tank, (2) bubble size, holdup, and the interfacial area between the gas and liquid over which transport occurs, and (3) the mass transfer coefficient $k_l a$. These, in turn, are determined by the impeller rotational speed, gassing rate, tank and impeller geometry, and fluid properties and rheology. A review of these different areas of interest and the state of the art follows.

Of the many impeller geometries, the six-blade disk style impeller is most often used in the standard configuration (Holland and

Chapman, 1966) with gas sparging occurring below the impeller in a fully baffled turbulent agitated tank. This configuration is the assumed geometry for discussion unless otherwise stated.

The works cited in this chapter are for noncavitating conditions and low pressure from 1 to 10 atm. The hydrodynamics of gas dispersion at higher pressures have not been studied and are likely very different than described here. Density differences are less severe between the gas and the liquid at high pressure and the effects of buoyancy less important.

It should also be noted that others have published reviews. These include, in books and contributed chapters, Nagata (1975), Oldshue (1983), Middleton (1985), Nienow and Ulbrecht (1985), and Smith (1985) and a review paper by Joshi et al. (1982). It is not the intent of this chapter to mimic these reviews, however. Gas dispersion in agitated tanks also falls under the general topic of multiphase flow for which there are many excellent reference texts, Brodkey (1967) being one.

Hydrodynamic Flow Regimes in Gas Dispersion

The hydrodynamic flow regimes occurring in an aerated, agitated tank did not receive significant attention until the late sixties and early seventies. The flow regimes in the bulk of the tank and in the impeller region require study to explain the complex phenomena that occur during gas dispersion. The work to date, although incomplete, has improved considerably our understanding of gas dispersion in agitated tanks.

The different flow processes during gas dispersion are complicated in detail but can be summarized as follows. Typically, gas is sparged below the disk style turbine, as shown in Fig. 6.1. It rises to the impeller and is dispersed by the impeller, either to rise to the top of the tank or recirculate to the impeller. An important dimensionless number in this process is the gas aeration number or gas flow number Q/ND^3 or Q_s/ND^3, which appears in the literature as N_A, N_Q, Fl, Fl_a, or Fl_g. The gassing rate from the sparger is Q or Q_s, the impeller rotational speed is N, and the impeller diameter is D. The aeration number includes the effects of gassing rate, impeller rotational speed, and impeller diameter, but it does not account for blade number and blade width effects. The aeration number is important in determining the flow phenomena occurring in the impeller region. Another dimensionless group is the Froude number Fr, N^2D/g, N^2T/g, which is the ratio of inertial forces to gravitational forces. The Froude number can be

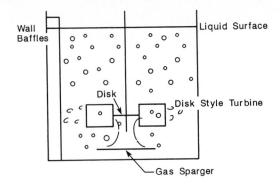

Figure 6.1   The gas dispersion process.

defined for the impeller or tank. However, unless otherwise stated, the Froude number will be for the impeller.

The amount of recirculated gas or recirculation is important. This is gas which has been dispersed by the impeller and has circulated back to the impeller for redispersion. This gas recirculation rate is difficult to determine, and its effect is not included in the aeration number.

Low-viscosity, Newtonian flow regimes for the disk style turbine

The many hydrodynamic studies published concerning flow over blunt objects indicate that vortex systems exist around impeller blades. However, the importance of such systems to the dispersion of gas was not fully understood or appreciated until recently. These vortex systems are low-pressure regions into which the sparged gas accumulates to form ventilated gas cavities. The word ventilated means that gas flows into the cavity and is dispersed from the cavity. There are unventilated cavities which form under laminar flow conditions, but these seldom occur.

The type of cavity formed is a function of the impeller geometry, gassing rate from the sparger, impeller rotational speed, and gas recirculation rate. Takeda and Hoshino (1966), Rennie and Valentin (1968), and Takashima and Mochizuki (1971) were among the first to publish photographs that clearly showed the presence of the trailing vortex systems occurring behind the blades of rotating impellers.

Single-phase vortex systems.   The flow field of an impeller consists of intake flows, vortices, and high-speed discharge flows as shown in Fig. 6.2 for the disk style turbine and the pitched blade turbine pumping upward. For gas dispersion, the trailing vortices are important. Stud-

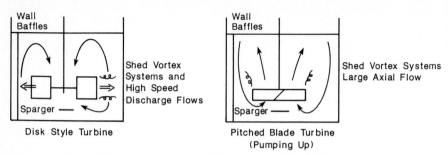

Figure 6.2   General flow patterns for the disk style and pitched blade turbines.

ies exist on impeller bound vortices as reviewed briefly here and in more detail in Chap. 4.

van't Riet and Smith (1973), using a rotating camera apparatus in single-phase flow, measured the angular velocities and calculated the circumferential velocities in the interior of the vortex systems as a function of vortex radius. The circumferential velocities in the vortices were found to be approximately the same order of magnitude as the impeller tip speed. van't Riet and Smith noted that, at such circumferential speeds, substantial centrifugal forces existed in the flow which significantly affected gas dispersion.

The centrifugal and pressure fields (van't Riet and Smith, 1974) were found more intense than described by the conventional vortex models. Dimensionless angular and axial velocities, centrifugal accelerations, and pressure coefficient data were given by van't Riet and Smith (1974). The shear rates that have considerable importance in multiphase processing occurred in such vortex systems. Popiolek et al. (1984) reported detailed laser-doppler turbulence measurements of the velocity field within the trailing vortices as well. A more thorough review of these is given in Chap. 4.

Cavity types.   Different cavity types and cavity systems have been identified for the disk style turbine and include vortex cavities, clinging cavities, large cavities, ragged cavities, and the 3-3 cavity structure as shown in Fig. 6.3. The 3-3 cavity structure consists of three large cavities and three clinging cavities in the form of a symmetrical pattern around a six-blade disk style turbine. These cavities form behind the impeller blades of a disk style turbine from which gas is dispersed.

van't Riet and Smith (1973) considered that the type of gas cavity formed behind the impeller blades depended on the gassing rate, the recirculation of gas from the bulk of the tank, and the level of the centrifugal forces of the vortex systems. At low rotational speeds, the

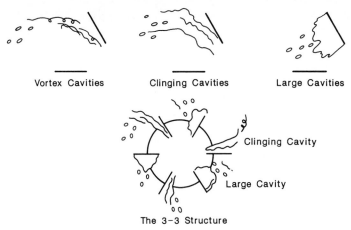

Vortex Cavities   Clinging Cavities   Large Cavities

Clinging Cavity

Large Cavity

The 3-3 Structure

Figure 6.3   The various cavities and cavity systems of the 3-3 structure.

buoyant force of the bubbles dominated over the centrifugal forces, and the gas was not captured by the vortices. At higher speeds, gas was drawn into the vortex systems to form cavities from which the gas was then dispersed. Gas dispersion occurred from the gas cavities and not by a turbulent eddy dispersion mechanism. Nienow and Wisdom (1974) and Bruijn et al. (1974) provided a fairly complete description of these gas cavities. The following summarizes the results of Bruijn et al.

At very low gassing rates, N_A less than about 0.01, gas was captured by the vortex systems to form vortex cavities. The properties of the single liquid phase flow and vortex systems remained very prevalent. Overall, the vortex cavity was the least important of the cavity types because of the low quantities of gas actually dispersed.

At very high gassing rates, the vortex systems were no longer present in the flow, and large cavities blanketed the backside of the impeller blades for N_A greater than about 0.03. The dispersion of gas occurred at the rear of the large cavity; however, the pressure in the large cavities was not low enough for effective bubble capture since the vortex and the accompanying centrifugal forces were either extremely weak or no longer present in the flow. The pressure in the large cavity was nearly equal to the static pressure near the blade.

The minimum speed at which large cavities formed was given by the Froude number value of 0.1. If the gassing rate was high and the Froude number was below 0.1, the impeller became flooded where the bubbles escaped from the large cavities because of buoyancy forces (Bruijn et al., 1974). Impeller flooding was the point at which gas was no longer dispersed effectively by the impeller.

Between the vortex cavity and the large cavity regimes, a clinging cavity or a 3-3 cavity regime occurred on the backside of the impeller blades. A 3-3 structure was briefly mentioned by Bruijn et al. where three large cavities and three clinging cavities alternated between blades of the impeller.

More recently, the literature has focused upon the 3-3 structure of the disk style turbine. Of all the cavity types or cavity systems, the 3-3 structure appears to be the most important because of the gas-handling capability of this system. Smith et al. (1985) noted that large cavities of the 3-3 structure cannot occur below a Froude number of 0.045. As a result, the 3-3 structure cannot form below this critical Froude number value.

Nienow et al. (1985) reported another cavity type called the ragged cavity. The ragged cavities were considered to be large cavities formed behind the impeller blades under flooded conditions. Six ragged cavities were found to draw more power than the three clinging and the three large cavities of the 3-3 structure.

Warmoeskerken and Smith (1989) reported flow regime data for the hollow blade disk style turbine. Due to its unique blade configuration, concave in the direction of rotation, over that of the flat blades of the standard disk style turbine, this particular impeller design had much better gas handling and dispersing capabilities than a standard disk style impeller. The transition from the vortex-clinging cavity regime to the large cavity regime occurred at higher superficial gas velocities than for the standard disk impeller. The hollow blade impeller was also more difficult to flood than a normal disk style impeller.

At higher gas flow rates and impeller rotational speeds, the gas dispersion process becomes nonobservable visually, and the exact gas dispersion mechanism(s) remain unclear. In some cases, a gas-liquid foam structure or a gas band exists. Gassed vortex ligaments that originate from the impeller region can easily be observed under high gassing rates at the wall. Whether these coherent gassed vortex ligaments or the foam structure are significant with regard to mass transfer has not been investigated.

The various cavity types are very important because they affect the power draw, the power ratio, P_g/P_0, the liquid pumping capacity of the impeller and the liquid hydrodynamics around the impeller blades. The various cavities also form the interfacial area necessary for transport.

Flow regime maps. Ismail et al. (1984) studied power characteristics and cavity formation for disk style turbines and identified three critical impeller rotational speeds, N_{c1}, N_m, and N_{c2}. At the lowest rotational speed N_{c1}, most of the bubbles entered the gas cavities, holdup in the tank increased, and large cavities were formed similar to those reported by Bruijn et al. (1974). Flooded conditions existed below N_{c1}. The power ratio P_g/P_0 decreased until a minimum was reached at N_m where all blades had large cavities. At higher rotational speeds, P_g/P_0 increased, and large and clinging cavities were observed. At N_{c2}, a maximum in power was observed where vortex cavities were formed. P_g/P_0 remained constant or decreased slightly after the maximum.

Ismail et al. correlated N_{c1}, N_m, and N_{c2} as functions of aeration number and impeller geometry, which included blade width W, blade number n_p, and blade length variable ($D^2 - I^2$), where $I/2$ was the inner radius where the blade initially started for the disk style turbine. The correlating parameter appeared as:

$$q = \left(\frac{Q}{ND^3}\right)\left(\frac{D}{n_pW}\right)\left(\frac{D^2}{D^2 - I^2}\right) \qquad (6.1)$$

which is an aeration number modified by impeller geometry. The region of large and clinging cavities collapsed to a single band on the plot, which indicated the region of operation for the 3-3 structure. A correlation for power draw was given in graphical form by Ismail et al. which included geometric effects of the impeller on power.

A flow regime map for a 1.2-m-diameter tank was given by Warmoeskerken and Smith (1985) for the three possible flow regimes of the disk style turbine which were those initially named by Bruijn et al. (1974): (1) vortex and clinging cavities, (2) the 3-3 structure which involves three clinging cavities and three large cavities (for a six-blade turbine), and (3) large or ragged cavities of flooding. In the resulting flow regime map, there were three different transition lines for flow regime changes of gas cavities residing on a six-blade disk style turbine: *A to B, B to C,* and *A to C* as shown in Fig. 6.4.

The *A-B* transition is from vortex-clinging cavities to large-ragged cavities of flooding. The ragged cavities were thought to be more streamlined than the vortex or clinging cavities, causing less drag on the blade. Pumping capacity dropped because the ragged cavity obstructed liquid pumping. Reduced liquid pumping and the streamline effect on drag caused power draw to drop at flooding. The power curve (to be discussed later) had an inflection point which cannot be associated with the formation of the 3-3 structure. The *B-C* transition occurs between the 3-3 cavity structure and the large-ragged cavities of the flooded condition. Typically, there was an increase in power draw as flooding was reached, probably because the 3-3 cavity structure was more streamlined and required less power than the ragged-large cavities of flooding. The *A-C* transition is between the clinging-vortex cavities and the 3-3 structure. The transition was gradual with no abrupt changes in power draw and was thought to be associated with the inflection point in the power draw curve.

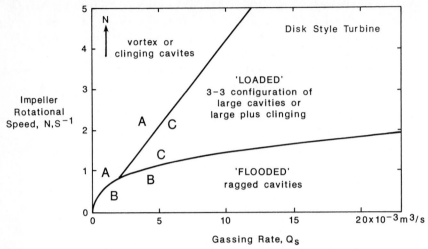

Figure 6.4   General flow map for a 1.2-m diameter tank with a 0.48-m diameter, six-blade turbine. *(From J. M. Smith and M. M. C. G. Warmoeskerken, Proc. 5th Eur. Conf. on Mixing, Wurzburg, Germany, BHRA Fluid Eng., Cranfield, England, 115, 1985. By permission.)*

The transitions *A-B, B-C,* and *A-C* were independent of whether the system was coalescing or noncoalescing, which agreed with the fact that cavities were independent of changes in surface tension (Bruijn et al., 1974). Clearance and liquid height had little effect on the various transitions. However, when the sparger diameter was larger than the disk of the turbine, the transition curves and the amount of gas recirculation changed. The lines for the various transitions will be given in a later section after flooding and recirculation are discussed more fully.

Various effects.   Several items concerning the behavior of gas cavities should be noted.

Concerning the cavity effect on power draw, Bruijn et al. (1974) found that the effect of vortex cavities on power draw was small. At higher gassing rates, the clinging cavity was present, which caused a reduction in power draw to 80 percent of ungassed power. Large cavities reduced the power to about 40 to 50 percent of the ungassed power level. As a result, power draw from the impeller was related to cavity structures that occurred behind the impeller blades. Power draw is discussed in more detail in a later section.

Concerning the cavity effect on liquid circulation, the large cavities (Bruijn et al., 1974) blocked the radial outflow of the impeller and reduced the radial liquid velocities in the bulk of the tank and, hence, recirculation of both gas and liquid to the impeller. For clinging and vortex cavities, liquid circulation appeared unaffected by the presence of gas. However, as discussed later, there is a steady increase in mixing and circulation times with increasing gassing rates for the liquid phase.

Concerning the effects of electrolytes, Bruijn et al. found that the bulk liquid properties such as liquid viscosity and liquid density were important only in determining the cavity shapes and the flow regime. Changes in surface tension and dissolved solutes did not affect the cavities. However, the closure of the gas cavities in dispersing the gas was affected by the surface properties. Electrolytes and electrolytic solutions (i.e., ionic solutions) caused a reduction in the bubble size and increased the amount of small bubbles recirculating into the impeller. Such tiny bubbles, because of their small size, acted as part of the continuous phase fluid. Such systems are called noncoalescing; noncoalescing in the bulk of the tank. Such bubbles can coalesce in cavities, as will be discussed later.

Concerning the effects of geometry, rapid reductions in power draw as a function of aeration number N_A indicated the formation of large gas cavities and poor gas dispersion. Such power studies can be used to judge various geometric effects. The number of blades (Bruijn et al., 1974) affected the type of cavity formed. More blades provided higher gas-handling capacity and higher liquid-pumping capacity and, hence, better gas dispersion and liquid mixing. Bruijn et al. found that a 12-blade impeller had over three times the gas-handling capability of a 6-blade impeller. The increased gas-handling capacity, because of increased number of blades, was also noted by Calderbank (1956b). Different blade geometries for the disk turbine (van't Riet et al., 1976) were studied, including concave and convex blades. The convex blade formed the large cavity structures at very low gassing rates (i.e., low N_A values), and the power draw for this blade geometry dropped very rapidly with aeration number. The concave impeller blade, because of its shape, resulted in lower reductions in power draw and higher gas-handling capacity. Apparently, this particular blade geometry hindered the formation of large cavities. Perforated blades were also studied and were observed to behave erratically. It was also suggested that the size of gas spargers should be less than the impeller diameter since at greater sizes bubbles tended to by-pass the impeller (Bruijn et al., 1974).

Recirculation.   The amount of gas recirculation partially determines the nature of the gas phase mixing that occurs in the tank. The amount of recirculation also determines, in part, the interfacial concentration for mass transfer. As a result, the impeller rotational speed for the initial onset of recirculation becomes important. The various gas flow rates which occur in an aerated, agitated tank are shown pictorially in Fig. 6.5. The amount of gas sparged Q_s plus the amount recirculated Q_r equals the amount of gas leaving the impeller Q_T.

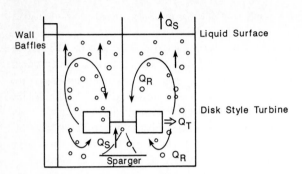

Q_S – Sparged gas flow rate
Q_R – Recirculated gas flow rate
Q_T – Total gas flow leaving the impeller

Figure 6.5 Gas dispersion showing the various circulation loops. Q_s: sparged gas flow; Q_r: recirculated gas flow; and Q_T: total gas flow leaving the impeller.

Gas bubbles represent mass and momentum deficits which are mixed in the tank. In a sense, they can be considered just as any quantity being mixed in the tank. Holmes et al. (1964) found that the average recirculation velocity V_r for single-phase liquid was related to the rotational speed, impeller, and tank diameters as:

$$V_r = \frac{cND^2}{T} \tag{6.2}$$

Recirculation velocity V_r should also be proportional to the tank diameter divided by the recirculation time. For the recirculation of gas bubbles, the bubble rise velocity V_t must be less than or equal to the recirculation velocity of the fluid, hence:

$$V_t < \frac{cND^2}{T} \tag{6.3}$$

If the rise velocity of bubbles is assumed constant:

$$V_t = K \tag{6.4}$$

and set proportional to the recirculation velocity, then the rotational speed for recirculation (Greaves and Kobbacy, 1981) varies as:

$$N_R \propto \frac{T}{D^2} \tag{6.5}$$

There is also an effect of gas holdup on this rotational speed. The more gas holdup or gas flow in the tank, the more difficult it is to recirculate gas bubbles through the impeller due, in part, to: (1)

longer circulation paths for the liquid because of the gas voids and (2) enhanced separation of gas from the liquid in the upper region of the tank because of the presence of more gas (Greaves and Kobbacy, 1981). These effects can be accounted for by using the gas flow rate Q. Bubble-size distributions are important in treating recirculation. Very small bubbles behave as fluid particles and recirculate continuously. Hence, there is always some recirculation.

Bruijn et al. (1974) reported a measure of recirculation through data on an external distribution coefficient α (not to be confused with the sign for proportionality), which was defined as the ratio of the amount of recirculated gas to that sparged. When α is large, the gas dispersed by the impeller is primarily recirculated gas. The magnitude of α indicates the amount of gas phase mixing that occurs.

van't Riet et al. (1976) studied recirculation of gas into the cavity structures which results in a special type of coalescence called impeller coalescence. Recirculation of gas resulted in lower power draw than with no recirculation since the impeller received more gas than the amount sparged. From the difference in power draw levels between conditions of recirculation and no recirculation, the amount of gas recirculation into the impeller was estimated. The nature of the bubble size affected recirculation. Ionic solutions had lower bubble size, higher holdup, higher impeller coalescence, and larger differences in recirculation power and nonrecirculating power (van't Riet et al., 1976).

van't Riet et al. (1976) found the distribution coefficient α to be a strong function of impeller rotational speed and gassing rate. At low gas sparging rates and/or high impeller rotational speeds, α was above 2, indicating (1) considerable recycle of gas into the impeller and (2) considerable gas mixing as a result of impeller coalescence. As rotational speed decreased or the gas sparging rate increased, α decreased. Values of α for ionic or electrolytic solutions were equal to or greater than nonionic solutions because of the increase in holdup for ionic liquids which caused higher impeller coalescence and better gas mixing. In ionic liquids, coalescence in the bulk of the tank was considered negligible.

Twelve and eighteen blade impellers had much higher α values than a six-blade impeller. The convex blade impeller had the lowest α values. Impellers with 12 or 18 flat or concave blades were shown to be most effective for gas dispersion and gas mixing applications.

Nienow et al. (1977) developed a correlation for the impeller rotational speed at the onset of recirculation based on observed maximums in gassed power curves. The correlation obtained was given as:

$$\frac{N_R D^2}{Q^{0.2} T} = 1.5 \left(\frac{m}{s^2}\right)^{0.4} \tag{6.6}$$

and follows the discussion given above. The reasoning behind this correlation was that the recirculated gas contributed to the streamlining of the cavities and reduced the power draw of the impeller. As a result, the onset of recirculation occurred at the maximums in the power curve. Nienow et al. (1977) rearranged the correlation to obtain:

$$(N_Q)_R^{-0.5}(\mathrm{Fr}_I)_R\left(\frac{D}{T}\right)^{2.5} = 0.28 \qquad (6.7)$$

Smith et al. (1986) also rearranged Nienow's correlation to obtain a different form:

$$\mathrm{Fl} = 13\mathrm{Fr}_I\left(\frac{T}{D}\right)^{-5} \qquad (6.8)$$

There are a number of ways in which the external distribution coefficient α can be estimated and used to describe the amount of gas recirculation through the impeller. Local void fraction or holdup measurements around the impeller can be used to determine α. Gas dilution methods can also be developed based on tracer studies.

The most popular methods have been based on power draw data and comparing N_Q before and after recirculation at constant power draw. The requirement "at constant power draw" implies that the gas cavities are the same shape and receiving the same amount of gas flow. However, the amount of recirculated gas cannot be determined with certainty from such a procedure since power data can only establish the point at which recirculated gas begins to affect impeller power draw.

The external distribution coefficient has been estimated from power data in two ways as summarized in Fig. 6.6: one based on power draw data obtained by varying Q_s at constant N; the other based on power draw data obtained by varying N at constant Q_s. van't Riet et al. (1976) related the external distribution to the change in gas flow number Q/ND^3 with recirculation, where Q to the impeller is the sum of Q_s and Q_r, and without recirculation, where Q to the impeller is Q_s at constant power draw. The subscripts s and r designate sparged and recirculated gas, respectively. The external distribution coefficient α was calculated as:

$$\alpha = \frac{Q_r}{Q_s} \qquad (6.9)$$

Nienow et al. (1977) used power data obtained by varying N at constant Q_s before and after recirculation and defined the external distribution coefficient α as:

$$\alpha = \frac{\Delta N_Q}{N_{Q_s}} \qquad (6.10)$$

ΔN_Q was obtained as the difference in N_Q before and after recirculation and contains the difference between Q_T and Q_s or, in other words, Q_r. N_{Q_s} (N_{Q_1} in Fig. 6.6) is the gas flow number before the onset of recirculation and includes only Q_s. The onset of recirculation was identified as the local maximum occurring in the power curve.

In other studies on the external distribution coefficient, Nienow et al. (1979) developed an ingenious way of studying the recirculation to the impeller by performing tracer balances around the impeller cavities and recirculation loops. The results reported the external distribution coefficient as shown in Fig. 6.7. Actual tracer concentrations inside the cavities were measured. Although there was recirculation of gas even below the rotational speeds for flooding N_F and recirculation N_R (as described by Nienow et al., 1977), the external distribution coefficient was very low as indicated by the data in Fig. 6.7. From a practical viewpoint, the rotational speeds N_F and N_R can be used to determine the onset of recirculation. The results of Nienow's study also indicated that a constant gas composition throughout a recirculation loop is not a completely valid assumption.

Warmoeskerken, Feijen, and Smith (1981) collapsed curves of power draw into one single curve up to the flooding transition using the external distribution coefficient α defined somewhat differently from that above. The amount of recirculated gas was related to the gas holdup ϕ and impeller pumping as:

$$Q_r = \alpha \phi N D^3 \qquad (6.11)$$

Rearranging in terms of a total gas flow number that includes gas sparged and gas recirculated, the relationship becomes:

$$N_{QT} = N_{Q_s} + \alpha \phi \qquad (6.12)$$

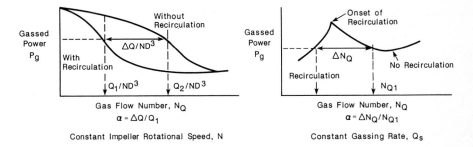

Figure 6.6   External distribution coefficient α from power data.

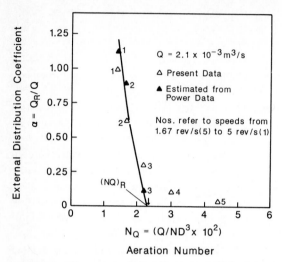

Figure 6.7 The external distribution coefficient as a function of aeration number. *(From A. W. Nienow, C. M. Chapman, and J. C. Middleton, Chem. Eng. J.*, 17, 111, 1979. *By permission.)*

Warmoeskerken et al. found that the external distribution coefficient α as defined above, equaled 0.2 up to 5 s^{-1} and 0.3 after that.

Recirculated gas does not have to pass through the impeller. Instead, gas can recirculate and be entrained in the impeller stream without passing through the impeller. Such secondary recirculated gas cannot significantly affect power draw of the impeller, and this secondary recirculation has not been studied in the literature.

Flooding transition. Flooding is defined as the flow regime where the impeller cannot disperse the gas properly and is represented pictorially in Fig. 6.8. The flooding transition is the demarcation between

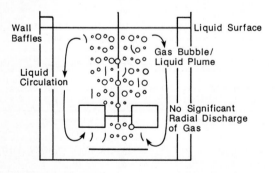

Figure 6.8 Flooding.

the impeller being flooded and a condition of good gas dispersion. Although there tends to be inconsistencies in the literature as to the proper definition and determination of the flooding transition, the transition is rather distinct with specific properties. Reviews of flooding for the disk style turbine are given by Warmoeskerken and Smith (1984; 1985) in which work by Rushton and Bimbinet (1968), Biesecker (1972), Wiedmann et al. (1981), and Wiedmann (1983) have been cited. Steiff (1985) has also discussed flooding.

At the transition between flooding and initial gas loading of the impeller, shown in Fig. 6.9, Warmoeskerken and Smith cited a balance between the pumping of the impeller and buoyant pumping of the bubbles. This balance reduced to:

$$Fl = aFr_I \qquad (6.13)$$

for the flooding transition where Fl is Q/ND^3 and Fr_I is the impeller Froude number ND^2/g. Biesecker (1972) obtained a similar relationship using a power balance. Data by Zwietering (1963), Mikulcova et al. (1967), and Biesecker (1972) showed the constant a to be roughly 0.6, but Warmoeskerken and Smith (1985) suggested a value of 1.2 as more appropriate. Warmoeskerken and Smith attributed the differences between the correlations to differences in geometry, stirrer type, and liquid height.

The balance, as presented, does not include the dependency on T/D ratio, which is known to be significant. Tatterson and Morrison (1987) accounted for this in the following manner. At the transition from flooding to a gas loading condition, power input from the impeller to the fluid P_g balances: (1) a portion of the power of the rising liquid plume written in terms of a multiple of the quantity $\rho_L Q_G g D$, and (2) the power increase P_C necessary initially to establish the increased radial liquid circulation. The power contained in the liquid plume is typ-

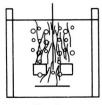

Flooding

High Gassing Rate
Low Rotational Speed

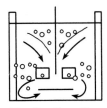

Flooding Transition

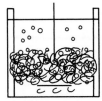

Gassed Band Visually
Covering Impeller

High Gassing Rate
High Rotational Speed

Figure 6.9 Flooding, flooding transition, and gassed band around the impeller.

ically written in terms of the gas flow rate Q_G (the gas sparging rate) because the liquid flow rate in the plume is not usually known. As a result, at the transition from flooding to loading, several events occur: (1) a portion of the liquid plume is stopped; (2) there is an increase in pumping from the impeller; and (3) gas is dispersed radially. The balance is:

$$P_g = \beta \rho_L Q_G g D + P_C \qquad (6.14)$$

where P_g is the power supplied by the agitator.

The first term on the right side has been used and often appears in models (Warmoeskerken and Smith, 1985). It is the power necessary to stop a portion of the rising liquid plume. After the transition from flooding to a condition of good gas dispersion, the liquid-bubble plume no longer exists.

The second term P_C is the initial power necessary to establish liquid circulation locally around the impeller and can be obtained from the following equation:

$$P_C = \int_0^{t'} \left(\frac{P_g}{V}\right) Q_L \, dt \qquad (6.15)$$

where t' is some time over which flooding is stopped and circulation is started (i.e., the time required for transition). Q_L is the volumetric pumping rate of the impeller. The term P_C, however, is not the total power to cause circulation throughout the tank.

The actual integration for P_C cannot be performed at the present time, but P_C can be scaled. The time t' was scaled according to the circulation time of the impeller-tank configuration of interest (Greaves et al., 1979; Greaves et al., 1983), and Q_L was scaled proportional to ND^3. As a result, the integral for P_C was scaled as:

$$P_C \propto \left[\left(\frac{P_g}{P_0}\right) N_P \rho_L N^3 D^2\right] [ND^3] \, t_C \qquad (6.16)$$

where the circulation time t_C (Holmes et al., 1964) is:

$$t_C = \frac{KT^2}{ND^2} \qquad (6.17)$$

Substituting these relations, using a constant K' to handle all constants involved with P_C, the following power balance was obtained:

$$\left(\frac{P_g}{P_0}\right) N_P(\rho_L N^3 D^5) = \beta \rho_L Q_G g D + K'\left(\frac{P_g}{P_0}\right) N_P(\rho_L N^3 D^2) ND^3 \frac{1}{N}\frac{T^2}{D^2} \qquad (6.18)$$

where K' is a constant. Rearranged, the equation became:

$$\mathrm{Fl} = \left(\frac{P_g N_P}{P_0 \beta}\right)\left[\mathrm{Fr}_I - K'\mathrm{Fr}_T\left(\frac{T}{D}\right)\right] \qquad (6.19)$$

where Fr_T is the tank Froude number. To test the equation, data given by Zwietering (1963) were used, and the correlation proved successful. The correlation also explained the differences in the constant a of 0.6 and 1.2 as arising from the nature of the bubble-liquid plume.

Other correlations are also available on the flooding transition which accounts for the D/T effects. Nienow et al. (1985a,b) presented an empirical correlation:

$$(\mathrm{Fl}_g)_F = 30\left(\frac{D}{T}\right)^{3.5}(\mathrm{Fr}_I)_F \qquad (6.20)$$

which fitted their data well for both coalescing and noncoalescing systems for D/T ranging from 0.22 to 0.5.

The flooding transition can be detected in a number of ways: (1) visual observations (e.g., Rushton and Bimbinet, 1968) in which the gas moves axially before rising; the gas no longer ignores the presence of the rotating impeller, (2) photographic observation where large cavities exist behind the impeller blades (Ismail et al., 1984), (3) conductivity probes measuring bubble void fractions (Mikulcova et al., 1967), (4) step increases or decreases in the power curve (Biesecker, 1972; Wiedmann et al., 1981; Nienow et al., 1985), (5) micropropeller rotation noting the increase in pumping (Roustan and Bruxelmane, 1981; Warmoeskerken and Smith, 1985), (6) vibrating vane with strain gages (Smith and Warmoeskerken, 1985), and (7) hydrophones (Sutter et al., 1987).

The detection of the flooding transition is important in that it occurs at the highest possible operating value of superficial gas velocity, which implies maximum transport between the gas and liquid, i.e., the highest mass transfer coefficient $k_l a$ values. However, further work is necessary to support this conclusion. The mass transfer coefficient $k_l a$ is discussed later.

Flooding has typically been associated with the impeller and the effect of impeller rotation in the dispersion of gas. Some have defined flooding differently. Nienow et al. (1977) defined flooding as being the region where gas was dispersed throughout the tank, the focus being placed on complete dispersion of the gas in the bottom portion of the tank. Visual observations indicated that the optimum occurred at C/D of 0.25. Although the complete dispersion of gas into the bottom of the tank is subjective, the rotational speed at which complete dispersion occurred was correlated as:

$$\frac{N_F D^2}{Q^{0.5} T^{0.2}} = 4 \text{ m}^{0.3} \text{s}^{-0.5} \tag{6.21}$$

which included the effect of scale (Nienow et al., 1977). The location of N_F was determined from minimums in the power curves. Nienow et al. (1985) redefined this flooding concept as one of complete dispersion denoted by the impeller rotational speed N_{CD}. There is another form of flooding, tank flooding, which has not been well studied in the literature. Blakebrough et al. (1961) mentioned tank flooding briefly as well as citing work by Steel and Brierley (1959) which indicated another occurrence of a tank flooded condition. In Blakebrough's work, tank flooding occurred when the volumetric flow rate of gas per minute was in excess of the liquid holdup in the tank, in excess of 1 vvm (vvm is volume of gas per volume of liquid per minute). Etchells, in unpublished work presented at an Engineering Foundation Conference on Mixing, also discussed tank flooding from 1 to 5 vvm.

The flooding transition for high-viscosity liquids has not been studied or well defined. The above development pertains to low-viscosity liquids and gas dispersion in the turbulent regime.

Complete dispersion.   Nienow et al. (1985) studied the complete dispersion condition N_{CD} for aerated tanks agitated by a disk style turbine. The correlation for dispersing air under a complete dispersal condition (Nienow et al., 1977) was given as:

$$(\text{Fl})_{CD} = 0.2 \left(\frac{D}{T}\right)^{0.5} (\text{Fr}_I)_{CD}^{0.5} \tag{6.22}$$

Desplanches et al. (1988) also discussed the concept of complete dispersion in an aerated system in the standard configuration using the disk style impeller. The aeration number at complete dispersion was correlated as:

$$N_A = 0.0021 \, \text{Fr}^{1.3} \left(\frac{D}{T}\right)^{1.3} \left(\frac{\rho_L^2 g D^3}{\mu_L^2}\right)^{0.35} \tag{6.23}$$

for $0.3 < T < 0.63$ m, $0.3 < D/T < 0.5$, $0.13 < $ vvm < 1.8, and $0.1 < \mu_L < 1.0$ Pa $\cdot$ s. The gassed power number for this zone was correlated as:

$$N_{Pg} = 0.34 (N_A)^{-0.15} \left(\frac{D}{T}\right)^{0.1} \left(\frac{\rho_L^2 g D^3}{\mu_L^2}\right)^{0.1} \tag{6.24}$$

The concept of complete dispersion throughout the tank, shown in

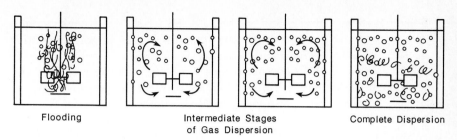

Flooding   Intermediate Stages   Complete Dispersion
                           of Gas Dispersion

Figure 6.10   Flow regimes in the bulk of the tank.

Fig. 6.10, is one of interfacial area. The gas is dispersed throughout
the tank, giving the impression of full utilization of the tank volume.
This may be misleading, and data on interfacial area as a function of
the condition of dispersion are required to support the concept. If the
effect of complete dispersion on interfacial area is significant, interfacial area should be a function of clearance. Typically, clearance does
not appear in correlations for interfacial area, as shown later. However, Hughmark (1980) used liquid volume in correlations for holdup
and interfacial area, which supports the concept of complete dispersion. Blakebrough et al. (1961) indicated that regions of low or no aeration, such as that occurring beneath the impeller under incomplete
dispersion, should be avoided because of the possible production of unwanted anaerobic products in such regions during aerobic fermentations. However, such a comment ignores mixing which takes place in
liquid phase.

Greaves and Kobbacy (1981) defined efficient gas contacting as the
condition where the sparged gas coalesced with impeller gas cavities
and was then dispersed without recirculation to the impeller. Greaves
and Kobbacy referred to this operation as the "efficient mixing zone."
The region was between the gas recirculation regime and the flooding
transition. To treat this concept, Greaves and Kobbacy defined an impeller dispersion efficiency as the ratio of rate of gas coalescence in the
impeller region to the rate of gas sparged. From their data, they developed a correlation for the rotational speed at the onset of recirculation as:

$$N_R = 0.57\left(\frac{T^{0.97}Q^{0.13}}{D^{2.34}}\right) \tag{6.25}$$

For the flooding transition:

$$N_F = 1.52\left(\frac{T^{0.20}Q^{0.29}}{D^{1.74}}\right) \tag{6.26}$$

Similar equations for N_R and N_F were given for ionic solutions.

Boiling systems. Smith and Verbeek (1988) studied the cavity development and cavity regimes around impeller blades in boiling liquids in the standard configuration using a Rushton impeller ($H = T$, $T/D = 2.5$ and 3.0). The generation of cavities was controlled by the relationship between the liquid temperature and the pressure drop occurring in the vortex systems behind the blade. Cavitation first occurred in the axis of the vortices.

An impeller Froude number of 0.5 was necessary to obtain cavities which were attached to the blade. The general nature of the gas cavities were similar to those generated in gassed systems. However, a variety of cavity arrangements were observed. The impeller rotational speeds were very high ranging between 1 and 60 s^{-1}. Different straight-line relationships between impeller Froude number and the dimensionless pressure ratio P_v/P were obtained for the different cavity types. P_v is the vapor pressure and P is the ambient pressure at the impeller level. The onset of cavitation occurred over a range of impeller Froude numbers, 0.0 to 4.0, for a range of pressure ratios P_v/P, 0.70 to 1.0. At incipient cavitation, a simple correlation was obtained as:

$$N^2 D^3 = \frac{0.35 m^3}{s^2}\left(1.02 - \frac{P_v}{P}\right) \tag{6.27}$$

which interrelates scale, impeller tip speed, and vapor pressure. Cavitation numbers, $2(P_v - P)/\rho V^2$, for two disk style impellers, $D = 5.4$ and 6.5 cm, were 3.5 and 4.6, respectively. Reduced pumping capacity and power draw also occurred in boiling systems. In a steadily boiling system, large cavities developed, and, in a rapidly boiling system, the impeller approached flooding. In simmering conditions, the gas loading on the impeller was greater than the boil off rate.

Smith and Smit (1988) also reported on the impeller hydrodynamics of boiling systems. Turbine impellers in such systems developed gas cavities similar to those in gas sparged systems. Smith and Smit reviewed the different types of cavities. The various transitions between the cavities types were a function of liquid temperature and impeller speed. The vapor evolution and agitation intensity were interrelated, and the transitions to large cavities and to flooding were functions of heat flux. At the transition to flooding, the heat transfer dropped because of the poor liquid circulation. This, in turn, reduced the tendency of the liquid to boil which tended to change the impeller flooding to a loaded state, giving rise to a hysteresis loop. Vapor loading of the impeller was determined by the impeller tip speed and was independent of the T/D ratio.

Summary. Three rotational speeds have been discussed: N_F, the rotational speed of the flooding transition; N_{CD}, the rotational speed for complete dispersion; and N_R, the rotational speed for recirculation. The equations for these were given above. By doing so, the different impeller rotational speeds have been emphasized. Smith and Warmoeskerken (1985) and Smith et al. (1986) provided an overview of the locations of the various flow regimes for the disk style turbine in the standard geometry ($H/T = 1$).

1. The minimum speed to hold any stable cavity structure occurred at a specific Froude number:

$$Fr_l = 0.045 \qquad (6.28)$$

2. The transition for the vortex-clinging cavities to the first large cavity occurred at:

$$Fl = 3.8 \times 10^{-3}\left(\frac{Re^2}{Fr_l}\right)^{0.067}\left(\frac{T}{D}\right)^{0.5} \qquad (6.29)$$

3. The maximum gas flow rate at which gas recirculation occurred at:

$$Fl = 13Fr_l^2\left(\frac{T}{D}\right)^{-5} \qquad (6.30)$$

which is a rearrangement of the correlation given by Nienow et al. (1977)

4. The maximum gassing rate before the transition to flooding (Nienow et al., 1985) occurred at:

$$Fl = 30Fr_l\left(\frac{T}{D}\right)^{-3.5} \qquad (6.31)$$

Tatterson and Morrison (1987) and Nienow et al. (1985) discussed the flooding transition as cited above. Warmoeskerken and Smith (1989) reported a flow regime map for hollow blade impellers.

High-viscosity, Newtonian, and non-Newtonian flow regimes in the impeller region

Most studies of gas dispersion have been done in low-viscosity Newtonian liquids; however, there are some studies of gas dispersion in highly viscous and non-Newtonian liquids. The impeller typically used is the disk style turbine. A general review in this area was given by Nienow and Ulbrecht (1985).

Bruijn et al. (1974) and Ranade and Ulbrecht (1977) have studied the gas cavities in viscous liquids. Gas dispersion in such liquids occurs at much lower impeller Reynolds numbers because of the high viscosities. The vortex systems and centrifugal forces occurring in high-viscosity liquids differ, as a result, from those occurring in low-viscosity liquids. The trailing vortices form at an impeller Reynolds number of about 100, indicating that gas cavities form at about this Reynolds number. Whether gas cavities form at all in laminar flow under creeping flow conditions remains unstudied. The most appropriate impeller for gas dispersion under creeping flow conditions has not been determined.

Bruijn et al. (1974) found that the cavities formed in viscous liquids were very stable and remained attached to the impeller blades even after the gas was turned off. The reduction in reduced power draw occurred at lower gassing rates after these stable cavities had been formed. Different cavity shapes were found by Bruijn et al. (1974). Below a Froude number of 0.17, bubbles escaped upward from the cavities because of their buoyancy. Under normal dispersion operations, cavities dispersed gas through long trailing filaments at the end of the cavities. Bruijn et al. did not mention viscoelastic and non-Newtonian effects.

Ranade and Ulbrecht (1977) investigated power draw and gas dispersion in a Newtonian fluid (corn syrup) and in various solutions of a viscoelastic fluid [polyacrylamide (PAA)] using the disk style turbine impeller. Mass transfer studies, discussed later, were performed in polyacrylamide (PAA). All studies were in the transition and initial turbulent regimes.

For single-phase mixing, the shear thinning nature of the fluids reduces power draw below that of a Newtonian fluid (Metzner and Otto, 1957; Metzner et al., 1961) in the transition region. As turbulent mixing regime is approached, the shear thinning nature of the fluid becomes unimportant, and the power number curve exhibits a minimum and returns to the turbulent power number levels of Newtonian fluids. In the study by Ranade and Ulbrecht (1977) of viscoelastic fluids, power draw was reduced by the elastic properties, resulting in substantially lower power numbers than Newtonian and shear thinning fluids. No minimums were observed in the power number curves, and the laminar-transition regime extended past the typical Newtonian laminar to turbulent transition impeller Reynolds number, as shown in the nonaerated curve in Fig. 6.11.

Under aerated conditions, Ranade and Ulbrecht (1977) found a substantial reduction in gas power draw from the impeller which occurred abruptly above a specific, critical, impeller Reynolds number. After this, power draw was constant and unchanged despite changes in gassing rate, Fig. 6.12. Ranade and Ulbrecht explained this in terms of the cavity structures that formed behind the impeller blades. At very low rotational speeds, the gas was evenly dispersed; large bubbles rose to the liquid surface because of buoyancy forces while the smaller bubbles recirculated into the impeller. These smaller bubbles agglomer-

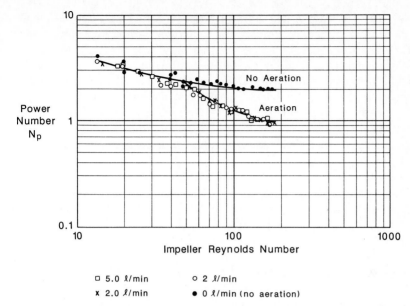

Figure 6.11 Influence of gas flow rate upon power numbers in 0.5 percent PAA solutions. *(From V. R. Ranade and J. J. Ulbrecht, Proc. 2nd Eur. Conf. on Mixing, St. John's College, Cambridge, England, BHRA Fluid Eng., Cranfield, England, F6–83, 1977. By permission.)*

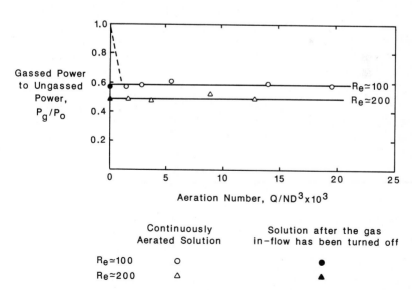

Figure 6.12 Relative power reduction in an aerated 0.5 percent PAA solution. *(From V. R. Ranade and J. J. Ulbrecht, Proc. 2nd Eur. Conf. on Mixing, St. John's College, Cambridge, England, BHRA Fluid Eng., Cranfield, England, F6–83, 1977. By permission.)*

ated to form cavities of discrete bubbles, called "small-bubble cavities," behind the impeller blades. With increasing impeller rotational speed, the small bubbles formed into a single large cavity, which remained fixed to the impeller blade even after the gas flow rate was shut off. The formation of the large cavities occurred at the point in the power number curve where aeration affected power draw (Fig. 6.11). The cavity apparently displaced the wake region of the blade through which power was transported from the blade to the fluid. The existence of the large cavity indicated a lower pressure drop across the blade than for the small-bubble cavity.

Ranade and Ulbrecht (1977) found that the power number curve in the large cavity regime appeared as a continuation of the laminar power number curve. This was attributed to a streamlining effect of the large cavity attached to the blades. At higher rotational speeds, a split cavity was formed, having extended tips through which gas was dispersed. However, the power number curve of split cavity was the same as that of the large cavity. The various cavity types observed by Ranade and Ulbrecht are shown in Fig. 6.13. Ranade and Ulbrecht also noted the possible segregation of gas in such liquids with large bubbles rising to the surface and small bubbles caught in closed recirculation loops with little impeller coalescence occurring in both the small-bubble and large cavity regimes.

Solomon et al. (1981) provided a discussion of flow patterns in agitated, aerated, shear thinning, and viscoelastic fluids for different impellers and combinations of impellers. Only large-diameter disk style turbines in combination with pitched blade turbines provided satisfactory mixing. A minimum impeller speed for gas dispersion was given. Nienow (1984) discussed gas dispersion in fermentation processes involving Xanthan gums.

Ozcan et al. (1988) studied gas dispersion in high-viscosity fluids in the transition region in a 0.49-m-diameter tank agitated by a disk style impeller in the standard configuration. Stable cavities occurred at Froude numbers greater than 0.41. Three regimes were noted. Be-

Small Bubble   Residual Cavity   Split Cavity
   Cavity   at Zero Gassing Rate

Figure 6.13   Gas cavities in highly viscous liquids. *(From V. R. Ranade and J. J. Ulbrecht, Proc. 2nd Eur. Conf. on Mixing, St. John's College, Cambridge, England, BHRA Fluid Eng., Cranfield, England, F6–83, 1977. By permission.)*

low a minimum impeller Reynolds number, regime I, gassed and ungassed power number curves were the same. Gassing had no affect on power consumption. The value of the minimum Reynolds number varied between 46 and 315 with viscosity and impeller diameter. Above the minimum Reynolds number and below an impeller Reynolds number of 400, regime II, the gassed power number passed through a minimum as a function of the impeller Reynolds number and behaved differently from the ungassed power number. Reduction in power number due to gassing in this range was dependent upon the Froude number and independent of the aeration number and gassing rate. Above an impeller Reynolds number of 400, regime III, power number was a function of the aeration number, the Froude number, and the impeller Reynolds number. At very low impeller speeds or at very low Froude numbers, no power reductions occurred.

Laminar regime. Very little work has been done in the dispersion of gas in high-viscosity liquids in the laminar regime. Most studies concerning high-viscosity liquids have been in the transition regime and the initial turbulent regime. The impeller typically used is the disk style turbine. In the laminar regime, this impeller is considered ineffective in single-phase bulk mixing, and applications of this impeller in gas dispersion under laminar flow conditions should be questioned. The most effective impeller geometry for the gas dispersion in the laminar regime has not been found. Nishikawa et al. (1983) reported that anchor and helical ribbon impellers were much more effective than the disk style turbine in gas dispersion in highly viscous liquids. Generally, however, very few studies have reported results comparing impeller designs for gas dispersion in highly viscous liquids in either laminar or transition mixing regimes.

Flow regimes in the bulk of the tank

The flow regimes in the bulk of the tank have been less studied than the flow regimes around the impeller. Flow in this region is less dramatic than in the impeller region and, therefore, more difficult to study. To characterize this region, liquid and bubble circulation velocities, bubble size, and bubble concentration distributions, holdup distributions and coalescence mechanisms are needed. Obviously, the flow properties are distributed and difficult to quantify.

Preen (1961), as cited by Westerterp et al. (1963), studied the dispersion occurring in the upper portion of an aerated, agitated tank. Preen found that very little dispersion occurred in this region and that coalescence occurred primarily between bubbles of small diameter ($< 0.5 \times 10^{-3}$m) and large bubbles ($> 3 \times 10^{-3}$m). Preen concluded that interfacial area was determined by coalescence and not by the dispersion, a result specific to this region of the tank. In the coalescence mechanism cited by Preen, small bubbles entered into the larger bubbles through the large trailing wake of the large bubbles. Under such conditions, coalescence was not because of any thinning of a liquid layer between bubbles.

Gal-Or and Resnick (1966) studied the bubble flow patterns, gas, and liquid velocities and the relative velocities between the gas bubbles and liquid in an aqueous solution of glycerin at low gassing rates. The flow patterns of the two phases were identical under nonflooded conditions and the velocities varied linearly with impeller rotational speed. The liquid velocities were higher than the gas velocities, and the differences between these increased with impeller rotational speed.

Bulk flow patterns.   Bulk flow patterns add to the basic information that describes the process behavior and can indirectly affect power draw substantially.

Chapman et al. (1983) reviewed the behavior of the bulk flow patterns occurring in the tank for different impellers: disk turbines, pitched blade turbines pumping up or down, marine impeller, and pitched blade turbines with disks. The flow patterns in the bulk of the tank and impeller clearance were found to affect the nature of the gas recirculation and, hence, power draw of the impellers. A summary of the results of Chapman et al. follows.

For the most part, the flow patterns in aerated, agitated tanks for the disk turbine resembled those of a single-phase flow and the bulk flow patterns reported by Nienow et al. (1977) and shown in Fig. 6.10. Among the more interesting effects, Chapman et al. found that, at a particular gas flow number at low clearance, the bulk flow pattern changed from a radial flow pattern to an axial one. This was an unexpected occurrence, indicating that care should be taken in making generalizations about flow patterns. This change was also reflected in the power curve (Chapman et al., 1983). Under such circumstances, the nature of the gas cavities changed as well since power draw changed. Blakebrough et al. (1961) also discussed bulk flow patterns for the disk style turbine.

One important flow pattern for the disk style turbine that has not been emphasized in the literature is the band of high gas content that forms around the impeller at very high gassing rates and very high rotational speeds. This can be considered an extension of the impeller region into the bulk of the tank (Rennie and Valentin, 1968) and is shown pictorially in Fig. 6.9.

Some observations for the pitched blade turbines (pumping up) were provided by Chapman et al. (1983). At low gassing rates, the power draw was low because of the liquid plume pumping induced by the gas. As the rotational speed increased, the liquid flow, generated by the impeller, dominated the flow field and the power draw increased. The impeller was no longer flooded. Further increases in rotational speed caused the power to drop but the variation of P_g/P_0 for pitched blade turbines was not nearly as great as for the disk style turbines. The rotational speed at complete dispersal N_{CD} was considered to occur at the maximum in the power curve. At this point, gas recirculated in the tank volume below the impeller.

For the pitched blade turbines (pumping down) and marine propellers, Chapman et al. (1983) made the following observations. At low rotational speeds, the gas rose through the impeller with little dispersion. At higher speeds, gas filled the vortex cores to form long streamers moving downward and out radially. Streamers were also formed with the pitched blade turbine pumping upward with gas rising along the shaft. Large fluctuations in torque and power draw were observed. Periodic flow pulsing in single-phase flow have been reported by Tatterson et al. (1980) as well. Overall, pulsing was associated with poor gas dispersion and was not desired in process operations. Marine propellers were also observed to have torque fluctuations. Smith (1985) has also commented upon the different flow regimes for the pitched blade turbine which requires review and comparison with Chapman et al. Operation at conditions where flow patterns and torque fluctuations occur should be avoided.

Comparison of impellers. Impeller types listed (Chapman et al., 1983) in order of increasing N_{CD} (i.e., the impeller rotational speed necessary for complete dispersion) were: (1) the standard disk turbine, (2) the pitched blade turbine with a disk (i.e., an angled blade disk style turbine), (3) the pitched blade turbine pumping upward, (4) the marine impeller, and (5) the pitched blade turbine pumping down. The impellers listed in order of increasing power draw at N_{CD} (Chapman et al., 1983) were: (1) the pitched blade turbine with a disk, (2) the standard disk turbine, marine impeller, and the pitched blade turbine pumping up, and (3) the pitched blade turbine pumping down. Ironically, the standard disk style turbine had the poorest performance of all the impellers at high gassing rates, with the exception of the pitched blade turbine pumping down (Chapman et al., 1983).

Central vortex. The central vortex, Fig. 6.14, entrains gas from the surface of the tank at high rotational speeds. The rotational speed for surface entrainment of gas is a strong function of impeller depth, impeller diameter, baffle geometry, and liquid height. The central vortex can play an important role in gas dispersion in an unbaffled or par-

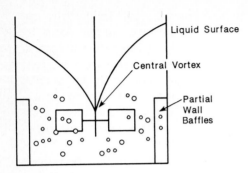

Figure 6.14 Surface entrainment of gas in a partially baffled tank showing the central vortex.

tially baffled tank (Boerma and Lankester, 1968). However, transient loading on the shaft can be expected, and, in multiphase systems, centrifugal separation may occur which is usually not desirable. Blakebrough et al. (1961) noted centrifugal separation in fermentations of filamentous organisms.

Studies of the central vortex include Zlokarnik (1971), LeLan and Angelino (1972), and Rieger et al. (1979) on the vortex depth, Nagata (1975) on velocity profiles, and Brennan (1975) on geometric effects on the behavior of the central vortex. Significant surface aeration or entrainment occurs when the central vortex depth reaches the impeller. Hence, correlations for vortex depth are often given or rearranged to provide correlations for the critical diameter-rotational speed at which significant entrainment occurs.

Clark and Vermeulen (1964) were among the first to provide a correlation for the point at which surface entrainment significantly affected power for various flat blade impellers. The rotational speed for surface entrainment was correlated using:

$$N_{Fr}\frac{D^2W}{T^2H}\left(\frac{C}{H}\right)^{2/3} = 5 \times 10^{-3} \tag{6.32}$$

Small quantities of air were entrained at lower rotational speeds than this with power unaffected. Matsumara et al. (1977) and Albal et al. (1983) discussed surface entrainment. Data by Albal et al. were for a dual pitched blade impeller system.

Greaves and Kobbacy (1981) have studied surface aeration in a 0.2-m-diameter tank and provided a detailed description of the fluid mechanics of surface entrainment. Two rotational speeds were defined: the rotational speed N_{SA1} for the onset of surface aeration where gas bubbles were first detected beneath the surface and the rotational speed N_{SA2} where surface entrainment affected the power draw of the impeller. N_{SA1} was correlated as:

$$N_{SA1} = c_5 \frac{(T^2 H^2)^{1/3}}{D^2} \left(1 - \frac{C}{H}\right)^{1/3} \left(\frac{p}{p_a}\right)^{-0.13} \tag{6.33}$$

where the constant c_5 varied from 0.375 for electrolytic solutions (0.11 M K_2SO_4) to 0.476 for distilled water. N_{SA2} was correlated as:

$$N_{SA2} = c_6 \frac{(T^2 H^2)^{1/3}}{D^2} \left(1 - \frac{C}{H}\right)^{1/3} \left(\frac{p}{p_a}\right)^{-0.13} \tag{6.34}$$

where the constant c_6 was a function of bubble rise velocity and varied between 0.62 for electrolytic solutions (0.11 M K_2SO_4) to 0.82 for distilled water. Greaves and Kobbacy reported data on bubble size versus tank position and impeller rotational speed. Such data would be useful in determining recirculation.

Surface aeration was studied by Heywood et al. (1985) in an impressive study of designs that minimized surface aeration in low-viscosity liquids in ungassed baffled systems. Five different impellers in liquids with different surface tensions were studied. Changes in impeller and tank diameters, clearance, eccentricity, and liquid height were investigated and correlations were developed to design mixing systems that minimized or eliminated surface aeration. Three different rotational speeds were defined where: (1) bubbles first appeared at the surface, (2) bubbles first entered the impeller flow, and (3) the entrained gas affected the power draw.

Surface aeration and entrainment are not typically desired in processing because of contamination, impeller loading, and other problems. Based on Clark and Vermeulen's work, the rotational speed for surface entrainment varied as:

$$N_{SE} \propto D^{-2} C^{-0.33} \tag{6.35}$$

On scaleup, N_{SE} becomes small because of the increase in size of the impellers. As a result, the clearance C should be reduced to offset the decrease in N_{SE} (Nienow et al., 1977).

Surface entrainment under gassing conditions and by solids. Matsumura et al. (1977) and Nienow et al. (1979), Fig. 6.15, have shown that surface aeration was negligible in gassed systems even at very low gassing rates. Apparently, the sparged gas hindered the entrainment of surface gas. Nienow et al. (1979) also found that surface aeration decreased rapidly with increases in gas sparging and was not important in terms of mass transfer under gas sparging conditions. Surface aeration studied under unsparged operations was found of little use in determining the importance of surface aeration under sparging condi-

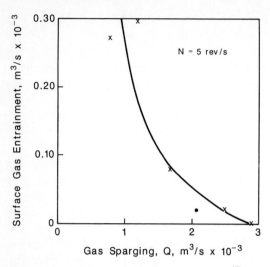

Figure 6.15   Surface gas entrainment rates. *(From A. W. Nienow, C. M. Chapman, and J. C. Middleton, Chem. Eng. J., 17, 111, 1979. By permission.)*

tions. Blakebrough et al. (1961) found that surface aeration had little effect on mass transfer.

Wilkinson (1975) studied the entrainment of air into liquids by solids. Entrainment occurred above a critical capillary number $\mu U/\sigma$ of 1.2.

Gas disengagement.   Mann et al. (1981) reported a study of gas disengagement from an aerated, agitated tank after the termination of gas sparging. Using a four-backmixed "zones-in-loops" model and a limited range of internal flow parameters (the liquid and gas split ratios at the wall and the external distribution coefficient), Mann et al. were able to closely match steady-state gas holdup and the dynamic disengagement gas holdup profiles for the tank. Estimates, based on the work, can be obtained for interfacial area, $k_l a$, and gas residence time distributions.

This particular article deserves additional examination since it emphasized additional parameters besides the external distribution coefficient. The split ratios of gas and liquid at the wall are very important in an aerated, agitated tank and are not typically studied. The split ratio for the gas flow would be useful in the characterization of complete dispersion.

Mixing and circulation times

In a study of the fluid motion and mixing in aerated, agitated tanks, mixing and circulation times are of interest. The following treats mixing in baffled aerated systems. Mixing in unbaffled aerated systems or partially baffled systems is not well studied. Mixing in such systems is expected to be much poorer than in baffled systems because of the formation of solid body rotation and an increased inability of the impeller to disperse gas.

Bryant and Sadeghzadeh (1979) studied circulation rates using a radio pill. Circulation time distributions were obtained and approximated by a log normal distribution. Both the circulation time t_C and the logarithmic standard deviation α_L varied linearly with gassing rate as:

$$\text{Circulation time:} \quad t_C = a + bQ = a\left(\frac{1 + b}{aQ}\right) \tag{6.36}$$

$$\text{Log standard deviation:} \quad \alpha_L = a' + b'Q = a'\left(\frac{1 + b'}{a'Q}\right) \tag{6.37}$$

where b/a was typically 0.1 to 0.14 and b'/a' was 0.043 to 0.053 with Q in m^3/s and t in s. Circulation times and the standard deviations were not strong functions of gassing rate. Mixing time was found to be roughly independent of gassing rate.

Nagase and Yuasi (1983) provided a qualitative understanding for the flow under aerated conditions using wall pressure and holdup measurements and estimated the effect of aeration on the mixing performance. The distributions of gas holdup throughout the tank were found similar to those given by Calderbank (1958). These distributions partially described the bubble flow in the bulk of the tank. Generally, the aerated liquid flow behaved in the same manner as the unaerated liquid flow in an agitated tank, but the mixing times under aeration conditions were longer by 20 to 30 percent. Liquid pumping of the impeller was reduced and circulation, and mixing times were increased because of the gas cavities and aeration. The bubble discharge from the impeller hindered the mixing of the liquid above the impeller with liquid below the impeller.

Prandit and Joshi (1983), following previous work by Joshi et al. (1982) provided the following correlation for mixing times in gas-liquid dispersions:

$$N\theta = 20.41\left(\frac{aH + T}{T}\right)\left(\frac{T}{D}\right)^{13/6}\left(\frac{W}{D}\right)\left(\frac{Q}{NV}\right)^{1/12}\left(\frac{N^2D^4}{gWV^{2/3}}\right)^{1/15} \tag{6.38}$$

where V and W are liquid volume and blade width, respectively. The variable a depended on the size of the recirculation loop and was equal to 1 for a centrally located impeller. The last two terms in the correlation are identical to those used by Hughmark (1980) in his correlation of power and interfacial area, which will be discussed later. There was no effect of sparger type on mixing times. The effect of electrolytes on mixing times was relatively small and was attributed to the viscosity changes because of the presence of the electrolytes. Data showing the effects of superficial velocity on mixing times were given for disk style turbines, propellers, and pitched blade turbines. The article included a study of bubble columns as well. A general review of mixing times in liquids is given in Chaps. 4 and 5.

Models

Residence time distributions (RTDs) of gas and liquid in agitated tanks have been studied to determine the internal nature of the flow and to develop models for the hydrodynamics. However, residence times distributions are not unique, and the results need to be interpreted in terms of the flow mechanisms. Although such work is outside the scope of this chapter, several of these studies are noteworthy.

Gas phase studies.   Hanhart et al. (1963) was perhaps the first to use gas phase residence time distributions to model gas phase mixing. In general, a considerable spread in the residence times was observed which increased with impeller rotational speed. The distributions were considered intermediate between a perfectly mixed condition and two perfectly mixed volumes of equal size in series. Hanhart et al. attributed the two well-mixed volumes to the tank volumes above and below the impeller. They also concluded that the driving force for mass transfer should be based on the composition of the gas leaving the dispersion and not on some average or logarithmic mean value.

Popovic et al. (1983) obtained gas RTD data in fermentation broths to determine the interrelationships between operating conditions, broths properties, and flow patterns. For low-viscosity systems, the gas holdup was divided into a well-mixed fraction m and a fraction p under plug flow conditions where:

$$m + p = 1 \qquad (6.39)$$

The determination of p and m was made from residence time distributions. The mixed fraction m was related to the gas recirculating back to the impeller and was found to decrease with increasing aeration number. Using work by Zlokarnick (1978), the ratio of m to p was correlated as:

$$\frac{m}{p} = K \left\{ \frac{P_g}{[Q \rho_L (\nu_L g)^{2/3}]} \right\}^a \qquad (6.40)$$

where a was roughly 0.25 and K was dependent on physical properties of the broth and coalescence.

For high-viscosity aerated fluids, Popovic et al., noting the appropriate literature, cited that in such systems: (1) the gas cavities behind the impeller blades were more stable than in low-viscosity liquids, (2) a large number of small gas bubbles were formed, and (3) the rate of bubble coalescence was lower. Popovic et al., for reasons not given, considered the rate of dispersion in high-viscosity liquids higher than in low-viscosity systems. For the model, a fraction d representing the small air bubbles, was added, which was typically 5 to 20 percent of the gas holdup. This fraction was included in the model volume balance, which appeared as:

$$d + m + p = 1 \qquad (6.41)$$

Because the well-mixed fraction m consisted mainly of the gas in the stable cavities, m remained fairly constant with aeration number. Under such conditions, the power number remained constant. Both m and p were correlated separately in terms of $Q\rho_L(\nu_L g)^{2/3}$. Increased power input was found to increase the liquid circulation rather than to increase gas dispersion. The fraction d was also correlated to the operating variables and the gas fraction exchanging between m and d. Overall, the models developed by Popovic et al. (1983) pointed to the significant differences between low-viscosity and high-viscosity fluids.

Liquid phase studies. Brown and Halsted (1979) developed a rather impressive liquid phase mixing model for aerated, agitated tanks. Their study and model were based on residence time distributions, but the model was developed to be physically realistic where the parameters of the model were related to physical quantities and known fluid mechanics occurring in the tank. The parameters of their model were given as functions of impeller rotational speed and gassing rate. Brown and Halsted checked their model against unaerated data in the literature. For the aerated cases studied, the parameters and model did not change significantly with gassing rate, and the effect of aeration on overall liquid mixing was very limited. This roughly agrees with the experimental work cited above on mixing and circulation times.

Gas and liquid phase studies. Mann and Hackett (1988) applied the network of zones model as discussed in Chap. 4 to gas-liquid dispersion in aerated agitated tanks. The model incorporated both gas and liquid flows between cells in the network loops and provided dispersion, circulation, disengagement, and recirculation information. Spatial distributions of holdup were obtained, and holdup data of Calderbank (1958) and Nagase and Yasui (1983) were predicted.

Power Draw and Power Consumption in Gas Dispersion

Gassed power draw is energy per time leaving the impeller and entering the fluid in the tank. The power primarily enters the liquid phase. The power dissipation occurs in the liquid phase as well since the volume fraction of the gas phase is relatively small in comparison. Gassed power consumption is the dissipation of this energy by viscosity in the tank. Under steady-state conditions, power draw and power consumption are equal. The mechanism by which power is transferred from the impeller to the fluid is unclear, but obviously the transfer mechanism occurs locally from the impeller by a drag force and relative velocity mechanism.

Power draw under gassing conditions is used as a parameter for the determination of a gas holdup, interfacial area, and mass transfer rates, and in the calculation of operating costs and comparison between different designs. As a result, correlations for power draw-consumption become important. Power consumption is also an independent design variable if so selected. Accurate correlations are hence needed to size agitator drives.

Power draw is an integral quantity dependent on impeller flow regime, which, in turn, is determined by gassing rate, impeller rotational speed, impeller geometry, and fluid properties and rheology. Integral quantities, as such, are easy to measure, but power draw under aerated conditions is an exception to this. The nature of the gas cavities and bubble dynamics in the tank affect the flow patterns and power draw. Typically, in much of the literature on power draw, impeller flow regimes were not established. These flow regimes cause significant changes in power draw from that of single-phase systems. Most power studies have been performed in the small-scale tanks, but such power measurements may not be accurate, given the difficulty in such measurements on small drives. In addition, in gas power studies, the power ratio P_g/P_0 is reported where P_0, the ungassed power number, is assumed to be constant. However, Greaves and Kobbacy (1981) noted that P_0 was not necessarily constant and varied with tank diameter and clearance. Greaves and Kobbacy felt that such an effect would also appear in gassed studies since it appeared in ungassed studies. Whether this is true or not and to what extent remains to be demonstrated.

The following reviews the gassed power draw literature. The division of the literature is based on whether flow regimes were also identified. All power correlations, unless otherwise stated, are for nonboiling liquid systems and noncavitating impellers.

Power draw correlations

Most early correlations for gassed power reported the ratio of gassed power to ungassed power P_g/P_0, a function of aeration number or flow number N_A, N_Q, Fl (or Q/ND^3) only, where Q is the volumetric gas flow rate from the sparger, N is the impeller rotational speed, and D is the impeller diameter.

A word of caution on the use of the aeration number is in order: Aeration number includes effects of impeller diameter D, gas sparging rate Q, and impeller rotational speed N. This parameter can lead to confusion in comparing studies, particularly when comparisons are made between varying impeller rotational speed and varying gassing rate. A low aeration number can mean low gassing rate or high impeller rotational speed which are different states. A high aeration number can mean high gassing rate or low impeller rotational speed, which also are different states. The power input P_g and the amount of recirculation can be quite different for the same N_A. Overall, the use of N_A is ambiguous and has left the power draw information underspecified.

The properties of the liquid and system geometry, including tank, impeller, and sparger geometry, also affect the gassed power input. These effects added further difficulties in the development of a general universal power correlation. As a result, correlations contain an assortment of dimensionless groups that included: impeller Weber number, gas holdup, and impeller Reynolds number, as well as geometric dimensionless groups.

Typically, for power draw data, P_g/P_0 drops rapidly from 1.0 to between 0.3 and 0.6 as N_A approaches 0.035 to 0.05. This provides the approximate order of magnitude that can be expected for such power data curves. Initial studies include those by Oyama and Endoh (1955) and Calderbank (1958). Calderbank provided two curves:

$$\frac{P_g}{P_0} = 1 - 12.6\left(\frac{Q}{ND^3}\right) \qquad \text{for } \frac{Q}{ND^3} < 0.035 \qquad (6.42)$$

and
$$\frac{P_g}{P_0} = 0.62 + 1.85\left(\frac{Q}{ND^3}\right) \qquad \text{for } \frac{Q}{ND^3} > 0.035 \qquad (6.43)$$

Data were also reported by Westerterp (1963) that showed P_g/P_0 as a function of impeller rotational speed and gassing rate.

Michel and Miller (1962) correlated gassed power for disk style turbines in tanks up to 0.3 m in diameter using the following correlation:

$$P_g = c\left(\frac{P_0^2 Q^{0.44}}{N_A}\right)^{0.45} = c\left(\frac{P_0^2 ND^3}{Q^{0.56}}\right)^{0.45} \qquad (6.44)$$

The constant c was dependent on the geometry of the impeller. The liquid specific gravities in the study ranged from 0.9 to 1.6, and the liquid viscosities varied from 0.001 to 0.1 Pa · s (1 to 100 cp). For the disk style turbine, c was found to be 0.08 when units of horsepower, revolutions per minute, feet, and cubic feet per minute were used. Interfacial tension, which ranged from 27 to 72 dyne/cm, had no effect on gassed power. This agrees with studies cited above that found no effect of surface tension and electrolytes on cavity shape. However, a D/T effect was noted by Michel and Miller, which implies an effect due to gas recirculation. This effect is not included in the correlation.

Nagata (1975) provided a gassed power correlation as:

$$\log\left(\frac{P_g}{P_0}\right) = -192\left(\frac{D}{T}\right)^{4.38}\left(\frac{\rho D^2 N}{\mu}\right)^{0.115}\left(\frac{N^2 D}{g}\right)^{1.96(D/T)}\left(\frac{Q}{ND^3}\right) \tag{6.45}$$

Nagata noted that the decrease of P_g/P_0 was larger with a nozzle than with a ring sparger, indicating an effect of sparger-impeller geometry.

Hassan and Robinson (1977) developed a correlation for the ratio of gassed power to ungassed power:

$$\frac{P_g}{P_0} = C_1 N_{We}^m N_A^n \left(\frac{\rho_L}{\rho_D}\right) \tag{6.46}$$

where N_{We} is the impeller Weber number, $N^2 D^3 \rho_L/\sigma$. The exponent n was found independent of impeller type and was about -0.38. The exponent m varied between -0.25 for a disk style turbine and -0.22 for a four-blade paddle, indicating a weak dependency on surface tension. The constant C_1 varied depending on impeller type, tank size at constant D/T, and the strength of the electrolytic solution. No correlating parameters were developed to incorporate ionic strength into the correlation. Hassan and Robinson based their correlation on a force balance between the dynamic forces of the impeller and the surface forces of the gas bubbles. This balance justified the use of the impeller Weber number in the correlations.

Loiseau et al. (1977) used the form of the Michel-Miller correlation for correlating gassed power data for foaming and nonfoaming systems. For nonfoaming systems:

$$P_g = 0.83\left(\frac{P_0^2 ND^3}{Q^{0.56}}\right)^{0.45} \tag{6.47}$$

For foaming systems:

$$P_g = 0.69\left(\frac{P_0^2 ND^3}{Q^{0.56}}\right)^{0.45} \quad \text{for } \frac{P_0^2 ND^3}{Q^{0.56}} < 2 \times 10^3 \tag{6.48}$$

and $$P_g = 1.88\left(\frac{P_0^2 ND^3}{Q^{0.56}}\right)^{0.31} \quad \text{for } \frac{P_0^2 ND^3}{Q^{0.56}} > 2 \times 10^3 \quad (6.49)$$

The units of the correlation are SI units. The accuracy of these correlations was about 20 percent. Data from a system containing a strong surfactant, lauric alcohol, were included in the development of the correlation and were contained within the 20 percent accuracy of the correlation. The system studied was the standard configuration with a constant liquid height. Agreement with power draw data given by Michel and Miller (1962), Pharamond et al. (1975), van't Riet et al. (1976), and Oyama and Endoh (1955) was within 30 percent over the ranges of $1 < P_0^2 ND^3/Q^{0.56} < 10^7$ and 0.0005 m/s $< V_s < 0.09$ m/s. The different correlations indicated that foaming systems affected P_g in a different manner than nonfoaming systems.

Yung et al. (1979) reviewed the literature on gassed power correlation and found that the Michel-Miller correlation was adequate to correlate their data for both a standard turbine (disk style turbine) and paddle (flat blade) impeller. The value of c for the Michel-Miller correlation was found slightly dependent on impeller type: 0.812 and 0.829 for the standard turbine (disk style turbine) and paddle (flat blade) impeller, respectively. No effect of liquid properties or ionic concentration was observed in the data and the gassed power was found to be independent of the geometry of the tank bottom.

Luong and Volesky (1979) critically reviewed the various correlations for P_g/P_0. For Newtonian fluids and a CMC solution, 0.2 percent by weight, they recommended the following correlations which are similar to the correlation by Hassan and Robinson (1977):

$$\frac{P_g}{P_0} = 0.497\left(\frac{Q}{ND^3}\right)^{-0.38}\left(\frac{\rho_L D^3 N^2}{\sigma}\right)^{-0.18} \quad (6.50)$$

and for higher weight percent CMC solutions of 0.4 and 0.6 percent by weight:

$$\frac{P_g}{P_0} = 0.514\left(\frac{Q}{ND^3}\right)^{-0.38}\left(\frac{\rho_L D^3 N^2}{\sigma}\right)^{-0.194} \quad (6.51)$$

The exponents of -0.18 and -0.194 indicate again a weak relationship between P_g/P_0 and the impeller Weber number.

Hughmark (1980) reviewed the various correlations for the gassed power ratio P_g/P_0, noting that most published work involved only the disk style and flat blade turbines. Data were analyzed for different fluid properties and a range of impeller and tank diameters and a correlation was developed for the disk style and flat blade turbines in Newtonian fluids without electrolytes. Data sets, 243 in all, were obtained from Michel and Miller (1962), Pharamond et al. (1975), and Luong and Volesky (1979) and were correlated for gassed power ratio P_g/P_0 as:

$$\frac{P_g}{P_0} = 0.10\left(\frac{Q}{NV}\right)^{-1/4}\left(\frac{N^2 D^4}{g W V^{2/3}}\right)^{-1/5} \tag{6.52}$$

where V and W are liquid volume and blade width, respectively. Hughmark noted that: (1) the relationship between P_g/P_0 and the impeller Weber number was not statistically significant, (2) the aeration number should be modified to include liquid volume, and (3) a Froude number, related to the velocity fluctuations in the impeller stream, reduced the residual variance in the correlation. The last term in the correlation is this Froude number. Using an additional 148 data sets from Bimbinet (1959) for a total of 391 data sets, Hughmark's correlation demonstrated an average absolute deviation of 11.7 percent.

Kudrna, as cited by Machon et al. (1980), proposed the following correlation for P_g/P_0 as:

$$\frac{1}{1 - P_g/P_0} = K_1 + K_2\left(\frac{P_g/P_0}{N_A^2}\right)^{0.5} \tag{6.53}$$

Machon et al. (1980) fitted power data for aerated CMC solutions using the empirical constants $K_1 = 1.26$ and $K_2 = 4.09 \times 10^{-2}$. For glycerol data from Calderbank (1958), the constants were $K_1 = 1.76$ and $K_2 = 3.81 \times 10^{-2}$. Machon et al. (1985) provided data for the constants of the Kudrna equation for both one impeller and dual impeller systems.

Greaves and Kobbacy (1981) developed an impeller dispersion efficiency η defined as the ratio of the gas coalescence rate in the impeller region to the gas sparging rate. From this, they developed a general gas power correlation given as:

$$P_g = 1007\left(\frac{N^{3.33} D^{6.33}}{(\eta Q)^{0.404}}\right) \tag{6.54}$$

where P_g is in watts and η was correlated depending on whether there was recirculation or not. This power correlation has the advantage over that by Michel and Miller because the entire gas flow rate to the impeller is used through the ηQ term. Efficiency index η was correlated as 1 when N was between N_F and N_R, as η_F for $N < N_F$, and as η_R for $N > N_R$. For recirculation:

$$\eta_R = 4.13\left(\frac{N^{2.53} D^{5.93}}{T^{2.45} Q^{0.33}}\right) \tag{6.55}$$

and for flooding:

$$\eta_F = 0.4A \qquad\qquad \text{for } A < 1 \qquad\qquad (6.56)$$

$$\eta_F = 0.25 + 0.26A \qquad \text{for } A > 1 \qquad\qquad (6.57)$$

where A is:

$$A = \left(\frac{N^{2.53}D^{4.40}}{T^{0.5}Q^{0.73}}\right) \qquad\qquad (6.58)$$

Similar equations were given for ionic solutions in the same form with different constants.

Brown (1981) found that wattmeters were useful in the measurement of power draw in production scale fermenters. The gassed power number was correlated as:

$$N_{Pg} = \zeta N_P e^{-\beta Q} \qquad\qquad (6.59)$$

where ζ and β were tabulated as functions of gassing rates. The data also followed the Michel and Miller correlation.

Gray et al. (1982), noting the limited gassing rates and geometry studied by Michel and Miller (1962) and Hassan and Robinson (1977), developed a more general correlation for the disk style impeller in a baffled tank. With data from Bimbinet (1959) and Hassan and Robinson (1977), their final equation for total power input P_T which included power input due to rising bubbles, was:

$$\frac{P_T g_c}{\rho_L N^3 D^5} = 0.75\left(\frac{g^2 TC}{(V_s/\phi)^2 N^2 D^2}\right)^{0.25} \qquad\qquad (6.60)$$

where C is the impeller clearance. This correlation contains two Froude numbers: one for the liquid with the liquid velocities based on the impeller rotational speed and one for the gas with gas velocities based on the superficial gas velocity and gas holdup. Gassed power varied with clearance as:

$$P_g \propto \left(\frac{C}{D}\right)^{0.29} \qquad\qquad (6.61)$$

This effect of clearance on power draw was the same as that found by Gray et al. for single-phase systems.

Baczkiewicz and Michalski (1988) studied power draw and aeration rates in acetic acid fermentation using a self-aspirating tubular agitator. The agitator had gas feed tubes, 5.72×10^{-3} m diameter, as impeller blades. The tank was 0.44 m in diameter and had a liquid volume of 0.1 m^3. The ranges of operating variables were: $6.4 < N < 23$

s^{-1}, $1.39 \times 10^{-4} < Q$ 6×10^{-4} m^3/s, $0.35 < D/T < 0.55$, and $2 < n_p < 6$. The power was measured with a wattmeter. The gassed power per volume was correlated as:

$$\frac{P_g}{V} = 0.62 \, (N)^{3.44} \left(\frac{Q}{V_L}\right)^{-0.31} \left(\frac{D}{T}\right)^{4.48} (n_p)^{1.07} \qquad (6.62)$$

which shows the effect of blade number. The flow rate of gas through the impeller was correlated as:

$$\frac{Q}{V_L} = 2.36 \times 10^{-4} (N)^{1.53} \left(\frac{D}{T}\right)^{1.33} (n_p)^{0.26} \qquad (6.63)$$

The units are SI units.

Power draw based on flow regime-impeller hydrodynamics

The hydrodynamics and power draw are interrelated and are combined in the more recent studies of gas dispersion. Power during gassing decreases because the streamlining of the impeller blade by the gas cavities causes a reduction in form drag or pressure drag of the blade. The pressure drop from the front of the blade to inside the gas cavities becomes less. The high-energy dissipative flows such as the trailing vortex systems become less dissipative at higher gas loadings to the point where they no longer exist. The gassed to ungassed power ratio, approached from drag analysis, is a shape measure of the cavity streamlining. Another view is that the gas cavities cause a reduction in the apparent density that the impeller experiences (Saito and Kamiwano, 1988). In any case, gas cavities interfere with the mechanism of power transfer from the blade to the liquid that occurs in single phase liquids. The manner in which the cavities interfere with the transfer of power and how they may change this mechanism of power transfer remain unclear.

It is more common to report the gassed power number and/or the gassed to ungassed power ratio than to report gassed power directly. The following summarizes the power correlations and data as a function of flow regime; however, there appears to be no particular way to collapse the power data into one correlation. A major difficulty is the determination of gas recirculation to the impeller which should be accounted for in the correlations. However, efforts are being made in that direction, e.g., Greaves and Kobbacy as cited above and those cited below.

Greaves and Barigou (1988) studied gassed power of a disk style turbine in a fully baffled 1.0-m-diameter tank ($C/T = 0.25$ and $D/T = 0.25$, 0.33, and 0.5) using air water and air electrolyte systems.

A single-point sparger was used. Gassing rate varied between 1.6 × 10^{-3} and 8.3 × 10^{-3} m³/s. Shaft power was measured using a precision calibrated torquemeter. Shaft speed was controlled to within 0.2 percent. The overall power correlation was given as:

$$P_g = 706.3 N^{3.01} Q^{-0.45} \left(\frac{D}{T}\right)^{5.38} \text{W/m}^3 \qquad (6.64)$$

Power data was also divided according to flow regime as:

Vortex-clinging cavity regime:

$$P_g = 441.4 N^{3.13} Q^{-0.50} \left(\frac{D}{T}\right)^{5.82} \text{W/m}^3 \qquad (6.65)$$

Large cavity regime:

$$P_g = 1737.1 N^{2.99} Q^{-0.31} \left(\frac{D}{T}\right)^{5.98} \text{W/m}^3 \qquad (6.66)$$

No effects of the ionic properties of the different solutions were found. The division of the data into flow regimes was significant at the 99.5 percent confidence level. The influence of gassing on power in the vortex clinging cavity regime was higher than in the large cavity regime.

Recirculation is a distributed process that depends greatly on the tank size and internal baffling arrangements. The transition to recirculation from a nonrecirculation condition is a gradual process, not nearly as abrupt as the flooding transition. As a result, it is difficult to model recirculation and correlate power draw.

Correlation requirements.   It is important to distinguish between what is needed in a gassed power correlation and present correlation methods. As noted, the correlation by Michel and Miller (1962) is quite successful and has been accepted by many. However, the correlation does not explicitly account for changes in impeller geometry for example. It has been shown that blade number, blade geometry, and impeller diameter change the gas-handling capability of the impeller. On close examination (Nienow et al., 1977), the success of the correlation by Michel and Miller can also be explained. Gassed power in a gas dispersion varies as:

$$P_g \propto N^3 D^5 \qquad (6.67)$$

and, if the gassing rate remains constant or is ignored [i.e., $(Q^{0.56})^{0.4}$ is a weak relationship], the right side of the correlation becomes:

$$(P_0^2 N D^3) \propto N^7 D^{13} \qquad (6.68)$$

Plotting the two quantities P_g and $P_0^2 N D^3$ versus each other results in an exponent of 0.4 for $(P_0^2 N D^3)$.

The inability of the given power correlations to determine gassed power draw can be attributed primarily to the complexity of the gas-liquid hydrodynamics occurring around the impeller blades and gas recirculation (Warmoeskerken and Smith, 1982) and the lack of use of such information in correlations. To say that P_g varies with N^3 is too simple since P_g is a complicated function of the impeller hydrodynamics (Warmoeskerken and Smith, 1982). Among the phenomena which require attention are the inflection point of P_g/P_0 and the final flooding-nonflooding transition. The inflection point in the power draw curves, which occurs around P_g/P_0 of 0.6 to 0.7, should be related to a hydrodynamic condition. Some order must also be given to P_g/P_0 at the flooding transition condition since this ratio can vary from 0.2 to 0.6. The gas flow number or aeration number $Q_s/N D^3$ is not the most appropriate parameter to correlated gassed power draw. Instead, a gas flow number or aeration number based on the total gas flow to the impeller would be more appropriate.

The correlation of P_g/P_0 with operating parameters is a classic example of flow regime characterization. Gassed power should be a function of the total gas it receives and the nature of its gas cavities and the impeller geometry. This requires knowledge of gas recirculation. As yet, there is no accepted correlation for the prediction of the amount of gas recirculated into the impeller.

To collapse the P_g/P_0 curves into one single correlation, the nature of the recirculation, impeller geometry, and gas cavity shape must be included in the correlation. Correlations should appear as functions of fluid properties, $Q/N D^3$, $Q_R/N D^3$, impeller and cavity shape parameters and P_g/P_0 at the inflection point and the flooding transition. The ratios of P_g/P_0 at the inflection point and at the flooding transition may uniquely define the changes in cavity shape, although further work is necessary to support this (Warmoeskerken and Smith, 1982). The cavity shape is thought to be fixed for a specific rotational speed and has a significant effect in determining P_g/P_0 (Warmoeskerken and Smith, 1982). The correlation of P_g/P_0 is indirectly an attempt to correlate the shape of the gas cavities since P_g/P_0 at the flooding transition and the inflection point or the formation region of the 3-3 structure depend directly on cavity shape (Warmoeskerken and Smith, 1982).

It may not be possible to develop a single power curve correlation since every case may be specific to the cavity shape and the nature of its growth as gassing rate is increased or rotational speed is decreased. Specific to rotational speed, there is a specific maximum cavity size before flooding. At the inflection point where the 3-3 structure is thought to form, there are specific sizes of cavities during the formation of the three large cavities (Warmoeskerken and Smith, 1982). Furthermore, to obtain a single correlation to predict across the flooding transition may be expecting too much. There is no reason why there can be a single correlation that spans from $P_g/P_0 = 1$ into the flooding state since there are substantial flow regime changes involved.

The nature of gas cavities under flooded conditions have not been studied, and the concept of blade-bound cavities may be inappropriate. The power draw of an impeller under flooded conditions may be better correlated using parameters that describe the bubble-liquid plume rather than those associated with cavities.

Power data

There are two power data curves typically reported as shown in Fig. 6.16: one with increasing gas flow rate at constant N, Fig. 6.16a, the other with increasing rotational speed at constant Q, Fig. 6.16b.

Single-disk style turbine. Gassed power curves, obtained by increasing the gassing rate Q (Warmoeskerken and Smith, 1985) show a smooth reduction in P_g/P_0 from 1 to approximately 0.45 as flooding is approached. As P_g/P_0 changes, various cavity systems are encountered: vortex cavities, clinging cavities, the 3-3 structure and the ragged cavities of flooding as shown in Fig. 6.17.

For $Fr_I > 0.045$ and at the transition to flooding, there is a step in-

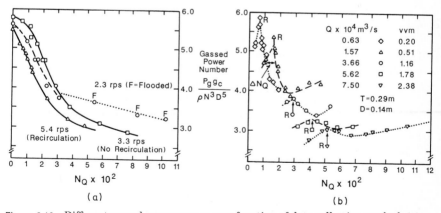

Figure 6.16   Different gassed power curves as a function of data collection method. (a) Increasing Q at constant N; (b) increasing N at constant Q. *(From A. W. Nienow, D. J. Wisdom, and J. C. Middleton, Proc. 2nd Eur. Conf. on Mixing, St. John's College, Cambridge, England, BHRA Fluid Eng., Cranfield, England, F1-1, 1–16, 1977. By permission.)*

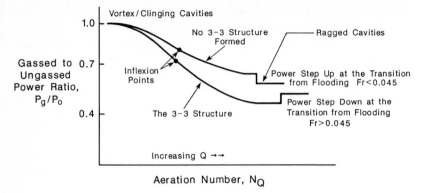

Figure 6.17 Gassed power ratio and different gas flow regimes-gas cavities. *(From A. W. Nienow, M. Konno, M. M. C. G. Warmoeskerken, and J. M. Smith, Proc. 5th Eur. Conf. on Mixing, Wurzburg, Germany, BHRA Fluid Eng., Cranfield, England, 143, 1985. By permission.)*

crease in P_g/P_0 of about 0.05 for the transition from the 3-3 structure to the large cavities of flooding. This step increase is attributed to the impeller pumping more gas under conditions of good gas dispersion than under flooding conditions. More gas pumped means more streamlining of the gas cavities than that occurring under flooded conditions. At P_g/P_0 of about 0.7 or around the inflection point in the curve, one or two large cavities form on the six-blade impeller, but this arrangement is unstable and the 3-3 structure appears. The 3-3 structure consists of three large cavities and three clinging cavities in the form of a symmetrical pattern around the six-blade impeller. Whether the transition to the 3-3 structure has anything to do with the properties of the inflection point with regard to the hydrodynamics remains unclear. The 3-3 structure remains coherent over a large range of gassing rates for P_g/P_0 between 0.7 and 0.4 if $Fr_I > 0.045$.

The 3-3 structure does not appear below an impeller Froude number of 0.045. The reason for this particular magnitude of Froude number is unclear, although it is attributed to buoyancy effects preventing the formation of large cavities in the 3-3 structure system. Below the critical Froude number, the inflection also occurs in the power curves; the transition to flooding occurs between vortex and clinging cavities to the ragged cavities. At the transition to flooding, there is a step decrease in power, indicating that: (1) the ragged cavities streamline the impeller blades more effectively than vortex-clinging cavities or (2) a drastic decrease in pumping occurs which draws less power. Since the 3-3 structure never forms, the inflection point cannot universally indicate the onset of the 3-3 structure. Between 1.0 to 0.7, the vortex and clinging cavities are most prevalent.

For both Froude number regimes and at higher gassing rates, the

P_g/P_0 curve levels off as flooding is approached which indicates that the cavities reach a fairly constant size (Warmoeskerken and Smith, 1982).

Warmoeskerken, Feijen, and Smith (1981) collapsed the power draw curves to one curve up until flooding, using a definition of recirculation based on α, ϕ, and impeller pumping to define an appropriate aeration number as given above. The data are shown in Fig. 6.18.

Gas power curves P_g/P_0 can also be plotted as a function of changing impeller rotational speed (Nienow et al., 1977; Nienow et al., 1985). Under such conditions, N_A decreases as rotational speed increases at a fixed gassing rate. The flow regime corresponding to the ratio P_g/P_0 moves from a flooded state associated with low values of the ratio, P_g/P_0. Above N_F, a local maximum in P_g/P_0 occurs which is considered as the onset of recirculation. Typical data are shown in Fig. 6.16b in terms of power number. Ismail et al. (1984) discussed the different flow regimes as P_g/P_0 varied with impeller rotational speed.

Warmoeskerken and Smith (1989) reported power data for the hollow blade disk style turbine. Power actually increased upon gassing and, at high relative aeration numbers, the relative power reached 0.6. The standard flat blade disk style impeller reaches values of 0.25 to 0.40 for relative power.

Power draw curves have not been reported for the case where the impeller diameter D was varied in the aeration number.

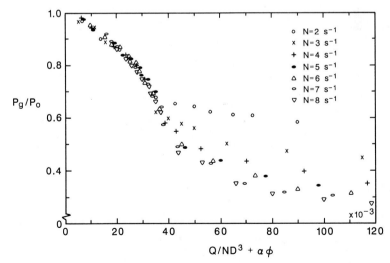

Figure 6.18   Collapsed gassed power curves as a function of the sum of N_Q and $\alpha\phi$.   *(From M. M. C. G. Warmoeskerken and J. M. Smith, Proc. 4th Eur. Conf. on Mixing, Noordwijkerhout, Netherlands, BHRA Fluid Eng., Cranfield, England, G1, 237, 1982. By permission.)*

Pitched blade turbine. The pitched blade turbine is not considered relevant to gas dispersion, but it is important as a pumping and recirculating impeller and as part of a multiple-impeller system. As with radial flow impellers, the pitched blade turbine can have a disk but, since this defeats the pumping of the impeller to some degree, pitched turbines are usually studied without disks. The pumping direction of the pitched turbine can be up or down, but pumping upward has been neglected in much of the early literature. Such a pumping direction does have advantages. In multiple arrangements, two pitched turbines can mimic a disk style turbine by having the top impeller pump down and the bottom impeller pump up. Other multiple impeller arrangements include a pitched blade turbine with a disk style turbine, with the pitched blade turbine pumping up and the disk turbine being the bottom impeller. Various other arrangements are possible, but the most optimum arrangements for gas dispersion are not necessarily what would be assumed from single-phase studies.

Generally, pitched blade turbines do not form the type of cavities that disk style turbines form under gassing conditions. Smith (1985) should be consulted for details concerning the various cavity types. Blade angle is important. Vortices are formed by the pitched turbine, one per blade, at the blade tip. For pitched blade impellers with bolted blades, vortices also form at the bolt joint. In one sense, the vortices are wing tip vortices, and, in another sense, the vortices are roll vortices or mixing layer vortices formed between two streams with different velocities (i.e., the large axial flow through the impeller and the flow in the bulk of the tank). These vortices and the subsequent cavities formed during gassing are much more coherent than those formed with the disk style turbine. The gas cavity structures for the pitched turbine have not been well studied and are not exactly cavities but resemble gassed vortex cores or ligaments instead. The gas or gas cavities cannot easily reside on the blade because of the hydrodynamics of the pitched blade turbine, the tilt of the blade, and the large axial flow through the center of the turbine. Work by Warmoeskerken et al. (1984) and Smith (1985) include other comments concerning gas dispersion using the pitched blade turbine.

The power draw from the pitched blade turbine, pumping up, changed very little as a function of gassing rate except at high gassing rates where large cavities reside on the blade (Smith, 1985). This particular pumping mode is advantageous. Studies reporting data on gassed power draw include Chapman et al. (1983) and Smith (1985). Some of the data by Chapman et al. are shown in Fig. 6.19. Generally, gassed power and power number remain in the vicinity of the ungassed values except when large cavities are present. The pitched blade turbine, pumping down, is not recommended since the power

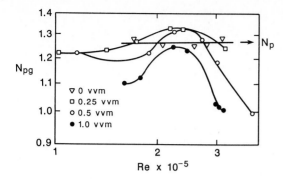

Figure 6.19 Power number versus impeller Reynolds number for a four-blade pitched blade turbine pumping up. *(From C. M. Chapman, A. W. Nienow, M. Cooke, and J. C. Middleton, Chem. Eng. Res. Des., 61, 82, 1983. By permission.)*

draw varied greatly in an unsteady-state manner (Chapman et al., 1983).

Bujalski et al. (1988) reported correlations for the aeration number at flooding and complete dispersion for the six blade pitched blade turbine, $D/T = 0.5$, in the up-pumping mode as:

At flooding:

$$(Fl_g)_F = 6000 \ (Fr)_F^{2.7}\left(\frac{T}{1.0 \text{ m}}\right)^{1.55} \tag{6.69}$$

And at complete dispersion:

$$(Fl_g)_{CD} = 12.1 \times 10^4 \ (Fr)_{CD}^{5.0}\left(\frac{T}{1.0 \text{ m}}\right)^{2.8} \tag{6.70}$$

Correlations and flow patterns regarding the pitched blade turbine in the down-pumping mode were also given.

Multiple-impeller systems. Work on dual- or multiple-impeller systems include: Hicks and Gates (1976), Nienow and Lilley (1979), Kuboi and Nienow (1982), Nishikawa et al. (1984), Machon et al. (1985), Roustan (1985), and Smith et al. (1986). The advantage of multiple-impeller systems is the capability of handling more gas. Such work is usually based on past operating knowledge obtained from single impellers but studies reported here have logically investigated a variety of arrangements of different impellers. Power draw, gas holdup, interfacial area, and $k_l a$ are of interest for multiple-impeller systems just as with single impellers.

The primary difficulty in studying multiple-impeller systems is the

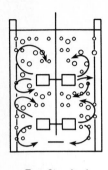
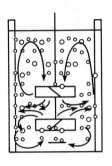
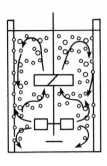

| Two Standard Disk Style Turbines | Two Pitched Blade Turbines Mimicking a Disk Style Turbine | A Disk Style Turbine with a Pitched Blade Turbine Pumping Up |

Figure 6.20   Different multiple impeller systems used in gas dispersion.

enormous variations in design. Different types of impellers at different positions can be used with differing impeller blade number and impeller spacings. Internal baffling arrangements can be varied. D/T effects can be present, and flow regimes on different impellers may interact. Recirculation effects can affect the flow, and sparging locations at different gassing rates for different impellers can be selected. Impellers are usually mounted on the same shaft and rotate at the same speed. If the impellers were on different shafts, the impeller rotational speed for each impeller is an additional design variable. Side- and bottom-entering agitators in conjunction with top-entering agitators are also design options. The cycling of the gas flow rate through individual impeller spargers is also possible.

Kuboi and Nienow (1982) studied eight combinations of dual impellers consisting of disk style turbines and pitched blade turbines. The gas dispersion was assessed visually. The impeller configurations which provided good dispersion characteristics were: (1) two standard disk style turbines, (2) two 45°, pitched blade turbines with the bottom turbine pumping up and top turbine pumping down, and (3) a bottom disk turbine with a pitched turbine above pumping up. The configurations are shown in Fig. 6.20.

The traditional mode of pumping downward for the pitched blade turbine showed torque and axial flow instabilities. The pitched blade turbine pumping upward also showed an inability to disperse gas at low rotational speeds typical of pitched blade turbines. The gas cavities formed on the disk and pitched turbines were similar to those formed on single turbines. Gassed power draw was high for the disk impeller system. In this system, the power for the top second impeller can be estimated for design purposes using the following equation:

$$\left(\frac{P_g}{P_0}\right)_2 = 0.5\left[\left(\frac{P_g}{P_0}\right)_1 + (1 + \phi)\right] \tag{6.71}$$

where $(P_g/P_0)_1$ is for the impeller on the bottom, estimated from single impeller correlations, and ϕ is the gas holdup.

Upward pumping (bottom position)-downward pumping (top position) pitched blade turbines dispersed gas at a low power draw which was insensitive to gassing rate. The disk turbine with a upward-pumping pitched turbine showed large hysteresis loops in flow patterns and significant step changes in power draw. As a result, the optimum geometries for a dual impeller system were: (1) two standard disk style turbines and (2) two 45°, pitched blade turbines with the bottom turbine pumping up and top turbine pumping down.

Roustan (1985) studied power draw in nonstandard vessels, $H = 2T$, using two and three disk style turbines in tanks with hemispherical bottoms. Without gas, two-impeller systems drew twice the power of one-impeller systems. When sparged from the bottom, the lower impeller became flooded. Power draw data curves were given for two- and three-impeller systems and D/T effects were noted. The presence of a third impeller did not increase holdup significantly.

Machon et al. (1985) provided power curves for three multiple impeller configurations: (1) two turbines, (2) one turbine and one pitched blade turbine pumping up, and (3) one turbine and one pitched blade turbine pumping down. For single disk turbine systems, Machon found that the flooding correlations all reduced to:

$$N^A Q^{-1} = k \tag{6.72}$$

where the literature values for A varied between 2.0 to 3.0. Machon et al. found A to be 3.0 for single turbines and 3.43 for dual turbines. The disk style turbine with the pitched blade turbine pumping upward was also found to be a reasonable configuration for gas dispersion. Apparently, in the pumping-up mode, recirculation of gas to the turbine was less, permitting the sparged gas to be dispersed more efficiently. The configuration with the turbine and pitched blade turbine pumping downward was an unsuitable configuration for dispersion. Where the impellers for different configurations were spaced about one vessel diameter apart, the power draw and power numbers of the dual configuration were found equal to the sum of the quantities for separate impellers.

Smith et al. (1986) established flow regimes maps for multiple-impeller designs and developed relationships between operating variables for standard disk turbine systems. In such systems, the bottommost turbine behaved as a single impeller under sparging conditions. The upper impellers behaved differently and received gas only on a recirculation basis unless they were sparged to directly. The behavior of the upper turbines was not similar to a single impeller under gas sparging conditions because of the larger amounts of recirculated gas received by the upper turbines.

The H/T ratio used in the Smith study was 3, with three disk style turbines being studied. For single-phase studies, when the impeller spacings were smaller than $1.5D$, the power draw dropped off rapidly to that of a single impeller. A spacing of $1.5D$ was the point where power draw was 90 percent of that of two separate impellers. Smith et al. used a spacing of $2.5D$ in the study. Sparge rings were also placed below each turbine, and the experimental equipment had the capability of detecting the formation of the 3-3 structure at each impeller.

Where gas was sparged to the bottom impeller, the transition to the 3-3 structure from the vortex-clinging cavity structures for the impeller occurred at approximately the same operating conditions as that for a single impeller. For this geometry, $Fl = 0.39$. The transition to the 3-3 structure from the vortex-clinging cavities occurred at much higher gassing rates for the upper impellers since these impellers received only recirculated gas. Under conditions for the 3-3 structure formation, the power draw for the upper impellers was higher than for the lower impeller. However, increasing gas recirculation at higher impeller speeds permitted the 3-3 structure to form on the upper impellers as well, which balanced power draw. The development of the 3-3 structure on the upper impellers occurred following the relationship:

$$Fr_i^* Fl = 0.019 \tag{6.73}$$

The 3-3 structure formation for all higher level impellers occurred at:

$$Fr_i^* Fl = 1.85\left(\frac{T}{D}\right)^{-5} \tag{6.74}$$

which accounts for the effect of scale.

Smith et al. (1986) studied gas supplied to the middle impeller as well. Cavity development was influenced by gas flow rates to the impellers. When gas was supplied only to the middle impeller, the transition to the 3-3 structure on the middle impeller behaved in a similar manner as that of a single impeller. With gas to both impellers, the transition lines were displaced.

It is well known that the bottom impeller floods first in a multiple-impeller system, long before the flooding of the upper impellers. In their study, Smith et al. found that (1) the gas flow number at the flooding transition for the upper impellers was approximately twice the gas flow number at flooding for the lowest impeller and (2) the two upper impellers behaved similarly over the entire range of gas flow numbers, drawing more power than the bottom impeller. Smith et al. also provided a fairly complete flow regime map for their multiple-impeller system.

Nocentini et al. (1988) studied the gas behavior of a multiple disk style impeller system using a two-phase axial dispersion model and residence time distribution curves of a tracer injected in the gas phase. The RTD curves were used to evaluate the axial dispersion coefficient and the overall volumetric mass transfer coefficient. The mixing in the gas phase was intermediate between perfect mixing and plug flow. The gas phase Peclet number $V_s H/\phi D_G$ ranged from 10 to 30 and decreased with increasing impeller rotational speed and decreasing gassing rate.

Abrardi et al. (1988) discussed flow regimes for a two disk style impeller system. Different flow regimes resided on the impellers, and flooding and loading coexisted together on the two impellers. Instabilities in flow regimes frequently occurred between the impellers. Abrardi et al. found very poor agreement between their data and the power equation by Kuboi and Nienow (1982) given above. For their two disk style impeller system, the power number was 10 and was practically independent of conditions. For two impeller systems, they suggested the use of the empirical correlation based upon the work of Michel and Miller and provided the following correlation:

$$P_g = 1.224 \left(\frac{P_0 N D^3}{Q^{0.56}} \right)^{0.432} \tag{6.75}$$

for their power calculations. The distance between their two impellers was $2D$.

Non-Newtonian power correlations and effects on gassed power draw-consumption. Drag-reducing additives, such as polyacrylamide (PAA), polyethylene-oxide (PEO), and fibers, cause a reduction in power draw and torque in agitated tanks. Quraishi et al. (1976) found the reduction was enhanced by the presence of the secondary gas phase under aeration, in some cases, by more than 70 percent. However, the torque suppression was not simply a sum of contributions from the drag-reducing material and aeration. Apparently, the drag-reducing materials affected the nature of the gas cavities by stabilizing cavities and the gas dispersion. The actual level in torque or power reduction was fairly independent of aeration number but was a function of concentration of the drag reducing material. Ranade and Ulbrecht (1977) found similar results.

The flooding point (Quraishi et al., 1976) was also suppressed because of the stabilization of the cavities. The torque suppression was greater for larger impellers as well, which requires consideration during scaleup. Machon et al. (1980) have also provided data on the effects of shear thinning CMC solutions on power draw in the correlation developed by Kudrna cited above.

Ranade and Ulbrecht (1977) reported power draw for gas dispersion in the transition and turbulent regimes for corn syrup and solutions of polyacrylamide (PAA). Power number, Fig. 6.11, was independent of aeration number after the initial gassing had formed stable cavities behind the impeller blades. Correlations for the power ratio P_g/P_0 appeared as:

$$\frac{P_g}{P_0} = K \tag{6.76}$$

as shown in Fig. 6.12, where the constant K depended on the impeller Reynolds number and the concentration and rheological properties of the PAA solution. P_g/P_0 was independent of aeration number in this regime.

Nienow et al. (1983) studied a variety of fluids including a shear thinning, viscoelastic fluid with a yield point. In the turbulent regime, Re > 900, the power draw in the rheologically complex solutions was not much different from that of water. Between 10 < Re < 900, the power draw was independent of gassing rate; however, higher levels of elasticity and the presence of a yield stress led to the lowest power numbers. For Re < 10, the aerated and unaerated power numbers were the same. Nienow (1984) reported similar data for Xanthan Gum with additional data on holdup and mass transfer coefficients.

Dynamic studies of power draw.   Greaves and Economides (1979) studied the dynamics of gas-liquid contacting in aerated vessels. Positive changes in gas flow rate were introduced, and the power response was measured. Different responses at fixed speeds were observed depending on impeller diameter, gas flow rate, and the rotational speed for recirculation N_R. Below N_R, step responses in power were very rapid and abrupt. In the recirculation regime, the step responses were slower. The responses appeared to be first order, and the time constants could be used directly to estimate the rate of recirculation. Greaves et al. (1983) also studied the dynamic response in power of the disk style turbine following step changes in gassing rates. The power change was modeled with a linear first-order model. Gas cavities noted above were used to explain the results. The dynamic changes in the cavities and power depended on impeller speed and gassing rate.

Interfacial Area, Holdup, and Bubble Diameter

Interfacial area, holdup, and bubble diameter describe gas dispersion. The proper approach in modeling a gas dispersion process is to use the equations of continuity and motion with bubble population balances including bubble size and concentration distributions, bubble coalescence, and dispersion mechanisms. To perform such modeling in a distributed fashion is extremely difficult since fundamental knowledge is unavailable. Work, however, is developing, e.g., Midoux and Charpentier (1979).

The integral quantities of interfacial area, holdup, and bubble diameter are reported in the technical literature and aid in the understanding of mass transfer in agitated tanks. Bubble concentrations are not discussed in the literature but are implied in data on the integral quantities of holdup ϕ, interfacial area a, and bubble diameter D_b.

Gas holdup ϕ is the dimensionless volume fraction of the gas phase in the dispersion and is estimated by a volume or area balance performed on a cross section of the tank, summed over all bubbles:

$$V_t \, \Sigma \, (\pi D_b^2) = V_s(\pi T^2) \tag{6.77}$$

where V_t is the rise velocity of the bubbles and V_s is the superficial gas velocity, $4Q/\pi T^2$ as shown in Fig. 6.21. Holdup is also the dimensionless area fraction of the bubbles or the ratio of $\Sigma \, (D_b^2)/T^2$, which is also the ratio of superficial velocity V_s, to the rise velocity V_t, of the bubbles or:

$$\phi = \frac{V_s}{V_t} \tag{6.78}$$

Air bubbles in water, ranging in size from 2 to 4 mm, rise at a con-

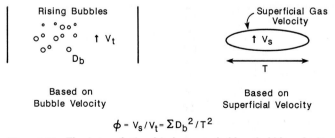

$$\phi = V_s/V_t = \Sigma D_b^2/T^2$$

Figure 6.21 The interrelationship between holdup, bubble velocity, and superficial velocity with bubble and tank diameters.

stant velocity of about 260 mm/s (Calderbank, 1958). The effects of the liquid phase viscosity are not important, and, as a result, viscosity is not included in holdup correlations for low viscosity liquids.

Holdup ϕ, interfacial area a, and bubble diameter D_b are related as:

$$D_b = \frac{6\phi}{a} \tag{6.79}$$

where the factor of 6 is based on the volume of a sphere divided by the surface area. Bubble diameter, holdup, and interfacial area can be measured independently and do not necessarily follow this relationship. Holdup values are typically between 0 and 25 percent; the diameter of air bubbles in water range from 0.5 to 20 mm; and interfacial area, 0.1 to 5.0 cm^{-1}.

Bubble diameter, interfacial area, and holdup can be measured by a number of experimental methods. Some experimental methods result in average quantities for the whole tank while other methods are specific to a local volume in the tank. Before using a particular correlation, the experimental methods used to obtain the data should be noted.

General correlations

In the dispersion of a gas bubble, viscous and interfacial forces of the bubble are present, but interfacial forces must be considered most important since viscous forces of a gas can be considered small. Any resistance to breakup by viscous forces is because of the continuous phase. Breakup is caused by dynamic forces of the continuous phase that impact on the bubble and are resisted by interfacial forces.

Interfacial area, holdup, and bubble diameter are integral quantities that are easily estimated from tank conditions. Correlations for these are available. Bubble size distributions and concentration and how these are distributed in the tank are more difficult to measure and correlate.

Vermeulen et al. (1955) studied interfacial area in an enclosed tank with a flat blade impeller using a light transmission technique. Bubble diameter varied with rotational speed to the − 1.5 power. There was no noticeable effect of blade width on bubble diameter. However, a small effect of liquid viscosity was noted, diameter varying to the + 0.25 power of the liquid viscosity. The effect of the gas viscosity was not determined, but the effect of the gas phase density was not significant. The final equation for specific interfacial area was given as:

$$a = \frac{1400 N^{1.5} D \rho_L^{0.5} \mu_G^{0.75} \phi}{\sigma \mu_L^{0.25} f_\phi} \tag{6.80}$$

The correlation requires holdup and the f_ϕ function to be known. Among the other items mentioned was the dynamic effects of trace impurities on the results and the use of simple shaker bottle tests to establish initially the effects of changes in impurities. Trace impurities can have a substantial impact on interfacial area.

Calderbank (1956a; 1956b; 1958), based on work by Hinze (1955), cited that a balance existed between interfacial forces and the turbulent fluctuations (dynamic forces) during breakup. From this balance, an equation was developed to correlate interfacial area and the Sauter mean diameter or the volume-surface diameter as:

Interfacial area:

$$a_0 = 1.44\left[\frac{(P_g/V)^{0.4}\rho_c^{0.2}}{\sigma^{0.6}}\right]\left(\frac{V_s}{V_t}\right)^{0.5} \tag{6.81}$$

Sauter mean diameter in centimeters:

$$D_{SM} = 4.15\left[\frac{(P_g/V)^{0.4}\rho_c^{0.2}}{\sigma^{0.6}}\right]^{-1}(\phi)^{0.5} + 0.09 \text{ cm} \tag{6.82}$$

For holdup ϕ, Calderbank developed:

$$\phi = \left(\frac{V_s\phi}{V_t}\right)^{0.5} + 0.0216\left[\frac{(P_g/V)^{0.4}\rho_c^{0.2}}{\sigma^{0.6}}\right]\left(\frac{V_s}{V_t}\right)^{0.5} \tag{6.83}$$

The term $[(P_g/V)^{0.4}\rho_c^{0.2}]/\sigma^{0.6}$ has to be in centimeters. These equations should not be used for high impeller Reynolds numbers and for large tank sizes greater than 100L. For $Re^{0.7}(ND/V_s)^{0.3} > 20{,}000$ following the data provided, the interfacial area was correlated by Calderbank (1958) as:

$$\log_{10}\left(\frac{2.3a}{a_0}\right) = 1.95 \times 10^{-4}Re^{0.7}\left(\frac{ND}{V_s}\right)^{0.3} \tag{6.84}$$

Given the present understanding of gas cavities cited above, the balance between interfacial and dynamic forces occurs in the cavities as gas is dispersed. Gas dispersion in the bulk of the tank by dynamic forces is unlikely. For the dispersion of air in electrolytic and alcoholic solutions, liquid viscosity effects were noted by Calderbank (1958). Calderbank also provided vertical and radial distributions of interfacial area for a 100L tank. The results indicated poor gas dispersion in the bottom of the tank.

Kawecki et al. (1967) studied the bubble size distribution in the impeller region. Impeller rotational speed and gassing rate influenced the bubble size distribution. The data were asymmetric about the mean and were well represented by the Erlang distribution.

The Erlang distribution used was:

$$\frac{A^*(r)\Delta r}{A_t^*} = \frac{m}{b(m-1)!}\left(\frac{mr}{b}\right)^{m-1}\exp\left(-\frac{mr}{b}\right)\Delta r \qquad (6.85)$$

$A^*(r)$ $\Delta r/A_t^*$ is the fraction of total surface area associated with bubbles having a radius between $r - \Delta r/2$ and $r + \Delta r/2$. Data on a/ϕ, given by Kawecki et al., varied linearly with impeller rotational speed and gassing rate. Midoux and Charpentier (1979) also cited a useful bubble size distribution and Brodkey (1967) discussed size distributions generally for multiphase flow.

Lee and Meyrick (1970) developed a parameter to correlate gas holdup in aqueous solutions of strong inorganic salts defined as:

$$\psi = \frac{2C}{RT}\left(\frac{d\sigma}{dC}\right)^2\left[1 + \frac{d\ln(f)}{d\ln(C)}\right]^{-1} \qquad (6.86)$$

where R is the universal gas constant, T is absolute temperature, C is concentration, f is the activity coefficient, and σ is surface tension. Machon et al. (1977) confirmed the use of this parameter. Holdup increased with increasing ψ for otherwise equal conditions, up to $\psi_{crit} = 3 \times 10^{-6}$ (kg^{2}m^3/kmol s^{-4}). Beyond this value, holdup remained constant for a given impeller rotational speed. For their geometry and salt solutions, Machon et al. found holdup to vary as:

$$\phi = 1.3 \times 10^{-4}We^{1.0}N_A^{0.36}\left[2 - \exp\left(\frac{-\psi}{\psi_{crit}}\right)\right] \qquad (6.87)$$

Unfortunately, the exponents varied with geometry.

Miller (1974) emphasized the importance of gas sparging rather than mechanical agitation. Various aspects of bubble behavior during sparging were discussed, including single-bubble behavior and laminar and turbulent chain bubbling. Miller defined holdup in a slightly different manner as:

$$\phi = \frac{V_s}{V_t + V_s} \qquad (6.88)$$

and modified the forms of Calderbank's correlations for holdup, interfacial area, and bubble diameter to include this definition. Miller incorporated the power input from sparging into Calderbank's correlations. His results indicated that at superficial gas velocities of 150 mm/s and higher, mechanical agitation did little to increase mass transfer, but high impeller rotational speeds were not studied. As a reference, the rise velocity of air bubbles in water is about 260 mm/s.

Loiseau et al. (1977) reported holdup data for foaming and

nonfoaming systems for the standard configuration using the disk style impeller for a fixed dispersion height system. At lower rotational speeds, gas dispersion by the impeller was inadequate, and gas passed directly through the tank. The minimum impeller rotational speed for gas dispersion is discussed in a later section (Westerterp et al., 1963). Above this minimum rotational speed, holdup increased with rotational speed but reached constant values at high impeller rotational speeds. Highest holdup values, e.g., 0.25, were obtained using foaming systems in which bubble coalescence was low. The smallest holdup values reached, e.g., 0.1, were obtained for nonfoaming viscous systems, $48 \times 10^{-3} \mathrm{Pa} \cdot \mathrm{s}$ in this study. For nonfoaming systems, holdup, using SI units, was correlated as:

$$\phi = 0.011 V_s^{0.36} \sigma^{-0.36} \mu^{-0.056} \left(\frac{P_g}{V} + \frac{P_s}{V} \right)^{0.27} \tag{6.89}$$

and for foaming systems:

$$\phi = 0.0051 V_s^{0.24} \left(\frac{P_g}{V} + \frac{P_s}{V} \right)^{0.57} \tag{6.90}$$

where power P_s from sparging was calculated, assuming an isothermal gas expansion, using:

$$\frac{P_s}{V} = \frac{Q \rho_g R T}{[(\mathrm{MW})(V) \ln (p_s/p_0)]} \tag{6.91}$$

where R = gas constant
$\quad\quad T$ = temperature
$\quad$ MW = molecular weight of the gas
$\quad p_s, p_0$ = pressures at sparger and liquid surfaces

Surface tension was not found useful in correlating holdup data for the foaming systems. Foaming is typically encountered in fermentation processes.

Yung et al. (1979) provided holdup correlations that accounted for D/T effects. The correlation was of the form:

$$\phi = \zeta N_A N_{\mathrm{We}} \left(\frac{D}{T} \right)^{1.4} \tag{6.92}$$

where ζ varied as a function of geometry of the tank bottom. For aqueous and nonelectrolytic solutions in hemispherical bottomed tanks, ζ was 0.0068 and, for flat-bottomed tanks, ζ was 0.0052. Yung noted that the effects of the tank bottom on holdup require further study since the holdup for flat-bottom tanks was 25 percent lower than for hemispherical bottom tanks. Holdup was not a function of impeller clearance from C/T of 0.225 to 0.45. The holdup results by Kawecki et al. (1967) were three times larger than those of Yung et al. which was attributed to differences in measurement techniques. Data by Yung et al. were average holdup data, whereas, Kawecki et al. obtained data in the impeller discharge region. Yung et al. (1979), following Calderbank (1958), did not consider viscosity effects on holdup in the correlations. For electrolytic solutions, Yung et al. correlated holdup dimensionally as varying with the group $Q^{0.5}N^{0.8}D^{1.4}$.

Hughmark (1980) applied modified forms of the aeration and Froude numbers to correlate holdup data from five literature sources. His final correlation was:

$$\phi = 0.74\left(\frac{Q}{NV}\right)^{1/2}\left(\frac{N^2D^4}{gWV^{2/3}}\right)^{1/2}\left(\frac{D_bN^2D^4}{\sigma V^{2/3}}\right)^{1/4} \tag{6.93}$$

which had an average absolute deviation of 11.5 percent between the calculated and experimental values of 208 data sets. The parameters V, W, and D_b are liquid volume, blade width, and bubble diameter, respectively. Applying the same in a correlation for interfacial area, Hughmark obtained:

$$a = 1.38\left(\frac{g\rho}{\sigma}\right)^{1/2}\left(\frac{Q}{NV}\right)^{1/3}\left(\frac{N^2D^4}{gWV^{2/3}}\right)^{0.592}\left(\frac{D_bN^2D^4}{\sigma V^{2/3}}\right)^{0.187} \tag{6.94}$$

The agreement between the correlation and interfacial area data from 12 literature studies was reasonable. The correlation holds for the standard disk style and flat blade impeller geometries $W = 0.2D$ in the standard configuration for gas dispersion where $0.75 < H/T < 1.87$ and $0.33 < C/H < 0.66$. Liquid height and impeller depth do not appear in the correlation. Hughmark also indicated that the correlation for interfacial area could be used in the calculation of k_la over a range of vessel sizes from $10L$ to $15,000L$ and superficial gas velocities to 0.053 m/s. Chapman et al. (1983) also cited work by Kiple suggesting the use of tank volume in correlating mass transfer coefficients.

Sridhar and Potter et al. (1980a) performed a verification study of Calderbank's initial correlations, presented in 1958, which were based on an assumed balance between surface tension forces and those caused by turbulent fluctuations. The various ways to measure interfacial area were reviewed and limitations of the techniques were also discussed. Sridhar and Potter used a fiber optic probe and light transmission to obtain local measurements in one quadrant of their tank.

Sridhar and Potter reported data on the arithmetic mean interfacial area versus radial distance averaged over liquid height as a function of superficial gas velocity. These curves were similar to those reported by Calderbank (1958) and Nagase and Yuasi (1983). Maximum interfacial area occurred at $r/R \cong 0.5$ for D/T of 0.35. The average interfacial area increased with impeller rotational speed but leveled off at higher impeller speeds. Agreement between value of Westerterp's minimum rotational speed (Westerterp et al., 1963, discussed in the next section), for gas dispersion and the initial increase in interfacial area with impeller speed was noted, 13.3 s^{-1} versus 11.6 s^{-1}. The average values for the interfacial area, found by Sridhar and Potter, were approximately one-half those by Kawecki et al. (1967) obtained in the impeller stream. Sridhar attributed this difference to local values versus average values and to different measurement methods.

The power input from gas sparging was divided into two mechanisms by Sridhar and Potter: (1) the dissipation of the kinetic energy of the gas jets, issuing from the sparger at the velocity u_s gave rise to jetting power P_j modeled as:

$$P_j = (\tfrac{1}{2})Q\rho_g(u_s)^2 \tag{6.95}$$

and (2) the rise of the gas through the liquid provided expansion power P_e (Lamont, 1958) modeled as:

$$P_e = Q\rho_L gH \tag{6.96}$$

This last equation has been discussed in detail in Chap. 4.

The two mechanisms are pictorially shown in Fig. 6.22. Jetting power P_j was considered unimportant by Miller (1974). Power input from expansion or the rise of the gas through the liquid has been noted before, but jetting power P_j has not. Power P_j is determined by the volumetric flow rate of the gas, density of the gas, and the sparging velocity which is fixed by Q and the diameter of the holes in the sparger.

Sridhar and Potter made several major conclusions. Calderbank's correlation for average interfacial area was valid under the conditions for which it was established. The relationship was modified for use in a wider range of gas velocities and was given as:

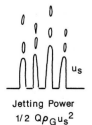

Jetting Power
$1/2\ Q\rho_G u_s^2$

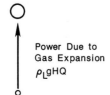

Power Due to
Gas Expansion
$\rho_L gHQ$

Figure 6.22 Various forms of power input in addition to mechanical agitation.

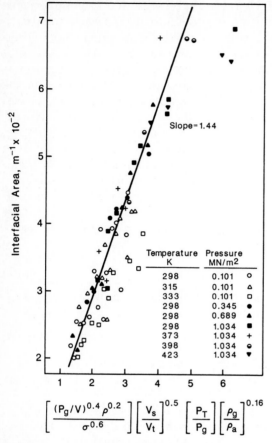

Figure 6.23 Correlation for interfacial area. *(From T. Sridhar and O. E. Potter, Chem. Eng. Sci., 35, 683, 1980. By permission.)*

$$a = 1.44 \left[\frac{(P_g/V)^{0.4} \rho_c^{0.2}}{\sigma^{0.6}} \right] \left(\frac{V_s}{V_t} \right)^{0.5} \left(\frac{P_T}{P_g} \right) \left(\frac{\rho_g}{\rho_a} \right)^{0.16} \qquad (6.97)$$

where P_g = gassed power input from agitation as it was in Calderbank's original correlation

P_T = total power input from agitation, sparging, and bubble rise

ρ_c = liquid density

The density ratio is the density of the sparging gas to that of air at operating conditions. The correlation, shown in Fig. 6.23, had an average deviation of 8 percent for P_T/P_g up to 1.4. The correlation assumed that the rise velocity for the gas bubbles was 265 mm/s as did Calderbank's correlation.

Chapman et al. (1983) provided the following holdup correlation for a number of different impeller geometries and pumping directions:

$$\phi = 1.97(\epsilon_T)^{0.31}(V_s)^{0.67} \tag{6.98}$$

where ϕ is dimensionless, ϵ_T is in watts per kilogram, and V_s is in meters per second.

Greaves and Barigou (1988) studied gas holdup of a disk style turbine in a fully baffled 1.0-m-diameter tank (C/T = 0.25 and D/T = 0.25, 0.33, and 0.5) using air water and air electrolyte systems. A single-point sparger was used. Gassing rate varied between 1.6×10^{-3} and 8.3×10^{-3} m³/s. Gas holdup was measured by the difference in height between ungassed and gassed conditions. General correlations were given for air water system as:

$$\phi = 4.07N^{0.62}Q^{0.64}\left(\frac{D}{T}\right)^{1.39} \tag{6.99}$$

and for the air electrolyte system as:

$$\phi = 4.20N^{0.79}Q^{0.52}\left(\frac{D}{T}\right)^{1.92} \tag{6.100}$$

Abrardi et al. (1988) provided the following holdup correlation for both single and dual disk style impeller systems as:

$$\phi = 0.210\left[\frac{P_g}{V_L}(1 - \phi)\right]^{0.270}(V_s)^{0.650} \tag{6.101}$$

Greaves and Barigou (1988) noted that, to develop a basic understanding of gas dispersion, a more detailed understanding of the structure of the gas dispersion was required. Most research to date reported global averages and provided empirical correlations. To determine the true mass transfer condition, measurements of bubble size and bubble size distributions were needed spatially in the tank. Greaves and Barigou developed a capillary suction probe to measure local bubble size distributions throughout a 1.0-m-diameter tank with a disk style turbine in the standard configuration and a single-point sparger. Two systems were studied: an air-water fast-coalescing system and an air-0.15M NaCl aqueous solution coalescence retarded system.

Reproducible Sauter mean diameters were obtained as a function of gassing rate and impeller rotational speed. Typically, at low gassing rates in the impeller stream, the Sauter diameter decreased with increasing rotational speed. At high gassing rates, the Sauter diameter remained constant and independent of impeller speed, indicating an equilibrium between coalescence and dispersion. At constant impeller speed, the Sauter diameter increased with increasing gassing rate. Under varying conditions in impeller rotational speed and gassing rate, the power input, the impeller hydrodynamics, the bubble density, bubble dispersion, and coalescence in the impeller discharge changed. In the bulk of the tank, the effect of impeller rotational speed on the bubble dispersion was insignificant.

Bubble size distributions were obtained as a function of spatial location and wide variations in distributions were noted. A very narrow log normal distribution, having a small average bubble size, occurred near the impeller in the impeller discharge. Close to the wall, a broader log normal distribution occurred which had a higher average bubble size. In the bulk of the tank, very broad distributions were measured. Global cumulative frequency distributions were attempted for the entire tank as a function of system type. For the air-water system, a Weibull cumulative frequency distribution was obtained as:

$$f(D_{SM}) = 0.222 D_{SM} \exp\left(\frac{D_{SM}}{3}\right)^2 \qquad (6.102)$$

For the electrolyte system, an exponential cumulative frequency distribution was obtained as:

$$f(D_{SM}) = 1.0 \exp\left(- D_{SM}\right) \qquad (6.103)$$

Pressure effects. Sridhar and Potter (1980) found that interfacial area increased with increasing pressure because of an increase in jetting power P_j caused by an increase in the density of the gas. The expansion power P_e remained constant under these conditions. At 10 atm, the power input from gas sparging P_j was substantial, up to 50 percent of the power input from mechanical agitation. Under such conditions, Sridhar and Potter found that increasing gassing rate (i.e., P_j) was more effective at increasing interfacial area than increasing the impeller rotational speed. The importance of P_j at high pressure levels indicates a need for further study on the effects of the sparger geometry on interfacial area under pressure. However, the increase in interfacial area with pressure was found to be no longer significant above 10 atm. The reason for this was not given by Sridhar and Potter. Obviously, at very high pressures, the density difference between a gas and a liquid becomes small, buoyant forces of the gas bubbles are not as significant, and the hydrodynamics change.

Sridhar and Potter (1980) reported correlations for gas holdup and bubble diameters for pressurized systems. They found the form of the correlations by Calderbank (1958) to hold generally. Sridhar et al. calculated the gassed power using the correlation by Michel and Miller (1962) and took exception to the way in which the total power input was calculated by Miller (1974). Sridhar et al. (1980) assumed that the total pressure of their system did not affect the power input from mechanical agitation. Their final correlations for bubble diameter and holdup were:

$$D_{SM} = 4.15 \left[\frac{(P_g/V)^{0.4} \rho_c^{0.2}}{\sigma^{0.6}} \right]^{-1} \left(\frac{P_g}{P_T} \right) \left(\frac{\rho_a}{\rho_g} \right)^{0.16} (\phi)^{0.5} + 0.09 \text{ cm} \qquad (6.104)$$

$$\phi = \left(\frac{V_s \phi}{V_t} \right)^{0.5} + 0.0216 \left[\frac{(P_g/V)^{0.4} \rho_c^{0.2}}{\sigma^{0.6}} \right] \left(\frac{V_s}{V_t} \right)^{0.5} \left(\frac{P_T}{P_g} \right) \left(\frac{\rho_g}{\rho_a} \right)^{0.16} \qquad (6.105)$$

where again the units for the bracketed term containing gassed power must result in centimeter. The total power input P_T was the sum of the power because of mechanical agitation, gas sparging, and gas expansion because of bubble rise. High pressures were found to cause: (1) increased power input because of sparging, (2) increased holdup values to 0.3 at 10 atm, and (3) decreased bubble diameter.

Temperature effects. Different temperature effects were noted by Sridhar and Potter (1980). At 298K, interfacial area increased with superficial velocity, then decreased moderately at higher superficial velocities. This reduction was attributed to changes in the dispersing capability of the impeller at the higher gassing rates. At higher temperatures of 315 and 333K, interfacial area continued to increase at high superficial gas velocities because of reductions in surface tension and liquid viscosity and an increase in power input because of gas sparging. The net effect of increasing temperature was to increase the power input to the liquid.

Holdup correlations based on flow regimes

Gas holdup is an important quantity in the evaluation of reactor performance. Unfortunately, no universal correlation can predict holdup with high reliability, and often correlations are based upon variables which are themselves difficult to predict. There is also a general lack of insight as to what the correlating variables should be. However, improvements in correlations can be made if the correlations incorporate flow regime information.

There is little work which bases holdup correlations on impeller flow regimes. Smith and Warmoeskerken (1985) provided dimensional correlations for holdup as:

Before the formation of the 3-3 structure:

$$\phi = 0.62Q^{0.45}N^{1.6} \qquad (6.106)$$

And for the 3-3 structure as:

$$\phi = 1.67Q^{0.75}N^{0.7} \qquad (6.107)$$

SI units are required for these correlations.

For a three disk style impeller system, Smith et al. (1986) provided a holdup correlation:

$$\phi = (Fr_I Fl)^{0.66} \qquad (6.108)$$

At the formation of the 3-3 structure on the lowest impeller, the holdup was 7.2 percent.

Greaves and Barigou (1988) classified their holdup data according to hydrodynamic flow regime and obtained the following correlations:

Air water system in the vortex-clinging cavity regime:

$$\phi = 3.85N^{0.73}Q^{0.62}\left(\frac{D}{T}\right)^{1.64} \qquad (6.109)$$

Air water system in the large cavity regime:

$$\phi = 1.33N^{0.60}Q^{0.44}\left(\frac{D}{T}\right)^{1.31} \qquad (6.110)$$

Air electrolyte system in the vortex-clinging cavity regime:

$$\phi = 3.86N^{0.92}Q^{0.41}\left(\frac{D}{T}\right)^{2.56} \qquad (6.111)$$

Air electrolyte system in the large cavity regime:

$$\phi = 2.86N^{0.76}Q^{0.31}\left(\frac{D}{T}\right)^{1.64} \qquad (6.112)$$

The classification according to flow regime was highly significant at the 99.5 percent confidence level. The level of gas holdup depended upon the hydrodynamic cavity regime and the type of system. Higher holdups were obtained for the vortex-clinging cavity regime because of the reduced pumping capacities of the large cavities. The sensitivity of gas holdup to impeller speed and D/T ratio were higher for ionic solutions. Gas holdup for nonionic solutions was more sensitive to gassing rate.

Non-Newtonian effects on interfacial area, holdup, and bubble diameter

The effects of viscous Newtonian and non-Newtonian fluids on interfacial area, holdup, and bubble diameters have been studied. Effects of drag-reducing additives are also included in this section.

In a study of polymer additives on gas-liquid mass transfer, Ranade and Ulbrecht (1978) included data on the holdup. Holdup was found dependent on the type of polymer additive and its concentration. Polymer additives primarily reduced the specific interfacial area and holdup, but the mechanism by which this occurred remained unclear.

Machon et al. (1980) studied the effects of the shear thinning nature of liquid on the holdup. The solutions used in their study showed no drag-reducing properties but were shear thinning. CMC solutions were studied in which holdup varied between 1 to 8 percent. For the same aeration number and agitation conditions, increasing the shear thinning nature of the fluid decreased holdup. This was attributed to an increase in the proportion of large bubbles which have a shorter residence time in the tank. However, it was also found that a very small amount of CMC in water (0.015 percent wt) increased holdup. Holdup was found proportional to gassed power P_g and gas flow rate Q as:

$$\phi \propto P_g^{0.3} Q^{0.7} \tag{6.113}$$

or to power consumption per unit mass ϵ and superficial velocity V_s as:

$$\phi \propto \epsilon^{0.3} V_s^{0.7} \tag{6.114}$$

At constant gassing rate conditions, holdup and power consumption behaved in a similar manner when plotted as functions of CMC concentration. CMC concentration lowered both power draw and gas holdup, apparently altering the cavity structure occurring around the impeller blades and affecting the dispersion of gas by the cavities.

At all CMC concentrations, two bubble sizes were observed: very tiny bubbles, having the size of a pin head, and large bubbles. This is in agreement with Ranade and Ulbrecht (1977) cited above. The tiny bubbles remained in solution for considerably longer times than the large bubbles. The tiny bubbles were thought to be associated with the presence of sodium ion in the CMC solution, causing the initial increase in holdup above deionized water at the lowest CMC weight percent studied. As the concentration of CMC increased, spherical cap bubbles were observed. These bubbles did not recirculate to the impeller but rose rapidly to the liquid surface. The large bubbles were present in the flow because of the inability of the impeller to disperse the gas. The large bubbles were not formed by coalescence. Apparently the CMC concentrations affected the nature of the gas cavities of the impeller.

Bubble coalescence behavior

The effects of electrolytes on bubble behavior are of interest. Lessard and Zieminski (1971) studied the dependence of bubble coalescence on ionic charge. The mechanism of coalescence was based on thinning and rupture of the liquid film between adjacent bubbles, Fig. 6.24. In their study, adjacent bubbles were formed redundantly in solution of various salts and the percentage of coalescing pairs was determined. The coalescence in pure water was 100 percent. It was found that the transition concentration from coalescence to noncoalescence was a function of electrolyte valence: 3-1 and 2-2 electrolytes substantially prevented coalescence at concentrations above 0.04 M, 2-1 and 1-2 electrolytes above 0.6 M, and 1-1 electrolytes above 0.2 to 0.3 M. The results were correlated using solution entropy and a self-diffusion parameter. Effects of ionic size, viscosity, and hydration were also discussed. The interface and the mass transfer coefficient k_l were also affected to some degree.

Bubble coalescence and coalescence times have also been studied by Marrucci and Nicodemo (1967), Marrucci (1969), Sagert and Quinn (1976; 1978a; 1978b), Keitel and Onken (1982) and Drogaris and Weiland (1983). Bubble coalescence can also occur when bubbles are drawn into the wakes of other bubbles, Fig. 6.24 (Preen, 1961). However, considerable work is needed with regard to its modeling. The frequency of coalescence by this mechanism is difficult to model because it requires detailed information concerning bubble sizes, wakes, frequency, and location of the surrounding bubbles. Bubble formation and coalescence can also occur by the collapse of gassed vortex ligaments.

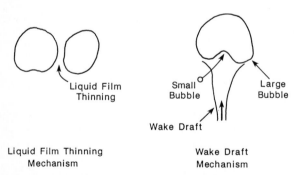

Liquid Film Thinning

Small Bubble   Large Bubble

Wake Draft

Liquid Film Thinning Mechanism   Wake Draft Mechanism

Figure 6.24 Bubble coalescence mechanisms.

Surface active agents. Meijboom and Vogtlander (1974*a*; 1974*b*) discussed the various effects of surface active agents, either from the gas phase or the liquid phase, on mass transfer from gas bubbles into a liquid. In particular, the effect of grease and oil, originating from the compressor, on bubble behavior was noted. Other major effects were classified in a scheme using bubble Reynolds number, a capillary number, and the Deborah number. Other references concerning surface agents include Zieminski et al. (1960), Zieminski and Hill (1962), and Zieminski and Lessard (1967). Patton (1979) has an excellent summary on surface active agents.

Mass Transfer Coefficients and Interfacial Area

Gas-liquid mass transfer is often the rate-controlling step in stirred tank reactors or aerobic fermenters. The rate of gas absorption becomes important as a result. Using the film mechanism to describe the transport, two mass transfer coefficients, the gas side coefficient $k_g a$ and the liquid side coefficient $k_l a$ appear. In comparison of these for a sphere and assuming that the Sherwood numbers $k_g L/D_G$ and $k_l L/D_L$ for both sides of the interface are roughly the same order of magnitude, the mass transfer coefficients k_g and k_l change only as the characteristic length L and the diffusivity D change. Given the high diffusivities in the gas phase and low diffusivities in the liquid phase that typically occur, mass transfer resistance and concentration gradients in the gas phase can be considered negligible. Hence, the mass transfer coefficient $k_l a$ becomes the most significant transfer coefficient. The mass transfer coefficient $k_l a$ contains both the mass transfer coefficient k_l, which is roughly independent of area, and interfacial area a. Of these, the interfacial area can change drastically in an agitated tank, much more so than k_l.

Correlations for mass transfer coefficients have not been divided according to impeller flow regimes, and most of the work on $k_l a$ has ignored the hydrodynamics of the impeller. However, there is a direct interrelationship between these flow regimes and mass transfer, which appears through gassed power P_g. For example, one method used to correlate $k_l a$ is the dimensional correlation:

$$k_l a = K \left(\frac{P_g}{V}\right)^{\alpha} (V_s)^{\beta} \tag{6.115}$$

where $k_l a$ is in reciprocal time units, P_g/V is gassed power per unit volume, V_s is superficial gas velocity, and the constant K has the necessary units. Flow regimes determine P_g which determines $k_l a$.

Mass transfer involves not only the mass transfer coefficient $k_l a$ but also the concentrations in the respective phases. In a gas dispersion process, the question of mixing arises in both phases. Typically, the liquid phase is considered well mixed (Nagase and Yasui, 1983). For continuous-flow processes this requires having the residence time of the material in the tank longer than the mixing time by some factor. The gas phase, being the less dense discrete phase, behaves differently. There remain considerable differences as to the nature of the gas phase mixing in the literature.

Gas can either leave the tank or recirculate in the tank. The amount of recirculation, impeller coalescence and redispersion and coalescence in the bulk of the tank and the nature of the gas and liquid hydrodynamics complicate the process. The amount of recirculation, for example, as that shown by van't Riet et al. (1976), depends on the flow conditions in the tank.

The concentration of the gas phase can be considered well mixed under some conditions and not under others. If recirculation is high, indicating good dispersion and considerable impeller coalescence, and the amount of sparged gas is low with no surface entrainment, the gas phase can be considered well mixed, having the composition of the exit stream. If the impeller is near flooding, indicating poor dispersion and little impeller coalescence, the gas phase cannot be considered well mixed, and a log mean driving force for the concentration difference may be necessary. Under flooding conditions, backmixing is not obtained in the gas phase. The liquid phase is being mixed by the bubble plume, so, under flooded conditions, the liquid phase is still under conditions of backmixing. However, circulation and mixing times increase (Nagase and Yasui, 1983) under such conditions.

In most studies to follow, the two phases are considered well mixed. However, the mixing of the gas phase is still under investigation, and further work is still needed which is similar to that of Nienow et al. (1979).

Studies of $k_l a$ often include studies on interfacial area. Such studies are discussed in this section. Also there is considerable confusion concerning the measurement of the mass transfer coefficient $k_l a$ in the literature. Flow regimes, ranges of variables, the effects of scale, and the methods used to acquire the $k_l a$ data require understanding before a particular correlation can be appropriately used.

Mass Transfer Coefficients and Interfacial Area Studies in Low-Viscosity Liquids

Chemical and physical methods.   There are two general methods by which mass transfer coefficients and interfacial area can be determined: (1) chemical methods and (2) physical methods, which can be subdivided into direct and indirect methods. Chemical methods are based on the determination of absorption rate per interfacial area, which can be predicted from the theory of gas absorption with chemical reaction in the liquid phase. Interfacial area can then be calculated from measurements of the total absorption rates. The kinetics of the reaction are necessary, which is the major difficulty with chemical methods. For fast chemical reactions, the rate of absorption is determined by the chemical reaction only and not by the physical absorption. The reactions are selected to be very fast, and, as a result, the rate of absorption is independent of k_l and the hydrodynamic conditions.

Another type of chemical method is based upon a steady-state reaction where the rate of addition of the reacting specie to the tank is related to $k_l a$. The chemical kinetics of the reaction, sophisticated models, and assumptions do not enter into the calculation of $k_l a$. The technique will be described later.

Physical methods determine $k_l a$ from gas absorption without chemical reaction. Such methods do not depend on kinetics and will be described later.

The major reason for the development of the chemical methods (Reith and Beek, 1973) was to eliminate the hydrodynamic effects occurring at or near the interface that changed k_l and affected interfacial area a. Westerterp et al. (1963) cited two problems with some of the competing physical methods: (1) probe interference of the flow that affects the measurements and (2) local measurements. Both of these tend to be unimportant as long as the fluid phases are well mixed and well mixed around the probe. Physical methods that do not require the use of probes do not have these limitations.

In comparing the general methods, the physical and the steady-state reaction methods are appropriate for the measurement of $k_l a$. The determination of mass transfer coefficients $k_l a$ by fast chemical reaction methods involves a considerable number of variables which have interdependencies. These interdependencies significantly influence the results. Considerable work has been done using chemical methods, and comparisons between results of the two general methods are important for fundamental understanding. Furthermore, not all processes involve mass transfer and physical transport. Reaction based chemical methods deal with the important area of mass transfer with chemical reaction which is outside the scope of this chapter.

Westerterp et al. (1963) studied interfacial areas using a fast chemical reaction to determine interfacial area. Results indicated that the absorption rate was proportional to the difference between the rotational speed minus the minimum rotational speed for gas absorption. However, the catalyst type affected the absorption rate and mass transfer, making the method dependent on the reliability of kinetic information.

The minimum impeller rotational speed N_0 for gas dispersion was correlated as:

$$N_0 D = \left(\frac{\sigma g}{\rho}\right)^{0.25}\left[A + B\left(\frac{T}{D}\right)\right]$$ (6.116)

where N_0 is in revolutions per second, A is 1.22, and B is 1.25 for the disk style turbine in the standard geometry. The interfacial area above the minimum rotational speed was correlated as:

$$a = c\left(\frac{1 - \phi}{H}\right)(ND - N_0 D)\left(\frac{\rho T}{\sigma}\right)^{0.5}$$ (6.117)

where c is a constant and H is the liquid height. The concept of a minimum rotational speed for effective mass transfer is similar to the concept of impeller flooding as cited above and the magnitudes of the impeller rotational speeds are approximately the same.

Westerterp et al. (1963) gave several examples concerning optimizing gas-liquid reactors and noted the effects of D/T on gas dispersion. At low D/T ratios, the tank geometry approached an unbaffled condition permitting solid body rotation of the liquid and the central gas plume at low impeller rotational speed. Westerterp (1963) recommended that, for equal interfacial area on scaleup, the procedure should be based on equal tip speed and D/T ratio. Bourne (1964) took exception to this and showed that scaleup was more appropriately done based on equal power per volume.

Westerterp et al. discussed the requirements of the chemical method of the oxidation of sulphite solutions used in his work. The method was difficult and more complex than physical methods developed since 1963. De Waal (1966) showed that the absorption in Westerterp's work was not chemically enhanced. From this, Mehta and Sharma (1971) and Reith and Beek (1973) concluded that the interfacial areas found by Westerterp et al. were far too small. However, the concept of the minimum rotational speed (Westerterp et al. 1963) is important.

Sideman et al. (1966) recommended the following correlation form for $k_l a$:

$$\frac{k_l a D^2}{D_L} = f\left(\frac{\rho N D^2}{\mu}, \frac{\mu}{\rho D_L}, \frac{\mu V_s}{\sigma}, \frac{\mu_G}{\mu}\right) \qquad (6.118)$$

which includes the Sherwood number, the Reynolds number, the Schmidt number, the reciprocal of a gas number, and a viscosity ratio of the gas and liquid. The liquid viscosity appears in every term on the right side of the correlation which has generated criticism (Nishikawa et al., 1981a). For low-viscosity, nonelectrolytic liquids, Calderbank found no effect of liquid viscosity on interfacial area.

Rennie and Valentin (1968) performed an interesting photographic study of gas absorption of ammonia in water followed by a fast chemical reaction. The ammonia reacted with dilute hydrochloric acid containing an indicator. Although the ammonia in the gas was well in excess of that absorbed by the liquid, the coloration of the indicator occurred initially only in the impeller region and not in the other regions of the tank. Rennie and Valentin concluded from this that the condition of the interfacial area was important. The gas-liquid interface in the vortices was most active in the transport of ammonia into the liquid. Rennie and Valentin also cited work by Wilhelm et al. (1966), who found that absorption rates were largely independent of the total liquid volume and depended mainly on the conditions of the impeller region.

Rennie and Valentin noted that most of the interfacial area was around the impeller. The generation of small bubbles occurred in the impeller region which then coalesced to form larger bubbles around the perimeter of the impeller. The addition of a surface active agent, nonyl alcohol, hindered coalescence of the small bubbles. These small bubbles were transported throughout the tank as part of the continuous phase.

Boerma and Lankester (1968) investigated the absorption of gas CO_2 from the surface in a solution of sodium carbonate in a small 194-mm-diameter tank. Two types of baffles were used, a long type extending to the liquid surface and a short type extending to approximately half the liquid height from the bottom. Below a certain rotational speed N_0 no absorption occurred. Above this minimum speed, absorption was proportional to $(N - N_0)$. Maximum absorption occurred at low impeller immersion depths, close to the liquid surface. The minimum rotational speed N_0 found by Boerma and Lankester, was comparable in magnitude to the minimum rotational speed found by Westerterp et al. (1963). The work by Westerterp et al. (1963) sparged gas below the impeller. At $N > 1.5 N_0$, the entire tank was filled with surface entrained gas. Boerma and Lankester noted that sparging gas from below the impeller was unnecessary and relatively large stirrers ($D/T = 0.6$) in combination with no baffles in the top half of the vessel was an optimal design for surface aeration. On the other hand, Blakebrough et al. (1961) did not recommend surface aeration for fermentation processes.

The similarity between the numerical values for the minimum rotational speeds for these two gas dispersion processes, given by Westerterp et al. (1963) and by Boerma and Lankester (1968), points to the flow regime where the dynamic forces, provided by the impeller, dominate the buoyant forces of the gas bubbles, irrespective of the source of the gas.

Mehta and Sharma (1971) studied the various effects of rotational speed, impeller type and diameter, height of liquid and impeller clearance, the viscosity and interfacial tension of the liquid phase, and ionic strength and nature of the ions in the liquid phase on gas absorption using a method based on a slow chemical reaction. The details of the procedure and theory were given by Sharma and Mashelkar (1968) and Sharma and Danckwerts (1970). The work assumed perfectly mixed conditions in both phases. Surface aeration and gas sparging were studied.

Specific results of Mehta and Sharma (1971) included some important design considerations. The interfacial area was found inversely proportional to liquid height for gas sparged conditions, similar to the results of Westerterp et al. (1963). For surface aeration, interfacial area dropped sharply as the ratio of liquid height to tank diameter H/T was increased. The $k_l a$ for surface aeration was 0.1 to 0.7 times the $k_l a$ for gas sparging. In the range of rotational speeds between 6 and 30 s^{-1} under gas sparging conditions, $k_l a$ ranged from 0.08 to 0.30 s^{-1} and interfacial area, from 1.5 to 5 cm^{-1}. Impeller clearance C/T between 0.34 and 0.5 had no effect on $k_l a$ for sparged systems. $k_l a$ and interfacial area increased linearly with rotational speed N above N_0 for a variety of vessel geometries and impeller types. There was no effect of gas flow rate on interfacial area or $k_l a$. The gassing rate was varied up to 0.12 m/s, indicating the impeller was effectively dominating the flow pattern; the lowest impeller tip speed studied was 1.3 m/s. At a constant impeller tip speed, the data indicated that the impeller to tank diameter ratio D/T should be 0.4 to 0.5 for design that balances the circulation loops in the top and bottom of the tank. With the impeller diameter and rotational speed fixed, increasing the tank diameter decreased $k_l a$ and interfacial area. Under such conditions, the dominance of the impeller was less in larger tanks, permitting the gas to leave the tank more easily. Both interfacial area and $k_l a$ varied linearly with $ND/(T)^{0.5}$. Six-blade curved and disk style turbines provided comparable $k_l a$ values. Axial flow turbines had the lowest $k_l a$ values. $k_l a$ values were also found to increase with blade number.

Mehta and Sharma (1971) noted the effects of ionic strength and nature of ions on gas dispersion. $k_l a$ and the liquid viscosity increased linearly with ionic strength. The type of ion also affected the interfacial area. Interfacial area was found to vary inversely with the one-third power of surface tension and to increase linearly with liquid viscosity. The increase in viscosity caused less coalescence and lower rise velocity of the gas phase. A secondary phase, whether liquid or solid, increased the interfacial area. Smaller bubble sizes were present because the secondary phase reduced rates of coalescence. The effect on $k_l a$ was less defined since small quantities of solids decreased $k_l a$; at higher solids, $k_l a$ increased. Mehta and Sharma also found k_l and interfacial area independently using a Danckwert's plot. The interfacial area a and $k_l a$ were found to increase with impeller rotational speed while k_l remained constant. The magnitude of interfacial area a was 3 cm^{-1} and k_l, 0.013 cm/s^{-1}.

Mehta and Sharma (1971) concluded that $k_l a$ and the interfacial area a cannot be predicted a priori for a design given all the possible effects that may occur under operating conditions.

Bossier et al. (1973) noted the need to develop a chemical method to measure interfacial areas for organic liquids. Although having the disadvantage of a high heat of reaction, the reaction of aluminum alkyls with oxygen was selected. Under the same agitation conditions, the mass transfer coefficient k_l appeared to be approximately the same for organic liquids and aqueous solutions. The interfacial area for the organic liquids, however, was about four times that of aqueous solutions.

Koetsier et al. (1973a) noted a design choice available in performing the mass transfer in a closed vessel in which holdup became, to some degree, an independent variable. $k_l a$ data were obtained in two different enclosed tanks, 0.19 and 0.6 m, using a transient physical method and a chemical method using sulphite. They found good agreement between the two methods, an average difference of 8.5 percent. For tap water:

$$k_l a = 0.05 \left(\frac{ND^2}{T^{1.5}} \right)^{1.95} \tag{6.119}$$

which can be rearranged in terms of power per unit volume as:

$$k_l a = 0.05 \left(\frac{N^3 D^5}{T^3} \right)^{0.65} \left(\frac{D}{T} \right)^{0.65} T^{-0.33} \tag{6.120}$$

For electrolytic solutions:

$$k_l a = 0.11 \left(\frac{ND^2}{T^{1.5}} \right)^{2.1} \tag{6.121}$$

which can be rearranged in terms of power per unit volume as:

$$k_l a = 0.11 \left(\frac{N^3 D^5}{T^3}\right)^{0.70} \left(\frac{D}{T}\right)^{0.70} T^{-0.35} \qquad (6.122)$$

The constants 0.05 and 0.11 have units of s/m. As shown in these equations, the mass transfer coefficient was a function of power consumption per unit volume, impeller to tank diameter ratio, and scale. Koetsier et al. cited that these relationships were limited to gas fraction of 1 percent by volume, which is very restrictive. Koetsier et al. described how a polargraphic cell measures the oxygen concentration, noting the restrictions on the use of polargraphic cells.

Koetsier et al. (1973b) obtained k_l from the mass transfer data $k_l a$ from their previous study and the correlation provided by Calderbank (1958) for interfacial area a. If such a procedure can be accepted, k_l for tap water was correlated as:

$$k_l = 0.002 \left(\frac{ND^2}{T^{1.6}} - 0.45\right) \qquad (6.123)$$

with the constant 0.45 having units of $m^{0.4}$/s and k_l with units of m/s. Koetsier et al. found that k_l varied between 0.5 to 4.0×10^{-3} m/s. The increase in k_l was attributed to an increase in surface renewal at the higher rotational speeds. For an electrolytic solution of 0.6 kmol NaCl/m^3, again using the correlation by Calderbank, k_l was correlated as:

$$k_l = \frac{0.11}{c_1} \left(\frac{N^3 D^5}{T^3}\right)^{0.30} \left(\frac{D}{T}\right)^{0.70} T^{-0.35} \qquad (6.124)$$

where the constant c_1 was:

$$c_1 = 5.93 \phi^{0.6} \left(\frac{\mu_L}{\mu_G}\right)^{0.25} \left(\frac{\rho_L}{\sigma}\right)^{0.6} \qquad (6.125)$$

For the solutions studied, the leading coefficient $0.11/c_1$ was 3.25×10^{-4} with units of $m^{0.75}/s^{1.9}$ for $\phi = 0.01$. Koetsier et al. concluded that large-diameter impellers resulted in higher k_l values for the same power consumption and that the gas fraction ϕ in the range of 0.01 to 0.2 had little effect on k_l. The upper limit of holdup for closed stirred tanks was not reported by Koetsier et al. (1973a; 1973b).

The work by Yagi and Yoshida (1975) correlated $k_l a$ data using the desorption of oxygen for two kinds of Newtonian fluids (glycerol and millet jelly). The results, obtained for non-Newtonian fluids, are reported later. For Newtonian fluids, Yagi and Yoshida assumed a general correlation based on dimensional analysis and correlated $k_l a$ data as:

$$\frac{k_l a D^2}{D_L} = 0.060 \left(\frac{\rho N D^2}{\mu}\right)^{1.5} \left(\frac{N^2 D}{g}\right)^{0.19} \left(\frac{\mu}{\rho D_L}\right)^{0.5} \left(\frac{\mu V_s}{\sigma}\right)^{0.6} \left(\frac{ND}{V_s}\right)^{0.32} \quad (6.126)$$

The first term in the correlation is a modified Sherwood number; the second, the Reynolds number; the third, the Froude number; the fourth, the Schmidt number; the fifth, the reciprocal of a gas number; and the last, the aeration number N_A. The viscosity in the correlation is the liquid phase viscosity. It is important to note the similarity of this correlation with Frossling (1938) and Higbie (1935) type correlations and the use of the aeration number. Power per unit volume-superficial gas velocity correlations for $k_l a$ can be obtained from this correlation. For geometrically similar vessels and the same fluid properties, the correlation reduces approximately to:

$$k_l a \propto (N^3 D^2)^{0.74} (V_s)^{0.28} \quad (6.127)$$

which indicates that $k_l a$ scales with power per unit volume to the 0.74 power for Newtonian fluids.

Smith et al. (1977) noted large discrepancies in $k_l a$ data and the lack of data when design departed from the single six-blade disk style turbine. Typically, such correlations appeared as:

$$k_l a = K \left(\frac{P_g}{V}\right)^\alpha (V_s)^\beta \quad (6.128)$$

α was 0.475 and β was 0.4 in their work. Although such correlations accounted for scale in P_g, V, and V_s in a gross fashion, Smith noted that the correlations were an oversimplification in which specific details important for optimization were lost. Actual $k_l a$ data for an air-water system are shown in Fig. 6.25 in which lines of constant impeller rotational speed and superficial gas velocity are shown. With regard to their data, Smith et al. noted the following. For coalescing systems:

1. $k_l a / V_s^{0.4}$ passed through a maximum value as P_g/V was varied at a fixed rotational speed N, directly implying an optimum impeller diameter that maximized $k_l a$ and minimized V_s.

2. At high P_g/V, $k_l a$ became less dependent on P_g/V which was attributed to depletion and gas recirculation.

For noncoalescing systems, similar comments were made:

1. $k_l a / V_s^{0.4}$ passed through a maximum value as P_g/V was varied at a fixed impeller tip speed πND, directly implying an optimum impeller rotational speed N.

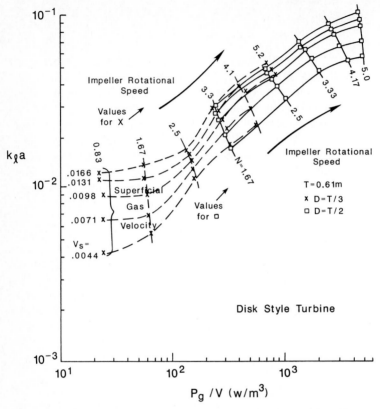

Figure 6.25 $k_l a$ versus P_g/V for an air water system in a 0.61-m diameter tank. *(From J. M. Smith, K. van't Riet, and J. C. Middleton, Proc. 2nd Eur. Conf. on Mixing, St. John's College, Cambridge, England, BHRA Fluid Eng., Cranfield, England, F4-51, 1977. By permission.)*

2. At high P_g/V values, the $k_l a$ correlation failed. Under such conditions, $k_l a$ was roughly constant.

3. Effects of scale on $k_l a$ data were very pronounced. $k_l a$ data for two different tank sizes did not match over the same ranges of P_g/V and V_s. The data for an air-electrolyte system are shown in Fig. 6.26.

Recirculation was shown to be more important for noncoalescing systems than for coalescing systems. Other comments by Smith et al. were:

1. Ungassed $k_l a$ data obtained by surface aeration cannot be used to determine $k_l a$ under gassed conditions.

2. Concave shaped blades and impellers with a large number of blades increased $k_l a$ values.

3. The concept of a minimum rotational speed for gas dispersion was

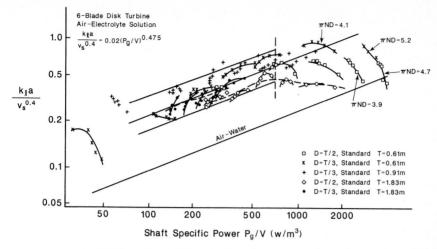

Figure 6.26   ($k_1a/V_s^{0.4}$) versus P_g/V for air electrolyte systems in various tanks.   *(From J. M. Smith, K. van't Riet, and J. C. Middleton, Proc. 2nd Eur. Conf. on Mixing, St. John's College, Cambridge, England, BHRA Fluid Eng., Cranfield, England, F4-51, 1977. By permission.)*

unnecessary if power input from gas sparging was incorporated into the $k_l a$ correlations. Total power input determined the nature of the gas dispersion.

DeMore et al. (1988) suggested that the maximums in the $k_l a/V_s^{0.4}$ were because the disk style turbines reached choked conditions which commonly occur in two-phase flow situations.

The article by Smith et al. (1977) is very important in its emphasis on $k_l a$ effects. In particular, typical P_g/V and V_s correlations in the literature are gross correlations that can be used to determine the approximate $k_l a$ level over several orders of magnitude for P_g/V. Such correlations cannot be used to optimize mass transfer for a particular design.

Legrys (1978) studied $k_l a$ at constant impeller rotational speed. He found that small increases in gas superficial velocity increased $k_l a$ and decreased total power input since $k_l a \cong V_s^{0.4}$. As flooding conditions were approached, increases in superficial velocity continued to increase $k_l a$ as total power input increased. Under these conditions, power input from sparging (P_e or P_s) was increased. P_g from mechanical agitation was fairly constant so that total power input increased. Power was plotted versus $k_l a$ and showed a minimum in the range of superficial velocities from 0.03 to 0.05 m/s. Power input because of kinetic energy of sparging was very small compared to gas expansion (i.e., P_s or P_e). Legry noted: (1) the inefficiencies in gas compression, (2) that mechanical agitation ensured efficient use of compressed gas, and (3) that the superficial velocity in large vessels increased with height, implying that $k_l a$ was lower around the sparger.

van't Riet (1979) provided correlations for $k_l a$ data divided according to the nature of the solution. The general correlation followed the form:

$$k_l a = K\left(\frac{P_g}{V}\right)^{\alpha}(V_s)^{\beta} \qquad (6.129)$$

For pure water:

$$k_l a = 2.6 \times 10^{-2}\left(\frac{P_g}{V}\right)^{0.4}(V_s)^{0.5} \ \ \mathrm{s}^{-1} \qquad (6.130)$$

up to volumes of 2600 L and $500 < P_g/V < 10,000$ W/m^3. For water with electrolytes:

$$k_l a = 2.0 \times 10^{-3}\left(\frac{P_g}{V}\right)^{0.7}(V_s)^{0.2} \ \mathrm{s}^{-1} \qquad (6.131)$$

for volumes between 2 and 4400 L and power per volume between 500 and 10,000 W/m^3. These correlations were accurate to within 20 to 40 percent. Both interfacial area and the mass transfer coefficient were dependent on power per volume.

For the electrolytic solutions, van't Riet noted that average bubble size was smaller than in nonelectrolytic solutions. This was attributed to a decrease in bubble coalescence caused by a more rigid interface; hence, the term noncoalescing. The gas phase was assumed to be well mixed because of impeller coalescence. Holdup ϕ increased as bubble size became smaller. van't Riet also noted that k_l was typically on the order of 4×10^{-4} m/s. Both flooding and power input from gas expansion had a substantial impact on $k_l a$. There was no effect of impeller type or number of impellers on the $k_l a$ correlations, but obviously some impellers can provide higher P_g/V values than other impellers. The choice of the sparger had little effect on $k_l a$ although spargers were recommended to be smaller than the impeller. The effect of fluid height was incorporated into P_g/V considerations, and the position and clearance of the impeller were unimportant.

These comments were made by van't Riet as gross effects of geometry on $k_l a$ correlations over a large span of P_g/V without specific reference to a particular geometry. In a particular arrangement, $k_l a$ can be improved by changing geometry at a constant P_g/V value.

Bryant and Sadeghzadeh (1979) obtained $k_l a$ data as a function of the reciprocals of circulation times with and without gas. In this work,

$k_l a$ was a function of a reduction in circulation rate of the liquid because of the gas. The correlation appeared as:

$$k_l a = 1.0 \left(\frac{1}{t_0} - \frac{1}{t} \right) + 0.03 \text{ s}^{-1} \tag{6.132}$$

where t_0 and t are the unaerated and aerated circulation times. This approach to $k_l a$ based on circulation times warrants further study.

Chandrasekharan and Calderbank (1981) suggested the following correlation:

$$k_l a = \zeta \left(\frac{P_g}{V} \right)^{\alpha} (V_s)^{\beta} \tag{6.133}$$

Their data indicated that $k_l a$ does not remain constant during scaleup at constant P_g/V and V_s. As a result, they suggested that ζ, α, and β should be a function of scale and correlated these parameters using power laws of tank diameter. One correlation was given as:

$$k_l a = \left(\frac{c}{T^4} \right) \left(\frac{P_g}{V} \right)^{A} (Q)^{A/\sqrt{T}} \tag{6.134}$$

where $c = 0.0248$ and $A = 0.551$ with SI units. They also suggested the use of tanks-in-series approach to model the gas phase mixing.

Davies et al. (1985) reported $k_l a$ correlations based on gassed power per volume and superficial gas velocity, which are given in a later section.

Hickman (1988) applied a steady-state $H_2 O_2$ technique described later to measure $k_l a$ in 0.6- and 2.0-m-diameter tanks in the standard configuration using disk style impellers. Combined data between the two systems provided the following correlation:

$$k_l a = 0.046 \left(\frac{P_g}{V} \right)^{0.54} V_s^{0.68} \tag{6.135}$$

Differences in scale were observed in the $k_l a$ data. A 1.4 percent CMC solution was also studied. $k_l a$ was found to vary as $\mu^{-0.6}$ and the CMC data were very much lower than the air water data.

Machon et al. (1988) obtained $k_l a$ data in a multiple-impeller system, 0.29 m in diameter and 0.8 m high, having two disk style impellers and 10 percent wall baffles. The $k_l a$ data for the air water system varied as:

$$k_l a \propto Q^{0.57} P_g^{0.49} C^{0.14} \tag{6.136}$$

which shows $k_l a$ increasing with increasing clearance of the bottom impeller. $k_l a$ data for carboxymethyl cellulose solutions of 1, 2, and 3 percent were also obtained. The 2 and 3 percent solutions were in the laminar regime and the 1 percent solution was in the transition regime which caused diverse changes in flow character. No single regression relation for the $k_l a$ data was obtained for these data. The $k_l a$ data for the high-viscosity CMC solutions were substantially less than that for the air water system.

Baczkiewicz and Michalski (1988) studied $k_l a$ in acetic acid fermentation using a self-aspirating tube agitator. The details of the system were given earlier under power draw. The acetic acid fermentation medium strongly suppressed bubble coalescence and $k_l a$ values were relatively high. $k_l a$ was correlated as:

$$k_l a = 0.17(N)^{1.68}\left(\frac{Q}{V_L}\right)^{0.59}\left(\frac{D}{T}\right)^{2.17}(n_p)^{0.46} \qquad (6.137)$$

which shows the effects of D/T and paddle number on $k_l a$. The correlation was rearranged to the traditional form as:

$$k_l a = 0.83\left(\frac{P_g}{V}\right)^{0.48}(V_s)^{0.74} \qquad (6.138)$$

At maximum aeration, $k_l a$ was correlated as:

$$k_l a = 4.0 \times 10^{-4}\left(\frac{P_g}{V}\right)^{0.76} \qquad (6.139)$$

Linek et al. (1988) were critical of most published $k_l a$ data and correlations. Several reasons for this included: (1) improper experimental techniques, used to determine $k_l a$, and (2) inadequate $k_l a$ correlations used to correlate the data. Linek et al. (1988) suggested that a proper $k_l a$ correlation should appear as:

$$k_l a = f(P_T/V, V_s, \text{fluid parameters, coalescence parameters}) \qquad (6.140)$$

Typically, the coalescence parameters are neglected. However, the coalescence parameters are necessary to obtain a general correlation applicable to both coalescing and noncoalescing systems. It is unlikely that there is a distinct abrupt transition between coalescence and noncoalescence in systems generally and the coalescence parameters would be used to account for the transition and the differences in coalescing and noncoalescing systems. Linek et al. reviewed the literature in the development of a coalescence parameter. The final correlation for $k_l a$, given by Linek et al., was:

$$k_l a = \exp\left(- 4.18n_1 - 4.36\right)\left(\frac{P_T}{V}\right)^{n_1} V_s^{0.4} \qquad (6.141)$$

where n_1 was given as:

$$n_1 = \frac{1.5X_2}{1/10^6 + 3.05X_2} + 0.56 \qquad (6.142)$$

where X_2 was defined as:

$$X_2 = C\left(\frac{d\sigma}{dC}\right)^2\left(\frac{1 + d\ln\sigma}{d\ln C}\right)^{-1} \qquad (6.143)$$

The fluid properties have not been included in this correlation. More work is necessary in this area. Hopefully, the coalescence parameter would account for the effect of impurities on $k_l a$.

Warmoeskerken and Smith (1989) cited $k_l a$ data and provided a $k_l a$ correlation for the hollow blade disk style turbine as:

$$\frac{k_l a}{V_s^{0.5}} = 0.5\left(\frac{P_g}{V}\right)^{0.45} \qquad (6.144)$$

It was suspected that the impeller will provide a 20 to 30 percent increase in $k_l a$ over that of a standard disk impeller.

Chemical methods versus physical methods. Differences in the magnitude of $k_l a$ and interfacial area and differences in the effects of operating variables such as impeller rotational speed on these quantities do appear in the literature. To a large degree, these differences depend on the type of flow regime in which the data were taken and the type of experimental technique used to obtain the data: chemical methods versus physical methods and local measurements versus average measurements. Although liquid phase can be considered well mixed, the gas phase and interfacial area are not necessarily well mixed. Hence, the assumption of well-mixed gas phase and interfacial area system and the simple weighing of $k_l a$ data, assumed according to the well-mixed assumption, are in serious question.

Prasher (1975) suggested that, although the interfacial area behaved the same throughout the vessel for fast chemical reactions, the nature of the interfacial area for physical absorption varied drastically because of the wide range of turbulence in the tank. Work by Rennie and Valentin (1968) supports the same conclusion.

Sridhar and Potter (1978) noted that the interfacial areas for the chemical reaction methods were larger than for physical methods. In their work, interfacial area determined by chemical methods varied linearly with N^3D^2 or power per unit volume which is not typical of such data.

Sridhar and Potter made other comparisons between a light transmission method and chemical methods for interfacial area. Typically, chemical methods yielded 1.5 to 2 times the interfacial area obtained using light transmission. Sridhar and Potter concluded that significant differences existed between the two methods. Making several estimations of actual tank conditions, they concluded that: (1) chemical reactions such as sulfite oxidation were still not fast enough to eliminate the hydrodynamic effects for the absorption rate and (2) the chemical methods weighed the interfacial area of the impeller region disproportionately higher than the interfacial area in the rest of the tank. In the impeller region, the rates of mass transfer are very high and may well match the fast chemical reactions used to determine mass transfer coefficients.

Effects of temperature in reaction methods are tied to fluid properties and their temperature dependencies. Assuming $k_l a D^2 / D_L$ equals a constant, $k_l a$ varies directly with liquid diffusivity and hence, directly with temperature (Smith et al., 1977). Zlokarnik (1978) and Linek and Vacek (1981) have also discussed the limitations of the chemical methods for the determination of $k_l a$. Blakebrough et al. (1961) discussed the application and shortcomings of chemical methods as applied to fermentation.

In any case, $k_l a$ is system specific because of all the variations of the phenomena and geometric effects that impact on the mass transport process. As a result, $k_l a$ cannot be calculated a priori in design with certainty (Mehta and Sharma, 1971; Nienow et al., 1985) without pilot data. Even with pilot data, the specific scaled process can provide only $k_l a$ data with absolute certainty, and tests should be performed on these units using steady-state methods to determine the actual $k_l a$. Using steady-state methods on actual units is perhaps the only way of determining the process $k_l a$, and even these data will vary according to disturbances and operating conditions.

Several studies have been performed in defense of fast reaction chemical methods in the determination of $k_l a$. Midoux and Charpentier (1979) developed a model for noncoalescing, gas-liquid dispersions. In their model, bubbles retained their identity until they left the reactor or dissolved completely into the liquid. Generalizations of single bubble behavior to bubble distributions and the use of residence time distributions led to the limiting and sufficient criteria for the validity of the chemical method in the determination of $k_l a$.

Ruchti et al. (1985) discussed some of the practical guidelines for the sulfite oxidation method for $k_l a$. Ruchti et al. noted the work by Linek and Vacek (1981) in which the sensitivity of the kinetics to trace impurities was emphasized, making reproducibility impossible under fast chemical reaction conditions. However, Ruchti et al. gave an empirical procedure by which the appropriate reaction conditions were determined.

In any case, different processes are being accomplished when chemical and physical methods are used to determine interfacial area. The different methods may easily weigh the interfacial area differently according to location. It is equally important that methods for the determination of $k_l a$ by fast chemical reactions be continued because of limitations of the dynamic methods. For example, probe dynamics in dynamic methods may not be sufficiently rapid to follow $k_l a$. Probes may not be suitable in systems under laminar or transitional aeration where liquid mixing becomes poor.

Mass transfer and interfacial area based on flow regime

van't Riet (1979) noted that the exponents α and β in the equations for $k_l a$ varied from 0.4 to 1.0 for the exponent α on power per unit volume term and from 0 to 0.7 for the exponent β on superficial velocity, partly, because of distinctly different flow regimes over which the studies have been made. Except for the flooding transition, changes in flow regimes are gradual, and no distinct values for the exponents α and β can be expected. When the extremes are compared, the differences in flow regimes and exponents can be noted.

Nishikawa et al. (1981a) developed $k_l a$ correlations for three flow regimes: (1) gas sparging or bubbling controlled regime occurring at high gassing rates and low impeller rotational speeds, (2) agitation controlled gas dispersion regime occurring at higher impeller rotational speeds and moderate to low gassing rates, and (3) an intermediate regime between the two extremes where both regimes 1 and 2 impact on $k_l a$. The two regimes can be observed in Fig. 6.25. Nishikawa et al. (1981a) proposed a general correlation that covered all three regimes based on power input per unit mass originating from both gas sparging and agitation. The gas phase was considered under plug-flow rather than a perfectly mixed condition typically assumed in the literature. The validity of this assumption requires investigation and depends on the amount of gas recirculated into the impeller.

The flow regime and recirculation affect the type of driving force for mass transfer. Under agitation controlled conditions, the gas phase can be considered perfectly mixed and under sparging controlled conditions, a log mean driving force may be more appropriate. Nishikawa et al. assumed a perfectly mixed condition for the liquid phase. In their investigation, flat blade and disk style turbines were used, and $k_l a$ was determined by using a dynamic gassing out-in method and a chemical method when the dynamics of the oxygen probe were too slow. Nishikawa et al. (1981a) also cited work by Yasunishi that found the two methods agreed.

For the correlations by Nishikawa et al. (1981a), power per unit mass by gas sparging was calculated as:

$$P_{sm} = V_s g \tag{6.145}$$

showing that the superficial velocity is actually a form of power input from gas sparging. The power per unit mass from agitation was calculated as:

$$P_m = \frac{N_P N^3 D^5}{V} \tag{6.146}$$

where N_P is the ungassed power number. For the gas sparging controlling regime where the power input from gas sparging was P_{sm}, the mass transfer coefficient was correlated as:

$$k_l a = 1.25 \times 10^{-5}(P_{sm})^{1.0} \tag{6.147}$$

where there was no effect from agitation of the rotating impeller. $k_l a$ is in units of s^{-1}, and P_{sm} is in units of cm^2/s^3.

For the agitation controlled regime where the power input per unit mass from agitation was P_m, the mass transfer coefficients were correlated as:

$$k_l a = 3.92 \times 10^{-6}(P_{sm})^{0.33}(P_m)^{0.8} \quad \text{(for the disk turbine)} \tag{6.148}$$

$$k_l a = 5.69 \times 10^{-6}(P_{sm})^{0.33}(P_m)^{0.75} \quad \text{(for the flat blade turbine)} \tag{6.149}$$

$k_l a$ is in units of s^{-1}, and P_{sm} is in units of cm^2/s^3. The $k_l a$ correlations for the intermediate regime were much more complicated:

$$k_l a = [3.92 \times 10^{-6}(P_m)^{0.8} + 1.25 \times 10^{-5}W_p(P_{sm})^{0.66}](P_{sm})^{0.33}$$

$$\text{(for the disk turbine)} \tag{6.150}$$

$$k_l a = [5.69 \times 10^{-6}(P_m)^{0.75} + 1.25 \times 10^{-5}W_p(P_{sm})^{0.66}](P_{sm})^{0.33}$$

$$\text{(for the flat blade turbine)} \tag{6.151}$$

where W_p is a weighting term equal to $P_{sm}/[P_m/N_P + P_{sm}]$ to match the two extreme regimes.

The power input per unit mass P_m used in the correlations given by Nishikawa et al. was obtained from the ungassed power number and was not the actual power input per mass under gassing conditions. The power input per unit mass P_m at which agitation regime controls at a particular P_{sm} was obtained from:

$$(P_m)^{0.8}\left(1 + \frac{P_m}{N_P P_{sm}}\right) = 12.7(P_{sm})^{0.66} \qquad (6.152)$$

The minimum value for P_m at which gas sparging dominates was given by:

$$P_m = 0.514(P_{sm})^{0.84} \qquad (6.153)$$

Nishikawa et al. (1981a) found that these relationships held for large-scale vessels up to 3000 L using data from Fuchs et al. (1971). The relationships overestimated $k_l a$ data of Fuchs et al. (1971) for a 51,000-L vessel, which were similar to the $k_l a$ data obtained from gas sparging alone. Nishikawa et al. (1981) also found the surface aeration contribution to $k_l a$ insignificant.

Nishikawa et al. (1981b) extended their study to include a broader range of fluid properties using aqueous solutions of millet jelly and non-Newtonian fluids, which is discussed in a later section. Their general approach was different in that $k_l a$ was based on correlations rather than power per mass. For the agitation controlled regime for disk style turbines, they used a correlation similar to that developed by Yagi and Yoshida (1975), given above, with several modifications:

$$\frac{(k_l a)_a T^2}{D_L} = 0.115\left(\frac{\rho N D^2}{\mu}\right)^{1.5}\left(\frac{N^2 D}{g}\right)^{0.367}$$

$$\times \left(\frac{\mu}{\rho D_L}\right)^{0.5}\left(\frac{\mu V_s}{\sigma}\right)^{0.5}\left(\frac{ND}{V_s}\right)^{0.167}\left(\frac{D}{T}\right)^{0.4} N_P^{0.8} \qquad (6.154)$$

A similar correlation was given for flat blade impellers. In the comparison of these correlations, Nishikawa et al. (1981b) indicated that the presence of the disk for the disk turbine had some detrimental effects in dispersing gas in viscous liquids.

For the sparging controlled regime, Nishikawa et al. (1981b) provided the following correlation:

$$\frac{(k_l a)_s T^2}{D_L} = 0.112\left(\frac{V_s^2}{gT}\right)^{0.5}\left(\frac{\mu}{\rho D_L}\right)^{0.5}\left(\frac{gT^2\rho}{\sigma}\right)^{0.66}\left(\frac{gT^3\rho^2}{\mu^2}\right)^{0.42} \qquad (6.155)$$

Nishikawa et al. (1981b) combined the two mass transfer coefficients into one, using a weighting function based on relative power consumption as used previously by Nishikawa et al (1981a):

$$k_l a = (k_l a)_a + \left[\frac{P_{sm}}{(P_m)/N_P + P_{sm}}\right](k_l a)_s \qquad (6.156)$$

where $P_{sm}/[P_m/N_P + P_{sm}]$ is the weighting parameter W_p, and the subscripts a and s indicate mass transfer from agitation and gas sparging, respectively. The power terms P_m and P_{sm} indicate power input from agitation and gas sparging, respectively, and the power number N_P is the ungassed power number. Work has also been performed for multistage impeller systems (Nishikawa et al., 1984).

Mass transfer because of surface aeration. Albal et al. (1983), using a pressurized system, studied surface aeration for gas dispersion. Under such conditions, gas flow rate was not an independent variable. They used a dual pitched blade configuration in which the pressure in the head space above the liquid surface was followed in time and used to determine absorption and $k_l a$. $k_l a$ was found independent of pressure between 2 and 8 MN/m^2, although no statements were made concerning the effect of pressure on surface entrainment. [Sridhar and Potter (1980) found a definite pressure effect on $k_l a$ at pressures up to 1 MN/ m^2 (10 atm) and no effect after that.] In the study by Albal et al., $k_l a$ was proportional to $(P_g/V)^{1.5}$ up to P_g/V of 4 kW/m^3 and $(P_g/V)^{0.8}$ after that. There was no comment on the effects of flow regime behavior in the bulk of the tank to explain the change in exponents. $k_l a$ values reported by Albal et al. were small because of the limitations of surface aeration and the lower power input by the pitched blade turbines. Data on $k_l a$ decreased with increasing viscosity and increasing surface tension. Albal et al. noted that the critical speed for the suspension of solids under aerated conditions was 20 to 30 percent over that for unaerated conditions given by Zwietering (1958). $k_l a$ increased initially and then decreased with increasing solid concentration which is the typical effect of a third phase. Three phase effects are discussed later. Power per volume increased continuously with solid concentration. The final correlation for $k_l a$ for the transfer of oxygen into water was given as:

$$\frac{(k_l a)D^2}{D_L} = 1.41 \times 10^{-3} \left(\frac{\mu}{\rho D_L}\right)^{0.5} \left(\frac{\rho N D^2}{\mu}\right)^{0.67} \left(\frac{\rho N^2 D^3}{\sigma}\right)^{1.29} \quad (6.157)$$

Mass transfer coefficients and correlations based on impeller flow regimes. Mass transfer coefficient correlations have been given according to the type of fluid: electrolytic and nonelectrolytic solutions. This was a division based on flow regime: small-bubble regime versus large-bubble regime. Correlations have also been given according to: (1) sparger controlled or agitator controlled contacting and (2) foaming and nonfoaming systems. The interrelationship between impeller flow regimes and mass transfer coefficients is far more difficult to establish.

The vortex and clinging cavities are so different from large cavities and the cavities of the 3-3 structure that different correlations may be developed for each of the major cavity types (Smith et al., 1986), particularly if the impeller region is the most active region in the tank (Rennie and Valentin, 1968). Opposing this view, $k_l a$ depends only on P_g/V and V_s, but this type of correlation is crude. Work in the direction of interrelating impeller flow regimes and mass transfer is appropriate.

Smith and Warmoeskerken (1985) provided correlations for $k_l a$ before the 3-3 structure as:

$$\frac{k_l a}{N} = 1.1 \times 10^{-7} \text{Fl}^{0.6} \text{Re}^{1.1} \qquad (6.158)$$

and after the formation of the large cavities of the 3-3 structure as:

$$\frac{k_l a}{N} = 1.6 \times 10^{-7} \text{Fl}^{0.42} \text{Re}^{1.02} \qquad (6.159)$$

Recent experimental techniques for the determination of $k_l a$

van't Riet (1979), in an excellent review of the various experimental techniques used in the determination of interfacial area and $k_l a$, cited the dynamic gassing out procedure and overall transfer rate as the most reliable and simple methods to obtain $k_l a$. Light transmission within its limitations (Sridhar and Potter, 1978) was cited as reliable and simple. Other techniques and their shortcomings were also given.

Various methods. The dynamic gassing out method is based on the absorption of a gas, usually oxygen, into solution. The dissolved gas concentration is measured with a polargraphic electrode, usually a Clark cell. A first-order model for the mass balance for the liquid phase can be written as:

$$\frac{dC_L(t)}{dt} = k_l a \left[\frac{C_G}{H_e} - C_L(t) \right] \qquad (6.160)$$

For time from t_1 to t_2, this equation integrates to:

$$k_l a = \frac{[C_G/H_e - C_L(t_1)]/[C_G/H_e - C_L(t_2)]}{t_2 - t_1} \qquad (6.161)$$

where H_e is the Henry's law constant. From $C_L(t)$ data, $k_l a$ can be calculated.

One difficulty with this method is the determination of the appropriate value for the Henry's law constant H_e, which is a function of gas composition and not recommended for partial pressures above 1 atm. Another difficulty is the question of whether the polargraphic electrode is detecting the gas concentration in the gas bubbles as well as in the liquid. In this case, the electrode should be placed to avoid contact with the bubbles. Dunn and Einsele (1975) showed that such simple first-order models underestimated $k_l a$ values because of improper assumptions about the gas phase concentrations and depletion. Specifically, Dunn and Einsele discussed the effect of gas phase depletion of the component being transported and showed how this affected measured $k_l a$ values.

Depletion has its greatest effect in large tanks at low superficial velocities and high holdups (Smith et al., 1977) and for noncoalescing systems. Smith et al. performed some sample calculations that showed $k_l a$ corrections may be 10 to 20 percent. Despite these difficulties, first-order models are most often assumed in the experimental determination of $k_l a$.

The overall transfer rate (OTR) is a steady-state measurement procedure in which the inlet and outlet gas specie concentrations, as well as the liquid specie concentration, are measured. $k_l a$ is then:

$$k_l a = \frac{\text{OTR}}{C_G/H_e - C_L} \tag{6.162}$$

van't Riet noted that this technique was not fully represented in the literature.

Linek et al. (1982a; 1982b) noted the two major types of dynamic methods available to determine $k_l a$: (1) simple interchange of sparged gasses, maintaining everything else constant, called procedure A by Linek et al., and (2) simultaneous startup of aeration and agitation, called procedure B. Such procedures require transport of one specie into solution and transport of one specie out of solution. Linek et al. noted that $k_l a$ data do not agree between the methods and with steady-state data. In procedure A, the specie leaving solution can interfere with the transport of the specie entering in the solution. This reduced $k_l a$ values if not appropriately accounted for in the model.

The objective of the study by Linek et al. was to check the agreement between procedure B and a steady-state method, the steady-state method being the most reliable method of the three. For procedure B to be equivalent to steady-state methods, the initial liquid must be totally degassed, and, if not done, the procedure was not equivalent to steady state procedures. Linek et al. found that: (1) $k_l a$ values obtained from different experimental techniques need not be the same; (2) significant effects on $k_l a$ resulted from the transport of counterdiffusing species; and (3) the flushing of the holdup of old gas with new gas in procedure A affected $k_l a$. Linek et al. (1982b) noted that dynamic methods led to lower $k_l a$ values.

In the general literature on $k_l a$, there is little categorization of studies reported in mass transfer research such as: "A through stagnant B" or statements concerning molar flux ratios. This is partly because of the poorly defined driving force for the mass transport and the fact that the gas phase mixing is so poorly understood and dependent on hydrodynamics.

Chapman et al. (1982), using material balances, established two equations for $k_l a$ that were independent of the gas phase concentration and the gas phase mixing. For the first method:

$$k_l a = \frac{dC_L^*/dt}{(Q/V_G) \int_0^t (1 - C_o^*) \, dt - C_L^*[(H_e V_L/V_G) + 1]} \tag{6.163}$$

where C_o^* and C_L^* are recorded concentrations divided by their new steady-state values. Using L'Hospital rule, Chapman et al. (1982) obtained the equation for the second method as:

$$k_l a = \frac{d^2 C_L^*/dt^2}{[(Q/V_G)(1 - C_o^*)] - (dC_L^*/dt)[(H_e V_L/V_G) + 1]} \quad \text{at } t = 0 \tag{6.164}$$

For $t = 0$, $C_o^* = 0$, and $dC_L^*/dt = 0$:

$$k_l a = \frac{V_G d^2 C_L^*}{Q \, dt^2} \quad \text{at } t = 0 \tag{6.165}$$

The advantages of the last equation were emphasized by Gibilaro et al. (1985).

Davies et al. (1985) discussed the same techniques to obtain $k_l a$ data which were referred to as: the double response method and the initial response method. For data obtained from the double response method, the mass transfer coefficient was correlated as:

$$k_l a = 0.17 \left(\frac{P_g}{V_L}\right)^{0.64} (V_s)^{0.22} \tag{6.166}$$

and for data obtained from the initial response method as:

$$k_l a = 0.49 \left(\frac{P_g}{V_L}\right)^{0.76} (V_s)^{0.45} \tag{6.167}$$

The double response method was found reliable, consistent, comparable to other methods, and insensitive to parameters. The initial response method was less reproducible than the double response method and had a greater dependency on volumetric gas flow rate and holdup which indicated different dependencies on operational conditions for the two techniques. Both methods required no assumptions as to the gas phase mixing. Results from these methods could be used in such a manner as to determine the gas phase mixing as a function of operating conditions.

As the equations show, $k_l a$ can be evaluated independent of the nature of the gas mixing. The liquid concentration C_L^* can be followed in time. A third-order derivative of the liquid concentration was also related to $k_l a$. Unfortunately, with these techniques, $k_l a$ becomes a function of the time constant of the measurement probe, which is a function of the local fluid velocity and the model assumed for the probe. The liquid mixing characteristics may become important in $k_l a$ measurements and may require consideration. Such questions require further investigation but should not distract from the novelty and potential advantages of these two relatively new methods.

Steady-state methods for $k_l a$. Steady-state methods for the measurement of $k_l a$ are very useful and often overlooked as an experimental procedure. Greaves and Loh (1985) used a steady-state method to determine $k_l a$ for two- and three-phase systems. To establish such a procedure, the liquid or slurry phase was removed at a constant rate, which established the lost term for the following balance:

$$k_l a V(C^* - C) = Q_L(C - C_i) \qquad (6.168)$$

where C^* is the interface concentration, C is the tank concentration, C_i is the inlet system concentration, V is the volume of the tank, and Q_L is the liquid flow rate in and out of the tank. The liquid phase was assumed well mixed. The results of the work agreed with van't Riet (1979), and the final $k_l a$ correlation obtained from the data was:

$$k_l a = 0.026\left(\frac{P_g}{V}\right)^{0.4}(V_s)^{0.5} \qquad (6.169)$$

Hickman (1988) presented a steady-state method for the determination of $k_l a$ using the decomposition of hydrogen peroxide:

$$2H_2O_2 \rightleftharpoons 2H_2O + O_2$$

At steady state, the rate of H_2O_2 decomposition r was proportional to the amount of H_2O_2 added to the process or:

$$\frac{Q_i C_{H_2O_2 i}}{V} = r \qquad (6.170)$$

where the subscript i stands for the inlet conditions. The amount of O_2 transferred across the interface was $0.5r$ or:

$$0.5r = k_l a\ \Delta C \qquad (6.171)$$

where $\Delta C = C_L - H_e C_G$. The equations show that the reaction rate and the steady-state concentration of H_2O_2 are not needed to determine $k_l a$. The technique was simple to use and steady state was reached quickly.

Mass transfer in viscous and non-Newtonian fluids

Viscous fluids. Mass transfer in viscous fluids requires consideration of the type of impeller used in the gas dispersion. Typically, the disk style impeller is inadequate in mixing viscous liquids. Concerning mass transfer, Nishikawa et al. (1983) showed the helical ribbon and anchor caused much higher $k_l a$ values in fluids above 2 poise (0.2 Pa · s) than did the disk style turbines. Further work is necessary in this area since effective laminar mixing in viscous fluids is usually accomplished by helical ribbon and screw impellers.

Non-Newtonian fluids. Ideally, mass transfer between a bubble and a non-Newtonian fluid requires consideration of the non-Newtonian viscosity and elastic effects, if any, as well as the non-Newtonian effects on flow regimes in the impeller region and tank, liquid and gas circulation, liquid and gas phase mixing, bubble size, interfacial area, holdup, and k_l. The simplest approach for mass transfer coefficients is to assume correlations for Newtonian fluids and to modify them to account for non-Newtonian effects mainly due to viscosity. For shear thinning fluids, the power law is assumed for viscosity, and apparent viscosity is given as a function of some average shear rate. Shear rate is then based on a flow quantity or mechanism. For an agitated tank, average shear rate is a constant times impeller rotational speed (Metzner and Otto, 1957). For viscoelastic phenomena, the first normal stress difference and the first normal stress coefficient are important fluid parameters modeled using a power law, relating deformation rate to impeller rotational speed and a characteristic time of the fluid to account for elastic effects.

Perez and Sandall (1974) correlated $k_l a$ for CO_2 absorption in water and Carbopol solutions using the apparent viscosity defined by Metzner and Otto (1957). They correlated their results with impeller rotation speed and the flow index n of the power law. Their final $k_l a$ correlation for impeller rotational speeds above $3.3\ s^{-1}$ was:

$$\frac{k_l a D^2}{D_L} = 21.2 \left(\frac{\rho N D^2}{\mu_a}\right)^{1.11} \left(\frac{\mu_a}{\rho D_L}\right)^{0.5} \left(\frac{V_s D}{\sigma}\right)^{0.447} \left(\frac{\mu_g}{\mu_a}\right)^{0.694} \qquad (6.172)$$

where the term DV_s/σ is not dimensionless, and μ_a is the apparent viscosity, calculated as:

$$\mu_a = K(11N)^{n-1}\left(\frac{3n + 1}{4n}\right)^n \qquad (6.173)$$

This correlation only correlated the data for $n = 1.0$ and 0.916. For the other n values studied, $n = 0.773$ and 0.594, the formation of foam apparently affected the results.

Yagi and Yoshida (1975), noting the lack of correlations for non-Newtonian fluids such as polymer solutions and fermentation broths, correlated $k_l a$ data obtained from the desorption of oxygen in two kinds of Newtonian fluids (glycerol and millet jelly) and non-Newtonian polymer solutions (sodium polyacrylate and sodium carboxyl methyl cellulose, CMC) having varying rheological properties. To take into account the shear thinning and the viscoelastic nature of the polymer solutions, the power law and a characteristic time, λ (Prest et al., 1970), were used. The apparent viscosity μ_a was based on the Metzner-Otto (1957) method where the average shear rate was 11.5 times the impeller rotational speed N. Using an apparent viscosity, Yagi and Yoshida assumed a general correlation form based on dimensional analysis and correlated $k_l a$ data as:

$$\frac{k_l a D^2}{D_L} = 0.060 \left(\frac{\rho N D^2}{\mu_a}\right)^{1.5} \left(\frac{N^2 D}{g}\right)^{0.19} \left(\frac{\mu_a}{\rho D_L}\right)^{0.5} \left(\frac{\mu_a V_s}{\sigma}\right)^{0.6} \left(\frac{ND}{V_s}\right)^{0.32} \qquad (6.174)$$

For viscoelastic fluids, an additional multiplication factor $[1 + 2.0(\lambda N)^{0.5}]^{-0.67}$ was added to the right side of the correlation to account for viscoelastic effects. The Deborah number used in the correlation was defined as λN, but no basic justification was given for this specific definition of the Deborah number. λ is the characteristic time of the fluid as discussed in Chap. 5. For the non-Newtonian fluids studied, increasing concentrations of polymer in the solution caused decreasing $k_l a$ values. The flow index n varied between 0.4 to 1.0 but was not a suitable parameter to correlate the effect of agitation. Much less dispersion occurred in the impeller region for the non-Newtonian fluids than for Newtonian fluids and, in one non-Newtonian case, a gas cavity structure appeared to cover the entire impeller. The normal stresses (i.e., the Weissenberg effect) generated in the polymer solution decreased the radial outward flow of the gas, resulting in less gas dispersion.

Ranade and Ulbrecht (1977) found substantial reductions in k_la in studies of viscoelastic solutions of PAA, from between 5 and 10 s^{-1} for water to between 1 and 3 s^{-1} for 0.1 percent PAA. These results were expected because of the flow behavior. Ranade and Ulbrecht indicated that mass transfer coefficients should be based on rheological properties for viscoelastic fluids.

Ranade and Ulbrecht (1978) studied the effect of polymer additives on gas-liquid mass transfer using a slow chemical reaction technique. Even a small addition of polymer substantially reduced the mass transfer coefficient, k_la; PAA in solution having a greater effect than CMC. The reductions in k_la were not attributed to surface active effects, diffusivity, solubility, or viscosity effects but to a direct reduction in the interfacial area a. Machon et al. (1980) found similar effects of CMC on gas holdup.

Ranade and Ulbrecht attributed the difference between PAA and CMC solutions to the elasticity of the PAA solutions. Viscoelasticity hindered coalescence but generally maintained the interfacial area intact once formed. In the study by Ranade and Ulbrecht, the elastic effects of PAA also hindered gas dispersion from the gas cavities of the impeller and hence, lowered the interfacial area and the mass transfer coefficient. Their final correlation for k_la was:

$$\frac{k_la T^2}{D_L} = 2.5 \times 10^{-4}\left(\frac{\rho ND^2}{\mu}\right)^{1.8}\left(\frac{\mu}{\mu_w}\right)^{1.39}(1 + 100\text{De})^{-0.67} \quad (6.175)$$

where μ_w is the viscosity for water and De is a modified Deborah number λN, λ being a characteristic fluid time. This correlation shows the effect of liquid viscosity on k_la as:

$$k_la \propto \mu^{-0.41} \quad (6.176)$$

which agrees with other correlations cited above.

Nishikawa et al. (1981b), as cited above, included non-Newtonian fluids in their combined correlation-power per unit mass approach by studying various CMC solutions. Elastic effects were not studied. As before, the two-regime model (i.e., gas sparged controlled and agitation controlled) was applied to correlated k_la using the power law to determine the apparent viscosity. The methods to calculate the average shear rates were different for the two regimes. For the agitation controlled regime and using the Metzner-Otto (1957) method, the average shear rate was related to impeller rotational speed (Nagata et al., 1971) as:

$$\gamma_{\text{av}} = 11.83N \quad (6.177)$$

This average shear rate gave rise to the apparent viscosity μ_{aa} due to agitation from the power law.

For the gas sparging regime, Nishikawa et al. related average shear rate to superficial gas velocity by:

$$\gamma_{av} = 50.0V_s \qquad (6.178)$$

where the constant 50 has units of cm^{-1}. This average shear rate provided an apparent viscosity μ_{as} due to gas sparging. Although elastic effects were not studied, Nishikawa et al. accounted for these in the two regimes using: (1) work by Yagi and Yoshida (1975) cited above for the agitation controlled regime and (2) work by Nakano (1976) for the sparging controlled regime. Nakano used the term $[1 + 0.18(\lambda V_t/D_{SM})^{0.45}]^{-1}$ for gas sparging, where V_t is the average rise velocity of the gas bubbles, and D_{SM} is the Sauter mean bubble diameter. The final correlation was:

$$\frac{k_l a T^2}{D_L} = 0.115\left(\frac{\rho ND^2}{\mu_{aa}}\right)^{1.5}\left(\frac{N^2D}{g}\right)^{0.367}\left(\frac{\mu_{aa}}{\rho D_L}\right)^{0.5}\left(\frac{\mu_{aa}V_s}{\sigma}\right)^{0.5}\left(\frac{ND}{V_s}\right)^{0.167}$$

$$\times \left(\frac{D}{T}\right)^{0.4}N_P^{0.8}[1 + 2.0(\lambda N)^{0.5}]^{-0.67} + 0.112\left[\frac{P_{sm}}{(P_m)/N_P + P_{sm}}\right]$$

$$\times \left(\frac{V_s^2}{gT}\right)^{0.5}\left(\frac{\mu_{as}}{\rho D_L}\right)^{0.5}\left(\frac{gT^2\rho}{\sigma}\right)^{0.66}\left(\frac{gT^3\rho^2}{\mu_{as}^2}\right)^{0.42}\left[1 + 0.18\left(\frac{\lambda V_t}{D_{SM}}\right)^{0.45}\right]^{-1} \qquad (6.179)$$

where the double subscripts, aa and as, are for "apparent, agitation" and "apparent, sparging," respectively. Again, the power number N_P was the ungassed non-Newtonian power number. The determination of power for non-Newtonian fluids under gassing conditions is still a matter of considerable research as noted above. Nishikawa et al. cited Hattori et al. (1972) as showing that $k_l a$ varied with viscosity to -0.45 power for fluids having viscosities less than 0.05 Pa · s (50 cP) and to the -1.03 power for fluids having viscosities greater than 0.05 Pa · s.

Nienow (1984), in a study on Xanthan gums and high-viscosity fermentations, found $k_l a$ to vary as:

$$k_l a \propto \mu_a^{-0.45}\left(\frac{P_g}{V}\right)^{0.66}(Q_G)^{0.34} \qquad (6.180)$$

Dynamic gassing out methods have been used in non-Newtonian fluids. However, there is no discussion in the literature of the flow behavior near probes used in these techniques. The effects of the flow on the probe dynamics, the measurements of dissolved oxygen concentration, and the state of the liquid phase mixing are of interest.

Viscoelastic fluids, in particular, have complex single-phase flow regimes that have not been studied in gas dispersion. The effects of viscoelasticity on probe measurements are not usually discussed in gas dispersion measurements.

Kawase and Moo-Young (1988) developed the following mass transfer correlation for Newtonian and non-Newtonian fluids:

$$k_l a = 0.675 \frac{\rho^{0.6}(P_T/V)^{(9+4n)/(10(1+n))}}{(K/\rho)^{0.5(1+n)}\sigma^{0.6}}(D_L)^{0.5}\left(\frac{V_s}{V_t}\right)^{0.5}\left(\frac{\mu_a}{\mu_w}\right)^{-0.25} \tag{6.181}$$

The correlation is for: $0.59 < n < 0.95$, 0.00355 Pa $\cdot$ s^n $< K < 10.8$ Pa $\cdot$ s^n, and $0.15 < T < 0.6$ m.

Scaleup of gas dispersion processes

Proper scaleup involves two major requirements: (1) a clear statement as to the criterion or criteria for scaleup and (2) the scaleup procedure itself.

Westerterp et al. (1963) recommended that, for equal interfacial area on scaleup, the scaleup procedure should be based on equal impeller tip speed and D/T ratio. Bourne (1964) showed that scaleup based on equal power per volume was more appropriate for equal interfacial area on scaleup. The difference points to differences in experimental methods used to study the gas dispersion process.

Nienow et al. (1977), in comments about different scaleup procedures, noted that superficial gas velocity V_s scales as $1/T^2$ at constant volumetric flow rate. At constant V_s on scaleup, Q must increase with scale since the gas flow rate varies with tank diameter squared. The mean residence time, τ or V/Q, at constant V_s on scaleup increases with tank diameter or $\tau \propto V/Q \propto T^3/T^2 \propto T$. Scaleup at constant mean residence time or constant vvm requires $Q \propto T^3$ or $V_s \propto T$, which requires the impeller to handle more gas and operate closer to flooding conditions.

Chandrasekharan and Calderbank (1981) commented that scaleup, which maintained geometric similarity, caused sacrifices in performance, and that $k_l a$ did not remain constant with constant P_g/V and V_s during scaleup (Smith et al., 1977). They recommended that, beyond a certain diameter, scaleup was best accomplished by increasing height. Legry (1978) provided counter arguments to this view.

Nishikawa et al. (1981a) summarized other general criteria for scaleup of a gas dispersion process:

1. Maintaining constant $k_l a$ on scaleup
2. Maintaining a specific value of the ratio of absorbing component

per unit volume in the large-scale vessel to that in the small-scale vessel on scaleup

3. Maintaining constant vvm, volume of gas per volume of liquid in the vessel per minute

Nishikawa et al. (1981*a*) gave examples of the different methods, indicating that maintaining constant $k_l a$ on scaleup may be inappropriate and that maintaining equal vvm on scaleup was the most rigorous method as was discussed above by Nienow et al. (1977).

Three-Phase Systems, Other Geometries, and Articles of Interest

Three-phase systems

The area of three-phase mixing is one of specialization. Many studies on three-phase systems have been accomplished as parts or extensions of two-phase studies. The basic problems of two-phase systems appear in three-phase systems with the additional complication of the changing flow behavior because of the presence of the third phase. Of these systems, gas-liquid-solid is probably the most studied because of the importance of three-phase slurry reactors. Gas-liquid-liquid systems have not been studied to the same extent. Gas-gas-liquid systems, which have not been studied extensively, can be used to study the mixing in the gas phase as in Nienow et al. (1979).

Although processing conditions for the two-phase systems may not match, they may be merged into a set of process conditions for a three-phase system. A gas-liquid-solid system is used as an example.

In gas-liquid systems, the quantities of interest are mass transfer, the flooding transition, the nature of gas recirculation, and perhaps the minimum impeller rotational speed for complete dispersion. Interfacial area and the mass transfer coefficient are functions of total power per volume and superficial gas velocity.

In solid-liquid processes typically minimum suspension is considered important since the entire solid-liquid interfacial area is available for transport at this condition. Above this speed, additional process requirements may be needed, for example, complete dispersion of the solid phase or avoidance of particle attrition.

On merging these, the process conditions for a three-phase system of gas, liquid, and solids become adequate gas dispersion and the suspension of the solids. Correlations are available which provide operating conditions for two-phase systems and are often used in calculations for three-phase systems. The assumption made is that the third phase does not alter the flow behavior significantly away from the two-phase behavior. For some processes, two-phase correlations are adequate; in others, the results can be very misleading. However, correlations for three-phase systems are developing. The requirement of treating three phases leads directly to the question of multiple impellers: the use of one impeller to treat one process objective or criterion and another impeller to treat another criterion. Laminar impellers described in Chaps. 3 and 5 could be used as well since they are full tank impellers (Litz, 1985).

Chapman et al. (1983b; 1983c), Greaves and Loh (1984), and Nienow et al. (1985) contain recommendations for design of gas-liquid-solid systems. Litz (1985) described the use of a helical screw impeller to entrain gas from the liquid surface. The same impeller configuration was useful in the suspension of solids. Brehm et al. (1985) reported mass transfer coefficient correlations for $k_l a$ for gas-liquid-solid systems based on power per volume P/V, superficial velocity V_s, and a ratio of dynamic viscosities of the solid free gas dispersion and the three-phase system. Greaves and Loh (1985) provided $k_l a$ data for three-phase systems as well. The work by Greaves and Loh is perhaps the most extensive study to date on $k_l a$ in gas-liquid-solid systems because of the weight fractions studied. At low solids concentrations, $k_l a$ did not differ from gas-liquids systems. Above the volume fraction of $X = 15$ percent vol/vol, $k_l a$ fell rapidly because of an increase in effective viscosity of the slurry and damping effects on the turbulence. At high weight fractions, solids 30 percent wt/wt to 40 percent wt/wt, $k_l a$ approached a limiting value that was not significantly affected by gassing rate or impeller speed. Gas holdup was dependent on particle density, and poor gas dispersion occurred in the lower portion of the vessel.

Das et al. (1985) studied the effect of an immiscible organic phase on mass transfer in aqueous gas dispersion. Interfacial area initially increased with increased organic fraction and then decreased. Apparently the inert phase initially prevented coalescence leading to an increase in interfacial area but then dampened turbulent fluctuations, causing decreased dispersion and lower interfacial area. Other more complex phenomena were also noted.

Cieszkowski and Dylag (1988) studied gas dispersion in gas-liquid-solid systems in a tank having a disk style turbine at low clearance in the standard configuration. Gas holdup was measured from the rise in height of the dispersion. A torque transducer was used to determine power, and the liquid interfacial area was determined by the sulfite method. For the most part, the gas dispersion in the three-phase system behaved similarly to that in the two-phase system. Holdup and interfacial area correlations were developed as:

$$\phi = 0.117\left(\frac{P_g}{V}\right)^{0.39} V_s^{0.58}(1 + \phi_s)^{-2.90} \qquad (6.182)$$

$$a = 32.77\left(\frac{P_g}{V}\right)^{0.66} V_s^{0.45}(1 + \phi_s)^{-2.11} \qquad (6.183)$$

Brehm and Oguz (1988) studied $k_l a$ in aqueous and organic slurries for various solids. Peculiar effects in some systems were noted. In some systems, fast-settling slurries prevented viscosity measurements; in other systems, the mass transfer was too fast, e.g., greater than 0.1 s^{-1}, for measurement by standard techniques used. The final $k_l a$ correlation for aqueous and organic systems was given as:

$$k_l a = 3.07 \times 10^{-3}\mu_{Sl}^{-0.34}\left(\frac{P_T}{V_{Sl}}\right)^{0.75} (V_s)^{(0.5\alpha)}\sigma_L^{-3.0}(D_L)^{0.5} \qquad (6.184)$$

where α is $(\sigma_{H_2O}/\sigma_L)^{0.5}$, $k_l a$ is in s^{-1} and the subscripts, Sl and L, are for the slurry phase and liquid phase, respectively. The various quantities in the correlation ranged as: $1.3 < 10^3\mu_{Sl} < 40.7$ Pa · s; $0.75 < P_T/V_{Sl} < 6.3$ kW/m^3; $0.84 < 10^3 V_s < 4.2$ m/s; $24.3 < 10^3\sigma_L < 71.8$ N/m; and $0.68 < 10^9 D_L < 2.41$ m^2/s. Different solids at low concentrations caused different effects, but at high concentrations, $\mu_{Sl}/\mu_L > 1.3$, the slurry $k_l a$ data behaved similarly to two-phase gas-liquid $k_l a$ data when the slurry properties were accounted for. The above correlation can be reduced to one similar to the Yagi-Yoshida correlation given above.

Saito and Kamiwano (1988) have also given comments on gas-liquid-solid systems. Raghav Rao and Joshi (1988) reported various effects of impeller design, impeller diameter and speed, solids loading, and tank diameter on mixing times for gas-liquid-solid systems. Mehta and Sharma (1971), Joosten et al. (1977), and Albal et al. (1983) also commented on the effects of the third phase on gas dispersion. Steiff (1985) discussed heat transfer in gas-liquid and gas-liquid-solid systems.

Other geometries

Very few studies of mechanically agitated gas dispersion deal with major changes in geometry away from the vertical agitator geometry.

Horizontal tanks and square tanks. Horizontally stirred vessels have been studied as to flow regimes and mass transfer capabilities. Andou et al. (1972) performed mass transfer studies in horizontally stirred vessels. The flow regimes in such vessels have been discussed by Ando et al. (1971). The major flow regime of interest was where the liquid remained on the tank bottom, not forming an annulus around the wall. In this regime, the impeller splashed the liquid violently, causing liquid drops in the vapor space, liquid film on the wall, and bubbles in the liquid. In such a system, power number was correlated with a Froude number, liquid fraction, and impeller diameter (Ando et al., 1971). Andou et al. found that the mass transfer coefficient k_la was proportional to the Froude and blade numbers and showed a maximum with the liquid fraction in the vicinity of 0.5 to 0.6. Since both k_la and power draw (and consumption) were proportional to Froude number, k_la was proportional to power per unit volume. Within the limits of the work, k_la was found independent of scale.

Levec and Pavko (1979) examined the dependency of the mass transfer coefficient on energy dissipation in experiments performed in tanks with square sections. The procedures developed by Mehta and Sharma (1971) were used, utilizing the reaction of NA-dithionite with aqueous solutions of NaOH (Juvekar and Sharma, 1973), which permitted the separate determination of the mass transfer coefficient k_la and interfacial area a. k_l was found proportional to about the 0.26 power of the energy dissipation rate and was considered independent of system geometry. The exponent on the energy dissipation rate varied slightly with impeller diameter.

Bauer and Moser (1985) noted that bubble columns and traditional stirred tanks have disadvantages on the large scale that are not recognized from small-scale studies. Low holdup and compression costs for the high-pressure head occur in large-scale units. Bauer and Moser noted that their particular horizontal agitated tank design provided high mass transfer coefficients at low power draw, performance being equal to or better than that of vertical aerated, agitated tanks. However, mixing times were dependent on power per volume and were much larger for the horizontally stirred tank because of poor axial circulation.

Roustan et al. (1988) studied liquid circulation and gas dispersion in a nonconventional tank having the impellers on a horizontal shaft. Three different types of impellers, i.e., propellers, a four-blade pitched blade turbine, and three-blade pitched blade turbine with a modified blade shape, were studied. The tank was a horizontal cylinder, 3 m long, with rounded ends and a central baffle plate, 2 m long. The impellers were placed below the baffle plate. Air was introduced upstream of the impeller blades. The flow was always turbulent. For single-phase liquid operation, power and pumping numbers for the impellers were similar to those in an agitated tank. For good dispersion of sparged gas, a critical impeller rotational speed was necessary which was dependent upon the impeller geometry and gassing rate. The four-blade pitched blade turbines performed the best of the impellers studied and had the highest liquid circulation under gassed conditions than the other impellers.

Surface aerators. Surface aeration can be enhanced with proper geometry. White and De Villiers (1977) investigated surface aeration using a stator (draft tube) and rotor (impeller) configuration. As the impeller rotated, a forced vortex was formed inside the stator, mixing the gas and liquid and causing the gas to enter the rotor region for dispersion. The amount of gas which was entrained under this operation was of interest.

The aeration number was modified to include the effect of the hydrostatic head h_R across the rotor as:

$$N'_A = \left(\frac{Q}{ND^3}\right)\left(\frac{D}{h_R}\right)^{0.5} \tag{6.185}$$

In such a form, the modified aeration number accounted for the effects of rotor depth. A balance between centrifugal forces and hydrostatic forces across the rotor indicated a critical rotational speed and critical Froude number, 0.230 ± 0.022, below which surface entrainment did not occur. The results depended on the type of solutions studied. For water, the entrainment rate was correlated as:

$$N'_A = 0.0231(\text{Fr} - \text{Fr}_c)^{1.84} \tag{6.186}$$

and for glycerin and Teepol solutions:

$$N'_A = 0.0977(\text{Fr} - \text{Fr}_c)^{2.33} \tag{6.187}$$

The glycerin and Teepol solutions had a much finer bubble size, higher holdup, and less bubble coalescence. Foam was present on the liquid surface for these studies.

Litz (1985) described the use of helical screw impeller as a surface aerator.

Other impellers and contacting methods. Keon and Pingaud (1977) studied self-inducing dispersers for gas-liquid applications. Such devices are useful in cases where fouling or plugging of sparger holes commonly occur. In their device, gas was introduced through a hollow shaft that supported the disperser. Other studies of such devices include Martin (1972) and Zlokarnik (1966; 1967).

Breucker et al. (1988) noted that large-diameter sparge rings, close to the wall hindered the flooding of the impeller. Various baffle arrangements were studied but did not significantly influence $k_l a$.

Suciu and Smigelschi (1976) studied the sizing of plunging jet systems. van de Donk et al. (1979) studied the effects of contaminants on the oxygen transfer rate achieved using a plunging liquid jet contactor. van Dierendonck et al. (1988) discussed the use of pump-around-loop reactors with a venturi system used for the gas-liquid contacting. The mass transfer coefficient $k_l a$ was found to be as high as 1 s^{-1} in such systems. The system was shown to have economic advantages for fast reacting systems and at high pressures. Warnecke and Hussmann (1989) used a steady-state method to determine $k_l a$ in a gas-liquid jet loop reactor.

Kawagoe et al. (1975) discussed $k_l a$ and bubble size in gas sparged contactors. Jackson and Shen (1978) studied aeration, mixing and $k_l a$ for deep tank fermentation systems. Bhavaraju et al. (1978) presented design procedures to predict both the mass transfer coefficient and the interfacial area in gas sparged tanks. Low, moderate, and high gassing rates and non-Newtonian power fluids were discussed. Two regions were considered in the design calculations: a gas sparged region and a bubble column region. $k_l a$ was based upon the calculation of k_l and area a separately, in which k_l was estimated from:

$$k_l = \left(\frac{4D_L V_t}{\pi D_b}\right)^{0.5} \tag{6.188}$$

and interfacial area a was from:

$$\phi = \frac{V_s}{V_s + V_t} \tag{6.189}$$

and
$$a = \frac{6\phi}{D_b} \tag{6.190}$$

The estimation of the bubble diameter was dependent upon flow regime and location in the tank. However, this procedure for $k_l a$ can be questioned as noted above.

Articles of interest

General studies. Laine and Soderman (1988) presented procedures for the design of gas-liquid systems based upon flow regime and other information as presented in this chapter.

Garner and Hammerton (1954) and Peebles and Garber (1953) provided discussions of bubble behavior and drag coefficients of bubbles in liquids. Brodkey (1967) also provided an exceptional review of bubble dynamics. Bubble formation in non-Newtonian and viscoelastic fluids has been studied by Shirotsuka and Kawase (1976) and Acharya et al. (1978).

Lehrer (1968) studied the gas agitation of liquids and performed mixing and heat transfer studies for paddles and spargers in nearly identical systems. Sparger configuration strongly influenced performance. The paddle stirrers provided shorter mixing times but the spargers provided higher heat transfer coefficients.

Mann et al. (1977) discussed the deteriorating mixing quality and yield losses on scaleup for gas dispersion processes.

Brucato et al. (1985) discussed the effects of errors in $k_l a$ and interfacial area on design results in a sensitivity analysis.

Older literature on gas-liquid contacting in agitated tanks includes Foust et al. (1944) on an "arrowhead" impeller design, Cooper et al. (1944) on a vaned disk impeller, and Friedman and Lightfoot (1957) on oxygen absorption for open flat blade impellers.

Aeration efficiency. Few studies have addressed the question of efficiency. Greaves and Kobbacy (1981) was cited above. Pollard (1978) defined aeration efficiency as the mass of oxygen absorbed per unit of net energy delivered to the dispersion. The efficiency index and $k_l a$ data were plotted as function of gassing rate and impeller rotational speed.

Other experimental techniques. Among some of the more interesting experimental techniques, Burgess and Calderbank (1975) developed a real-time probe to measure bubble size, velocity, and bubble orientation. Stravs and von Stockar (1985) applied ultrasonic pulse transmission to the measurement of interfacial area, and Hsi et al. (1985) used hydrophones to study gas dispersion around the impeller, in the tank bulk, and near a sparge ring. Chuang et al. (1984) developed an optical technique to measure bubble coalescence times. DeMore et al. (1988) measured the sound produced by the cavities during dispersion. Koh and Batterham (1989) studied liquid splashing during gas sparging.

Gas side mass transfer coefficients. A number of studies concern the gas side mass transfer coefficient and include Fukuda et al. (1980), Yadav and Sharma (1979), Matheron and Sandall (1979), and Alper et al. (1980).

Heat transfer. Heat transfer studies of gas-liquid or three-phase systems include Steiff (1985), Man (1985), Strek and Karcz (1988), and Brain and Man (1989). Desplanches et al. (1988) also discussed heat transfer for aerated Newtonian and highly viscous liquids in the transition regime.

Nomenclature

a	specific interfacial surface area, exponent, constant
A	constant, exponent

A^*	area fraction
B	constant
b	parameter in Erlang distribution, exponent
C	clearance, impeller height off bottom, constant, concentration
C^*	liquid interface concentration
C_G	gas phase concentration
C_L	liquid phase concentration
C_i	inlet liquid concentration
C_L^*	normalized liquid phase concentration (dimensionless)
C_o^*	normalized outlet concentration (dimensionless)
c, c_1, c_5, c_6	constants
D	impeller diameter
D_b	bubble diameter
De	Deborah number, λN
D, D_L, D_G	liquid diffusivity, gas diffusivity, dispersion coefficient
D_{SM}	Sauter mean bubble diameter
d	disperse phase: small bubble fraction
f	activity coefficient
Fl, Fl_G, Fl_a, Fl_g	gas flow number, Q/ND^3
Fr	Froude number, N^2D/g
Fr_c	critical Froude number, N_c^2D/g
Fr_I	impeller Froude number, N^2D/g
Fr_T	tank Froude number, N^2T/g
g	acceleration of gravity
g_c	gravitational constant
H	liquid height in tank
H_e	Henry's law constant
h_R	hydrostatic head across rotor
I	inside diameter of disk style turbine
K_1, K_2, K'	constants
K	constant, consistency index in power law
k_g	mass transfer coefficient, gas side
k_l	mass transfer coefficient, liquid side
$k_l a$	mass transfer coefficient, liquid side
$(k_l a)_a$	mass transfer coefficient because of agitation
$(k_l a)_s$	mass transfer coefficient because of sparging
L	characteristic length
MW	molecular weight

m	fraction mixed, parameter in Erlang distribution
N	impeller rotational speed, usually revolutions per time
N_c	critical impeller rotational speed
N_{CD}	impeller rotational speed for complete dispersion
N_0	minimum impeller rotational speed
N_A, N_Q, N_{Qs}	aeration number, Q/ND^3 or Q_s/ND^3
N_{c1}	impeller rotational speed at the formation of large cavities; also equivalent to rotational speed at flooding
N_{c2}	impeller rotational speed at maximum power, vortex cavities are present
N'_A	modified aeration number, $Q/ND^3(D/h_R)^{0.5}$
N_F	impeller rotational speed at flooding
N_{Fr}	Froude number, N^2D/g
N_m	impeller rotational speed at minimum power, all blades have large cavities
N_P	ungassed power number, $P_o g_c/\rho N^3 D^5$
N_{Pg}	gassed power number, $P_g g_c/\rho N^3 D^5$
N_{QT}	aeration number based on total gas flow rate to the impeller
N_R	impeller rotational speed for gas recirculation
N_{SA}	impeller rotational speed for surface aeration
N_{SA1}	impeller rotational speed where gas bubbles first appear below surface
N_{SA2}	impeller rotational speed where entrained gas affects power draw of the impeller
N_{SE}	impeller rotational speed for surface entrainment
N_{We}	impeller Weber number, $N^2D^3\rho_L/\sigma$
n	power law exponent, flow index
$\dot{n}$	molar gas flow rate
n_1	coalescence parameter
n_p	paddle number
OTR	overall transfer rate
P, P_0	ungassed power, $N_P\rho N^3 D^5/g_c$, pressure
P_C	power to initiate circulation
P_e	power because of bubble expansion
P_g	gassed power
P_j	power because of jetting at sparger
P_m	power per unit mass due to agitation

P_s	power because of sparging, bubble expansion, or bubble rise
P_{sm}	power per unit mass due to gas sparging
P_T	total power input
P_v	vapor pressure
p	fraction under plug flow, tank pressure
p_a	atmosphere pressure
p_s, p_0	pressure at sparger, at liquid surface
Q, Q_G, Q_s	gas sparging rate, volumetric flow rate of gas
Q_i	liquid flow rate in tank
Q_L	liquid pumping rate of the impeller KND^3; liquid flow rate in and out of tank
Q_R	recirculated gas flow rate
Q_T	total gas flow rate to impeller
R	gas constant
Re	impeller Reynolds number, $\rho D^2 N/\mu$
r	bubble radius, reaction rate
T	tank diameter, temperature
t, t'	time, time of flooding
t_C	circulation time, $(K/N)(T/D)^2$
u_s	gas velocity exiting sparger holes
V	volume of liquid or dispersion
V_G	volume of gas
V_L	volume of liquid
V_r	recirculation velocity
V_s	superficial gas velocity
V_t	terminal or rise velocity of bubbles
W	blade width
W_p	weighting parameter
X	volume fraction
X_2	coalescence parameter

Greek symbols

α	external distribution coefficient, exponent, proportionality symbol
α_L	log standard deviation
β	exponent, multiplication factor
λ	characteristic time

ϕ, ϕ_s	gas phase holdup, solid void fraction
σ	interfacial surface tension
ϵ, $\bar{\epsilon}$	energy dissipation rate per unit volume or mass, average value
ζ	constant
τ	mean residence time
θ	mixing time
γ, γ_{av}	shear rate, average shear rate
ρ	density
ρ_a	density of air at operating conditions
ρ_g, ρ_G	gas phase density, gas density at sparging conditions
ρ_L	density of the continuous liquid phase
ρ_D	density of the dispersion
μ	liquid viscosity, apparent viscosity
μ_a	apparent viscosity
μ_g, μ_G	gas phase viscosity
μ_w	viscosity of water
ν	kinematic viscosity, liquid phase kinematic viscosity
ν_L	liquid phase kinematic viscosity
ψ	modified aeration number of Ismail et al. (1984), holdup parameter of Lee and Meyrick (1970)

Subscripts

a	apparent, air
a, a	apparent, agitation
a, s	apparent, sparging
c	continuous phase, critical
CD	complete dispersion
D	dispersed phase
d	dispersion
F	flooding
G, g	gas phase
L	liquid phase
0, o	unaerated
p	paddle
r	recirculated
R	recirculation, rotor
s	superficial, sparging
sl	slurry phase

SA surface aeration

SE surface entrainment

T total

w water

1, 2   first, second

References

Abrardi, V., G. Rovero, S. Sicardi, G. Baldi, and R. Conti, *Proc. 6th Eur. Conf. on Mixing, Pavia, Italy*, BHRA Fluid Eng., Cranfield, England, 329, 1988.
Acharya, A., R. A. Mashelkar, and J. J. Ulbrecht, *IE&C: Fundam.*, 17, 230, 1978.
Albal, R. S., Y. T. Shah, A. Schumpe, and N. L. Carr, *Chem. Eng. J.*, 27, 61, 1983.
Alper, E., W. D. Deckwer, and P. V. Danckwerts, *Chem. Eng. Sci.*, 35, 1263, 1268.
Ando, K., H. Hara, and K. Endoh, *International Chem. Eng.*, 11, 735, 1971.
Andou, K., H. Tabo, and K. Endoh, *J. Chem. Eng. Japan*, 5, 193, 1972.
Baczkiewicz, J., and H. Michalski, *Proc. 6th Eur. Conf. on Mixing, Pavia, Italy*,BHRA Fluid Eng., Cranfield, England, 473, 1988.
Bauer, A., and A. Moser, *Proc. 5th Eur. Conf. on Mixing, Wurzburg, Germany*, BHRA Fluid Eng., Cranfield, England, 171, 1985.
Bhavaraju, S. M., T. W. F. Russell, and H. W. Blanch, *AIChEJ.*, 24, 454, 1978.
Biesecker, B. O., *VDI Forschungsheft*, 554, 1972.
Bimbinet, J. J., M.S. thesis, Purdue University, Lafayette, Ind., 1959.
Blakebrough, N., G. Hamer, and M. W. Walker, *The Chem. Engr.*, A71, October 1961.
Boerma, H., and J. H. Lankester, *Chem. Eng. Sci.*, 23, 799, 1968.
Bossier, J. A., R. E. Farritor, G. A. Hughmark, and J. T. F. Kao, *AIChEJ.*, 19, 1065, 1973.
Bourne, J. R., *Chem. Eng. Sci.*, 19, 513, 1964.
Brain, T. J. S., and K. L. Man, *Chem. Eng. Prog.*, 76, July 1989.
Brehm, A., H. Oguz, and B. Kisakurek, *Proc. 5th Eur. Conf. on Mixing, Wurzburg, Germany*, BHRA Fluid Eng., Cranfield, England, 419, 1985.
Brehm, A., and H. Oguz, *Proc. 6th Eur. Conf. on Mixing, Pavia, Italy*, BHRA Fluid Eng., Cranfield, England, 413, 1988.
Brennan, D. J., *Trans. Instn. Chem. Engrs.*, 54, 209, 1976.
Breucker, C., A. Steiff, and P.-M. Weinspach, *Proc. 6th Eur. Conf. on Mixing, Pavia, Italy*, BHRA Fluid Eng., Cranfield, England, 399, 1988.
Brodkey, R. S., *The Phenomena of Fluid Motions*, The Ohio State Bookstore, Columbus, Ohio, 43210, 1967.
Brown, D. E., and D. J. Halsted, *Chem. Eng. Sci.*, 34, 853, 1979.
Brown, D. E., *Fluid Mixing, I. Chem. Eng. Sym. Ser. No. 64, I. Chem. Eng.*, Rugby, Warks, England, N1, 1981.
Brucato, A., V. Brucato, and L. Rizzuti, *Proc. 5th Eur. Conf. on Mixing, Wurzburg, Germany*, BHRA Fluid Eng., Cranfield, England, 427, 1985.
Bruijn, W., K. van't Riet, and J. M. Smith, *Trans. Instn. Chem. Engrs.*, 52, 88, 1974.
Bryant, J., and S. Sadeghzadeh, *Proc. 3rd. Eur. Conf. on Mixing, University of York, York, England*, BHRA Fluid Eng., Cranfield, England, F3, 325, 1979.
Bujalski, W., M. Konno, and A. W. Nienow, *Proc. 6th Eur. Conf. on Mixing, Pavia, Italy*, BHRA Fluid Eng., Cranfield, England, 389, 1988.
Burgess, J. M., and P. H. Calderbank, *Chem. Eng. Sci.*, 30, 743, 1975.
Calderbank, P. H., *Brit. Chem. Eng.*, 206, August 1956a.
Calderbank, P. H., *Brit. Chem. Eng.*, 267, September 1956b.
Calderbank, P. H., *Trans. Instn. Chem. Engrs.*, 36, 443, 1958.
Calderbank, P. H., *Trans. Instn. Chem. Engrs.*, 37, 173, 1959.
Chandrasekharan, K., and P. H. Calderbank, *Chem. Eng. Sci.*, 36, 819, 1981.
Chapman, C. M., L. G. Gibilaro, and A. W. Nienow, *Chem. Eng. Sci.*, 37, 891, 1982.

Chapman, C. M., A. W. Nienow, M. Cooke, and J. C. Middleton, *Chem. Eng. Res. Des.*, **61**, 82, 1983a.

Chapman, C. M., A. W. Nienow, M. Cooke, and J. C. Middleton, *Chem. Eng. Res. Des.*, **61**, 167, 1983b.

Chapman, C. M., A. W. Nienow, M. Cooke, and J. C. Middleton, *Chem. Eng. Res. Des.*, **61**, 182, 1983c.

Chuang, K. T., A. J. Stirling, and J. C. Baker, *IEC: Fundam.*, **23**, 109, 1984.

Cieszkowski, J., and M. Dylag, *Proc. 6th Eur. Conf. on Mixing, Pavia, Italy*, BHRA Fluid Eng., Cranfield, England, 421, 1988.

Clark, M. W., and T. Vermeulen, *AIChEJ.*, **10**, 420, 1964.

Cooper, C. M., G. A. Fernstrom, and S. A. Miller, *Ind. Eng. Chem.*, **36**, 504, 1944.

Das, T. R., A. Bandopadhyay, R. Parthasarathy, and R. Kumar, *Chem. Eng. Sci.*, **40**, 209, 1985.

Davies, S. N., L. G. Gibilaro, J. C. Middleton, M. Cooke, and P. M. Lynch, *Proc. 5th Eur. Conf. on Mixing, Wurzburg, Germany*, BHRA Fluid Eng., Cranfield, England, 27, 1985.

DeMore, L. S., W. F. Pafford, and G. B. Tatterson, *AIChEJ.*, **34**, 1922, 1988.

Desplanches, H., F. Essayem, J. L. Chevalier, M. Bruxelmane, and C. Delvosalle, *Proc. 6th Eur. Conf. on Mixing, Pavia, Italy*, BHRA Fluid Eng., Cranfield, England, 479, 1988.

DeWaal, K. J. A., and J. C. Okeson, *Chem. Eng. Sci.*, **21**, 559, 1966.

Drogaris, G., and P. Weiland, *Chem. Eng. Sci.*, **38**, 1501, 1983.

Dunn, I. J., and A. J. Einsele, *J. Applied Chem. and Biotech.*, **25**, 707, 1975.

Foust, H. C., D. E. Mack, and J. H. Rushton, *Ind. Eng. Chem.*, **36**, 517, 1944.

Friedman, A. M., and E. N. Lightfoot, *Ind. Eng. Chem.*, **49**, 1227, 1957.

Frossling, N., *Gerlands. Beitr. Geophys.*, **52**, 170, 1938.

Fuchs, R., D. D. Y. Ryu, and A. E. Humphrey, *I&EC: Proc. Des. Dev.*, **10**, 190, 1971.

Fukuda, T., K. Idogawa, K. Ikeda, K. Ando, and K. Endoh, *J. Chem. Eng. Japan*, **13**, 298, 1980.

Garner, F. H., and D. Hammerton, *Chem. Eng. Sci.*, **3**, 1, 1954.

Gibilaro, L. G., S. N. Davies, M. Cooke, P. M. Lynch, and J. C. Middleton, *Chem. Eng. Sci.*, **40**, 1811, 1985.

Gray, D. J., R. E. Treybal, and S. M. Barnet, *AIChEJ.*, **28**, 195, 1982.

Greaves, M., and C. A. Economides, *Proc. 3rd. Eur. Conf. on Mixing, University of York, York, England*, BHRA Fluid Eng., Cranfield, England, F5, 357, 1979.

Greaves, M., and K. A. H. Kobbacy, *Fluid Mixing, I. Chem. Eng. Sym. Ser. No. 64, I. Chem. E.*, Rugby, Warks, England, H1, 1981.

Greaves, M., and K. A. H. Kobbacy, *Fluid Mixing, I. Chem. Eng. Sym. Ser. No. 64, I. Chem. E.*, Rugby, Warks, England, L1, 1981.

Greaves, M., K. A. H. Kobbacy, and G. C. Millington, *Chem. Eng. Sci.*, **38**, 1909, 1983.

Greaves, M., and V. Y. Loh, *Fluid Mixing II, Symposium Series No. 89, I. Chem. Eng.*, Rugby, Warks, England, 69, 1984.

Greaves, M., and V. Y. Loh, *Proc. 5th Eur. Conf. on Mixing, Wurzburg, Germany*, BHRA Fluid Eng., Cranfield, England, 451, 1985.

Greaves, M., and M. Barigou, *Fluid Mixing III, I. Chem. Eng. Sym. Ser. No. 108*, 235, 1988.

Greaves, M., and M. Barigou, *Proc. 6th Eur. Conf. on Mixing, Pavia, Italy*, BHRA Fluid Eng., Cranfield, England, 313, 1988.

Hassan, I. T. M., and C. W. Robinson, *AIChEJ.*, **23**, 48, 1977.

Hanhart, J., H. Kramers, and K. R. Westerterp, *Chem. Eng. Sci.*, **18**, 503, 1963.

Hattori, K., S. Yokoo, and O. Imada, *J. Ferment. Technol.*, **50**, 737, 1972.

Heywood, N. I., P. Madhvi, and M. McDonagh, *Proc. 5th Eur. Conf. on Mixing, Wurzburg, Germany*, BHRA Fluid Eng., Cranfield, England, 243, 1985.

Hickman, A. D., *Proc. 6th Eur. Conf. on Mixing, Pavia, Italy*, BHRA Fluid Eng., Cranfield, England, 369, 1988.

Hicks, R. W., and L. E. Gates, *Chem. Eng.*, July 19, 1976, p. 141.

Higbie, R., *Trans. Amer. Inst. Chem. Engr.*, **31**, 365, 1935.

Hinze, J. O., *AIChEJ.*, **1**, 289, 1955.

Holland, F. A., and F. S. Chapman, *Liquid Mixing and Processing in Stirred Tanks*, Reinhold, New York, 1966.
Holmes, D. B., R. M. Voncken, and J. A. Dekker, *Chem. Eng. Sci.*, **19**, 201, 1964.
Hsi, R., M. Tay, D. Bukur, G. B. Tatterson, and G. L. Morrison, *Chem. Eng. J.*, **31**, 153, 1985.
Hughmark, G. A., *I&EC: Proc. Des. Dev.*, **19**, 638, 1980.
Ismail, A. F., Y. Nagase, and J. Imon, *AIChEJ.*, **30**, 487, 1984.
Jackson, M. L., and C. C. Shen, *AIChEJ.*, **24**, 63, 1978.
Joosten, G. E. H., J. G. M. Schilder, and J. J. Janssen, *Chem. Eng. Sci.*, **32**, 563, 1977.
Joshi, J. B., A. B. Pandit, and M. M. Sharma, *Chem. Eng. Sci.*, **37**, 813, 1982.
Juvekar, V. A., and M. M. Sharma, *Chem. Eng. Sci.*, **28**, 976, 1973.
Kawagoe, M., K. Nakao, and T. Otake, *J. Chem. Eng. Japan*, **8**, 254, 1975.
Kawase, Y., and M. Moo-Young, *Chem. Eng. Res. Des.*, **66**, 284, 1988.
Kawecki, W., T. Reith, J. W. van Heuven, and W. J. Beek, *Chem. Eng. Sci.*, **22**, 1519, 1967.
Keitel, G., and U. Onken, *Chem. Eng. Sci.*, **37**, 1635, 1982.
Koen, C., and Pingaud, B., *Proc. 2th Eur. Conf. on Mixing, St. John's College, Cambridge, England*, BHRA Fluid Eng., Cranfield, England, F5, 67, 1977.
Koetsier, W. T., D. Thoenes, and J. F. Frankena, *Chem. Eng. J.*, **5**, 61, 1973.
Koetsier, W. T., and J. F. Frankena, *Chem. Eng. J.*, **5**, 71, 1973.
Koh, P. T. L., and R. J. Batterham, *Chem. Eng. Res. Des.*, **67**, 211, 1989.
Kuboi, R., and A. W. Nienow, *Proc. 4th Eur. Conf. on Mixing, Noordwijkerhout, Netherlands*, BHRA Fluid Eng., Cranfield, England, G2, 247, 1982.
Laine, J., and J. Soderman, *Proc. 6th Eur. Conf. on Mixing, Pavia, Italy*, BHRA Fluid Eng., Cranfield, England, 305, 1988.
Lamont, A. G. W., *Canad. J. Chem. Eng.*, 153, August 1958.
Le Lan, A., and H. Angelino, *Chem. Eng. Sci.*, **27**, 1969, 1972.
Lee, J. C., and D. L. Meyrick, *Trans. Instn. Chem. Eng.*, **157**, T37, 1970.
Legrys, G. A., *Chem. Eng. Sci.*, **33**, 83, 1978.
Lehrer, L. H., *I&EC Proc. Des. Dev.*, **7**, 226, 1968.
Lessard, R. R., and S. A. Zieminski, *I&EC: Fundam.*, **10**, 260, 1971.
Levec, J., and S. Pavko, *Chem. Eng. Sci.*, **34**, 1159, 1979.
Linek, V., and V. Vacek, *Chem. Eng. Sci.*, **36**, 1747, 1981.
Linek, V., P. Benes, V. Vacek, and F. Hovorka, *Chem. Eng. J.*, **25**, 77, 1882.
Linek, V., and V. Vacek, *Chem. Eng. Sci.*, **37**, 1425, 1982.
Linek, V., P. Benes, and O. Holecek, *Biotech. Bioeng.*, **32**, 482, 1988.
Litz, L. M., *Chem. Eng. Prog.*, **81**, 11, 36, November 1985.
Loiseau, B., N. Midoux, and J. C. Charpentier, *AIChEJ.*, **23**, 931, 1977.
Luong, H. T., and B. Volesky, *AIChEJ.*, **25**, 894, 1979.
Machon, V., J. Vlcek, and V. Kudrna, *Proc. 2nd Eur. Conf. on Mixing, St. John's College, Cambridge, England*, BHRA Fluid Eng., Cranfield, England, F2, 17, 1977.
Machon, V., J. Vlcek, A. W. Nienow, and J. Solomon, *Chem. Eng. J.*, **19**, 67, 1980.
Machon, V., J. Vlcek, and J. Skrivanek, *Proc. 5th Eur. Conf. on Mixing, Wurzburg, Germany*, BHRA Fluid Eng., Cranfield, England, 155, 1985.
Machon, V., J. Vlcek, and V. Hudcova, *Proc. 6th Eur. Conf. on Mixing, Pavia, Italy*, BHRA Fluid Eng., Cranfield, England, 351, 1988.
Man, K. L., *Proc. 5th Eur. Conf. on Mixing, Wurzburg, Germany*, BHRA Fluid Eng., Cranfield, England, 221, 1985.
Mann, R., J. C. Middleton, and I. B. Parker, *Proc. 2nd Eur. Conf. on Mixing, St. John's College, Cambridge, England*, BHRA Fluid Eng., Cranfield, England, F3, 35, 1977.
Mann, R., P. P. Mavros, and J. C. Middleton, *Fluid Mixing, I. Chem. Eng. Sym. Ser. No. 64*, G1, 1981.
Mann, R., and L. A. Hackett, *Proc. 6th Eur. Conf. on Mixing, Pavia, Italy*, BHRA Fluid Eng., Cranfield, England, 321, 1988.
Marrucci, G., and L. Nicodemo, *Chem. Eng. Sci.*, **22**, 1257, 1967.
Marrucci, G., *Chem. Eng. Sci.*, **24**, 975, 1969.
Martin, G. Q., *I&EC Proc. Des. Dev.*, **11**, 397, 1972.
Matheron, E. R., and O. C. Sandall, *AIChEJ.*, **25**, 332, 1979.

Matsumara, M., H. Masunaga, K. Haraya, and J. Kobayashi, *J. Ferment. Technol.*, **55**, 388, 1977.

Mehta, V. D., and M. M. Sharma, *Chem. Eng. Sci.*, **26**, 461, 1971.

Meijboom, F. W., and J. G. Vogtlander, *Chem. Eng. Sci.*, **29**, 857, 1974a.

Meijboom, F. W., and J. G. Vogtlander, *Chem. Eng. Sci.*, **29**, 949, 1974b.

Metzner, A. B., and R. E. Otto, *AIChEJ.*, **3**, 3, 1957.

Metzner, A. B., R. H. Feehs, H. L. Ramos, R. E. Otto, and J. D. Tuthill, *AIChEJ.*, **7**, 1, 3, 1961.

Michel, B. J., and S. A. Miller, *AIChEJ.*, **8**, 262, 1962.

Middleton, J. C., "Gas Liquid Dispersion and Mixing," *Mixing in the Process Industries*, edited by N. Harnby, M. F. Edwards, and A. W. Nienow, Butterworths, London, 1985, chap. 17, pp. 322–355.

Midoux, N., and J. C. Charpentier, *Proc. 3rd. Eur. Conf. on Mixing, University of York, York, England*, BHRA Fluid Eng., Cranfield, England, F4, 337, 1979.

Mikulcova, E., V. Kudrna, and J. Vlcek, *Scient. Pap., Inst. Chem. Tech.*, Prague, **k1**, 167, 183, 1967.

Miller, D. N., *AIChEJ.*, **20**, 445, 1974.

Nagase, Y., and H. Yasui, *Chem. Eng. J.*, **27**, 37, 1983.

Nagata, S., M. Nishikawa, H. Tada, and S. Gotoh, *J. Chem. Eng. Japan*, **4**, 72, 1971.

Nagata, S., "Agitation in Gas Liquid Systems," *Mixing: Principles and Applications*, Kodansha, Ltd., Tokyo, and Wiley, 1975, pp. 335–368.

Nagata, S., *Mixing: Principles and Applications*, Kodansha Ltd., Tokyo, and Halsted Press, New York, 1975, p. 60.

Nakano, S., M. Eng. Thesis, Kyoto Univ., Kyoto, 1976.

Nienow, A. W., and D. J. Wisdom, *Chem. Eng. Sci.*, **29**, 1994, 1974.

Nienow, A. W., D. J. Wisdom, and J. C. Middleton, *Proc. 2nd Eur. Conf. on Mixing, St. John's College, Cambridge, England*, BHRA Fluid Eng., Cranfield, England, F1, F1-16, 1977.

Nienow, A. W., C. M. Chapman, and J. C. Middleton, *Chem. Eng. J.*, **17**, 111, 1979.

Nienow, A. W., and M. D. Lilly, *Biotech. Bioeng.*, **21**, 2341, 1979.

Nienow, A. W., D. J. Wisdom, J. Solomon, V. Machon, and J. Vlcek, *Chem. Eng. Commun.*, **19**, 273, 1983.

Nienow, A. W., *The World Biotech Report*, Online Publication, London, England, **1**, 293, 1984.

Nienow, A. W., M. Konno, M. M. C. G. Warmoeskerken, and J. M. Smith, *Proc. 5th Eur. Conf. on Mixing, Wurzburg, Germany*, BHRA Fluid Eng., Cranfield, England, 143, 1985a.

Nienow, A. W., M. Konno, and W. Bujalski, *Proc. 5th Eur. Conf. on Mixing, Wurzburg, Germany*, BHRA Fluid Eng., Cranfield, England, 1, 1985b.

Nienow, A. W., and J. J. Ulbrecht, "Gas Liquid Mixing and Mass Transfer in High Viscosity Liquids," *Mixing of Liquids by Mechanical Agitation*, edited by J. J. Ulbrecht and G. K. Patterson, Gordon & Breach, New York, 1985, Chap. 6.

Nishikawa, M., M. Nakamura, H. Yagi, and K. Hashimoto, *J. Chem. Eng. Japan*, **14**, 219, 1981a.

Nishikawa, M., M. Nakamura, and K. Hashimoto, *J. Chem. Eng. Japan*, **14**, 227, 1981b.

Nishikawa, M., S. Nishioka, and S. Fujida, *Kagaku Kogaku Ronbunshu*, **9**, 76, 1983.

Nishikawa, M., S. Nishioka, and T. Kayama, *J. Chem. Eng. Japan*, **17**, 541, 1984.

Nocentini, M., F. Magelli, and G. Pasquali, *Proc. 6th Eur. Conf. on Mixing, Pavia, Italy*, BHRA Fluid Eng., Cranfield, England, 337, 1988.

Oldshue, J. Y., *Fluid Mixing Technology, Chemical Engineering*, McGraw-Hill, New York, 1983.

Oyama, Y., and K. Endoh, *Chem. Eng. Japan*, **19**, 2, 1955.

Ozcan, G., M. Decloux, and M. Bruxelmane, *Proc. 6th Eur. Conf. on Mixing, Pavia, Italy*, BHRA Fluid Eng., Cranfield, England, 361, 1988.

Pandit, A. B., and J. B. Joshi, *Chem. Eng. Sci.*, **38**, 1189, 1983.

Patton, T. C., *Paint Flow and Pigment Dispersion*, Wiley Interscience, Wiley, New York, 1979.

Peebles, F. N., and H. J. Garber, *Chem. Eng. Progr.*, February 1953, p. 88.

Perez, J. F., and O. C. Sandall, *AIChEJ.*, **20**, 770, 1974.

Pharamond, J. C., M. Roustan, and H. Roques, *Chem. Eng. Sci.*, **30**, 907, 1975.
Pollard, G. J., *Proceedings of the International Symposium on Mixing*, Mons, C4, 1, 1978.
Popovic, M., A. Papalexiou, and M. Peuss, *Chem. Eng. Sci.*, **38**, 2015, 1983.
Popiolek, Z., J. H. Whitelaw, M. Yianneskis, *Proc. 2nd International Conf. on Applications of Laser Doppler Anemometry to Fluid Mechanics*, Lisbon, Portugal, paper, 17.1, July 1984, pp. 1–6.
Prasher, B. D., *AIChEJ.*, **21**, 407, 1975.
Preen, B. V., Ph.D. dissertation, Univ. of Durham, South Africa, 1961.
Prest, W. M., R. S. Porter, and J. W. O'Reilly, *J. Appl. Polym. Sci.*, **14**, 2697, 1970.
Quaraishi, A. Q., R. A. Mashelkar, and J. J. Ulbrecht, *J. Non-Newtonian Fluid Mech.*, **1**, 223, 1976.
Raghav Rao, K. S. M. S., and J. B. Joshi, *Proc. 6th Eur. Conf. on Mixing, Pavia, Italy*, BHRA Fluid Eng., Cranfield, England, 427, 1988.
Ranade, V. R., and J. J. Ulbrecht, *Proc. 2nd Europ. Conf. on Mixing, St. John's College, Cambridge, England*, BHRA Fluid Eng., Cranfield, England, F6-83, 1977.
Ranade, V. R., and J. J. Ulbrecht, *AIChEJ.*, **24**, 796, 1978.
Reith, T., and W. J. Beek, *Chem. Eng. Sci.*, **28**, 1331, 1973.
Rennie, J., and F. H. H. Valentin, *Chem. Eng. Sci.*, **23**, 663, 1968.
Rieger, F., P. Ditl, and V. Novak, *Chem. Eng. Sci.*, **34**, 397, 1979.
Roustan, M., and M. Z. A. Bruxelmane, *7th Int. Congress of Chem. Eng.*, CHISA, Prague, Czech., **B3.3**, 1981, pp. 1–8.
Roustan, M., *Proc. 5th Eur. Conf. on Mixing, Wurzburg, Germany*, BHRA Fluid Eng., Cranfield, England, 127, 1985.
Roustan, M., A. Meziane, and G. Faup, *Proc. 6th Eur. Conf. on Mixing, Pavia, Italy*, BHRA Fluid Eng., Cranfield, England, 381, 1988.
Ruchti, G., I. J. Dunn, J. R. Bourne, and U. von Stockar, *Chem. Eng. J.*, **30**, 29, 1985.
Rushton, J. H., and J. J. Bimbinet, *Canadian J. Chem. Eng.*, **46**, 16, 1968.
Sagert, N., and M. Quinn, *Canad. J. Chem. Eng.*, **54**, 392, 1976.
Sagert, N., and M. Quinn, *J. Colloid Interface Sci.*, **65**, 415, 1978*a*.
Sagert, N., and M. Quinn, *Chem. Eng. Sci.*, **33**, 1087, 1978*b*.
Saito, F., and M. Kamiwano, *Proc. 6th Eur. Conf. on Mixing, Pavia, Italy*, BHRA Fluid Eng., Cranfield, England, 407, 1988.
Sharma, M. M. and R. A. Mashelkar, *Proc. of a Symp. presented the Tripartite Chemical Engineering Conference, Montreal, Canada*, Instn. of Chem. Engrs., United Kingdom, 1968, p. 10.
Sharma, M. M., and P. V. Danckwerts, *Brit. Chem. Eng.*, **15**, 522, 1970.
Shirotsuka, T., and Y. Kawase, *J. Chem. Eng. Japan*, **9**, 234, 1976.
Sideman, S., H. Hortacsu, and J. W. Fulton, *Ind. Eng. Chem.*, **58**, 32, 1966.
Smith, J. M., K. van't Riet, and J. C. Middleton, *Proc. of the 2nd Eur. Conf. on Mixing, Cambridge, England*, BHRA Fluid Eng., Cranfield, England, F4-51, 1977.
Smith, J. M., "Dispersion of Gases in Liquids," *Mixing of Liquids by Mechanical Agitation*, edited by J. J. Ulbrecht and G. K. Patterson, Gordon and Breach, New York, 1985, pp. 139–202.
Smith, J. M., and M. M. C. G. Warmoeskerken, *Proc. 5th Eur. Conf. on Mixing, Wurzburg, Germany*, BHRA Fluid Eng., Cranfield, England, 115, 1985.
Smith, J. M., M. M. C. G. Warmoeskerken, and E. Zeef, AIChE Annual Meeting, Miami Beach, paper 134*b*, November 1986.
Smith, J. M., and D. G. F. Verbeek, *Chem. Eng. Res. Des.*, **66**, 39, 1988.
Smith, J. M., and L. Smit, *Proc. 6th Eur. Conf. on Mixing, Pavia, Italy*, BHRA Fluid Eng., Cranfield, England, 297, 1988.
Solomon, J., A. W. Nienow, and G. W. Pace, *Fluid Mixing, I. Chem. Eng. Sym. Ser. No. 64, I. Chem. E.*, Rugby, Warks, England, A1, 1981.
Sridhar, T., and O. E. Potter, *Chem. Eng. Sci.*, **33**, 1347, 1978.
Sridhar, T., and O. E. Potter, *Chem. Eng. Sci.*, **35**, 683, 1980*a*.
Sridhar, T., and O. E. Potter, *I&EC: Funds.*, **19**, 21, 1980*b*.
Steel, R., and M. R. Brierley, *Applied Microbiology*, **7**, 51, 1959.
Steiff, A., *Proc. 5th Eur. Conf. on Mixing, Wurzburg, Germany*, BHRA Fluid Eng., Cranfield, England, **209**, 1985.
Stravs, A. A., and U. von Stockar, *Chem. Eng. Sci.*, **40**, 1169, 1985.

Strek, F., and J. Karcz, *Proc. 6th Eur. Conf. on Mixing, Pavia, Italy*, BHRA Fluid Eng., Cranfield, England, 375, 1988.

Suciu, G. D., and O. Smigelschi, *Chem. Eng. Sci.*, **31**, 1217, 1976.

Sutter, T. A., G. L. Morrison, and G. B. Tatterson, *AIChEJ.*, **33**, 668, 1987.

Takashima, I., and M. Mochizuki, *J. Chem. Eng. Japan*, **4**, 66, 1971.

Takeda, K., and T. Hoshino, *Kagaku Kogaku*, **4**, 394, 1966.

Tatterson, G. B., H. H. S. Yuan, and R. S. Brodkey, *Chem. Eng. Sci.*, **35**, 1369, 1980.

Tatterson, G. B., and G. L. Morrison, *AIChEJ.*, **33**, 1751, 1987.

van de Donk, J. A. C., R. G. J. M. van der Lans, and J. M. Smith, *Proc. 5th Eur. Conf. on Mixing, Wurzburg, Germany*, BHRA Fluid Eng., Cranfield, England, F1, 289, 1985.

van Dierendonck, L. L., G. W. Meindersma, and G. M. Leuteritz, *Proc. 6th Eur. Conf. on Mixing, Pavia, Italy*, BHRA Fluid Eng., Cranfield, England, 287, 1988.

van't Riet, K., and J. M. Smith, *Chem. Eng. Sci.*, **28**, 1073, 1973.

van't Riet, K., J. M. Boom, and J. M. Smith, *Trans. Instn. Chem. Engrs.*, **54**, 124, 1976.

van't Riet, K., *I&EC: Pro. Des. Dev.*, **18**, 357, 1979.

van't Riet, K., and J. M. Smith, *Proc. 1st Eur. Conf. on Mixing and Centrifugal Sep.*, *Churchill College, Cambridge, England*, BHRA Fluid Eng., Cranfield, England, B2, 19, 1974.

Vermeulen, T., G. M. Williams, and G. E. Langlois, *Chem. Eng. Progr.*, **51**, 2, 85F, 1955.

Warmoeskerken, M. M. C. G., J. Feijen, and J. M. Smith, *Fluid Mixing, I. Chem. Eng. Sym. Ser. No. 64, I. Chem. Eng.*, Rugby, Warks, England, J1, 1981.

Warmoeskerken, M. M. C. G., and J. M. Smith, *Proc. 4th Eur. Conf. on Mixing, Noordwijkerhout, Netherlands*, BHRA Fluid Eng., Cranfield, England, G1, 237, 1982.

Warmoeskerken, M. M. C. G., J. Speur, and J. M. Smith, *Chem. Eng. Communs.*, **25**, 11, 1984.

Warmoeskerken, M. M. C. G., and J. M. Smith, *Fluid Mixing II, Sym. Series No. 89, I. Chem. Eng.*, Rugby, Warks, England, 59, 1984.

Warmoeskerken, M. M. C. G., and J. M. Smith, *Chem. Eng. Sci.*, **40**, 2063, 1985.

Warmoeskerken, M. M. C. G., and J. M. Smith, *Chem. Eng. Res. Des.*, **67**, 193, 1989.

Warnecke, H. J., and P. Hussmann, *Chem. Eng. Commun.*, **78**, 131, 1989.

Westerterp, K. R., L. L. van Dierendonck, and J. A. DeKraa, *Chem. Eng., Sci.*, **18**, 157, 1963.

Westerterp, K. R., *Chem. Eng. Sci.*, **18**, 495, 1963.

White, D. A., and J. U. De Villiers, *Chem. Eng. J.*, **14**, 113, 1977.

Wiedmann, J. A., *Chem. Ing. Tech.*, **55**, 689, 1983.

Wiedmann, J. A., A. Steiff, and P. M. Weinspach, 7th Int. Congress of Chem. Eng., CHISA, Prague, Czech., B3.2, pp. 1–16, 1981.

Wilhelm, R. R., W. A. Donohue, D. J. Valêsano, and G. A. Brown, *Biotech and Bioeng.*, **8**, 55, 1966.

Wilkinson, W. L., *Chem. Eng. Sci.*, **30**, 1227, 1975.

Yadav, G. D., and M. M. Sharma, *Chem. Eng. Sci.*, **34**, 1423, 1979.

Yagi, H., and F. Yoshida, *Ind. Eng. Chem. Proc. Des. Dev.*, **14**, 488, 1975.

Yung, C. N., C. W. Wong, and C. L. Chang, *Canad. J. Chem. Eng.*, **57**, 672, 1979.

Zieminski, S., M. Caron, and R. Blackmore, *I&EC: Fundam.*, **6**, 233, 1967.

Zieminski, S., C. Goodwin, and R. Hill, *TAPPI*, **43**, 1029, 1960.

Zieminski, S., and R. Hill, *J. Chem. Eng. Data*, **7**, 51, 1962.

Zieminski, S., and R. Lessard, *I&EC: Proc. Des. Dev.*, **8**, 69, 1969.

Zlokarnik, M., *Chem. Ing. Techn.*, **38**, 357, 1966.

Zlokarnik, M., *Chem. Ing. Techn.*, **39**, 1163, 1967.

Zlokarnik, M., *Chem. Ing. Techn.*, **43**, 1028, 1971.

Zlokarnik, M., *Adv. Biochem. Eng.*, **8**, 133, 1978.

Zwietering, T. N., *Chem. Eng. Sci.*, **8**, 244, 1958.

Zwietering, Th. N., *De Ingenieur (Chemische Techniek 6)*, **42**, October 18, 1963, Chap. 60–61.

Problems

The following problems are a combination of applied design problems and fundamental problems. Their solution will help in the understanding of material presented in their respective chapters. Most of the problems are not difficult; however, some are very lengthy.

The applied design problems have been obtained from Dr. Arthur W. Etchells, Principal Consultant, and Dr. James N. Tilton, Senior Consulting Engineer, who work in the Engineering Services Division of E.I. du Pont de Nemour & Co. These design problems are meant to be open-ended and have many reasonably correct solutions. They are also representative of the types of problems encountered in the mixing practice and may not contain all the information necessary for their solution.

Chapter 2

1   How does power number scale with liquid height, baffle number, clearance, blade width, blade number, vessel diameter, vessel bottom, and impeller number for both baffled and unbaffled tanks under turbulent conditions? How significant is the effect of feed location on power number?

2   In a plant setting, how would you measure power usage of a mixing unit?

3   Why is power number independent of impeller Reynolds number in a highly turbulent flow in a baffled tank and dependent on impeller Reynolds number in an unbaffled tank under the same operating conditions?

4   Changes of 1 and 0.1 percent in impeller diameter cause what changes in the respective power draw?

5   You have received four bids from different manufacturers for an agitator to be installed in your company's latest polymerization reactor.
   The process specification was

   Vessel volume: 10,000 gal (50 m^3)
   Tank diameter: 12 ft (4 m)
   Tank liquid level: 4 m

Fluid density: 1.34 s.g. (1340 kg/m^3)

Initial viscosity: 1 cP

Final viscosity: 250 cP

Good motion is to be maintained during the 100-min batch polymerization. Below are the bid summaries. Which would you recommend buying and
why?

Manufacturer	AGITCO	BEATCO	CHEMIX	DYFLO
Power, hp/W	10/7.5	5/3.6	5/3.6	3/2.3
Speed, rpm	68	45	37	45
Imp. dia., in	47	56	59	65
Cost, $	7500	5000	10000	15000

Power cost: $400 per horsepower year or $500 per kilowatt year

All manufacturers except DYFLO recommend a 45° pitched blade turbine
(N_p = 1.5, flow number = 0.6). DYFLO recommends their new hydrofoil impeller (N_p = 0.3, flow number = 0.3). All impellers are mounted 48 in off the
tank bottom.

The plant proposes running this unit continuously. What changes would
you make, and what rate do you suggest they run at?

6 List five negative aspects of the use of power laws.

7 For an impeller in mercury at room temperature, D = 1.0 ft, N = 2 s^{-1},
what is the impeller Reynolds number and power draw if N_p = 4.0?

8 How does volume scale for geometrically similar systems?

9 Given what you now know, how would you define N_P for turbulent flow
conditions? For laminar flow conditions? The definitions do not have to be the
same. What adjustments would you have to make for multiphase liquid systems using these definitions?

10 Would the flow patterns around the impeller be the same between these
situations:
 a A stationary centered impeller with the tank being rotated
 b A rotating centered impeller with a stationary tank?

 Please explain your answer.

Chapter 3

1 How does power number scale with liquid height, baffle number, clearance, blade width, blade number, vessel diameter, vessel bottom, and impeller
number for both baffled and unbaffled tanks under laminar flow conditions?
How significant is the effect of feed location on power number?

2   Use Eq. (3.20) to calculate the power draw for a off-centered screw impeller for $D = 1.0$ ft, $T = 1.7$ ft, $a = 12$, $E = 0.1$ and 0.2, $p = 0.75$ ft, $C_b = 0.5$ ft, and $N = 0.1$ s^{-1} in water, 100-cP fluid and 10,000-cP fluid. What are the torque levels at these conditions?

3   Show how the Carreau rheological model reduces to the power law. What are the implications of this in power draw correlations?

4   Show how Eq. (3.87) was obtained.

5   A new water-based automotive finish is to be stored in a flat-bottomed cylindrical vessel 5 by 5 ft. Color adjustments will be made in this tank. The material is a Bingham plastic with the following properties.

Density: 1.2 g/cm^3

Yield stress: 300 dyne/cm^2

Infinite shear viscosity: 100 cP

Design an agitator system for this vessel. Use multiple impellers as needed.

6   Calculate the function $4n/(3n + 1)$, for $n = 0.1$ to $n = 2.0$ and plot it. Do you think the term significantly improves your answers?

7   Show how Eq. (3.56) was obtained.

8   Obtain Eq. (3.64) from the coaxial cylinder analogy.

9   Explain why, upon scaleup, it is impossible to maintain dynamic and elastic similarity.

10   For geometrically similar systems, scaling factor = k, in laminar mixing, how does power scale in which different fluids are used?

11   Does the Metzner-Otto postulation hold for power law fluids having $n > 1.4$? Explain.

12   Since Godleski and Smith (1962) found that the Metzner-Otto postulation held for variable K and n, were the results of Godfrey et al. (1974) and Edwards et al. (1976) surprising? Explain.

13   For the polyethylene data in Fig. 3.14, what are the natural and characteristic times?

14   Show the development for $F(\Psi, \mu)$ used by Hiraoka et al. (1979).

Chapter 4

1   Hiraoka and Ito (1979) referred to a friction velocity in their mixing time study. Please explain the concept of friction velocity as applied to a mixing tank and its use in the calculation of mixing time.

2 The times required for convective ("distributive" and "dispersive") mixing and diffusive mixing both decrease with increasing intensity of agitation. The latter is the time required for uniformity on the molecular level, while the former is the time required for macroscopic uniformity, often called the *blend time.* Consider a liquid being agitated in the turbulent flow regime in a conventionally equipped stirred tank. What will be the effect on the macroscopic convective mixing time and on the diffusive mixing time for each of the following changes?

  a Doubling the agitator speed.
  b Increasing the impeller diameter by 20 percent, with all other variables fixed (including tank diameter).
  c Changing the solute with twice the molecular diffusivity.
  d Scaling up the vessel using geometric similarity for blending and motion. The vessel volume doubles on scaleup.

3 Corrsin (1957) obtained an equation for the mean-square fluctuation in concentration. Explain how the equation was developed.

4 Manning and Wilhelm (1963) provided spectra on concentration fluctuations. Please interpret these spectra and their meaning with regard to mixing and the presence of the trailing vortex systems of the disk style turbine.

5 An existing jacketed tank is proposed for use as a batch chemical reactor to make synthetic detergent. The reactants are initially all low-viscosity liquids. The batch is then heated with steam through the jacket. A reaction begins, and a precipitate is formed. This precipitate is a fine powder. After sufficient "cooking," the batch is cooled by adding water to the jacket and the slurry emptied and filtered and the solids dried. The controlling steps are associated with heat transfer. In the first production runs, the heat-up time was about 20 percent longer than desirable. The cool-down step took many hours. You have been asked the following questions because your résumé said you had taken a mixing course. To increase the heat transfer rate during heating by 20 percent, is it better to increase the agitator speed or to add an auxiliary coil? Why is the cooling time long, and what can be done to shorten the cooling time?
  Some information

  Vessel: tank diameter: 120 in; tank straight side: 120 in

  Impeller: one four-blade, 45° pitched blade turbine; diameter 40 in mounted 40 in off bottom; N_p = 1.7, speed 68 rpm, motor, 7.5 hp

  Fluids (initial): density 1.0 s.g.; viscosity 0.5 cP; thermal conductivity, 0.2 Btu/h/ft^2(°F/ft); specific heat, 0.5, Prandtl No., 9.2

  Fluids (final): slurry density, 1.0 s.g.; percent solids, 35 percent; particle size, 1 to 10 μm; rheology, non-Newtonian yield stress, 300 dyne/cm^2; infinite shear viscosity, 100 cP, apparent viscosity 30,000/γ + 100 cP

6 Justify the various equations used by Manning et al. (1965) to establish their micro-macro mixed models, and comment on the effect of reaction rate constant on outlet concentration.

7   Khang and Levenspiel (1976*b*) observed that the pulse response for the propeller was not as smooth as those of the disk style turbine. Please explain why in terms of the local flow patterns of the two impellers. Which impeller is more effective at micromixing?

8   It has been commented on that the momentum in a jet is constant. Is this statement correct? Please provide an explanation.

9   Calculate impeller Reynolds number, power consumption, pumping capacity, tip speed, and bulk velocity for the following arrangements:
   a Fluid with density 1.2 gm/cm^3 viscosity 45 cP; tank: 10-ft diameter by 10-ft straight side; volume about 6000 gal
      Impellers: (one- to four-bladed): 45° 36-in-diameter pitched blade turbine; blade width 6-in turning at 30 and 56 rpm;
      Impellers: (one- to six-bladed): 60-in-diameter flat blade turbine; blade width 10 in turning at 30 and 56 rpm
   b Fluid with density 0.9 gm/cm^3 and viscosity 100 cP: Tank 20 by 10 ft high about 23,000 gal and impellers as above
   c What are the impeller sizes which provide 5 hp/1000 gal in the above two tanks, assuming only the pitched blade turbine and then the combination of the pitched blade turbine with the flat blade turbine?
   Consider setting up a simple calculator program, spreadsheet, or a math solver for this problem.

10   For conditions and arrangements in the problem above, calculate mixing time and determine the scaleup relation for constant bulk velocity per volume. Calculate power/volume, torque/volume, and tip speed for the different arrangements.

11   Prove that the mean residence time τ is equal to V_T/Q_F. (See Zwietering, 1959.)

12   Patterson (1974) developed a model for the determination of the $\overline{c^2}/\overline{c_0^2}$. Review the various assumptions in his model, and explain how the model can be incorporated into a general k-ϵ model for a reactor having a second-order reaction.

13   A large tank of water is stirred by a impeller at 1.0 rps. The impeller is 0.5 m in diameter and has a turbulent power number of 2. Estimate the characteristic size and velocity fluctuations of the energy-containing eddies and the Kolmogorov eddies in the impeller region. The density and viscosity are 1000 kg/m^3 and 0.001 Pa · s, respectively. What are the corresponding values if the impeller speed is doubled?
   Compare these eddy characteristics to those for the flow of water in a 5-cm-ID pipe at a velocity of 2 m/s.

14   Energy-containing eddies transfer most of their energy to smaller eddies, at a rate on the order of the power input. Only a small fraction of their energy is dissipated directly by viscosity. Obtain an order-of-magnitude estimate for the ratio of the energy dissipated by viscosity directly from an energy-containing eddy with a fluctuating velocity u' and size l to the total dissipation rate. Try to express your answer in terms of the Reynolds number for the eddy.

15 The kinetics of the competitive-consecutive coupling reactions between 1-naphthol (A) and diazotized sulfonanilic acid (B) to produce 4-(4'-sulfophenylazo)-1-naphthol $(R,$ monoazo dyestuff) and 2, 4-bis (4'-sulfuphenylazo)-1-naphthol $(S,$ bisazo dyestuff) are well understood.

$$A + B \rightarrow R$$

$$R + B \rightarrow S$$

Suppose R is the desired product and S is undesired. The rate constants k_1 and k_2 are such that we should be able to make minimal amounts of S in alkline solution at 298 K, pH 10:

$$r_1 = k_1 C_A C_B \qquad k_1 = 7300 \text{ m}^3/\text{mol sec}$$

$$r_2 = k_2 C_R C_B \qquad k_2 = 3.5 \text{ m}^3/\text{mol sec}$$

The reaction will be carried out in a laboratory reactor. There is no volume change on mixing.

   a Assuming instantaneous complete mixing of the feed streams, calculate the composition of an effective combined feed stream.
   b Assuming ideal CSTR behavior with perfect micromixing (no segregation), calculate the product composition. What is the fraction of the feed B which ends up in secondary undesired product S? Call this fraction X.
   c Because of the fast reaction kinetics, the selectivity is not equal to that calculated in b above for the ideal CSTR. It is found that a stirrer speed of 180 rpm is required to keep X below 10 percent. Based on this information, specify an agitator and speed that would maintain the same selectivity in a plant reactor. The plant reactor volume and feed rates are 100 times those for the lab reactor. There are many possible "correct" answers. What will be the required power? Do not worry about picking standard speeds or powers.

A feed: $Q_A = 3.18 \times 10^{-5}$ m^3/s   Feed $C_A = 0.0578$ mol/m^3
B feed: $Q_B = 3.18 \times 10^{-6}$ m^3/s   Feed $C_B = 0.55$ mol/m^3
$\rho = 1000$ kg/m^3

Reactor dimensions:

   Volume: 0.063 m^3

   Tank diameter: 0.440 m

   Diameter of six-bladed Rushton turbine: 0.153 m

   Liquid volume above turbine: 0.041 m

   Submersion of turbine: 0.275 m

   Height and width of turbine blades: 0.030 and 0.037 m

Diameters of feed pipes for A and B solution: 0.018 and 0.014 m

Spacing between feed pipes and turbine: 0.061 m

Spacing between feed pipes: 0.048 m

Width of baffles (four): 0.040 m

Spacing between baffles and wall: 0.045 m

16 Laufhutte and Mersmann (1985) presented a model for the power number in the transition region using an analogy to pipe flow. Review the basis for the model, and make comments on the model as to whether it is fundamentally correct.

17 Derive the k equation in cylindrical coordinates.

18 Three competitive-consecutive reactions are carried out in a liquid phase batch laboratory reactor, where initial compositions have been optimized to give the desired product distribution to meet market needs. The batch reactor was well agitated, and it was found that increasing agitator speed had no effect on the results.

$$A + B \rightarrow R$$
$$R + B \rightarrow S$$
$$S + B \rightarrow T$$

B is mostly depleted in 50 s and A in 100 s. As A and B are depleted, the concentrations of R, S, and T increase with $C_R > C_S > C_T$. A commercial scale continuous-flow reactor is to be built to handle to a total feed rate of 0.015 m^3/s (238 GPM). The proposed design is a 10-m^3 reactor, 2 m in diameter with a liquid depth of 2.5 m and standard baffling. A standard four-blade axial flow impeller ($D = 0.6$ m, $W = 0.12$ m) turning at 45 rpm has been specified. The following physical property data may be assumed:

Liquid density: 1000 kg/m^3

Liquid viscosity: 0.001 Pa · s

Molecular diffusivity: 10^{-9} m^2/s (all species)

Based on this information, answer the following questions as quantitatively as possible.

   a What power will be required to drive the impeller?

   b What will be the Kolmogorov eddy size based on the average power dissipation in the vessel? Based on assuming that most of the power is dissipated in half the volume swept out by the impeller?

   c Will the commercial reactor be well mixed so that it may be modeled as an ideal CSTR? Answer this question carefully, considering both macro-convective mixing and micromixing-segregation effects.

   d Will the optimized inlet concentrations determined in the lab apply to the commercial reactor?

19 Lauder (1954) and Sandborn (1955) were cited by Brodkey (1967) as providing data concerning the Reynolds stress $\overline{u_r u_z}$. How did this $\overline{u_r u_z}$ correlation data vary with r, θ, and z? How was this correlation related to wall shear stress and frictional velocity?

Chapter 5

1   What were the basic assumptions used by Solomon et al. (1981) to establish a relationship for cavern size? Compare their analysis with that of Wichterle and Wein (1981), and explain the similarities and differences in the analyses. Also review the more recent articles on the subject and make comparisons.

2   What physical reasoning can be given which justifies the relationship between pressure drop, due to pumping, and μN for the helical screw impeller?

3   A cylindrical vessel 10 by 10 ft is to be used as a spinning solution storage tank for a polymer solution used to make synthetic fibers. Its capacity is about 50,000 lb of solution. Feed rate into and out of the vessel is 10,000 lb/h.

Provide a double helical ribbon impeller design and a screw impeller design for this vessel. Provide mixing times. The screw design should consider the use of a draft tube.

Physical properties: Density: 1.3 gm/cm^3; viscosity 500,000 cP (millipascal second)

If the tank were twice as high, how would your designs change?

4   In Nagata et al. (*Intern. Chem. Eng.*, **12**, 180, 1972), please explain why the addition of a screw to a helical ribbon impeller increased power consumption but, in some cases, did not improve the k_m value.

5   Review and discuss the similarities and differences between the work on pumping capacities of a screw by reviewing work by Seichter (1971; 1981), and Chavan and Ulbrecht (1973) and that by Rieger, Ditl, and Novak (1974) and Rieger and Novak (1985). Which of the reported correlations should be used for design purposes and why?

6   Cheng (1979) has discussed three simple flow fields: simple shear, uniaxial extension, and planar extension. Draw these flow fields, and comment on their various properties. Explain the difference between extensional viscosity and shear viscosity. How do these change as a function of fluid type?

7   Figure 5.1 shows the direction of rotation and pumping direction for the helical ribbon impeller, the screw impeller, and the helical ribbon screw impeller. Examine the direction of rotation and the pumping direction, and convince yourself that these are correctly shown in the figure.

8   Why do viscoelastic fluids climb an impeller shaft?

9   How would you determine the natural and characteristic times for a viscoelastic liquid?

10   In laminar mixing, the highest shear rates are associated with the blade tip. True or False?

11   Toroidal vortices are a major hindrance to mixing in the laminar regime. True or False?

12   For $N = 1.0 \text{ s}^{-1}$, what are the pumping capacities, i.e., k_Q values, for a helical ribbon impeller with $D = 0.7$ ft? For a helical screw impeller with $D = 0.5$ ft? For a turbulent impeller with $D = 0.25$ ft? Which impeller pumps more?

13   Show that the units of μN, i.e., $[(\text{lb/ft}^3)\text{ ft}^2/\text{s}]\text{s}^{-1}$ can be converted into units of pound-force per square inch.

14   List three mixing performance criteria, and explain their meaning.

15   Chavan (*AIChEJ.*, **29**, 2, 177, 1983) used a term F_g in a force balance on a helical impeller. Please explain the origin of this term; explain why it is necessary.

Chapter 6

1   In design, it is important to know the source and reliability of design correlations. Which correlations would you use for:
  a   Gas power draw
  b $k_l a$ mass transfer coefficient
  c   Holdup
  Explain why for each case.

2   What are the cavity types which form in laminar situations?

3   Why does the rotational speed for recirculation vary with $(T/D)^2$?

4   What is the external distribution coefficient? Typically, what are its values?

5   The chemical *monochloro-stuff* is made by a liquid phase chlorination reaction in an agitated tank. Chlorine gas is sparged into the tank, where it dissolves in the liquid and reacts with the organic feedstock. Monochloro-stuff can react with dissolved chlorine to make *dichloro-stuff* and even *trichloro-stuff*. These by-products are undesirable.

A bench-scale reactor is built to determine optimum operating conditions. The reactor operates with continuous liquid flow. It is found that good conversion to monochloro-stuff, with acceptably low production of by-product, is obtained with the following design:

Reactor diameter	0.30 m (0.98 ft)
Liquid depth	0.30 m (0.98 ft)
Baffles	4

Impeller diameter (six-blade flat blade turbine)	0.15 m (0.49 ft)
Speed	155 rpm
Power number	5 (ungassed)
Chlorine feed rate	1.13×10^{-4} m^3/s (0.24 cfm)
Liquid density	800 kg/m^3 (50 lb/ft^3)

This design is scaled up for commercial operation at 1000 times the lab liquid flow rate. The plant design has complete geometric similarity to the lab unit, with 1000 times the lab volume. The chlorine gas rate used is 1000 times the lab chlorine rate. Keeping constant tip speed would require 15.5 rpm on the plant scale. Since this is less than the minimum standard speed, a speed of 30 rpm is chosen.

It is found that the plant reactor does not work nearly as well as the lab reactor. There is much more production of undesired by-product.

What went wrong? Determine the likely problem, or problems, with the plant reactor, and recommend a fix.

6 Gas sparging significantly reduces surface entrainment. True or False?

7 Mixing and circulation times in the liquid phase of a gas-sparged tank is pretty much unaffected by gas sparging. Explain this statement.

8 Recently, your company hired a contract engineer from Engineers R Us to design a gas-sparged, mechanically agitated vessel to oxygenate an aqueous stream. The stream contains a hazardous chemical that reacts quickly with oxygen to form a product safe to release to the environment. The oxygen transfer demand is 25 lb/h (10^{-4} kgmole/s).

Tank diameter T	3 m (9.8 ft)
Clear liquid depth H	2 m (6.6 ft)
Impeller diameter D	1 m (3.3 ft)
Impeller type	six-blade flat blade radial flow
Impeller speed N	68 rpm
Motor size	7.5 hp
Impeller clearance	0.3 m (1 ft) off bottom
Baffles	four standard width and offset, full tank height
Ring sparger	diameter = 0.9 m, centered on bottom
Air sparging rate	superficial velocity, V_s = 0.1 ft/s

The vessel is installed as designed. After startup, you receive a panic call from the process engineer running the unit. It is not performing as designed. Even though the air rate has been increased 50 percent above the design rate, the oxygen transfer is still only 30 percent of the requirement. You are asked to determine why the unit is not meeting the design rate and what can be done to correct the situation. What is your response?

The following additional information may be useful. The Henry's law constant for oxygen in water at the process temperature is H_e = 4.01 × 10⁴ atm/ mol fr. You may assume this value is a good approximation for the process liquid. Because the fast chemical reaction consumes oxygen, it may be assumed that the bulk oxygen concentration in the liquid is approximately zero. The reaction is not fast enough to affect the mass transfer coefficient. A clever operator has used an ammeter to determine the current drawn by the agitator motor. The motor is drawing "much less" than 7.5 hp.

Physical properties for water at 20°C may be assumed. (ρ_L = 1000 kg/m³, μ_L = 0.001 cP)

Conversion factors and constants:

1 hp = 745.7 W

1 ft³ = 7.48 gal

Gas constant 8314 J/kgmole K

No evidence has been found of any chemical contaminants interfering with the mass transfer.

9 How is holdup related to superfacial gas velocity and gas bubble terminal velocity?

10 The aeration number typically used in correlations inadequately specifies the gas dispersion process. What would you recommend as an improvement in the definition of the aeration number?

11 Calderbank (1956a; 1956b; 1958) developed a correlation for bubble diameter based on balance between interfacial tension and dynamic forces. Please explain development of this correlation.

12 Gray et al. (1982), in his development of total power input under gassing conditions, used two Froude numbers. Please explain why two Froude numbers could be used, and provide a rationalization for the correlation.

13 The holdup results by Kawecki et al. (1967) were three times larger than those of Yung et al. (1979). Review these articles and their respective experimental techniques for determining holdup, and give reasons for the factor of three difference.

14 Oxygen absorption from dry air into water is to be studied in a 1-m-diameter stirred tank. The tank is operated at atmospheric pressure and 20°C. At this temperature, oxygen solubility is given by a Henry's law constant H_e = 4.01 × 10⁴ atm. The unaerated liquid depth is 0.8 m. The tank is equipped with standard baffles and a six-blade radial flow turbine. The impeller diameter is 0.3 m. Gas is sparged at 0.01 std m³/s (21 SCFM). The tank operates in batch mode with respect to the liquid. The water is initially oxygen-free.

Consider a range of realistic impeller speeds. For each, estimate the required agitator power and the time required to raise the oxygen mole fraction in the liquid to 10⁻⁶.

You may find it a useful simplification to assume that the fraction of the feed oxygen transferred is small. If you make this assumption, justify it.

Additional data:

Oxygen diffusivity in water 20°C: 1.8×10^{-5} cm^2/s

Surface tension of water at 20°C: 73 dyn/cm

15 Miller and Sridhar and Potter exchanged correspondence concerning their work in *Ind. Eng. Chem. Funds.*, **20**, 105–109, 1981. Study the correspondence and related articles and comment on the debate.

16 Compare and comment on the similarities between Ismail et al. (1984) and Warmoeskerken and Smith (1985a; 1985b).

17 Derive the equations for $k_l a$ provided by Chapman et al. (1982).

Index